# THE 33 STRATEGIES OF WAR

**ROBERT GREENE**, author of *The 48 Laws of Power*, *The Art of Seduction* and *Mastery*, has a degree in classical studies. He lives in Los Angeles.

**JOOST ELFFERS** is the producer of *The 48 Laws of Power*, *The Art of Seduction* and a number of other bestselling books. He lives in New York.

OTHER TITLES BY ROBERT GREENE

*The Art of Seduction* (A Joost Elffers Production)

*The 48 Laws of Power* (A Joost Elffers Production)

*Mastery*

THE 33 STRATEGIES OF

# WAR

ROBERT GREENE

A JOOST ELFFERS PRODUCTION

P

PROFILE BOOKS

This paperback edition published in 2007

First published in Great Britain in 2006 by
PROFILE BOOKS LTD
3A Exmouth House
Pine Street
London EC1R 0JH
*www.profilebooks.com*

First published in the United States in 2006 by
Viking Penguin, a member of Penguin Group (USA) Inc.



5 7 9 10 8 6 4

Printed and bound in China
through Asia Pacific Offset
Set in Baskerville and Times Ten
Designed by Joost Elffers and Amy Hill



A CIP catalogue record for this book is available from the British Library.

ISBN-10: 1 86197 978 9
ISBN-13: 978 1 86197 978 0

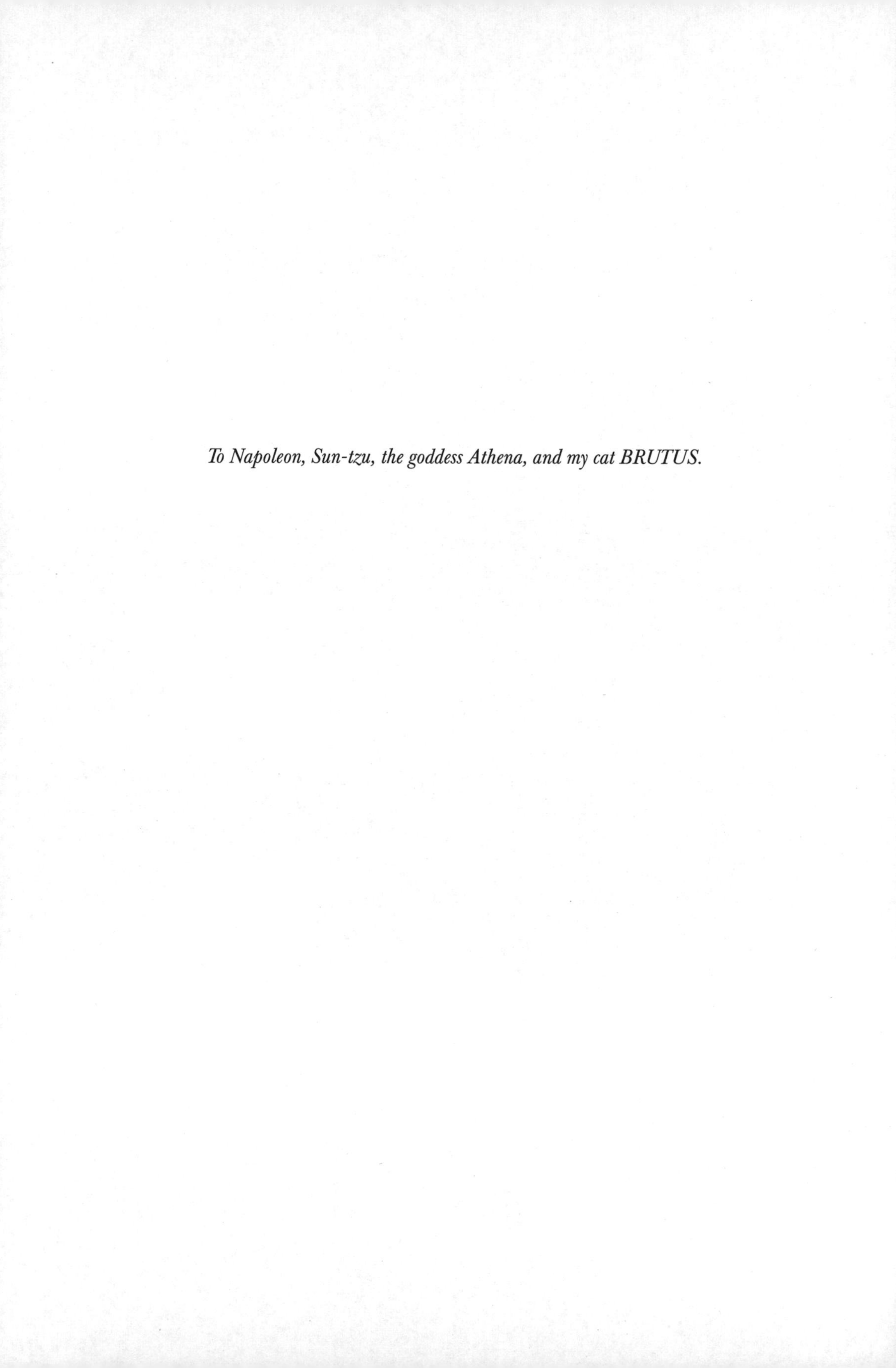

*To Napoleon, Sun-tzu, the goddess Athena, and my cat BRUTUS.*

## CONTENTS

## PART II ORGANIZATIONAL (TEAM) WARFARE



## PART III DEFENSIVE WARFARE

# PREFACE

*The life of man upon earth is a warfare.*

JOB 7:1

*Qui desiderat pacem, praeparet bellum (let him who wants peace prepare for war)*

VEGETIUS, A.D. FOURTH CENTURY

We live in a culture that promotes democratic values of being fair to one and all, the importance of fitting into a group, and knowing how to cooperate with other people. We are taught early on in life that those who are outwardly combative and aggressive pay a social price: unpopularity and isolation. These values of harmony and cooperation are perpetuated in subtle and not-so-subtle ways—through books on how to be successful in life; through the pleasant, peaceful exteriors that those who have gotten ahead in the world present to the public; through notions of correctness that saturate the public space. The problem for us is that we are trained and prepared for peace, and we are not at all prepared for what confronts us in the real world—war.

This war exists on several levels. Most obviously, we have our rivals on the other side. The world has become increasingly competitive and nasty. In politics, business, even the arts, we face opponents who will do almost anything to gain an edge. More troubling and complex, however, are the battles we face with those who are supposedly on our side. There are those who outwardly play the team game, who act very friendly and agreeable, but who sabotage us behind the scenes, use the group to promote their own agenda. Others, more difficult to spot, play subtle games of passive aggression, offering help that never comes, instilling guilt as a secret weapon. On the surface everything seems peaceful enough, but just below it, it is every man and woman for him- or herself, this dynamic infecting even families and relationships. The culture may deny this reality and promote a gentler picture, but we know it and feel it, in our battle scars.

It is not that we and our colleagues are ignoble creatures who fail to live up to ideals of peace and selflessness, but that we cannot help the way we are. We have aggressive impulses that are impossible to ignore or repress. In the past, individuals could expect a group—the state, an extended family, a company—to take care of them, but this is no longer the case, and in this uncaring world we have to think first and foremost of ourselves

*[Strategy] is more than a science: it is the application of knowledge to practical life, the development of thought capable of modifying the original guiding idea in the light of ever-changing situations; it is the art of acting under the pressure of the most difficult conditions.*

HELMUTH VON MOLTKE, 1800–1891

and our interests. What we need are not impossible and inhuman ideals of peace and cooperation to live up to, and the confusion that brings us, but rather practical knowledge on how to deal with conflict and the daily battles we face. And this knowledge is not about how to be more forceful in getting what we want or defending ourselves but rather how to be more rational and strategic when it comes to conflict, channeling our aggressive impulses instead of denying or repressing them. If there is an ideal to aim for, it should be that of the strategic warrior, the man or woman who manages difficult situations and people through deft and intelligent maneuver.

Many psychologists and sociologists have argued that it is through conflict that problems are often solved and real differences reconciled. Our successes and failures in life can be traced to how well or how badly we deal with the inevitable conflicts that confront us in society. The common ways that people deal with them—trying to avoid all conflict, getting emotional and lashing out, turning sly and manipulative—are all counterproductive in the long run, because they are not under conscious and rational control and often make the situation worse. Strategic warriors operate much differently. They think ahead toward their long-term goals, decide which fights to avoid and which are inevitable, know how to control and channel their emotions. When forced to fight, they do so with indirection and subtle maneuver, making their manipulations hard to trace. In this way they can maintain the peaceful exterior so cherished in these political times.

This ideal of fighting rationally comes to us from organized warfare, where the art of strategy was invented and refined. In the beginning, war was not at all strategic. Battles between tribes were fought in a brutal manner, a kind of ritual of violence in which individuals could display their heroism. But as tribes expanded and evolved into states, it became all too apparent that war had too many hidden costs, that waging it blindly often led to exhaustion and self-destruction, even for the victor. Somehow wars had to be fought more rationally.

The word "strategy" comes from the ancient Greek word *strategos,* meaning literally "the leader of the army." Strategy in this sense was the art of generalship, of commanding the entire war effort, deciding what formations to deploy, what terrain to fight on, what maneuvers to use to gain an edge. And as this knowledge progressed, military leaders discovered that the more they thought and planned ahead, the more possibilities they had for success. Novel strategies could allow them to defeat much larger armies, as Alexander the Great did in his victories over the Persians. In facing savvy opponents who were also using strategy, there developed an upward pressure: to gain an advantage, a general had to be even more strategic, more indirect and clever, than the other side. Over time the arts of generalship became steadily more sophisticated, as more strategies were invented.

Although the word "strategy" itself is Greek in origin, the concept appears in all cultures, in all periods. Solid principles on how to deal with the inevitable accidents of war, how to craft the ultimate plan, how to best organize the army—all of this can be found in war manuals from ancient China to modern Europe. The counterattack, the flanking or enveloping maneuver, and the arts of deception are common to the armies of Genghis Khan, Napoleon, and the Zulu king Shaka. As a whole, these principles and strategies indicate a kind of universal military wisdom, a set of adaptable patterns that can increase the chances for victory.

Perhaps the greatest strategist of them all was Sun-tzu, author of the ancient Chinese classic *The Art of War*. In his book, written probably the fourth century B.C., can be found traces of almost all the strategic patterns and principles later developed over the course of centuries. But what connects them, in fact what constitutes the art of war itself in Sun-tzu's eyes, is the ideal of winning without bloodshed. By playing on the psychological weaknesses of the opponent, by maneuvering him into precarious positions, by inducing feelings of frustration and confusion, a strategist can get the other side to break down mentally before surrendering physically. In this way victory can be had at a much lower cost. And the state that wins wars with few lives lost and resources squandered is the state that can thrive over greater periods of time. Certainly most wars are not waged so rationally, but those campaigns in history that have followed this principle (Scipio Africanus in Spain, Napoleon at Ulm, T. E. Lawrence in the desert campaigns of World War I) stand out above the rest and serve as the ideal.

War is not some separate realm divorced from the rest of society. It is an eminently human arena, full of the best and the worst of our nature. War also reflects trends in society. The evolution toward more unconventional, dirtier strategies—guerrilla warfare, terrorism—mirrors a similar evolution in society, where almost anything goes. The strategies that succeed in war, whether conventional or unconventional, are based on timeless psychology, and great military failures have much to teach us about human stupidity and the limits of force in any arena. The strategic ideal in war—being supremely rational and emotionally balanced, striving to win with minimum bloodshed and loss of resources—has infinite application and relevance to our daily battles.

Inculcated with the values of our times, many will argue that organized war is inherently barbaric—a relic of man's violent past and something to be overcome for good. To promote the arts of warfare in a social setting, they will say, is to stand in the way of progress and to encourage conflict and dissension. Isn't there enough of that in the world? This argument is very seductive, but not at all reasonable. There will always be those in society and in the world at large who are more aggressive than we are, who find ways to get what they want, by hook or by crook. We must be vigilant and must know how to defend ourselves against such

*"Well, then, my boy,*
*develop your strategy*
*So that prizes in games*
*won't elude your grasp.*
*Strategy makes a better*
*woodcutter than strength.*
*Strategy keeps a pilot's*
*ship on course*
*When crosswinds blow*
*it over the wine-blue sea.*
*And strategy wins races*
*for charioteers.*
*One type of driver trusts*
*his horses and car*
*And swerves mindlessly*
*this way and that,*
*All over the course, without reining his horses.*
*But a man who knows*
*how to win with lesser*
*horses*
*Keeps his eye on the post*
*and cuts the turn close,*
*And from the start keeps*
*tension on the reins*
*With a firm hand as he*
*watches the leader."*

*THE ILIAD*, HOMER, CIRCA NINTH CENTURY B.C.

*The self is the friend of a man who masters himself through the self, but for a man without self-mastery, the self is like an enemy at war.*

THE *BHAGAVAD GITA*, INDIA, CIRCA A.D. FIRST CENTURY

types. Civilized values are not furthered if we are forced to surrender to those who are crafty and strong. In fact, being pacifists in the face of such wolves is the source of endless tragedy.

Mahatma Gandhi, who elevated nonviolence into a great weapon for social change, had one simple goal later on in his life: to rid India of the British overlords who had crippled it for so many centuries. The British were clever rulers. Gandhi understood that if nonviolence were to work, it would have to be extremely strategic, demanding much thought and planning. He went so far as to call nonviolence a new way of waging war. To promote any value, even peace and pacifism, you must be willing to fight for it and to aim at results—not simply the good, warm feeling that expressing such ideas might bring you. The moment you aim for results, you are in the realm of strategy. War and strategy have an inexorable logic: if you want or desire anything, you must be ready and able to fight for it.

Others will argue that war and strategy are primarily matters that concern men, particularly those who are aggressive or among the power elite. The study of war and strategy, they will say, is a masculine, elitist, and repressive pursuit, a way for power to perpetuate itself. Such an argument is dangerous nonsense. In the beginning, strategy indeed belonged to a select few—a general, his staff, the king, a handful of courtiers. Soldiers were not taught strategy, for that would not have helped them on the battlefield. Besides, it was unwise to arm one's soldiers with the kind of practical knowledge that could help them to organize a mutiny or rebellion. The era of colonialism took this principle further: the indigenous peoples of Europe's colonies were conscripted into the Western armies and did much of the police work, but even those who rose to the upper echelons were rigorously kept ignorant of knowledge of strategy, which was considered far too dangerous for them to know. To maintain strategy and the arts of war as a branch of specialized knowledge is actually to play into the hands of the elites and repressive powers, who like to divide and conquer. If strategy is the art of getting results, of putting ideas into practice, then it should be spread far and wide, particularly among those who have been traditionally kept ignorant of it, including women. In the mythologies of almost all cultures, the great gods of war are women, including Athena of ancient Greece. A woman's lack of interest in strategy and war is not biological but social and perhaps political.

Instead of resisting the pull of strategy and the virtues of rational warfare or imagining that it is beneath you, it is far better to confront its necessity. Mastering the art will only make your life more peaceful and productive in the long run, for you will know how to play the game and win without violence. Ignoring it will lead to a life of endless confusion and defeat.

The following are six fundamental ideals you should aim for in transforming yourself into a strategic warrior in daily life.

**Look at things as they are, not as your emotions color them.** In strategy you must see your emotional responses to events as a kind of disease that must be remedied. Fear will make you overestimate the enemy and act too defensively. Anger and impatience will draw you into rash actions that will cut off your options. Overconfidence, particularly as a result of success, will make you go too far. Love and affection will blind you to the treacherous maneuvers of those apparently on your side. Even the subtlest gradations of these emotions can color the way you look at events. The only remedy is to be aware that the pull of emotion is inevitable, to notice it when it is happening, and to compensate for it. When you have success, be extra wary. When you are angry, take no action. When you are fearful, know you are going to exaggerate the dangers you face. War demands the utmost in realism, seeing things as they are. The more you can limit or compensate for your emotional responses, the closer you will come to this ideal.

*Although a goddess of war, [Athena] gets no pleasure from battle . . . but rather from settling disputes, and upholding the law by pacific means. She bears no arms in time of peace and, if ever she needs any, will usually borrow a set from Zeus. Her mercy is great. . . . Yet, once engaged in battle, she never loses the day, even against Ares himself, being better grounded in tactics and strategy than he; and wise captains always approach her for advice.*

*The Greek Myths vol. 1*, Robert Graves, 1955

**Judge people by their actions.** The brilliance of warfare is that no amount of eloquence or talk can explain away a failure on the battlefield. A general has led his troops to defeat, lives have been wasted, and that is how history will judge him. You must strive to apply this ruthless standard in your daily life, judging people by the results of their actions, the deeds that can be seen and measured, the maneuvers they have used to gain power. What people say about themselves does not matter; people will say anything. Look at what they have done; deeds do not lie. You must also apply this logic to yourself. In looking back at a defeat, you must identify the things you could have done differently. It is your own bad strategies, not the unfair opponent, that are to blame for your failures. You are responsible for the good and bad in your life. As a corollary to this, look at everything other people do as a strategic maneuver, an attempt to gain victory. People who accuse you of being unfair, for example, who try to make you feel guilty, who talk about justice and morality, are trying to gain an advantage on the chessboard.

**Depend on your own arms.** In the search for success in life, people tend to rely on things that seem simple and easy or that have worked before. This could mean accumulating wealth, resources, a large number of allies, or the latest technology and the advantage it brings. This is being materialistic and mechanical. But true strategy is psychological—a matter of intelligence, not material force. Everything in life can be taken away from you and generally will be at some point. Your wealth vanishes, the latest gadgetry suddenly becomes passé, your allies desert you. But if your mind is armed with the art of war, there is no power that can take that away. In the middle of a crisis, your mind will find its way to the right solution. Having superior strategies at your fingertips will give your maneuvers irresistible force. As Sun-tzu says, "Being unconquerable lies with yourself."

*And Athena, whose eyes were as grey as owls: "Diomedes, son of Tydeus . . . You don't have to fear Ares or any other Of the immortals. Look who is here beside you. Drive your horses directly at Ares And when you're in range, strike. Don't be in awe of Ares. He's nothing but A shifty lout . . ." . . . And when Diomedes thrust next, She drove his spear home to the pit Of Ares' belly, where the kilt-piece covered it. . . . [Ares] quickly scaled the heights of Olympus, Sat down sulking beside Cronion Zeus, Showed him the immortal blood oozing From his wound, and whined these winged words: "Father Zeus, doesn't it infuriate you To see this violence? We gods Get the worst of it from each other Whenever we try to help out men . . ." And Zeus, from under thunderhead brows: "Shifty lout. Don't sit here by me and whine. You're the most loathsome god on Olympus. You actually like fighting and war. You take after your hardheaded mother, Hera. I can barely control her either. . . . Be that as it may, I cannot tolerate you're being in pain . . ." And he called Paieon to doctor his wound . . .*

**Worship Athena, not Ares.** In the mythology of ancient Greece, the cleverest immortal of them all was the goddess Metis. To prevent her from outwitting and destroying him, Zeus married her, then swallowed her whole, hoping to incorporate her wisdom in the process. But Metis was pregnant with Zeus's child, the goddess Athena, who was subsequently born from his forehead. As befitting her lineage, she was blessed with the craftiness of Metis and the warrior mentality of Zeus. She was deemed by the Greeks to be the goddess of strategic warfare, her favorite mortal and acolyte being the crafty Odysseus. Ares was the god of war in its direct and brutal form. The Greeks despised Ares and worshipped Athena, who always fought with the utmost intelligence and subtlety. Your interest in war is not the violence, the brutality, the waste of lives and resources, but the rationality and pragmatism it forces on us and the ideal of winning without bloodshed. The Ares figures of the world are actually quite stupid and easily misled. Using the wisdom of Athena, your goal is to turn the violence and aggression of such types against them, making their brutality the cause of their downfall. Like Athena, you are always one step ahead, making your moves more indirect. Your goal is to blend philosophy and war, wisdom and battle, into an unbeatable blend.

**Elevate yourself above the battlefield.** In war, strategy is the art of commanding the entire military operation. Tactics, on the other hand, is the skill of forming up the army for battle itself and dealing with the immediate needs of the battlefield. Most of us in life are tacticians, not strategists. We become so enmeshed in the conflicts we face that we can think only of how to get what we want in the battle we are currently facing. To think strategically is difficult and unnatural. You may imagine you are being strategic, but in all likelihood you are merely being tactical. To have the power that only strategy can bring, you must be able to elevate yourself above the battlefield, to focus on your long-term objectives, to craft an entire campaign, to get out of the reactive mode that so many battles in life lock you into. Keeping your overall goals in mind, it becomes much easier to decide when to fight and when to walk away. That makes the tactical decisions of daily life much simpler and more rational. Tactical people are heavy and stuck in the ground; strategists are light on their feet and can see far and wide.

**Spiritualize your warfare.** Every day you face battles—that is the reality for all creatures in their struggle to survive. But the greatest battle of all is with yourself—your weaknesses, your emotions, your lack of resolution in seeing things through to the end. You must declare unceasing war on yourself. As a warrior in life, you welcome combat and conflict as ways to prove yourself, to better your skills, to gain courage, confidence, and experience. Instead of repressing your doubts and fears, you must face them down, do battle with them. You want more challenges, and you invite more war. You are forging the warrior's spirit, and only constant practice will lead you there.

*The 33 Strategies of War* is a distillation of the timeless wisdom contained in the lessons and principles of warfare. The book is designed to arm you with practical knowledge that will give you endless options and advantages in dealing with the elusive warriors that attack you in daily battle.

Each chapter is a strategy aimed at solving a particular problem that you will often encounter. Such problems include fighting with an unmotivated army behind you; wasting energy by battling on too many fronts; feeling overwhelmed by friction, the discrepancy between plans and reality; getting into situations you cannot get out of. You can read the chapters that apply to the particular problem of the moment. Better still, you can read all of the strategies, absorb them, allowing them to become part of your mental arsenal. Even when you are trying to avoid a war, not fight one, many of these strategies are worth knowing for defensive purposes and for making yourself aware of what the other side might be up to. In any event, they are not intended as doctrine or formulas to be repeated but as aids to judgment in the heat of battle, seeds that will take root in you and help you think for yourself, developing the latent strategist within.

The strategies themselves are culled from the writings and practices of the greatest generals in history (Alexander the Great, Hannibal, Genghis Khan, Napoleon Bonaparte, Shaka Zulu, William Techumseh Sherman, Erwin Rommel, Vo Nguyen Giap) as well as the greatest strategists (Sun-tzu, Miyamoto Musashi, Carl von Clausewitz, Ardant du Picq, T. E. Lawrence, Colonel John Boyd). They range from the basic strategies of classical warfare to the dirty, unconventional strategies of modern times. The book is divided into five parts: self-directed war (how to prepare your mind and spirit for battle); organizational war (how to structure and motivate your army); defensive war; offensive war; and unconventional (dirty) war. Each chapter is illustrated with historical examples, not only from warfare itself but from politics (Margaret Thatcher), culture (Alfred Hitchcock), sports (Muhammad Ali), business (John D. Rockefeller), showing the intimate connection between the military and the social. These strategies can be applied to struggles of every scale: organized warfare, business battles, the politics of a group, even personal relationships.

Finally, strategy is an art that requires not only a different way of thinking but an entirely different approach to life itself. Too often there is a chasm between our ideas and knowledge on the one hand and our actual experience on the other. We absorb trivia and information that takes up mental space but gets us nowhere. We read books that divert us but have little relevance to our daily lives. We have lofty ideas that

*Then back to the palace of great Zeus*
*Came Argive Hera and Athena the Protector,*
*Having stopped brutal Ares from butchering men.*

*THE ILIAD*, HOMER, CIRCA NINTH CENTURY B.C.

*Against war it can be said: it makes the victor stupid, the defeated malicious. In favour of war: through producing these two effects it barbarizes and therefore makes more natural; it is the winter or hibernation time of culture, mankind emerges from it stronger for good and evil.*

FRIEDRICH NIETZSCHE, 1844–1900

*Without war human beings stagnate in comfort and affluence and lose the capacity for great thoughts and feelings, they become cynical and subside into barbarism.*

FYODOR DOSTOYEVSKY, 1821–1881

*Nature has made up her mind that what cannot defend itself shall not be defended.*

RALPH WALDO EMERSON, 1803–1882

we do not put into practice. We also have many rich experiences that we do not analyze enough, that do not inspire us with ideas, whose lessons we ignore. Strategy requires a constant contact between the two realms. It is practical knowledge of the highest form. Events in life mean nothing if you do not reflect on them in a deep way, and ideas from books are pointless if they have no application to life as you live it. In strategy all of life is a game that you are playing. This game is exciting but also requires deep and serious attention. The stakes are so high. What you know must translate into action, and action must translate into knowledge. In this way strategy becomes a lifelong challenge and the source of constant pleasure in surmounting difficulties and solving problems.

*In this world, where the game is played with loaded dice,*
*a man must have a temper of iron, with armor proof to the*
*blows o fate, and weapons to make his way against men.*
*Life is one long battle; we have to fight at every step;*
*and Voltaire very rightly says that if we succeed, it is*
*at the point of the sword, and that we die*
*with the weapon in our hand.*
*—Arthur Schopenhauer,* Counsels and Maxims, *1851*

PART

# I

# SELF-DIRECTED WARFARE

War, or any kind of conflict, is waged and won through strategy. Think of strategy as a series of lines and arrows aimed at a goal: at getting you to a certain point in the world, at helping you to attack a problem in your path, at figuring out how to encircle and destroy your enemy. Before directing these arrows at your enemies, however, you must first direct them at yourself.

Your mind is the starting point of all war and all strategy. A mind that is easily overwhelmed by emotion, that is rooted in the past instead of the present, that cannot see the world with clarity and urgency, will create strategies that will always miss the mark.

To become a true strategist, you must take three steps. First, become aware of the weakness and illness that can take hold of the mind, warping its strategic powers. Second, declare a kind of war on yourself to make yourself move forward. Third, wage ruthless and continual battle on the enemies within you by applying certain strategies.

The following four chapters are designed to make you aware of the disorders that are probably flourishing in your mind right now and to arm you with specific strategies for eliminating them. These chapters are arrows to aim at yourself. Once you have absorbed them through thought and practice, they will serve as a self-corrective device in all your future battles, freeing the grand strategist within you.

# 1

# DECLARE WAR ON YOUR ENEMIES

## THE POLARITY STRATEGY

*Life is endless battle and conflict, and you cannot fight effectively unless you can identify your enemies. People are subtle and evasive, disguising their intentions, pretending to be on your side. You need clarity. Learn to smoke out your enemies, to spot them by the signs and patterns that reveal hostility. Then, once you have them in your sights, inwardly declare war. As the opposite poles of a magnet create motion, your enemies—your opposites—can fill you with purpose and direction. As people who stand in your way, who represent what you loathe, people to react against, they are a source of energy. Do not be naïve: with some enemies there can be no compromise, no middle ground.*

## THE INNER ENEMY

In the spring of 401 B.C., Xenophon, a thirty-year-old country gentleman who lived outside Athens, received an intriguing invitation: a friend was recruiting Greek soldiers to fight as mercenaries for Cyrus, brother of the Persian king Ataxerxes, and asked him to go along. The request was somewhat unusual: the Greeks and the Persians had long been bitter enemies. Some eighty years earlier, in fact, Persia had tried to conquer Greece. But the Greeks, renowned fighters, had begun to offer their services to the highest bidder, and within the Persian Empire there were rebellious cities that Cyrus wanted to punish. Greek mercenaries would be the perfect reinforcements in his large army.

Xenophon was not a soldier. In fact, he had led a coddled life, raising dogs and horses, traveling into Athens to talk philosophy with his good friend Socrates, living off his inheritance. He wanted adventure, though, and here he had a chance to meet the great Cyrus, learn war, see Persia. Perhaps when it was all over, he would write a book. He would go not as a mercenary (he was too wealthy for that) but as a philosopher and historian. After consulting the oracle at Delphi, he accepted the invitation.

Some 10,000 Greek soldiers joined Cyrus's punitive expedition. The mercenaries were a motley crew from all over Greece, there for the money and the adventure. They had a good time of it for a while, but a few months into the job, after leading them deep into Persia, Cyrus admitted his true purpose: he was marching on Babylon, mounting a civil war to unseat his brother and make himself king. Unhappy to be deceived, the Greeks argued and complained, but Cyrus offered them more money, and that quieted them.

The armies of Cyrus and Ataxerxes met on the plains of Cunaxa, not far from Babylon. Early in the battle, Cyrus was killed, putting a quick end to the war. Now the Greeks' position was suddenly precarious: having fought on the wrong side of a civil war, they were far from home and surrounded by hostile Persians. They were soon told, however, that Ataxerxes had no quarrel with them. His only desire was that they leave Persia as quickly as possible. He even sent them an envoy, the Persian commander Tissaphernes, to provision them and escort them back to Greece. And so, guided by Tissaphernes and the Persian army, the mercenaries began the long trek home—some fifteen hundred miles.

A few days into the march, the Greeks had new fears: their supplies from the Persians were insufficient, and the route that Tissaphernes had chosen for them was problematic. Could they trust these Persians? They started to argue among themselves.

The Greek commander Clearchus expressed his soldiers' concerns to Tissaphernes, who was sympathetic: Clearchus should bring his captains to a meeting at a neutral site, the Greeks would voice their grievances, and the two sides would come to an understanding. Clearchus agreed and appeared the next day with his officers at the appointed time

*Then [Xenophon] got up, and first called together the under-officers of Proxenos. When they were collected he said: "Gentlemen, I cannot sleep and I don't think you can; and I can't lie here when I see what a plight we are in. It is clear that the enemy did not show us open war until they thought they had everything well prepared; and no-one among us takes the pains to make the best possible resistance. "Yet if we give way, and fall into the king's power, what do we expect our fate will be? When his own half-brother was dead, the man cut off his head and cut off his hand and stuck them up on a pole. We have no-one to plead for us, and we marched here to make the king a slave or to kill him if we could, and what do you think our fate will be? Would he not go to all extremes of torture to make the whole world afraid of making war on him? Why, we must do anything to keep out of his power! While the truce lasted, I never ceased pitying ourselves, I never ceased congratulating the king and his army. What a vast country I saw, how large, what endless provisions, what crowds of servants, how many cattle and sheep, what gold, what raiment! But when I thought of these our soldiers—we had no share in all these* and place—where, however, a large contingent of Persians surrounded and arrested them. They were beheaded that same day.

One man managed to escape and warn the Greeks of the Persian treachery. That evening the Greek camp was a desolate place. Some men argued and accused; others slumped drunk to the ground. A few considered flight, but with their leaders dead, they felt doomed.

That night Xenophon, who had stayed mostly on the sidelines during the expedition, had a dream: a lightning bolt from Zeus set fire to his father's house. He woke up in a sweat. It suddenly struck him: death was staring the Greeks in the face, yet they lay around moaning, despairing, arguing. The problem was in their heads. Fighting for money rather than for a purpose or cause, unable to distinguish between friend and foe, they had gotten lost. The barrier between them and home was not rivers or mountains or the Persian army but their own muddled state of mind. Xenophon didn't want to die in this disgraceful way. He was no military man, but he knew philosophy and the way men think, and he believed that if the Greeks concentrated on the enemies who wanted to kill them, they would become alert and creative. If they focused on the vile treachery of the Persians, they would grow angry, and their anger would motivate them. They had to stop being confused mercenaries and go back to being Greeks, the polar opposite of the faithless Persians. What they needed was clarity and direction.

Xenophon decided to be Zeus's lightning bolt, waking the men up and illuminating their way. He called a meeting of all the surviving officers and stated his plan: We will declare war without parley on the Persians—no more thoughts of bargaining or debate. We will waste no more time on argument or accusation among ourselves; every ounce of our energy will be spent on the Persians. We will be as inventive and inspired as our ancestors at Marathon, who fought off a vastly larger Persian army. We will burn our wagons, live off the land, move fast. We will not for one second lay down our arms or forget the dangers around us. It is us or them, life or death, good or evil. Should any man try to confuse us with clever talk or with vague ideas of appeasement, we will declare him too stupid and cowardly to be on our side and we will drive him away. Let the Persians make us merciless. We must be consumed with one idea: getting home alive.

The officers knew that Xenophon was right. The next day a Persian officer came to see them, offering to act as an ambassador between them and Ataxerxes; following Xenophon's counsel, he was quickly and rudely driven away. It was now war and nothing else.

Roused to action, the Greeks elected leaders, Xenophon among them, and began the march home. Forced to depend on their wits, they quickly learned to adapt to the terrain, to avoid battle, to move at night. They successfully eluded the Persians, beating them to a key mountain pass and moving through it before they could be caught. Although many enemy tribes still lay between them and Greece, the dreaded Persian

*good things unless we bought them, and few had anything left to buy with; and to procure anything without buying was debarred by our oaths. While I reasoned like this, I sometimes feared the truce more than the war now. "However, now they have broken the truce, there is an end both to their insolence and to our suspicion. There lie all these good things before us, prizes for whichever side prove the better men; the gods are the judges of the contest, and they will be with us, naturally. . . . "When you have appointed as many commanders as are wanted, assemble all the other soldiers and encourage them; that will be just what they want now. Perhaps you have noticed yourselves how crestfallen they were when they came into camp, how crestfallen they went on guard; in such a state I don't know what you could do with them. . . . But if someone could turn their minds from wondering what will happen to them, and make them wonder what they could do, they will be much more cheerful. You know, I am sure, that not numbers or strength brings victory in war; but whichever army goes into battle stronger in soul, their enemies generally cannot withstand them."*

*Anabasis: The March Up Country*,
Xenophon, 430?–355? B.C.

*Political thought and political instinct prove themselves theoretically and practically in the ability to distinguish friend and enemy. The high points of politics are simultaneously the moments in which the enemy is, in concrete clarity, recognized as the enemy.*

CARL SCHMITT, 1888–1985

army was now behind them. It took several years, but almost all of them returned to Greece alive.

Interpretation

Life is battle and struggle, and you will constantly find yourself facing bad situations, destructive relationships, dangerous engagements. How you confront these difficulties will determine your fate. As Xenophon said, your obstacles are not rivers or mountains or other people; your obstacle is yourself. If you feel lost and confused, if you lose your sense of direction, if you cannot tell the difference between friend and foe, you have only yourself to blame.

Think of yourself as always about to go into battle. Everything depends on your frame of mind and on how you look at the world. A shift of perspective can transform you from a passive and confused mercenary into a motivated and creative fighter.

We are defined by our relationship to other people. As children we develop an identity by differentiating ourselves from others, even to the point of pushing them away, rejecting them, rebelling. The more clearly you recognize who you do *not* want to be, then, the clearer your sense of identity and purpose will be. Without a sense of that polarity, without an enemy to react against, you are as lost as the Greek mercenaries. Duped by other people's treachery, you hesitate at the fatal moment and descend into whining and argument.

Focus on an enemy. It can be someone who blocks your path or sabotages you, whether subtly or obviously; it can be someone who has hurt you or someone who has fought you unfairly; it can be a value or an idea that you loathe and that you see in an individual or group. It can be an abstraction: stupidity, smugness, vulgar materialism. Do not listen to people who say that the distinction between friend and enemy is primitive and passé. They are just disguising their fear of conflict behind a front of false warmth. They are trying to push you off course, to infect you with the vagueness that inflicts them. Once you feel clear and motivated, you will have space for true friendship and true compromise. Your enemy is the polar star that guides you. Given that direction, you can enter battle.

*He that is not with me is against me.*

—Luke 11:23

## THE OUTER ENEMY

In the early 1970s, the British political system had settled into a comfortable pattern: the Labour Party would win an election, and then, the next time around, the Conservatives would win. Back and forth the power went, all fairly genteel and civilized. In fact, the two parties had come to resemble one another. But when the Conservatives lost in 1974, some of them had had enough. Wanting to shake things up, they proposed Margaret Thatcher as their leader. The party was divided that year, and Thatcher took advantage of the split and won the nomination.

No one had ever seen a politician quite like Thatcher. A woman in a world run by men, she was also proudly middle class—the daughter of a grocer—in the traditional party of the aristocracy. Her clothes were prim, more like a housewife's than a politician's. She had not been a player in the Conservative Party; in fact, she was on its right-wing fringes. Most striking of all was her style: where other politicians were smooth and conciliatory, she confronted her opponents, attacking them directly. She had an appetite for battle.

Most politicians saw Thatcher's election as a fluke and didn't expect her to last. And in her first few years leading the party, when Labour was in power, she did little to change their opinion. She railed against the socialist system, which in her mind had choked all initiative and was largely responsible for the decline of the British economy. She criticized the Soviet Union at a time of détente. Then, in the winter of 1978–79, several public-sector unions decided to strike. Thatcher went on the warpath, linking the strikes to the Labour Party and Prime Minister James Callaghan. This was bold, divisive talk, good for making the evening news—but not for winning elections. You had to be gentle with the voters, reassure them, not frighten them. At least that was the conventional wisdom.

In 1979 the Labour Party called a general election. Thatcher kept on the attack, categorizing the election as a crusade against socialism and as Great Britain's last chance to modernize. Callaghan was the epitome of the genteel politician, but Thatcher got under his skin. He had nothing but disdain for this housewife-turned-politician, and he returned her fire: he agreed that the election was a watershed, for if Thatcher won, she would send the economy into shock. The strategy seemed partly to work; Thatcher scared many voters, and the polls that tracked personal popularity showed that her numbers had fallen well below Callaghan's. At the same time, though, her rhetoric, and Callaghan's response to it, polarized the electorate, which could finally see a sharp difference between the parties. Dividing the public into left and right, she charged into the breach, sucking in attention and attracting the undecided. She won a sizable victory.

Thatcher had bowled over the voters, but now, as prime minister, she would have to moderate her tone, heal the wounds—according to the polls, at any rate, that was what the public wanted. But Thatcher as usual did the opposite, enacting budget cuts that went even deeper than she had proposed during the campaign. As her policies played out, the economy did indeed go into shock, as Callaghan had said it would, and unemployment soared. Men in her own party, many of whom had by that point been resenting her treatment of them for years, began publicly to question her abilities. These men, whom she called the "wets," were

*I am by nature warlike. To attack is among my instincts. To be able to be an enemy, to be an enemy—that presupposes a strong nature, it is in any event a condition of every strong nature. It needs resistances, consequently it* seeks *resistances. . . . The strength of one who attacks has in the opposition he needs a kind of gauge; every growth reveals itself in the seeking out of a powerful opponent—or problem: for a philosopher who is warlike also challenges problems to a duel. The undertaking is to master,* not *any resistances that happen to present themselves, but those against which one has to bring all one's strength, suppleness and mastery of weapons—to master* equal *opponents.*

FRIEDRICH NIETZSCHE, 1844–1900

*[Salvador Dalí] had no time for those who did not agree with his principles, and took the war into the enemy camp by writing insulting letters to many of the friends he had made in the Residencia, calling them pigs. He happily compared himself to a clever bull avoiding the cowboys and generally had a great deal of fun stirring up and scandalizing almost every Catalan intellectual worthy of the name. Dalí was beginning to burn his bridges with the zeal of an arsonist. . . . "We [Dalí and the filmmaker Luis Buñuel] had resolved to send a poison pen letter to one of the great celebrities of Spain," Dalí later told his biographer Alain Bosquet. "Our goal was pure subversion. . . . Both of us were strongly influenced by Nietzsche. . . . We hit upon two names: Manuel de Falla, the composer, and Juan Ramón Jiménez, the poet. We drew straws and Jiménez won. . . . So we composed a frenzied and nasty letter of incomparable violence and addressed it to Juan Ramón Jiménez. It read: 'Our Distinguished Friend: We believe it is our duty to inform you—disinterestedly—that your work is deeply repugnant to us because of its immorality, its hysteria, its arbitrary quality. . . .' It caused Jiménez great pain. . . ."*

THE PERSISTENCE OF MEMORY: A BIOGRAPHY OF DALÍ, MEREDITH ETHERINGTON-SMITH, 1992

the most respected members of the Conservative Party, and they were in a panic: she was leading the country into an economic disaster that they were afraid they would pay for with their careers. Thatcher's response was to purge them from her cabinet. She seemed bent on pushing everyone away; her legion of enemies was growing, her poll numbers slipping still lower. Surely the next election would be her last.

Then, in 1982, on the other side of the Atlantic, the military junta that ruled Argentina, needing a cause to distract the country from its many problems, invaded the Falkland Islands, a British possession to which, however, Argentina had a historical claim. The officers of the junta felt certain that the British would abandon these islands, barren and remote. But Thatcher did not hesitate: despite the distance—eight thousand miles—she sent a naval task force to the Falklands. Labour leaders attacked her for this pointless and costly war. Many in her own party were terrified; if the attempt to retake the islands failed, the party would be ruined. Thatcher was more alone than ever. But much of the public now saw her qualities, which had seemed so irritating, in a new light: her obstinacy became courage, nobility. Compared to the dithering, pantywaisted, careerist men around her, Thatcher seemed resolute and confident.

The British successfully won back the Falklands, and Thatcher stood taller than ever. Suddenly the country's economic and social problems were forgotten. Thatcher now dominated the scene, and in the next two elections she crushed Labour.

### Interpretation

Margaret Thatcher came to power as an outsider: a middle-class woman, a right-wing radical. The first instinct of most outsiders who attain power is to become insiders—life on the outside is hard—but in doing so they lose their identity, their difference, the thing that makes them stand out in the public eye. If Thatcher had become like the men around her, she would simply have been replaced by yet another man. Her instinct was to stay an outsider. In fact, she pushed being an outsider as far as it could go: she set herself up as one woman against an army of men.

At every step of the way, to give her the contrast she needed, Thatcher marked out an opponent: the socialists, the wets, the Argentineans. These enemies helped to define her image as determined, powerful, self-sacrificing. Thatcher was not seduced by popularity, which is ephemeral and superficial. Pundits might obsess over popularity numbers, but in the mind of the voter—which, for a politician, is the field of battle—a dominating presence has more pull than does likability. Let some of the public hate you; you cannot please everyone. Your enemies, those you stand sharply against, will help you to forge a support base that will not desert you. Do not crowd into the center, where everyone else is; there is no room to fight in a crowd. Polarize people, drive some of them away, and create a space for battle.

Everything in life conspires to push you into the center, and not just politically. The center is the realm of compromise. Getting along with other people is an important skill to have, but it comes with a danger: by always seeking the path of least resistance, the path of conciliation, you forget who you are, and you sink into the center with everyone else. Instead see yourself as a fighter, an outsider surrounded by enemies. Constant battle will keep you strong and alert. It will help to define what you believe in, both for yourself and for others. Do not worry about antagonizing people; without antagonism there is no battle, and without battle, there is no chance of victory. Do not be lured by the need to be liked: better to be respected, even feared. Victory over your enemies will bring you a more lasting popularity.

*Don't depend on the enemy not coming; depend rather on being ready for him.*
*—Sun-tzu,* The Art of War *(fourth century B.C.)*

## KEYS TO WARFARE

We live in an era in which people are seldom directly hostile. The rules of engagement—social, political, military—have changed, and so must your notion of the enemy. An up-front enemy is rare now and is actually a blessing. People hardly ever attack you openly anymore, showing their intentions, their desire to destroy you; instead they are political and indirect. Although the world is more competitive than ever, outward aggression is discouraged, so people have learned to go underground, to attack unpredictably and craftily. Many use friendship as a way to mask aggressive desires: they come close to you to do more harm. (A friend knows best how to hurt you.) Or, without actually being friends, they offer assistance and alliance: they may seem supportive, but in the end they're advancing their own interests at your expense. Then there are those who master moral warfare, playing the victim, making you feel guilty for something unspecified you've done. The battlefield is full of these warriors, slippery, evasive, and clever.

Understand: the word "enemy"—from the Latin *inimicus,* "not a friend"—has been demonized and politicized. Your first task as a strategist is to widen your concept of the enemy, to include in that group those who are working against you, thwarting you, even in subtle ways. (Sometimes indifference and neglect are better weapons than aggression, because you can't see the hostility they hide.) Without getting paranoid, you need to realize that there are people who wish you ill and operate indirectly. Identify them and you'll suddenly have room to maneuver. You can stand back and wait and see or you can take action, whether aggressive or just evasive, to avoid the worst. You can even work to turn this enemy into a friend. But whatever you do, do not be the naïve victim. Do not find yourself constantly retreating, reacting to your enemies' maneuvers. Arm yourself with prudence, and never completely lay down your arms, not even for friends.

*The opposition of a member to an associate is no purely negative social factor, if only because such opposition is often the only means for making life with actually unbearable people at least possible. If we did not even have the power and the right to rebel against tyranny, arbitrariness, moodiness, tactlessness, we could not bear to have any relation to people from whose characters we thus suffer. We would feel pushed to take desperate steps—and these, indeed, would end the relation but do* not, *perhaps, constitute "conflict." Not only because of the fact that . . . oppression usually increases if it is suffered calmly and without protest, but also because opposition gives us inner satisfaction, distraction, relief . . . Our opposition makes us feel that we are not completely victims of the circumstances.*

GEORG SIMMEL, 1858–1918

*As one travels up any one of the large rivers [of Borneo], one meets with tribes that are successively more warlike. In the coast regions are peaceful communities which never fight save in self-defense, and then with but poor success, whereas in the central regions, where the rivers take their rise, are a number of extremely warlike tribes whose raids have been a constant source of terror to the communities settled in the lower reaches of the rivers. . . . It might be supposed that the peaceful coast people would be found to be superior in moral qualities to their more warlike neighbors, but the contrary is the case. In almost all respects the advantage lies with the warlike tribes. Their houses are better built, larger, and cleaner; their domestic morality is superior; they are physically stronger, are braver, and physically and mentally more active and in general are more trustworthy. But, above all, their social organization is firmer and more efficient because their respect for and obedience to their chiefs and their loyalty to their community are much greater; each man identifies himself with the whole community and accepts and loyally performs the social duties laid upon him.*

WILLIAM MCDOUGALL, 1871–1938

People are usually good at hiding their hostility, but often they unconsciously give off signals showing that all is not what it seems. One of the closest friends and advisers of the Chinese Communist Party leader Mao Tse-tung was Lin Biao, a high-ranking member of the Politburo and possible successor to the chairman. In the late 1960s and early '70s, though, Mao detected a change in Lin: he had become effusively friendly. Everyone praised Mao, but Lin's praise was embarrassingly fervent. To Mao this meant that something was wrong. He watched Lin closely and decided that the man was plotting a takeover, or at the very least positioning himself for the top spot. And Mao was right: Lin was plotting busily. The point is not to mistrust all friendly gestures but to notice them. Register any change in the emotional temperature: unusual chumminess, a new desire to exchange confidences, excessive praise of you to third parties, the desire for an alliance that may make more sense for the other person than for you. Trust your instincts: if someone's behavior seems suspicious, it probably is. It may turn out to be benign, but in the meantime it is best to be on your guard.

You can sit back and read the signs or you can actively work to uncover your enemies—beat the grass to startle the snakes, as the Chinese say. In the Bible we read of David's suspicion that his father-in-law, King Saul, secretly wanted him dead. How could David find out? He confided his suspicion to Saul's son Jonathan, his close friend. Jonathan refused to believe it, so David suggested a test. He was expected at court for a feast. He would not go; Jonathan would attend and pass along David's excuse, which would be adequate but not urgent. Sure enough, the excuse enraged Saul, who exclaimed, "Send at once and fetch him unto me—he deserves to die!"

David's test succeeded because it was ambiguous. His excuse for missing the feast could be read in more than one way: if Saul meant well toward David, he would have seen his son-in-law's absence as no more than selfish at worst, but because he secretly hated David, he saw it as effrontery, and it pushed him over the edge. Follow David's example: say or do something that can be read in more than one way, that may be superficially polite but that could also indicate a slight coolness on your part or be seen as a subtle insult. A friend may wonder but will let it pass. The secret enemy, though, will react with anger. Any strong emotion and you will know that there's something boiling under the surface.

Often the best way to get people to reveal themselves is to provoke tension and argument. The Hollywood producer Harry Cohn, president of Universal Pictures, frequently used this strategy to ferret out the real position of people in the studio who refused to show what side they were on: he would suddenly attack their work or take an extreme position, even an offensive one, in an argument. His provoked directors and writers would drop their usual caution and show their real beliefs.

Understand: people tend to be vague and slippery because it is safer than outwardly committing to something. If you are the boss, they will mimic your ideas. Their agreement is often pure courtiership. Get them emotional; people are usually more sincere when they argue. If you pick an argument with someone and he keeps on mimicking your ideas, you may be dealing with a chameleon, a particularly dangerous type. Beware of people who hide behind a façade of vague abstractions and impartiality: no one is impartial. A sharply worded question, an opinion designed to offend, will make them react and take sides.

*Man exists only in so far as he is opposed.*

GEORG HEGEL, 1770–1831

Sometimes it is better to take a less direct approach with your potential enemies—to be as subtle and conniving as they are. In 1519, Hernán Cortés arrived in Mexico with his band of adventurers. Among these five hundred men were some whose loyalty was dubious. Throughout the expedition, whenever any of Cortés's soldiers did something he saw as suspicious, he never got angry or accusatory. Instead he pretended to go along with them, accepting and approving what they had done. Thinking Cortés weak, or thinking he was on their side, they would take another step. Now he had what he wanted: a clear sign, to himself and others, that they were traitors. Now he could isolate and destroy them. Adopt the method of Cortés: if friends or followers whom you suspect of ulterior motives suggest something subtly hostile, or against your interests, or simply odd, avoid the temptation to react, to say no, to get angry, or even to ask questions. Go along, or seem to turn a blind eye: your enemies will soon go further, showing more of their hand. Now you have them in sight, and you can attack.

An enemy is often large and hard to pinpoint—an organization, or a person hidden behind some complicated network. What you want to do is take aim at one part of the group—a leader, a spokesman, a key member of the inner circle. That is how the activist Saul Alinsky tackled corporations and bureaucracies. In his 1960s campaign to desegregate Chicago's public-school system, he focused on the superintendent of schools, knowing full well that this man would try to shift the blame upward. By taking repeated hits at the superintendent, he was able to publicize his struggle, and it became impossible for the man to hide. Eventually those behind him had to come to his aid, exposing themselves in the process. Like Alinsky, never aim at a vague, abstract enemy. It is hard to drum up the emotions to fight such a bloodless battle, which in any case leaves your enemy invisible. Personalize the fight, eyeball to eyeball.

Danger is everywhere. There are always hostile people and destructive relationships. The only way to break out of a negative dynamic is to confront it. Repressing your anger, avoiding the person threatening you, always looking to conciliate—these common strategies spell ruin. Avoidance of conflict becomes a habit, and you lose the taste for battle. Feeling guilty is pointless; it is not your fault you have enemies. Feeling wronged or victimized is equally futile. In both cases you are looking inward, concentrating on yourself and your feelings. Instead of internalizing a bad situation, externalize it and face your enemy. It is the only way out.

*The frequent hearing of my mistress reading the bible—for she often read aloud when her husband was absent—soon awakened my curiosity in respect to this mystery of reading, and roused in me the desire to learn. Having no fear of my kind mistress before my eyes, (she had given me no reason to fear,) I frankly asked her to teach me to read; and without hesitation, the dear woman began the task, and very soon, by her assistance, I was master of the alphabet, and could spell words of three or four letters . . . Master Hugh was amazed at the simplicity of his spouse, and, probably for the first time, he unfolded to her the true philosophy of slavery, and the peculiar rules necessary to be observed by masters and mistresses, in the management of their human chattels. Mr. Auld promptly forbade the continuance of her* [*reading*] *instruction; telling her, in the first place, that the thing itself was unlawful; that it was also unsafe, and could only lead to mischief. . . . Mrs. Auld evidently felt the force of his remarks; and, like an obedient wife, began to shape her course in the direction indicated by her husband. The effect of his words, on me, was neither slight nor transitory. His iron sentences—cold and*

The child psychologist Jean Piaget saw conflict as a critical part of mental development. Through battles with peers and then parents, children learn to adapt to the world and develop strategies for dealing with problems. Those children who seek to avoid conflict at all cost, or those who have overprotective parents, end up handicapped socially and mentally. The same is true of adults: it is through your battles with others that you learn what works, what doesn't, and how to protect yourself. Instead of shrinking from the idea of having enemies, then, embrace it. Conflict is therapeutic.

Enemies bring many gifts. For one thing, they motivate you and focus your beliefs. The artist Salvador Dalí found early on that there were many qualities he could not stand in people: conformity, romanticism, piety. At every stage of his life, he found someone he thought embodied these anti-ideals—an enemy to vent on. First it was the poet Federico García Lorca, who wrote romantic poetry; then it was André Breton, the heavy-handed leader of the surrealist movement. Having such enemies to rebel against made Dalí feel confident and inspired.

Enemies also give you a standard by which to judge yourself, both personally and socially. The samurai of Japan had no gauge of their excellence unless they fought the best swordsmen; it took Joe Frazier to make Muhammad Ali a truly great fighter. A tough opponent will bring out the best in you. And the bigger the opponent, the greater your reward, even in defeat. It is better to lose to a worthy opponent than to squash some harmless foe. You will gain sympathy and respect, building support for your next fight.

Being attacked is a sign that you are important enough to be a target. You should relish the attention and the chance to prove yourself. We all have aggressive impulses that we are forced to repress; an enemy supplies you with an outlet for these drives. At last you have someone on whom to unleash your aggression without feeling guilty.

Leaders have always found it useful to have an enemy at their gates in times of trouble, distracting the public from their difficulties. In using your enemies to rally your troops, polarize them as far as possible: they will fight the more fiercely when they feel a little hatred. So exaggerate the differences between you and the enemy—draw the lines clearly. Xenophon made no effort to be fair; he did not say that the Persians weren't really such a bad lot and had done much to advance civilization. He called them barbarians, the antithesis of the Greeks. He described their recent treachery and said they were an evil culture that could find no favor with the gods. And so it is with you: victory is your goal, not fairness and balance. Use the rhetoric of war to heighten the stakes and stimulate the spirit.

What you want in warfare is room to maneuver. Tight corners spell death. Having enemies gives you options. You can play them off against each other, make one a friend as a way of attacking the other, on and on. Without enemies you will not know how or where to maneuver, and you will lose a sense of your limits, of how far you can go. Early on, Julius Caesar identified Pompey as his enemy. Measuring his actions and calculating carefully, he did only those things that left him in a solid position in relation to Pompey. When war finally broke out between the two men, Caesar was at his best. But once he defeated Pompey and had no more such rivals, he lost all sense of proportion—in fact, he fancied himself a god. His defeat of Pompey was his own undoing. Your enemies force on you a sense of realism and humility.

Remember: there are always people out there who are more aggressive, more devious, more ruthless than you are, and it is inevitable that some of them will cross your path. You will have a tendency to want to conciliate and compromise with them. The reason is that such types are often brilliant deceivers who see the strategic value in charm or in seeming to allow you plenty of space, but actually their desires have no limit, and they are simply trying to disarm you. With some people you have to harden yourself, to recognize that there is no middle ground, no hope of conciliation. For your opponent your desire to compromise is a weapon to use against you. Know these dangerous enemies by their past: look for quick power grabs, sudden rises in fortune, previous acts of treachery. Once you suspect you are dealing with a Napoleon, do not lay down your arms or entrust them to someone else. You are the last line of your own defense.

*harsh—sunk deep into my heart, and stirred up not only my feelings into a sort of rebellion, but awakened within me a slumbering train of vital thought. It was a new and special revelation, dispelling a painful mystery, against which my youthful understanding had struggled, and struggled in vain, to wit: the white man's power to perpetuate the enslavement of the black man. "Very well," thought I; "knowledge unfits a child to be a slave." I instinctively assented to the proposition; and from that moment I understood the direct pathway from slavery to freedom. This was just what I needed; and got it at a time, and from a source, whence I least expected it. . . . Wise as Mr. Auld was, he evidently underrated my comprehension, and had little idea of the use to which I was capable of putting the impressive lesson he was giving to his wife. . . . That which he most loved I most hated; and the very determination which he expressed to keep me in ignorance, only rendered me the more resolute in seeking intelligence.*

*MY BONDAGE AND MY FREEDOM*, FREDERICK DOUGLASS, 1818–1895

**Image:**
The Earth. The enemy is the ground beneath your feet. It has a gravity that holds you in place, a force of resistance. Root yourself deep in this earth to gain firmness and strength. Without an enemy to walk upon, to trample, you lose your bearings and all sense of proportion.

**Authority:** If you count on safety and do not think of danger, if you do not know enough to be wary when enemies arrive, this is called a sparrow nesting on a tent, a fish swimming in a cauldron—they won't last the day. —*Chuko Liang (A.D. 181–234)*

## REVERSAL

Always keep the search for and use of enemies under control. It is clarity you want, not paranoia. It is the downfall of many tyrants to see an enemy in everyone. They lose their grip on reality and become hopelessly embroiled in the emotions their paranoia churns up. By keeping an eye on possible enemies, you are simply being prudent and cautious. Keep your suspicions to yourself, so that if you're wrong, no one will know. Also, beware of polarizing people so completely that you cannot back off. Margaret Thatcher, usually brilliant at the polarizing game, eventually lost control of it: she created too many enemies and kept repeating the same tactic, even in situations that called for retreat. Franklin Delano Roosevelt was a master polarizer, always looking to draw a line between himself and his enemies. Once he had made that line clear enough, though, he backed off, which made him look like a conciliator, a man of peace who occasionally went to war. Even if that impression was false, it was the height of wisdom to create it.

# 2

# DO NOT FIGHT THE LAST WAR

## THE GUERRILLA-WAR-OF-THE-MIND STRATEGY

*What most often weighs you down and brings you misery is the past, in the form of unnecessary attachments, repetitions of tired formulas, and the memory of old victories and defeats. You must consciously wage war against the past and force yourself to react to the present moment. Be ruthless on yourself; do not repeat the same tired methods. Sometimes you must force yourself to strike out in new directions, even if they involve risk. What you may lose in comfort and security, you will gain in surprise, making it harder for your enemies to tell what you will do. Wage guerrilla war on your mind, allowing no static lines of defense, no exposed citadels—make everything fluid and mobile.*

## THE LAST WAR

*Theory cannot equip the mind with formulas for solving problems, nor can it mark the narrow path on which the sole solution is supposed to lie by planting a hedge of principles on either side. But it can give the mind insight into the great mass of phenomena and of their relationships, then leave it free to rise into the higher realms of action. There the mind can use its innate talents to capacity, combining them all so as to seize on what is right and true as though this were a single idea formed by their concentrated pressure—as though it were a response to the immediate challenge rather than a product of thought.*

*On War*, Carl von Clausewitz, 1780–1831

No one has risen to power faster than Napoleon Bonaparte (1769–1821). In 1793 he went from captain in the French revolutionary army to brigadier general. In 1796 he became the leader of the French force in Italy fighting the Austrians, whom he crushed that year and again three years later. He became first consul of France in 1801, emperor in 1804. In 1805 he humiliated the Austrian and Russian armies at the Battle of Austerlitz.

For many, Napoleon was more than a great general; he was a genius, a god of war. Not everyone was impressed, though: there were Prussian generals who thought he had merely been lucky. Where Napoleon was rash and aggressive, they believed, his opponents had been timid and weak. If he ever faced the Prussians, he would be revealed as a great fake.

Among these Prussian generals was Friedrich Ludwig, prince of Hohenlohe-Ingelfingen (1746–1818). Hohenlohe came from one of Germany's oldest aristocratic families, one with an illustrious military record. He had begun his career young, serving under Frederick the Great (1712–86) himself, the man who had single-handedly made Prussia a great power. Hohenlohe had risen through the ranks, becoming a general at fifty—young by Prussian standards.

To Hohenlohe success in war depended on organization, discipline, and the use of superior strategies developed by trained military minds. The Prussians exemplified all of these virtues. Prussian soldiers drilled relentlessly until they could perform elaborate maneuvers as precisely as a machine. Prussian generals intensely studied the victories of Frederick the Great; war for them was a mathematical affair, the application of timeless principles. To the generals Napoleon was a Corsican hothead leading an unruly citizens' army. Superior in knowledge and skill, they would outstrategize him. The French would panic and crumble in the face of the disciplined Prussians; the Napoleonic myth would lie in ruins, and Europe could return to its old ways.

In August 1806, Hohenlohe and his fellow generals finally got what they wanted: King Friedrich Wilhelm III of Prussia, tired of Napoleon's broken promises, decided to declare war on him in six weeks. In the meantime he asked his generals to come up with a plan to crush the French.

Hohenlohe was ecstatic. This campaign would be the climax of his career. He had been thinking for years about how to beat Napoleon, and he presented his plan at the generals' first strategy session: precise marches would place the army at the perfect angle from which to attack the French as they advanced through southern Prussia. An attack in oblique formation—Frederick the Great's favorite tactic—would deliver a devastating blow. The other generals, all in their sixties and seventies, presented their own plans, but these too were merely variants on the tactics of Frederick the Great. Discussion turned into argument; several weeks went by. Finally the king had to step in and create a compromise strategy that would satisfy all of his generals.

A feeling of exuberance swept the country, which would soon relive the glory years of Frederick the Great. The generals realized that Napoleon knew about their plans—he had excellent spies—but the Prussians had a head start, and once their war machine started to move, nothing could stop it.

*He [Baron Antoine-Henri de Jomini]—often quite arbitrarily—presses [the deeds of Napoleon] into a system which he foists on Napoleon, and, in doing so, completely fails to see what, above all, really constitutes the greatness of this captain—namely, the reckless boldness of his operations, where, scoffing at all theory, he always tried to do what suited each occasion best.*

FRIEDRICH VON BERNHARDI, 1849–1930

On October 5, a few days before the king was to declare war, disturbing news reached the generals. A reconnaissance mission revealed that divisions of Napoleon's army, which they had believed was dispersed, had marched east, merged, and was massing deep in southern Prussia. The captain who had led the scouting mission reported that the French soldiers were marching with packs on their backs: where the Prussians used slow-moving wagons to provision their troops, the French carried their own supplies and moved with astonishing speed and mobility.

Before the generals had time to adjust their plans, Napoleon's army suddenly wheeled north, heading straight for Berlin, the heart of Prussia. The generals argued and dithered, moving their troops here and there, trying to decide where to attack. A mood of panic set in. Finally the king ordered a retreat: the troops would reassemble to the north and attack Napoleon's flank as he advanced toward Berlin. Hohenlohe was in charge of the rear guard, protecting the Prussians' retreat.

On October 14, near the town of Jena, Napoleon caught up with Hohenlohe, who finally faced the battle he had wanted so desperately. The numbers on both sides were equal, but while the French were an unruly force, fighting pell-mell and on the run, Hohenlohe kept his troops in tight order, orchestrating them like a corps de ballet. The fighting went back and forth until finally the French captured the village of Vierzehnheiligen.

Hohenlohe ordered his troops to retake the village. In a ritual dating back to Frederick the Great, a drum major beat out a cadence and the Prussian soldiers, their colors flying, re-formed their positions in perfect parade order, preparing to advance. They were in an open plain, though, and Napoleon's men were behind garden walls and on the house roofs. The Prussians fell like ninepins to the French marksmen. Confused, Hohenlohe ordered his soldiers to halt and change formation. The drums beat again, the Prussians marched with magnificent precision, always a sight to behold—but the French kept shooting, decimating the Prussian line.

Never had Hohenlohe seen such an army. The French soldiers were like demons. Unlike his disciplined soldiers, they moved on their own, yet there was method to their madness. Suddenly, as if from nowhere, they rushed forward on both sides, threatening to surround the Prussians. The prince ordered a retreat. The Battle of Jena was over.

Like a house of cards, the Prussians quickly crumbled, one fortress

THE BAT AND THE HOUSE-FERRETS

*A bat fell to the ground and was caught by a house-ferret. Realizing that she was on the point of being killed, she begged for her life. The house-ferret said to her that she couldn't let her go, for ferrets were supposed to be natural enemies to all birds. The bat replied that she herself was not a bird, but a mouse. She managed to extricate herself from her danger by this means. Eventually, falling a second time, the bat was caught by another house-ferret. Again she pleaded to the ferret not to eat her. The second ferret declared that she absolutely detested all mice. But the bat positively affirmed that she was not a mouse but a bat. And so she was released again. And that was how she saved herself from death twice by a mere change of name.*
This fable shows that it is not always necessary to confine ourselves to the same tactics. But, on the contrary, if we are adaptable to circumstances we can better escape danger.

*FABLES*, AESOP, SIXTH CENTURY B.C.

falling after another. The king fled east. In a matter of days, virtually nothing remained of the once mighty Prussian army.

### Interpretation

The reality facing the Prussians in 1806 was simple: they had fallen fifty years behind the times. Their generals were old, and instead of responding to present circumstances, they were repeating formulas that had worked in the past. Their army moved slowly, and their soldiers were automatons on parade. The Prussian generals had many signs to warn them of disaster: their army had not performed well in its recent engagements, a number of Prussian officers had preached reform, and, last but not least, they had had ten years to study Napoleon—his innovative strategies and the speed and fluidity with which his armies converged on the enemy. Reality was staring them in the face, yet they chose to ignore it. Indeed, they told themselves that Napoleon was the one who was doomed.

You might find the Prussian army just an interesting historical example, but in fact you are likely marching in the same direction yourself. What limits individuals as well as nations is the inability to confront reality, to see things for what they are. As we grow older, we become more rooted in the past. Habit takes over. Something that has worked for us before becomes a doctrine, a shell to protect us from reality. Repetition replaces creativity. We rarely realize we're doing this, because it is almost impossible for us to see it happening in our own minds. Then suddenly a young Napoleon crosses our path, a person who does not respect tradition, who fights in a new way. Only then do we see that our ways of thinking and responding have fallen behind the times.

Never take it for granted that your past successes will continue into the future. Actually, your past successes are your biggest obstacle: every battle, every war, is different, and you cannot assume that what worked before will work today. You must cut yourself loose from the past and open your eyes to the present. Your tendency to fight the last war may lead to your final war.

> *When in 1806 the Prussian generals . . . plunged into the open jaws of disaster by using Frederick the Great's oblique order of battle, it was not just a case of a style that had outlived its usefulness but the most extreme poverty of the imagination to which routine has ever led. The result was that the Prussian army under Hohenlohe was ruined more completely than any army has ever been ruined on the battlefield.*
>
> —*Carl von Clausewitz,* ON WAR *(1780–1831)*

## THE PRESENT WAR

In 1605, Miyamoto Musashi, a samurai who had made a name for himself as a swordsman at the young age of twenty-one, was challenged to a duel. The challenger, a young man named Matashichiro, came from the Yoshioka family, a clan itself renowned for swordsmanship. Earlier that year Musashi had defeated Matashichiro's father, Genzaemon, in a duel. Days later he had killed Genzaemon's younger brother in another duel. The Yoshioka family wanted revenge.

*I never read any treatises on strategy. . . . When we fight, we do not take any books with us.*

MAO TSE-TUNG, 1893–1976

Musashi's friends smelled a trap in Matashichiro's challenge and offered to accompany him to the duel, but Musashi went alone. In his earlier fights with the Yoshiokas, he had angered them by showing up hours late; this time, though, he came early and hid in the trees. Matashichiro arrived with a small army. Musashi would "arrive way behind schedule as usual," one of them said, "but that trick won't work with us anymore!" Confident in their ambush, Matashichiro's men lay down and hid in the grass. Suddenly Musashi leaped out from behind his tree and shouted, "I've been waiting long enough. Draw your sword!" In one swift stroke, he killed Matashichiro, then took a position at an angle to the other men. All of them jumped to their feet, but they were caught off guard and startled, and instead of surrounding him, they stood in a broken line. Musashi simply ran down the line, killing the dazed men one after another in a matter of seconds.

*REFRESHING THE MIND When you and your opponent are engaged in combat which is dragging on with no end in sight, it is crucial that you should come up with a completely different technique. By refreshing your mind and techniques as you continue to fight your opponent, you will find an appropriate rhythm-timing with which to defeat him. Whenever you and your opponent become stagnant, you must immediately employ a different method of dealing with him in order to overcome him.*

*THE BOOK OF FIVE RINGS*, MIYAMOTO MUSASHI, 1584–1645

Musashi's victory sealed his reputation as one of Japan's greatest swordsmen. He now roamed the country looking for suitable challenges. In one town he heard of an undefeated warrior named Baiken whose weapons were a sickle and a long chain with a steel ball at the end of it. Musashi wanted to see these weapons in action, but Baiken refused: the only way he could see them work, Baiken said, was by fighting a duel.

Once again Musashi's friends chose the safe route: they urged him to walk away. No one had come close to defeating Baiken, whose weapons were unbeatable: swinging his ball in the air to build up momentum, he would force his victim backward with a relentless charge, then hurl the ball at the man's face. His opponent would have to fend off the ball and chain, and while his sword arm was occupied, in that brief instant Baiken would slash him with the sickle across his neck.

Ignoring the warnings of his friends, Musashi challenged Baiken and showed up at the man's tent with two swords, one long, one short. Baiken had never seen someone fight with two swords. Also, instead of letting Baiken charge him, Musashi charged first, pushing his foe back on his heels. Baiken hesitated to throw the ball, for Musashi could parry it with one sword and strike him with the other. As he looked for an opening, Musashi suddenly knocked him off balance with a blow of the short sword and then, in a split second, followed with a thrust of the long one, stabbing him through and killing the once undefeated master Baiken.

A few years later, Musashi heard about a great samurai named Sasaki Ganryu, who fought with a very long sword—a startlingly beautiful weapon, which seemed possessed of some warlike spirit. This fight would be Musashi's ultimate test. Ganryu accepted his challenge; the duel would take place on a little island near the samurai's home.

*It is a disease to be obsessed by the thought of winning. It is also a disease to be obsessed by the thought of employing your swordsmanship. So it is to be obsessed by the thought of using everything you have learned, and to be obsessed by the thought of attacking. It is also a disease to be obsessed and stuck with the thought of ridding yourself of any of these diseases. A disease here is an obsessed mind that dwells on one thing. Because all these diseases are in your mind, you must get rid of them to put your mind in order.*

TAKUAN, JAPAN, 1573–1645

On the morning of the duel, the island was packed. A fight between such warriors was unprecedented. Ganryu arrived on time, but Musashi was late, very late. An hour went by, then two; Ganryu was furious. Finally a boat was spotted approaching the island. Its passenger was lying down, half asleep, it seemed, whittling at a long wooden oar. It was Musashi. He seemed lost in thought, staring into the clouds. When the boat came to shore, he tied a dirty towel around his head and jumped out of the boat, brandishing the long oar—longer than Ganryu's famous sword. This strange man had come to the biggest fight of his life with an oar for a sword and a towel for a headband.

Ganryu called out angrily, "Are you so frightened of me that you have broken your promise to be here by eight?" Musashi said nothing but stepped closer. Ganryu drew his magnificent sword and threw the sheath onto the sand. Musashi smiled: "Sasaki, you have just sealed your doom." "Me? Defeated? Impossible!" "What victor on earth," replied Musashi, "would abandon his sheath to the sea?" This enigmatic remark only made Ganryu angrier.

Then Musashi charged, aiming his sharpened oar straight for his enemy's eyes. Ganryu quickly raised his sword and struck at Musashi's head but missed, only cutting the towel headband in two. He had never missed before. In almost the same instant, Musashi brought down his wooden sword, knocking Ganryu off his feet. The spectators gasped. As Ganryu struggled up, Musashi killed him with a blow to the head. Then, after bowing politely to the men officiating over the duel, he got back into the boat and left as calmly as he had arrived.

From that moment on, Musashi was considered a swordsman without peer.

*Anyone can plan a campaign, but few are capable of waging war, because only a true military genius can handle the developments and circumstances.*

NAPOLEON BONAPARTE, 1769–1821

### Interpretation

Miyamoto Musashi, author of *The Book of Five Rings,* won all his duels for one reason: in each instance he adapted his strategy to his opponent and to the circumstances of the moment. With Matashichiro he decided it was time to arrive early, which he hadn't done in his previous fights. Victory against superior numbers depended on surprise, so he leaped up when his opponents lay down; then, once he had killed their leader, he set himself at an angle that invited them to charge at him instead of surrounding him, which would have been much more dangerous for him. With Baiken it was simply a matter of using two swords and then crowding his space, giving him no time to react intelligently to this novelty. With Ganryu he set out to infuriate and humiliate his haughty opponent—the wooden sword, the nonchalant attitude, the dirty-towel headband, the enigmatic remark, the charge at the eyes.

Musashi's opponents depended on brilliant technique, flashy swords, and unorthodox weapons. That is the same as fighting the last war: instead of responding to the moment, they relied on training, technology, and what had worked before. Musashi, who had grasped the essence of strategy when he was still very young, turned their rigidity into their downfall. His first thought was of the gambit that would take this particular opponent most by surprise. Then he would anchor himself in the moment: having set his opponent off balance with something unexpected, he would watch carefully, then respond with another action, usually improvised, that would turn mere disequilibrium into defeat and death.

In preparing yourself for war, you must rid yourself of myths and misconceptions. Strategy is not a question of learning a series of moves or ideas to follow like a recipe; victory has no magic formula. Ideas are merely nutrients for the soil: they lie in your brain as possibilities, so that in the heat of the moment they can inspire a direction, an appropriate and creative response. Let go of all fetishes—books, techniques, formulas, flashy weapons—and learn to become your own strategist.

> *Thus one's victories in battle cannot be repeated—they take their form in response to inexhaustibly changing circumstances.*
> —*Sun-tzu (fourth century* B.C.*)*

*Thunder and wind: the image of DURATION. Thus the superior man stands firm And does not change his direction. Thunder rolls, and the wind blows; both are examples of extreme mobility and so are seemingly the very opposite of duration, but the laws governing their appearance and subsidence, their coming and going, endure. In the same way the independence of the superior man is not based on rigidity and immobility of character. He always keeps abreast of the time and changes with it. What endures is the unswerving directive, the inner law of his being, which determines all his actions.*

*THE I CHING*, CHINA, CIRCA EIGHTH CENTURY B.C.

## KEYS TO WARFARE

In looking back on an unpleasant or disagreeable experience, the thought inevitably occurs to us: if only we had said or done *x* instead of *y*, if only we could do it over. Many a general has lost his head in the heat of battle and then, looking back, has thought of the one tactic, the one maneuver, that would have changed it all. Even Prince Hohenlohe, years later, could see how he had botched the retaking of Vierzehnheiligen. The problem, though, is not that we think of the solution only when it is too late. The problem is that we imagine that knowledge is what was lacking: if only we had known more, if only we had thought it through more thoroughly. That is precisely the wrong approach. What makes us go astray in the first place is that we are unattuned to the present moment, insensitive to the circumstances. We are listening to our own thoughts, reacting to things that happened in the past, applying theories and ideas that we digested long ago but that have nothing to do with our predicament in the present. More books, theories, and thinking only make the problem worse.

*My policy is to have no policy.*

ABRAHAM LINCOLN, 1809–1865

Understand: the greatest generals, the most creative strategists, stand out not because they have more knowledge but because they are able, when necessary, to drop their preconceived notions and focus intensely on the present moment. That is how creativity is sparked and opportunities are seized. Knowledge, experience, and theory have limitations: no amount of thinking in advance can prepare you for the chaos of life, for the infinite possibilities of the moment. The great philosopher of war Carl von Clausewitz called this "friction": the difference between our plans and what actually happens. Since friction is inevitable, our minds have to be capable of keeping up with change and adapting to the unexpected. The better we can adapt our thoughts to changing circumstances, the more realistic our responses to them will be. The more we lose ourselves in predigested theories and past experiences, the more inappropriate and delusional our response.

*If you put an empty gourd on the water and touch it, it will slip to one side. No matter how you try, it won't stay in one spot. The mind of someone who has reached the ultimate state does not stay with anything, even for a second. It is like an empty gourd on the water that is pushed around.*

TAKUAN, JAPAN, 1573–1645

It can be valuable to analyze what went wrong in the past, but it is far more important to develop the capacity to think in the moment. In that way you will make far fewer mistakes to analyze.

Think of the mind as a river: the faster it flows, the better it keeps up with the present and responds to change. The faster it flows, also the more it refreshes itself and the greater its energy. Obsessional thoughts, past experiences (whether traumas or successes), and preconceived notions are like boulders or mud in this river, settling and hardening there and damming it up. The river stops moving; stagnation sets in. You must wage constant war on this tendency in the mind.

The first step is simply to be aware of the process and of the need to fight it. The second is to adopt a few tactics that might help you to restore the mind's natural flow.

**Reexamine all your cherished beliefs and principles.** When Napoleon was asked what principles of war he followed, he replied that he followed none. His genius was his ability to respond to circumstances, to make the most of what he was given—he was the supreme opportunist. Your only principle, similarly, should be to have no principles. To believe that strategy has inexorable laws or timeless rules is to take up a rigid, static position that will be your undoing. Of course the study of history and theory can broaden your vision of the world, but you have to combat theory's tendency to harden into dogma. Be brutal with the past, with tradition, with the old ways of doing things. Declare war on sacred cows and voices of convention in your own head.

Our education is often a problem. During World War II, the British fighting the Germans in the deserts of North Africa were well trained in tank warfare; you might say they were indoctrinated with theories about it. Later in the campaign, they were joined by American troops who were much less educated in these tactics. Soon, though, the Americans began to fight in a way that was equal if not superior to the British style; they adapted to the mobility of this new kind of desert combat. According to Field Marshal Erwin Rommel himself, the leader of the German army in North Africa, "The Americans . . . profited far more than the British from their experience in Africa, thus confirming the axiom that education is easier than reeducation."

What Rommel meant was that education tends to burn precepts into the mind that are hard to shake. In the midst of combat, the trained mind may fall a step behind—focusing more on learned rules than on the changing circumstances of battle. When you are faced with a new situation, it is often best to imagine that you know nothing and that you need to start learning all over again. Clearing your head of everything you thought you knew, even your most cherished ideas, will give you the mental space to be educated by your present experience—the best school of all. You will develop your own strategic muscles instead of depending on other people's theories and books.

**Erase the memory of the last war.** The last war you fought is a danger, even if you won it. It is fresh in your mind. If you were victorious, you will tend to repeat the strategies you just used, for success makes us lazy and complacent; if you lost, you may be skittish and indecisive. Do not think about the last war; you do not have the distance or the detachment. Instead do whatever you can to blot it from your mind. During the Vietnam War, the great North Vietnamese general Vo Nguyen Giap had a simple rule of thumb: after a successful campaign, he would convince himself that it had actually been a failure. As a result he never got drunk on his success, and he never repeated the same strategy in the next battle. Rather he had to think through each situation anew.

Ted Williams, perhaps baseball's greatest pure hitter, made a point of always trying to forget his last at-bat. Whether he'd gotten a home run or a strikeout, he put it behind him. No two at-bats are the same, even against the same pitcher, and Williams wanted an open mind. He would not wait for the next at-bat to start forgetting: the minute he got back to the dugout, he started focusing on what was happening in the game taking place. Attention to the details of the present is by far the best way to crowd out the past and forget the last war.

**Keep the mind moving.** When we were children, our minds never stopped. We were open to new experiences and absorbed as much of them as possible. We learned fast, because the world around us excited us. When we felt frustrated or upset, we would find some creative way to get what we wanted and then quickly forget the problem as something new crossed our path.

All the greatest strategists—Alexander the Great, Napoleon, Musashi—were childlike in this respect. Sometimes, in fact, they even acted like children. The reason is simple: superior strategists see things as they are. They are highly sensitive to dangers and opportunities. Nothing stays the same in life, and keeping up with circumstances as they change requires a great deal of mental fluidity. Great strategists do not act according to preconceived ideas; they respond to the moment, like children. Their minds are always moving, and they are always excited and curious. They quickly forget the past—the present is much too interesting.

The Greek thinker Aristotle thought that life was defined by movement. What does not move is dead. What has speed and mobility has more possibilities, more life. We all start off with the mobile mind of a Napoleon, but as we get older, we tend to become more like the Prussians. You may think that what you'd like to recapture from your youth is

*Defeat is bitter. Bitter to the common soldier, but trebly bitter to his general. The soldier may comfort himself with the thought that, whatever the result, he has done his duty faithfully and steadfastly, but the commander has failed in his duty if he has not won victory—for that is his duty. He has no other comparable to it. He will go over in his mind the events of the campaign. "Here," he will think, "I went wrong; here I took counsel of my fears when I should have been bold; there I should have waited to gather strength, not struck piecemeal; at such a moment I failed to grasp opportunity when it was presented to me." He will remember the soldiers whom he sent into the attack that failed and who did not come back. He will recall the look in the eyes of men who trusted him. "I have failed them," he will say to himself, "and failed my country!" He will see himself for what he is—a defeated general. In a dark hour he will turn in upon himself and question the very foundations of his leadership and manhood.*

*And then he must stop! For if he is ever to command in battle again, he must shake off these regrets, and stamp on them, as they claw at his will and his self-confidence. He must beat off these attacks he delivers*

*against himself, and cast out the doubts born of failure. Forget them, and remember only the lessons to be learned from defeat—they are more than from victory.*

DEFEAT INTO VICTORY, WILLIAM SLIM, 1897–1970

your looks, your physical fitness, your simple pleasures, but what you really need is the fluidity of mind you once possessed. Whenever you find your thoughts revolving around a particular subject or idea—an obsession, a resentment—force them past it. Distract yourself with something else. Like a child, find something new to be absorbed by, something worthy of concentrated attention. Do not waste time on things you cannot change or influence. Just keep moving.

**Absorb the spirit of the times.** Throughout the history of warfare, there have been classic battles in which the past has confronted the future in a hopeless mismatch. It happened in the seventh century, when the Persians and Byzantines confronted the invincible armies of Islam, with their new form of desert fighting; or in the first half of the thirteenth century, when the Mongols used relentless mobility to overwhelm the heavy armies of the Russians and Europeans; or in 1806, when Napoleon crushed the Prussians at Jena. In each case the conquering army developed a way of fighting that maximized a new form of technology or a new social order.

You can reproduce this effect on a smaller scale by attuning yourself to the spirit of the times. Developing antennae for the trends that have yet to crest takes work and study, as well as the flexibility to adapt to those trends. As you get older, it is best to periodically alter your style. In the golden age of Hollywood, most actresses had very short careers. But Joan Crawford fought the studio system and managed to have a remarkably long career by constantly changing her style, going from siren to noir heroine to cult queen. Instead of staying sentimentally attached to some fashion of days gone by, she was able to sense a rising trend and go with it. By constantly adapting and changing your style, you will avoid the pitfalls of your previous wars. Just when people feel they know you, you will change.

**Reverse course.** The great Russian novelist Fyodor Dostoyevsky suffered from epilepsy. Just before a seizure, he would experience a moment of intense ecstasy, which he described as a feeling of being suddenly flooded with reality, a momentary vision of the world exactly as it is. Later he would find himself getting depressed, as this vision was crowded out by the habits and routines of daily life. During these depressions, wanting to feel that closeness to reality again, he would go to the nearest casino and gamble away all his money. There reality would overwhelm him; comfort and routine would be gone, stale patterns broken. Having to rethink everything, he would get his creative energy back. This was the closest he could deliberately come to the sense of ecstasy he got through epilepsy.

Dostoyevsky's method was a little extreme, but sometimes you have to shake yourself up, break free from the hold of the past. This can take the form of reversing your course, doing the opposite of what you would normally do in any given situation, putting yourself in some unusual circumstance, or literally starting over. In those situations the mind has to deal with a new reality, and it snaps to life. The change may be alarming, but it is also refreshing—even exhilarating.

Relationships often develop a certain tiresome predictability. You do what you usually do, other people respond the way they usually do, and around it goes. If you reverse course, act in a novel manner, you alter the entire dynamic. Do this every so often to break up the relationship's stale patterns and open it to new possibilities.

Think of your mind as an army. Armies must adapt to the complexity and chaos of modern war by becoming more fluid and maneuverable. The ultimate extension of this evolution is guerrilla warfare, which exploits chaos by making disorder and unpredictability a strategy. The guerrilla army never stops to defend a particular place or town; it wins by always moving, staying one step ahead. By following no set pattern, it gives the enemy no target. The guerrilla army never repeats the same tactic. It responds to the situation, the moment, the terrain where it happens to find itself. There is no front, no concrete line of communication or supply, no slow-moving wagon. The guerrilla army is pure mobility.

That is the model for your new way of thinking. Apply no tactic rigidly; do not let your mind settle into static positions, defending any particular place or idea, repeating the same lifeless maneuvers. Attack problems from new angles, adapting to the landscape and to what you're given. By staying in constant motion you show your enemies no target to aim at. You exploit the chaos of the world instead of succumbing to it.

**Image:** Water.
Adapting its shape
to wherever it
moves in
the stream,
pushing
rocks out of
its way, smoothing
boulders,
it never stops,
is never the same.
The faster it moves
the clearer
it gets.

*To know that one is in a certain condition, in a certain state, is already a process of liberation; but a man who is not aware of his condition, of his struggle, tries to be something other than he is, which brings about habit. So, then, let us keep in mind that we want to examine what is, to observe and be aware of exactly what is the actual, without giving it any slant, without giving it an interpretation. It needs an extraordinarily astute mind, an extraordinarily pliable heart, to be aware of and to follow what is; because what is is constantly moving, constantly undergoing a transformation, and if the mind is tethered to belief, to knowledge, it ceases to pursue, it ceases to follow the swift movement of what is. What is is not static, surely—it is constantly moving, as you will see if you observe it very closely. To follow it, you need a very swift mind and a pliable heart—which are denied when the mind is static, fixed in a belief, in a prejudice, in an identification; and a mind and heart that are dry cannot follow easily, swiftly, that which is.*

JIDDU KRISHNAMURTI, 1895–1986

**Authority:** Some of our generals failed because they worked out everything by rule. They knew what Frederick did at one place, and Napoleon at another. They were always thinking about what Napoleon would do. . . . I don't underrate the value of military knowledge, but if men make war in slavish observance to rules, they will fail. . . . War is progressive. —*Ulysses S. Grant (1822–85)*

## REVERSAL

There is never any value in fighting the last war. But while you're eliminating that pernicious tendency, you must imagine that your enemy is trying to do the same—trying to learn from and adapt to the present. Some of history's worst military disasters have come not out of fighting the last war but out of assuming that that's what your opponent will do. When Saddam Hussein of Iraq invaded Kuwait in 1990, he thought the United States had yet to recover from "Vietnam syndrome"—the fear of casualties and loss that had been so traumatic during the Vietnam period—and that it would either avoid war altogether or would fight in the same way it had, trying to win the fight from the air instead of on the ground. He did not realize that the American military was ready for a new kind of war. Remember: the loser in any battle may be too traumatized to fight again but may also learn from the experience and move on. Err on the side of caution; be ready. Never let your enemy surprise you in war.

# 3

# AMIDST THE TURMOIL OF EVENTS, DO NOT LOSE YOUR PRESENCE OF MIND

## THE COUNTERBALANCE STRATEGY

*In the heat of battle, the mind tends to lose its balance. Too many things confront you at the same time—unexpected setbacks, doubts and criticisms from your own allies. There's a danger of responding emotionally, with fear, depression, or frustration. It is vital to keep your presence of mind, maintaining your mental powers whatever the circumstances. You must actively resist the emotional pull of the moment—staying decisive, confident, and aggressive no matter what hits you. Make the mind tougher by exposing it to adversity. Learn to detach yourself from the chaos of the battlefield. Let others lose their heads; your presence of mind will steer you clear of their influence and keep you on course.*

## THE HYPERAGGRESSIVE TACTIC

Vice Admiral Lord Horatio Nelson (1758–1805) had been through it all. He had lost his right eye in the siege of Calvi and his right arm in the Battle of Tenerife. He had defeated the Spanish at Cape St. Vincent in 1797 and had thwarted Napoleon's Egyptian campaign by defeating his navy at the Battle of the Nile the following year. But none of his tribulations and triumphs prepared him for the problems he faced from his own colleagues in the British navy as they prepared to go to war against Denmark in February 1801.

*[Presence of mind] must play a great role in war, the domain of the unexpected, since it is nothing but an increased capacity of dealing with the unexpected. We admire presence of mind in an apt repartee, as we admire quick thinking in the face of danger. . . . The expression "presence of mind" precisely conveys the speed and immediacy of the help provided by the intellect.*

ON WAR, CARL VON CLAUSEWITZ, 1780–1831

Nelson, England's most glorious war hero, was the obvious choice to lead the fleet. Instead the Admiralty chose Sir Hyde Parker, with Nelson his second-in-command. This war was a delicate business; it was intended to force the disobedient Danes to comply with a British-led embargo on the shipping of military goods to France. The fiery Nelson was prone to lose his cool. He hated Napoleon, and if he went too far against the Danes, he would produce a diplomatic fiasco. Sir Hyde was an older, more stable, even-tempered man who would do the job and nothing more.

Nelson swallowed his pride and took the assignment, but he saw trouble ahead. He knew that time was of the essence: the faster the navy sailed, the less chance the Danes would have to build up their defenses. The ships were ready to sail, but Parker's motto was "Everything in good order." It wasn't his style to hurry. Nelson hated his casualness and burned for action: he reviewed intelligence reports, studied maps, and came up with a detailed plan for fighting the Danes. He wrote to Parker urging him to seize the initiative. Parker ignored him.

*More life may trickle out of men through thought than through a gaping wound.*

THOMAS HARDY, 1840–1928

At last, on March 11, the British fleet set sail. Instead of heading for Copenhagen, however, Parker anchored well to the north of the city's harbor and called a meeting of his captains. According to intelligence reports, he explained, the Danes had prepared elaborate defenses for Copenhagen. Boats anchored in the harbor, forts to the north and south, and mobile artillery batteries could blast the British out of the water. How to fight this artillery without terrible losses? Also, pilots who knew the waters around Copenhagen reported that they were treacherous, places of sandbars and tricky winds. Navigating these dangers under bombardment would be harrowing. With all of these difficulties, perhaps it was best to wait for the Danes to leave harbor and then fight them in open sea.

Nelson struggled to control himself. Finally he let loose, pacing the room, the stub of his lost arm jerking as he spoke. No war, he said, had ever been won by waiting. The Danish defenses looked formidable "to those who are children at war," but he had worked out a strategy weeks earlier: he would attack from the south, the easier approach, while Parker and a reserve force would stay to the city's north. Nelson would use his mobility to take out the Danish guns. He had studied the maps: sandbars were no threat. As for the wind, aggressive action was more important than fretting over wind.

Nelson's speech energized Parker's captains. He was by far their most successful leader, and his confidence was catching. Even Sir Hyde was impressed, and the plan was approved.

The next morning Nelson's line of ships advanced on Copenhagen, and the battle began. The Danish guns, firing on the British at close range, took a fierce toll. Nelson paced the deck of his flagship, HMS *Elephant*, urging his men on. He was in an excited, almost ecstatic state. A shot through the mainmast nearly hit him: "It is warm work, and this day may be the last to any of us at any moment," he told a colonel, a little shaken up by the blast, "but mark you, I would not be elsewhere for thousands."

Parker followed the battle from his position to the north. He now regretted agreeing to Nelson's plan; he was responsible for the campaign, and a defeat here could ruin his career. After four hours of back-and-forth bombardment, he had seen enough: the fleet had taken a beating and had gained no advantage. Nelson never knew when to quit. Parker decided it was time to hoist signal flag 39, the order to withdraw. The first ships to see it were to acknowledge it and pass the signal on down the line. Once acknowledged there was nothing else to do but retreat. The battle was over.

On board the *Elephant*, a lieutenant told Nelson about the signal. The vice-admiral ignored it. Continuing to pound the Danish defenses, he eventually called to an officer, "Is number sixteen still hoisted?" Number 16 was his own flag; it meant "Engage the enemy more closely." The officer confirmed that the flag was still flying. "Mind you keep it so," Nelson told him. A few minutes later, Parker's signal still flapping in the breeze, Nelson turned to his flag captain: "You know, Foley, I have only one eye—I have a right to be blind sometimes." And raising his telescope to his blind eye, he calmly remarked, "I really do not see the signal."

Torn between obeying Parker and obeying Nelson, the fleet captains chose Nelson. They would risk their careers along with his. But soon the Danish defenses started to crack; some of the ships anchored in the harbor surrendered, and the firing of the guns began to slow. Less than an hour after Parker's signal to stop the battle, the Danes surrendered.

The next day Parker perfunctorily congratulated Nelson on the victory. He did not mention his subordinate's disobedience. He was hoping the whole affair, including his own lack of courage, would be quietly forgotten.

### Interpretation

When the Admiralty put its faith in Sir Hyde, it made a classical military error: it entrusted the waging of a war to a man who was careful and methodical. Such men may seem calm, even strong, in times of peace, but their self-control often hides weakness: the reason they think things through so carefully is that they are terrified of making a mistake and of

*So Grant was alone; his most trusted subordinates besought him to change his plans, while his superiors were astounded at his temerity and strove to interfere. Soldiers of reputation and civilians in high places condemned, in advance, a campaign that seemed to them as hopeless as it was unprecedented. If he failed, the country would concur with the Government and the Generals. Grant knew all this, and appreciated his danger, but was as invulnerable to the apprehensions of ambition as to the entreaties of friendship, or the anxieties even of patriotism. That quiet confidence in himself which never forsook him, and which amounted indeed almost to a feeling of fate, was uninterrupted. Having once determined in a matter that required irreversible decision, he never reversed, nor even misgave, but was steadily loyal to himself and his plans. This absolute and implicit faith was, however, as far as possible from conceit or enthusiasm; it was simply a consciousness or conviction, rather, which brought the very strength it believed in; which was itself strength, and which inspired others with a trust in him, because he was able thus to trust himself.*

MILITARY HISTORY OF ULYSSES S. GRANT, ADAM BADEAU, 1868

what that might mean for them and their career. This doesn't come out until they are tested in battle: suddenly they cannot make a decision. They see problems everywhere and defeat in the smallest setback. They hang back not out of patience but out of fear. Often these moments of hesitation spell their doom.

Lord Nelson operated according to the opposite principle. Slight of build, with a delicate constitution, he compensated for his physical weakness with fierce determination. He forced himself to be more resolute than anyone around him. The moment he entered battle, he ratcheted up his aggressive impulses. Where other sea lords worried about casualties, the wind, changes in the enemy's formation, he concentrated on his plan. Before battle no one strategized or studied his opponent more thoroughly. (That knowledge helped Nelson to sense when the enemy was ready to crumble.) But once the engagement began, hesitation and carefulness were dropped.

Presence of mind is a kind of counterbalance to mental weakness, to our tendency to get emotional and lose perspective in the heat of battle. Our greatest weakness is losing heart, doubting ourselves, becoming unnecessarily cautious. Being more careful is not what we need; that is just a screen for our fear of conflict and of making a mistake. What we need is double the resolve—an intensification of confidence. That will serve as a counterbalance.

In moments of turmoil and trouble, you must force yourself to be more determined. Call up the aggressive energy you need to overcome caution and inertia. Any mistakes you make, you can rectify with more energetic action still. Save your carefulness for the hours of preparation, but once the fighting begins, empty your mind of doubts. Ignore those who quail at any setback and call for retreat. Find joy in attack mode. Momentum will carry you through.

*The senses make a more vivid impression on the mind than systematic thought. . . . Even the man who planned the operation and now sees it being carried out may well lose confidence in his earlier judgment. . . . War has a way of masking the stage with scenery crudely daubed with fearsome apparitions. Once this is cleared away, and the horizon becomes unobstructed, developments will confirm his earlier convictions—this is one of the great chasms between planning and execution.*

—*Carl von Clausewitz,* ON WAR *(1780–1831)*

*There was once a man who may be called the "generalissimo" of robbers and who went by the name of Hakamadare. He had a strong mind and a powerful build. He was swift of foot, quick with his hands, wise in thinking and plotting. Altogether there was no one who could compare with him. His business was to rob people of their possessions when they were off guard. Once, around the tenth month of a year, he needed clothing and decided to get hold of some. He went to prospective spots and walked about, looking. About midnight when people had gone to sleep and were quiet, under a somewhat blurry moon he saw a man dressed in abundant clothes sauntering about on a boulevard. The man, with his trouser-skirt tucked up with strings perhaps and in a formal hunting robe which gently covered his body, was playing the flute, alone, apparently in no hurry to go to any particular place. Wow, here's a fellow who's shown up just to give me his clothes, Hakamadare thought. Normally he would have gleefully run up and beaten his quarry down and robbed him of his clothes. But this time, unaccountably, he felt something fearsome about the man, so he followed him for a couple of hundred yards.*

## THE DETACHED-BUDDHA TACTIC

Watching the movie director Alfred Hitchcock (1899–1980) at work on a film set was often quite a surprise to those seeing it for the first time. Most filmmakers are wound-up balls of energy, yelling at the crew and barking out orders, but Hitchcock would sit in his chair, sometimes dozing, or at least with his eyes half closed. On the set of *Strangers on a Train*, made in 1951, the actor Farley Granger thought Hitchcock's behavior meant he was angry or upset and asked him if anything was wrong. "Oh," Hitchcock replied sleepily, "I'm so bored." The crew's complaints, an actor's tantrums—nothing fazed him; he would just yawn, shift in his chair, and ignore the problem. "Hitchcock . . . didn't seem to direct us at all," said the actress Margaret Lockwood. "He was a dozing, nodding Buddha with an enigmatic smile on his face."

It was hard for Hitchcock's colleagues to understand how a man doing such stressful work could stay so calm and detached. Some thought it was part of his character—that there was something inherently cold-blooded about him. Others thought it a gimmick, a put-on. Few suspected the truth: before the filmmaking had even begun, Hitchcock would have prepared for it with such intense attention to detail that nothing could go wrong. He was completely in control; no temperamental actress, no panicky art director, no meddling producer could upset him or interfere with his plans. Feeling such absolute security in what he had set up, he could afford to lie back and fall asleep.

Hitchcock's process began with a storyline, whether from a novel or an idea of his own. As if he had a movie projector in his head, he would begin to visualize the film. Next, he would start meeting with a writer, who would soon realize that this job was unlike any other. Instead of taking some producer's half-baked idea and turning it into a screenplay, the writer was simply there to put on paper the dream trapped in Hitchcock's mind. He or she would add flesh and bones to the characters and would of course write the dialog, but not much else. When Hitchcock sat down with the writer Samuel Taylor for the first script meeting on the movie *Vertigo* (1958), his descriptions of several scenes were so vivid, so intense, that the experiences seemed almost to have been real, or maybe something he had dreamed. This completeness of vision foreclosed creative conflict. As Taylor soon realized, although he was writing the script, it would remain a Hitchcock creation.

Once the screenplay was finished, Hitchcock would transform it into an elaborate shooting script. Blocking, camera positions, lighting, and set dimensions were spelled out in detailed notes. Most directors leave themselves some latitude, shooting scenes from several angles, for example, to give the film editor options to work with later on. Not Hitchcock: he essentially edited the entire film in the shooting script. He knew exactly what he wanted and wrote it down. If a producer or actor tried to add or change a scene, Hitchcock was outwardly pleasant—he could afford to pretend to listen—but inside he was totally unmoved.

Nothing was left to chance. For the building of the sets (quite elaborate in a movie like *Rear Window*), Hitchcock would present the production designer with precise blueprints, floor plans, incredibly detailed lists of props. He supervised every aspect of set construction. He was particularly attentive to the clothes of his leading actresses: according to Edith Head, costumer on many Hitchcock movies, including *Dial M for Murder*

*The man himself didn't seem to think, Somebody's following me. On the contrary, he continued to play the flute with what appeared to be greater calm.*
*Give him a try, Hakamadare said to himself, and ran up close to the man, making as much clatter as he could with his feet. The man, however, looked not the least disturbed. He simply turned to look, still playing the flute. It wasn't possible to jump on him. Hakamadare ran off.*
*Hakamadare tried similar approaches a number of times, but the man remained utterly unperturbed. Hakamadare realized he was dealing with an unusual fellow. When they had covered about a thousand yards, though, Hakamadare decided he couldn't continue like this, drew his sword, and ran up to him. This time the man stopped playing the flute and, turning, said, "What in the world are you doing?" Hakamadare couldn't have been struck with greater fear even if a demon or a god had run up to attack him when he was walking alone. For some unaccountable reason he lost both heart and courage. Overcome with deathly fear and despite himself, he fell on his knees and hands.*
*"What are you doing?" the man repeated. Hakamadare felt he*

*couldn't escape even if he tried. "I'm trying to rob you," he blurted out. "My name is Hakamadare." "I've heard there's a man about with that name, yes. A dangerous, unusual fellow, I'm told," the man said. Then he simply said to Hakamadare, "Come with me," and continued on his way, playing the flute again. Terrified that he was dealing with no ordinary human being, and as if possessed by a demon or a god, Hakamadare followed the man, completely mystified. Eventually the man walked into a gate behind which was a large house. He stepped inside from the verandah after removing his shoes. While Hakamadare was thinking, He must be the master of the house, the man came back and summoned him. As he gave him a robe made of thick cotton cloth, he said, "If you need something like this in the future, just come and tell me. If you jump on somebody who doesn't know your intentions, you may get hurt." Afterward it occurred to Hakamadare that the house belonged to Governor of Settsu Fujiwara no Yasumasa. Later, when he was arrested, he is known to have observed, "He was such an unusually weird, terrifying man!" Yasumasa was not a warrior by family tradition because he*

in 1954, "There was a reason for every color, every style, and he was absolutely certain about everything he settled on. For one scene he saw [Grace Kelly] in pale green, for another in white chiffon, for another in gold. He was really putting a dream together in the studio." When the actress Kim Novak refused to wear a gray suit in *Vertigo* because she felt it made her look washed out, Hitchcock told her he wanted her to look like a woman of mystery who had just stepped out of the San Francisco fog. How could she argue with that? She wore the suit.

Hitchcock's actors found working with him strange yet pleasant. Some of Hollywood's best—Joseph Cotten, Grace Kelly, Cary Grant, Ingrid Bergman—said that he was the easiest director to work for: his nonchalance was catching, and since his films were so carefully staged as not to depend on the actor's performance in any particular scene, they could relax. Everything went like clockwork. As James Stewart told the cast of *The Man Who Knew Too Much* (1956), "We're in the hands of an expert here. You can lean on him. Just do everything he tells you and the whole thing will be okay."

As Hitchcock sat calmly on the set, apparently half asleep, the cast and crew could see only the small part each one played. They had no idea how everything fit into his vision. When Taylor saw *Vertigo* for the first time, it was like seeing another man's dream. The film neatly duplicated the vision Hitchcock had expressed to him many months before.

### Interpretation

The first film Hitchcock directed was *The Pleasure Garden*, a silent he made in 1925. The production went wrong in every conceivable way. Hitchcock hated chaos and disorder; unexpected events, panicky crew members, and any loss of control made him miserable. From that point on, he decided, he would treat filmmaking like a military operation. He would give his producers, actors, and crew no room to mess up what he wanted to create. He taught himself every aspect of film production: set design, lighting, the technicalities of cameras and lenses, editing, sound. He ran every stage of the film's making. No shadow could fall between the planning and the execution.

Establishing control in advance the way Hitchcock did might not seem like presence of mind, but it actually takes that quality to its zenith. It means entering battle (in Hitchcock's case a film shoot) feeling calm and ready. Setbacks may come, but you will have foreseen them and thought of alternatives, and you are ready to respond. Your mind will never go blank when it is that well prepared. When your colleagues barrage you with doubts, anxious questions, and slipshod ideas, you may nod and pretend to listen, but really you're ignoring them—you've outthought them in advance. And your relaxed manner will prove contagious to other people, making them easier to manage in turn.

It is easy to be overwhelmed by everything that faces you in battle, where so many people are asking or telling you what to do. So many vital matters press in on you that you can lose sight of your goals and plans; suddenly you can't see the forest for the trees. Understand: presence of mind is the ability to detach yourself from all that, to see the whole battlefield, the whole picture, with clarity. All great generals have this quality. And what gives you that mental distance is preparation, mastering the details beforehand. Let people think your Buddha-like detachment comes from some mysterious source. The less they understand you the better.

*For the love of God, pull yourself together and do not look at things so darkly: the first step backward makes a poor impression in the army, the second step is dangerous, and the third becomes fatal.*

*—Frederick the Great (1712–86), letter to a general*

*was a son of Munetada. Yet he was not the least inferior to anyone who was a warrior by family tradition. He had a strong mind, was quick with his hands, and had tremendous strength. He was also subtle in thinking and plotting. So even the imperial court did not feel insecure in employing him in the way of the warrior. As a result, the whole world greatly feared him and was intimidated by him.*

*Legends of the Samurai*, Hiroaki Sato, 1995

## KEYS TO WARFARE

We humans like to see ourselves as rational creatures. We imagine that what separates us from animals is the ability to think and reason. But that is only partly true: what distinguishes us from animals just as much is our capacity to laugh, to cry, to feel a range of emotions. We are in fact emotional creatures as well as rational ones, and although we like to think we govern our actions through reason and thought, what most often dictates our behavior is the emotion we feel in the moment.

We maintain the illusion that we are rational through the routine of our daily affairs, which helps us to keep things calm and apparently controlled. Our minds seem rather strong when we're following our routines. But place any of us in an adverse situation and our rationality vanishes; we react to pressure by growing fearful, impatient, confused. Such moments reveal us for the emotional creatures we are: under attack, whether by a known enemy or unpredictably by a colleague, our response is dominated by feelings of anger, sadness, betrayal. Only with great effort can we reason our way through these periods and respond rationally—and our rationality rarely lasts past the next attack.

Understand: your mind is weaker than your emotions. But you become aware of this weakness only in moments of adversity—precisely the time when you need strength. What best equips you to cope with the heat of battle is neither more knowledge nor more intellect. What makes your mind stronger, and more able to control your emotions, is internal discipline and toughness.

No one can teach you this skill; you cannot learn it by reading about it. Like any discipline, it can come only through practice, experience, even a little suffering. The first step in building up presence of mind is to see the need for it—to want it badly enough to be willing to work for it. Historical figures who stand out for their presence of mind—Alexander the Great, Ulysses S. Grant, Winston Churchill—acquired it through adversity, through trial and error. They were in positions of responsibility in which they had to develop this quality or sink. Although these men may have been blessed with an unusual amount of personal fortitude, they had to work hard to strengthen this into presence of mind.

The ideas that follow are based on their experience and hard-won victories. Think of these ideas as exercises, ways to toughen your mind, each a kind of counterbalance to emotion's overpowering pull.

*The first quality of a General-in-Chief is to have a cool head which receives exact impressions of things, which never gets heated, which never allows itself to be dazzled, or intoxicated, by good or bad news. The successive simultaneous sensations which he receives in the course of a day must be classified, and must occupy the correct places they merit to fill, because common sense and reason are the results of the comparison of a number of sensations each equally well considered. There are certain men who, on account of their moral and physical constitution, paint mental pictures out of everything: however exalted be their reason, their will, their courage, and whatever good qualities they may possess, nature has not fitted them to command armies, nor to direct great operations of war.*

NAPOLEON BONAPARTE, 1769–1821

**Expose yourself to conflict.** George S. Patton came from one of America's most distinguished military families—his ancestors included generals and colonels who had fought and died in the American Revolution and the Civil War. Raised on stories of their heroism, he followed in their footsteps and chose a career in the military. But Patton was also a sensitive young man, and he had one deep fear: that in battle he would turn coward and disgrace the family name.

Patton had his first real taste of battle in 1918, at the age of thirty-two, during the Allied offensive on the Argonne during World War I. He commanded a tank division. At one point during the battle, Patton managed to lead some American infantrymen to a position on a hilltop overlooking a key strategic town, but German fire forced them to take cover. Soon it became clear that they were trapped: if they retreated, they would come under fire from positions on the sides of the hill; if they advanced, they would run right into a battery of German machine guns. If they were all to die, as it seemed to Patton, better to die advancing. At the moment he was to lead the troops in the charge, however, Patton was stricken by intense fear. His body trembled, and his legs turned to jelly. In a confirmation of his deepest fears, he had lost his nerve.

At that instant, looking into the clouds beyond the German batteries, Patton had a vision: he saw his illustrious military ancestors, all in their uniforms, staring sternly down at him. They seemed to be inviting him to join their company—the company of dead war heroes. Paradoxically, the sight of these men had a calming effect on the young Patton: calling for volunteers to follow him, he yelled, "It is time for another Patton to die!" The strength had returned to his legs; he stood up and charged toward the German guns. Seconds later he fell, hit in the thigh. But he survived the battle.

From that moment on, even after he became a general, Patton made a point of visiting the front lines, exposing himself needlessly to danger. He tested himself again and again. His vision of his ancestors remained a constant stimulus—a challenge to his honor. Each time it became easier to face down his fears. It seemed to his fellow generals, and to his own men, that no one had more presence of mind than Patton. They did not know how much of his strength was an effort of will.

The story of Patton teaches us two things. First, it is better to confront your fears, let them come to the surface, than to ignore them or tamp them down. Fear is the most destructive emotion for presence of mind, but it thrives on the unknown, which lets our imaginations run wild. By deliberately putting yourself in situations where you have to face fear, you familiarize yourself with it and your anxiety grows less acute. The sensation of overcoming a deep-rooted fear in turn gives you confidence and presence of mind. The more conflicts and difficult situations you put yourself through, the more battle-tested your mind will be.

Second, Patton's experience demonstrates the motivating power of a sense of honor and dignity. In giving in to fear, in losing your presence of mind, you disgrace not only yourself, your self-image, and your reputation but your company, your family, your group. You bring down the communal spirit. Being a leader of even the smallest group gives you something to live up to: people are watching you, judging you, depending on you. To lose your composure would make it hard for you to live with yourself.

*There was a fox who had never seen a lion. But one day he happened to meet one of these beasts face to face. On this first occasion he was so terrified that he felt he would die of fear. He encountered him again, and this time he was also frightened, but not so much as the first time. But on the third occasion when he saw him, he actually plucked up the courage to approach him and began to chat.*

This fable shows that familiarity soothes our fears.

FABLES, AESOP, SIXTH CENTURY B.C.

**Be self-reliant.** There is nothing worse than feeling dependent on other people. Dependency makes you vulnerable to all kinds of emotions—betrayal, disappointment, frustration—that play havoc with your mental balance.

Early in the American Civil War, General Ulysses S. Grant, eventual commander in chief of the Northern armies, felt his authority slipping. His subordinates would pass along inaccurate information on the terrain he was marching through; his captains would fail to follow through on his orders; his generals were criticizing his plans. Grant was stoical by nature, but his diminished control over his troops led to a diminished control over himself and drove him to drink.

Grant had learned his lesson by the time of the Vicksburg campaign, in 1862–63. He rode the terrain himself, studying it firsthand. He reviewed intelligence reports himself. He honed the precision of his orders, making it harder for his captains to flout them. And once he had made a decision, he would ignore his fellow generals' doubts and trust his convictions. To get things done, he came to rely on himself. His feelings of helplessness dissolved, and with them all of the attendant emotions that had ruined his presence of mind.

Being self-reliant is critical. To make yourself less dependent on others and so-called experts, you need to expand your repertoire of skills. And you need to feel more confident in your own judgment. Understand: we tend to overestimate other people's abilities—after all, they're trying hard to make it look as if they knew what they were doing—and we tend to underestimate our own. You must compensate for this by trusting yourself more and others less.

It is important to remember, though, that being self-reliant does not mean burdening yourself with petty details. You must be able to distinguish between small matters that are best left to others and larger issues that require your attention and care.

*In the words of the ancients, one should make his decisions within the space of seven breaths. Lord Takanobu said, "If discrimination is long, it will spoil." Lord Naoshige said, "When matters are done leisurely, seven out of ten will turn out badly. A warrior is a person who does things quickly." When your mind is going hither and thither, discrimination will never be brought to a conclusion. With an intense, fresh and undelaying spirit, one will make his judgments within the space of seven breaths. It is a matter of being determined and having the spirit to break right through to the other side.*

*HAGAKURE: THE BOOK OF THE SAMURAI*, YAMAMOTO TSUNETOMO, 1659–1720

**Suffer fools gladly.** John Churchill, the Duke of Marlborough, is one of history's most successful generals. A genius of tactics and strategy, he had tremendous presence of mind. In the early eighteenth century, Churchill was often the leader of an alliance of English, Dutch, and German armies against the mighty forces of France. His fellow generals were timid, indecisive, narrow-minded men. They balked at the duke's bold plans, saw dangers everywhere, were discouraged at the slightest setback, and promoted their own country's interests at the expense of the alliance. They had no vision, no patience: they were fools.

*On a famous occasion during the civil war, Caesar tripped when disembarking from a ship on the shores of Africa and fell flat on his face. With his talent for improvisation, he spread out his arms and embraced the earth as a symbol of conquest. By quick thinking he turned a terrible omen of failure into one of victory.*

CICERO: THE LIFE AND TIMES OF ROME'S GREATEST POLITICIAN, ANTHONY EVERITT, 2001

The duke, an experienced and subtle courtier, never confronted his colleagues directly; he did not force his opinions on them. Instead he treated them like children, indulging them in their fears while cutting them out of his plans. Occasionally he threw them a bone, doing some minor thing they had suggested or pretending to worry about a danger they had imagined. But he never let himself get angry or frustrated; that would have ruined his presence of mind, undermining his ability to lead the campaign. He forced himself to stay patient and cheerful. He knew how to suffer fools gladly.

Understand: you cannot be everywhere or fight everyone. Your time and energy are limited, and you must learn how to preserve them. Exhaustion and frustration can ruin your presence of mind. The world is full of fools—people who cannot wait to get results, who change with the wind, who can't see past their noses. You encounter them everywhere: the indecisive boss, the rash colleague, the hysterical subordinate. When working alongside fools, do not fight them. Instead think of them the way you think of children, or pets, not important enough to affect your mental balance. Detach yourself emotionally. And while you're inwardly laughing at their foolishness, indulge them in one of their more harmless ideas. The ability to stay cheerful in the face of fools is an important skill.

*We mean the ability to keep one's head at times of exceptional stress and violent emotion. . . . But it might be closer to the truth to assume that the faculty known as self-control—the gift of keeping calm even under the greatest stress—is rooted in temperament. It is itself an emotion which serves to balance the passionate feelings in strong characters without destroying them, and it is this balance alone that assures the dominance of the intellect. The counterweight we mean is simply the sense of human dignity, the noblest pride and deepest need of all: the urge to act rationally at all times. Therefore we would argue that a strong character is one that will not be unbalanced by the most powerful emotions.*

ON WAR, CARL VON CLAUSEWITZ, 1780–1831

**Crowd out feelings of panic by focusing on simple tasks.** Lord Yamanouchi, an aristocrat of eighteenth-century Japan, once asked his tea master to accompany him on a visit to Edo (later Tokyo), where he was to stay for a while. He wanted to show off to his fellow courtiers his retainer's skill in the rituals of the tea ceremony. Now, the tea master knew everything there was to know about the tea ceremony, but little else; he was a peaceful man. He dressed, however, like a samurai, as his high position required.

One day, as the tea master was walking in the big city, he was accosted by a samurai who challenged him to a duel. The tea master was not a swordsman and tried to explain this to the samurai, but the man refused to listen. To turn the challenge down would disgrace both the tea master's family and Lord Yamanouchi. He had to accept, though that meant certain death. And accept he did, requesting only that the duel be put off to the next day. His wish was granted.

In panic, the tea master hurried to the nearest fencing school. If he were to die, he wanted to learn how to die honorably. To see the fencing master ordinarily required letters of introduction, but the tea master was so insistent, and so clearly terrified, that at last he was given an interview. The fencing master listened to his story.

The swordsman was sympathetic: he would teach the poor visitor the art of dying, but first he wanted to be served some tea. The tea master proceeded to perform the ritual, his manner calm, his concentration perfect. Finally the fencing master yelled out in excitement, "No need for you to learn the art of death! The state of mind you're in now is enough for you to face any samurai. When you see your challenger, imagine you're about to serve tea to a guest. Take off your coat, fold it up carefully, and lay your fan on it just as you do at work." This ritual completed, the tea master was to raise his sword in the same alert spirit. Then he would be ready to die.

The tea master agreed to do as his teacher said. The next day he went to meet the samurai, who could not help but notice the completely calm and dignified expression on his opponent's face as he took off his coat. Perhaps, the samurai thought, this fumbling tea master is actually a skilled swordsman. He bowed, begged pardon for his behavior the day before, and hurried away.

When circumstances scare us, our imagination tends to take over, filling our minds with endless anxieties. You need to gain control of your imagination, something easier said than done. Often the best way to calm down and give yourself such control is to force the mind to concentrate on something relatively simple—a calming ritual, a repetitive task that you are good at. You are creating the kind of composure you naturally have when your mind is absorbed in a problem. A focused mind has no room for anxiety or for the effects of an overactive imagination. Once you have regained your mental balance, you can then face the problem at hand. At the first sign of any kind of fear, practice this technique until it becomes a habit. Being able to control your imagination at intense moments is a crucial skill.

**Unintimidate yourself.** Intimidation will always threaten your presence of mind. And it is a hard feeling to combat.

During World War II, the composer Dmitry Shostakovich and several of his colleagues were called into a meeting with the Russian ruler Joseph Stalin, who had commissioned them to write a new national anthem. Meetings with Stalin were terrifying; one misstep could lead you into a very dark alley. He would stare you down until you felt your throat tighten. And, as meetings with Stalin often did, this one took a bad turn: the ruler began to criticize one of the composers for his poor arrangement of his anthem. Scared silly, the man admitted he had used an arranger who had done a bad job. Here he was digging several graves: Clearly the poor arranger could be called to task. The composer was responsible for the hire, and he, too, could pay for the mistake. And what of the other composers, including Shostakovich? Stalin could be relentless once he smelled fear.

*However, he perceived now that it did not greatly matter what kind of soldiers he was going to fight, so long as they fought, which fact no one disputed. There was a more serious problem. He lay in his bunk pondering upon it. He tried to mathematically prove to himself that he would not run from a battle. . . . A little panic-fear grew in his mind. As his imagination went forward to a fight, he saw hideous possibilities. He contemplated the lurking menaces of the future, and failed in an effort to see himself standing stoutly in the midst of them. He recalled his visions of broken-bladed glory, but in the shadow of the impending tumult he suspected them to be impossible pictures. He sprang from the bunk and began to pace nervously to and fro. "Good Lord, what's th' matter with me?" he said aloud. He felt that in this crisis his laws of life were useless. Whatever he had learned of himself was here of no avail. He was an unknown quantity. He saw that he would again be obliged to experiment as he had in early youth. He must accumulate information of himself, and meanwhile he resolved to remain close upon his guard lest those qualities of which he knew nothing should everlastingly disgrace him. "Good Lord!" he repeated in dismay. . . .*

*For days he made ceaseless calculations, but they were all wondrously unsatisfactory. He found that he could establish nothing. He finally concluded that the only way to prove himself was to go into the blaze, and then figuratively to watch his legs to discover their merits and faults. He reluctantly admitted that he could not sit still and with a mental slate and pencil derive an answer. To gain it, he must have blaze, blood, and danger, even as a chemist requires this, that, and the other. So he fretted for an opportunity.*

THE RED BADGE OF COURAGE, STEPHEN CRANE, 1871–1900

Shostakovich had heard enough: it was foolish, he said, to blame the arranger, who was mostly following orders. He then subtly redirected the conversation to a different subject—whether a composer should do his own orchestrations. What did Stalin think on the matter? Always eager to prove his expertise, Stalin swallowed the bait. The dangerous moment passed.

Shostakovich maintained his presence of mind in several ways. First, instead of letting Stalin intimidate him, he forced himself to see the man as he was: short, fat, ugly, unimaginative. The dictator's famous piercing gaze was just a trick, a sign of his own insecurity. Second, Shostakovich faced up to Stalin, talking to him normally and straightforwardly. By his actions and tone of voice, the composer showed that he was not intimidated. Stalin fed off fear. If, without being aggressive or brazen, you showed no fear, he would generally leave you alone.

The key to staying unintimidated is to convince yourself that the person you're facing is a mere mortal, no different from you—which is in fact the truth. See the person, not the myth. Imagine him or her as a child, as someone riddled with insecurities. Cutting the other person down to size will help you to keep your mental balance.

**Develop your *Fingerspitzengefühl* (fingertip feel).** Presence of mind depends not only on your mind's ability to come to your aid in difficult situations but also on the speed with which this happens. Waiting until the next day to think of the right action to take does you no good at all. "Speed" here means responding to circumstances with rapidity and making lightning-quick decisions. This power is often read as a kind of intuition, what the Germans call "*Fingerspitzengefühl*" (fingertip feel). Erwin Rommel, who led the German tank campaign in North Africa during World War II, had great fingertip feel. He could sense when the Allies would attack and from what direction. In choosing a line of advance, he had an uncanny feel for his enemy's weakness; at the start of a battle, he could intuit his enemy's strategy before it unfolded.

To Rommel's men their general seemed to have a genius for war, and he did possess a quicker mind than most. But Rommel also did things to enhance his quickness, things that reinforced his feel for battle. First, he devoured information about the enemy—from details about its weaponry to the psychological traits of the opposing general. Second, he made himself an expert in tank technology, so that he could get the most out of his equipment. Third, he not only memorized maps of the North African desert but would fly over it, at great risk, to get a bird's-eye view of the battlefield. Finally, he personalized his relationship with his men. He always had a sense of their morale and knew exactly what he could expect from them.

Rommel didn't just study his men, his tanks, the terrain, and the enemy—he got inside their skin, understood the spirit that animated them, what made them tick. Having felt his way into these things, in battle he entered a state of mind in which he did not have to think consciously of the situation. The totality of what was going on was in his blood, at his fingertips. He had *Fingerspitzengefühl.*

Whether or not you have the mind of a Rommel, there are things you can do to help you respond faster and bring out that intuitive feel that all animals possess. Deep knowledge of the terrain will let you process information faster than your enemy, a tremendous advantage. Getting a feel for the spirit of men and material, thinking your way into them instead of looking at them from outside, will help to put you in a different frame of mind, less conscious and forced, more unconscious and intuitive. Get your mind into the habit of making lightning-quick decisions, trusting your fingertip feel. Your mind will advance in a kind of mental blitzkrieg, moving past your opponents before they realize what has hit them.

Finally, do not think of presence of mind as a quality useful only in periods of adversity, something to switch on and off as you need it. Cultivate it as an everyday condition. Confidence, fearlessness, and self-reliance are as crucial in times of peace as in times of war. Franklin Delano Roosevelt showed his tremendous mental toughness and grace under pressure not only during the crises of the Depression and World War II but in everyday situations—in his dealings with his family, his cabinet, his own polio-racked body. The better you get at the game of war, the more your warrior frame of mind will do for you in daily life. When a crisis does come, your mind will already be calm and prepared. Once presence of mind becomes a habit, it will never abandon you.

**Image:**
The Wind. The
rush of unexpected
events, and the doubts and
criticisms of those around you, are like
a fierce wind at sea. It can come from any point
of the compass, and there is no place to go to escape
from it, no way to predict when and in what direction it will
strike. To change direction with each gust of wind will
only throw you out to sea. Good pilots do not waste
time worrying about what they cannot control.
They concentrate on themselves, the skill
and steadiness of their hand, the
course they have plotted, and
their determination to
reach port, come
what may.

*The man with centre has calm, unprejudiced judgment. He knows what is important, what unimportant. He meets realilty serenely and with detachment keeping his sense of proportion. The* Hara no aru hito [*man with centre*] *faces life calmly, is tranquil, ready for anything.... Nothing upsets him. If suddenly fire breaks out and people begin to shout in wild confusion* [*he*] *does the right thing immediately and quietly, he ascertains the direction of the wind, rescues what is most important, fetches water, and behaves unhesitatingly in the way the emergency demands. The* Hara no nai hito *is the opposite of all this. The* Hara no nai hito *applies to the man without calm judgment. He lacks the measure which should be second nature. Therefore he reacts haphazardly and subectively, arbitrarily and capriciously. He cannot distinguish between important and unimportant, essential and unessential. His judgment is not based upon facts but on temporary conditions and rests on subjective foundations, such as moods, whims, "nerves."* The Hara no nai hito *is easily startled, is nervous, not because he is particularly sensitive but because he lacks that inner axis which would prevent his being thrown off centre and which would enable him to deal with situations realistically....*

**Authority:** A great part of courage is the courage of having done the thing before.
*—Ralph Waldo Emerson (1803–82)*

## REVERSAL

It is never good to lose your presence of mind, but you can use those moments when it is under threat to know how to act in the future. You must find a way to put yourself in the thick of battle, then watch yourself in action. Look for your own weaknesses, and think about how to compensate for them. People who have never lost their presence of mind are actually in danger: someday they will be taken by surprise, and the fall will be harsh. All great generals, from Julius Caesar to Patton, have at some point lost their nerve and then have been the stronger for winning it back. The more you have lost your balance, the more you will know about how to right yourself.

You do not want to lose your presence of mind in key situations, but it is a wise course to find a way to make your enemies lose theirs. Take what throws you off balance and impose it on them. Make them act before they are ready. Surprise them—nothing is more unsettling than the unexpected need to act. Find their weakness, what makes them emotional, and give them a double dose of it. The more emotional you can make them, the farther you will push them off course.

Hara [*centre, belly*] *is only in slight measure innate. It is above all the result of persistent self-training and discipline, in fact the fruit of responsible, individual development. That is what the Japanese means when he speaks of the* Hara no dekita hito, *the man who has accomplished or finished his belly, that is, himself: for he is mature. If this development does not take place, we have the* Hara no dekita inai hito, *someone who has not developed, who has remained immature, who is too young in the psychological sense. The Japanese also say* Hara no dekita inai hito wa hito no ue ni tatsu koto ga dekinai*: the man who has not finished his belly cannot stand above others (is not fit for leadership).*

*HARA: THE VITAL CENTRE*, KARLFRIED GRAF VON DÜRCKHEIM, 1962

# 4

# CREATE A SENSE OF URGENCY AND DESPERATION

## THE DEATH-GROUND STRATEGY

*You are your own worst enemy. You waste precious time dreaming of the future instead of engaging in the present. Since nothing seems urgent to you, you are only half involved in what you do. The only way to change is through action and outside pressure. Put yourself in situations where you have too much at stake to waste time or resources—if you cannot afford to lose, you won't. Cut your ties to the past; enter unknown territory where you must depend on your wits and energy to see you through. Place yourself on "death ground," where your back is against the wall and you have to fight like hell to get out alive.*

*Cortés ran all that aground with the ten ships. Cuba, to be sure, was still there, in the blue sea, with its farms, its cows and its tame Indians; but the way to Cuba was no longer through sunny blue waves, rocked in soft idleness, oblivious of danger and endeavor; it was through Moteçuçuma's court, which had to be conquered by ruse, by force, or by both; through a sea of warlike Indians who ate their prisoners and donned their skins as trophies; at the stroke of their chief's masterly hand, the five hundred men had lost that flow of vital memories and hopes which linked up their souls with their mother-island; at one stroke, their backs had been withered and had lost all sense of life. Henceforward, for them, all life was ahead, towards those forbidding peaks which rose gigantically on the horizon as if to bar all access to what was now not merely their ambition, but their only possible aim—Mexico, mysterious and powerful behind the conflicting tribes.*

HERNÁN CORTÉS: CONQUEROR OF MEXICO, SALVADOR DE MADARIAGA, 1942

## THE NO-RETURN TACTIC

In 1504 an ambitious nineteen-year-old Spaniard named Hernán Cortés gave up his studies in law and sailed for his country's colonies in the New World. Stopping first in Santo Domingo (the island today comprising Haiti and the Dominican Republic), then in Cuba, he soon heard about a land to the west called Mexico—an empire teeming with gold and dominated by the Aztecs, with their magnificent highland capital of Tenochtitlán. From then on, Cortés had just one thought: someday he would conquer and settle the land of Mexico.

Over the next ten years, Cortés slowly rose through the ranks, eventually becoming secretary to the Spanish governor of Cuba and then the king's treasurer for the island. In his own mind, though, he was merely biding his time. He waited patiently while Spain sent other men to Mexico, many of them never to return.

Finally, in 1518, the governor of Cuba, Diego de Velázquez, made Cortés the leader of an expedition to discover what had happened to these earlier explorers, find gold, and lay the groundwork for the country's conquest. Velázquez wanted to make that future conquest himself, however, so for this expedition he wanted a man he could control, and he soon developed doubts about Cortés—the man was clever, perhaps too much so. Word reached Cortés that the governor was having second thoughts about sending him to Mexico. Deciding to give Velázquez no time to nurse his misgivings, he managed to slip out of Cuba in the middle of the night with eleven ships. He would explain himself to the governor later.

The expedition landed on Mexico's east coast in March 1519. Over the next few months, Cortés put his plans to work—founding the town of Veracruz, forging alliances with local tribes who hated the Aztecs, and making initial contact with the Aztec emperor, whose capital lay some 250 miles to the west. But one problem plagued the conquistador: among the 500 soldiers who had sailed with him from Cuba were a handful who had been placed there by Velázquez to act as spies and make trouble for him if he exceeded his authority. These Velázquez loyalists accused Cortés of mismanaging the gold that he was collecting, and when it became clear that he intended to conquer Mexico, they spread rumors that he was insane—an all-too-convincing accusation to make about a man planning to lead 500 men against half a million Aztecs, fierce warriors known to eat their prisoners' flesh and wear the skins as trophies. A rational man would take the gold they had, return to Cuba, and come back later with an army. Why stay in this forbidding land, with its diseases and its lack of creature comforts, when they were so heavily outnumbered? Why not sail for Cuba, back home where their farms, their wives, and the good life awaited them?

Cortés did what he could with these troublemakers, bribing some, keeping a close eye on others. Meanwhile he worked to build a strong enough rapport with the rest of his men that the grumblers could do no harm. All seemed well until the night of July 30, when Cortés was awoken by a Spanish sailor who, begging for mercy, confessed that he had joined in a plot to steal a ship and return that very evening to Cuba, where the conspirators would tell Velázquez about Cortés's goal of conquering Mexico on his own.

Cortés sensed that this was the decisive moment of the expedition. He could easily squash the conspiracy, but there would be others. His men were a rough lot, and their minds were on gold, Cuba, their families—anything but fighting the Aztecs. He could not conquer an empire with men so divided and untrustworthy, but how to fill them with the energy and focus for the immense task he faced? Thinking this through, he decided to take swift action. He seized the conspirators and had the two ringleaders hanged. Next, he bribed his pilots to bore holes in all of the ships and then announce that worms had eaten through the boards of the vessels, making them unseaworthy.

Pretending to be upset at the news, Cortés ordered what was salvageable from the ships to be taken ashore and then the hulls to be sunk. The pilots complied, but not enough holes had been bored, and only five of the ships went down. The story of the worms was plausible enough, and the soldiers accepted the news of the five ships with equanimity. But when a few days later more ships were run aground and only one was left afloat, it was clear to them that Cortés had arranged the whole thing. When he called a meeting, their mood was mutinous and murderous.

This was no time for subtlety. Cortés addressed his men: he was responsible for the disaster, he admitted; he had ordered it done, but now there was no turning back. They could hang him, but they were surrounded by hostile Indians and had no ships; divided and leaderless, they would perish. The only alternative was to follow him to Tenochtitlán. Only by conquering the Aztecs, by becoming lords of Mexico, could they get back to Cuba alive. To reach Tenochtitlán they would have to fight with utter intensity. They would have to be unified; any dissension would lead to defeat and a terrible death. The situation was desperate, but if the men fought desperately in turn, Cortés guaranteed that he would lead them to victory. Since the army was so small in number, the glory and riches would be all the greater. Any cowards not up to the challenge could sail the one remaining ship home.

No one accepted the offer, and the last ship was run aground. Over the next months, Cortés kept his army away from Veracruz and the coast. Their attention was focused on Tenochtitlán, the heart of the Aztec empire. The grumbling, the self-interest, and the greed all disappeared. Understanding the danger of their situation, the conquistadors fought ruthlessly. Some two years after the destruction of the Spanish ships, and with the help of their Indian allies, Cortés's army laid siege to Tenochtitlán and conquered the Aztec empire.

*Meditation on inevitable death should be performed daily. Every day when one's body and mind are at peace, one should meditate upon being ripped apart by arrows, rifles, spears and swords, being carried away by surging waves, being thrown into the midst of a great fire, bring struck by lightning, being shaken to death by a great earthquake, falling from thousand-foot cliffs, dying of disease or committing seppuku at the death of one's master. And every day without fail one should consider himself as dead.*

*HAGAKURE: THE BOOK OF THE SAMURAI*, YAMAMOTO TSUNETOMO, 1659–1720

*There is something in war that drives so deeply into you that death ceases to be the enemy, merely another participant in a game you don't wish to end.*

*PHANTOM OVER VIETNAM*, JOHN TROTTI, USMC, 1984

*"You don't have time for this display, you fool," he said in a severe tone. "This, whatever you're doing now, may be your last act on earth. It may very well be your last battle. There is no power which could guarantee that you are going to live one more minute. . . ." ". . . Acts have power," he said, "Especially when the person acting knows that those acts are his last battle. There is a strange consuming happiness in acting with the full knowledge that whatever one is doing may very well be one's last act on earth. I recommend that you reconsider your life and bring your acts into that light. . . . Focus your attention on the link between you and your death, without remorse or sadness or worrying. Focus your attention on the fact you don't have time and let your acts flow accordingly. Let each of your acts be your last battle on earth. Only under those conditions will your acts have their rightful power. Otherwise they will be, for as long as you live, the acts of a timid man." "Is it so terrible to be a timid man?" "No. It isn't if you are going to be immortal, but if you are going to die there is not time for timidity, simply because timidity makes you cling to something that exists only in your thoughts. It soothes you*

Interpretation

On the night of the conspiracy, Cortés had to think fast. What was the root of the problem he faced? It was not Velázquez's spies, or the hostile Aztecs, or the incredible odds against him. The root of the problem was his own men and the ships in the harbor. His soldiers were divided in heart and mind. They were thinking about the wrong things—their wives, their dreams of gold, their plans for the future. And in the backs of their minds there was always an escape route: if this conquest business went badly, they could go home. Those ships in the harbor were more than just transportation; they represented Cuba, the freedom to leave, the ability to send for reinforcements—so many possibilities.

For the soldiers the ships were a crutch, something to fall back on if things got ugly. Once Cortés had identified the problem, the solution was simple: destroy the ships. By putting his men in a desperate place, he would make them fight with utmost intensity.

A sense of urgency comes from a powerful connection to the present. Instead of dreaming of rescue or hoping for a better future, you have to face the issue at hand. Fail and you perish. People who involve themselves completely in the immediate problem are intimidating; because they are focusing so intensely, they seem more powerful than they are. Their sense of urgency multiplies their strength and gives them momentum. Instead of five hundred men, Cortés suddenly had the weight of a much larger army at his back.

Like Cortés you must locate the root of your problem. It is not the people around you; it is yourself, and the spirit with which you face the world. In the back of your mind, you keep an escape route, a crutch, something to turn to if things go bad. Maybe it is some wealthy relative you can count on to buy your way out; maybe it is some grand opportunity on the horizon, the endless vistas of time that seem to be before you; maybe it is a familiar job or a comfortable relationship that is always there if you fail. Just as Cortés's men saw their ships as insurance, you may see this fallback as a blessing—but in fact it is a curse. It divides you. Because you think you have options, you never involve yourself deeply enough in one thing to do it thoroughly, and you never quite get what you want. Sometimes you need to run your ships aground, burn them, and leave yourself just one option: succeed or go down. Make the burning of your ships as real as possible—get rid of your safety net. Sometimes you have to become a little desperate to get anywhere.

*The ancient commanders of armies, who well knew the powerful influence of necessity, and how it inspired the soldiers with the most desperate courage, neglected nothing to subject their men to such a pressure.*

*—Niccolò Machiavelli (1469–1527)*

## THE DEATH-AT-YOUR-HEELS TACTIC

In 1845 the writer Fyodor Dostoyevsky, then twenty-four, shook the Russian literary world with the publication of his first novel, *Poor Folk*. He became the toast of St. Petersburg society. But something about his early fame seemed empty to him. He drifted into the fringes of left-wing politics, attending meetings of various socialist and radical groups. One of these groups centered on the charismatic Mikhail Petrashevsky.

Three years later, in 1848, revolution broke out all across Europe. Inspired by what was happening in the West, Russian radical groups like Petrashevsky's talked of following suit. But agents of Czar Nicholas I had infiltrated many of these groups, and reports were written about the wild things being discussed at Petrashevsky's house, including talk of inciting peasant revolts. Dostoyevsky was fervent about freeing the serfs, and on April 23, 1849, he and twenty-three other members of the Petrashevsky group were arrested.

*while everything is at a lull, but then the awesome, mysterious world will open its mouth for you, as it will open for every one of us, and then you will realize that your sure ways were not sure at all. Being timid prevents us from examining and exploiting our lot as men."*

*Journey to Ixtlan: The Lessons of Don Juan*, Carlos Castaneda, 1972

After eight months of languishing in jail, the prisoners were awakened one cold morning and told that today they would finally hear their sentences. A few months' exile was the usual punishment for their crime; soon, they thought, their ordeal would be over.

They were bundled into carriages and driven through the icy streets of St. Petersburg. Emerging from the carriages into Semyonovsky Square, they were greeted by a priest; behind him they could see rows of soldiers and, behind the soldiers, thousands of spectators. They were led toward a scaffold covered in black cloth at the center of the square. In front of the scaffold were three posts, and to the side was a line of carts laden with coffins.

*Lord Naoshige said, "The Way of the Samurai is in desperateness. Ten men or more cannot kill such a man. Common sense will not accomplish great things. Simply become insane and desperate."*

*Hagakure: The Book of the Samurai*, Yamamoto Tsunetomo, 1659–1720

Dostoyevsky could not believe what he saw. "It's not possible that they mean to execute us," he whispered to his neighbor. They were marched to the scaffold and placed in two lines. It was an unbelievably cold day, and the prisoners were wearing the light clothes they'd been arrested in back in April. A drumroll sounded. An officer came forward to read their sentences: "All of the accused are guilty as charged of intending to overthrow the national order, and are therefore condemned to death before a firing squad." The prisoners were too stunned to speak.

As the officer read out the individual charges and sentences, Dostoyevsky found himself staring at the golden spire of a nearby church and at the sunlight bouncing off it. The gleams of light disappeared as a cloud passed overhead, and the thought occurred to him that he was about to pass into darkness just as quickly, and forever. Suddenly he had another thought: If I do *not* die, if I am *not* killed, my life will suddenly seem endless, a whole eternity, each minute a century. I will take account of everything that passes—I will not waste a second of life again.

The prisoners were given hooded shirts. The priest came forward to read them their last rites and hear their confessions. They said good-bye to one another. The first three to be shot were tied to the posts, and the hoods were pulled over their faces. Dostoyevsky stood in the front, in the

*It had long been known, of course, that a man who, through disciplined training, had relinquished any desire or hope for survival and had only one goal—the destruction of his enemy—could be a redoubtable opponent and a truly formidable fighter who neither asked nor offered any quarter once his weapon had been unsheathed. In this way, a seemingly ordinary man who, by the force of circumstances rather than by profession, had been placed in the position of having to make a desperate choice, could prove dangerous, even to a skilled fencing master. One famous episode, for example, concerns a teacher of swordsmanship who was asked by a superior to surrender a servant guilty of an offense punishable by death. This teacher, wishing to test a theory of his concerning the power of that condition we would call "desperation," challenged the doomed man to a duel. Knowing full well the irrevocability of his sentence, the servant was beyond caring one way or the other, and the ensuing duel proved that even a skilled fencer and teacher of the art could find himself in great difficulty when confronted by a man who, because of his acceptance of imminent death, could go to the limit (and even beyond)*

next group to go. The soldiers raised their rifles, took aim—and suddenly a carriage came galloping into the square. A man got out with an envelope. At the last second, the czar had commuted their death sentences.

Later that morning, Dostoyevsky was told his new sentence: four years hard labor in Siberia, to be followed by a stint in the army. Barely affected, he wrote that day to his brother, "When I look back at the past and think of all the time I squandered in error and idleness, . . . then my heart bleeds. Life is a gift . . . every minute could have been an eternity of happiness! If youth only knew! Now my life will change; now I will be reborn."

A few days later, ten-pound shackles were put on Dostoyevsky's arms and legs—they would stay there for the length of his prison term—and he was carted off to Siberia. For the next four years, he endured the most abysmal prison conditions. Granted no writing privileges, he wrote novels in his head, memorized them. Finally, in 1857, still serving the army period of his sentence, he was allowed to start publishing his work. Where before he would torture himself over a page, spend half a day idling it away in thought, now he wrote and wrote. Friends would see him walking the streets of St. Petersburg mumbling bits of dialogue to himself, lost in his characters and plots. His new motto was "Try to get as much done as possible in the shortest time."

Some pitied Dostoyevsky his time in prison. That made him angry; he was grateful for the experience and felt no bitterness. But for that December day in 1849, he felt, he would have wasted his life. Right up until his death, in 1881, he continued writing at a frantic pace, churning out novel after novel—*Crime and Punishment, The Possessed, The Brothers Karamazov*—as if each one were his last.

### Interpretation

Czar Nicholas had decided to sentence the Petrashevsky radicals to hard labor soon after their arrest. But he wanted to teach them a harsher lesson as well, so he dreamed up the cruel theater of the death sentence, with its careful details—the priest, the hoods, the coffins, the last-second pardon. This, he thought, would really humble and humiliate them. In fact, some of the prisoners were driven insane by the events of that day. But the effect on Dostoyevsky was different: he had been afflicted for years with a sense of wandering, of feeling lost, of not knowing what to do with his time. An extremely sensitive man, that day he literally felt his own death deep in his bones. And he experienced his "pardon" as a rebirth.

The effect was permanent. For the rest of his life, Dostoyevsky would consciously bring himself back to that day, remembering his pledge never to waste another moment. Or, if he felt he had grown too comfortable and complacent, he would go to a casino and gamble away all his money. Poverty and debt were for him a kind of symbolic death, throwing him back on the possible nothingness of his life. In either case he would have to write, and not the way other novelists wrote—as if it were a pleasant little artistic career, with all its attendant delights of salons, lectures, and other frills. Dostoyevsky wrote as if his life were at stake, with an intense feeling of urgency and seriousness.

Death is impossible for us to fathom: it is so immense, so frightening, that we will do almost anything to avoid thinking about it. Society is organized to make death invisible, to keep it several steps removed. That distance may seem necessary for our comfort, but it comes with a terrible price: the illusion of limitless time, and a consequent lack of seriousness about daily life. We are running away from the one reality that faces us all.

As a warrior in life, you must turn this dynamic around: make the thought of death something not to escape but to embrace. Your days are numbered. Will you pass them half awake and halfhearted or will you live with a sense of urgency? Cruel theaters staged by a czar are unnecessary; death will come to you without them. Imagine it pressing in on you, leaving you no escape—for there *is* no escape. Feeling death at your heels will make all your actions more certain, more forceful. This could be your last throw of the dice: make it count.

*in his strategy, without a single hesitation or distracting consideration. The servant, in fact, fought like a man possessed, forcing his master to retreat until his back was almost to the wall. At last the teacher had to cut him down in a final effort, wherein the master's own desperation brought about the fullest coordination of his courage, skill, and determination.*

*Secrets of the Samurai*, Oscar Ratti and Adele Westbrook, 1973

*While knowing that we will die someday, we think that all the others will die before us and that we will be the last to go. Death seems a long way off. Is this not shallow thinking? It is worthless and is only a joke within a dream. . . . Insofar as death is always at one's door, one should make sufficient effort and act quickly.*

—Hagakure: The Book of the Samurai, *Yamamoto Tsunetomo (1659–1720)*

## KEYS TO WARFARE

Quite often we feel somewhat lost in our actions. We could do this or that—we have many options, but none of them seem quite necessary. Our freedom is a burden—what do we do today, where do we go? Our daily patterns and routines help us to avoid feeling directionless, but there is always the niggling thought that we could accomplish so much more. We waste so much time. Upon occasion all of us have felt a sense of urgency. Most often it is imposed from outside: we fall behind in our work, we inadvertently take on more than we can handle, responsibility for something is thrust into our hands. Now everything changes; no more freedom. We have to do this, we have to fix that. The surprise is always how much more spirited and more alive this makes us feel; now everything we do seems necessary. But eventually we go back to our normal patterns. And when that sense of urgency goes, we really do not know how to get it back.

Leaders of armies have thought about this subject since armies existed: how can soldiers be motivated, be made more aggressive, more desperate? Some generals have relied on fiery oratory, and those

*Taking advantage of the opportunity, they began to question Han Hsin. "According to* The Art of War, *when one fights he should keep the hills to his right or rear, and bodies of water in front of him or to the left," they said. "Yet today you ordered us on the contrary to draw up ranks with our backs to the river, saying 'We shall defeat Chao and feast together!' We were opposed to the idea, and yet it has ended in victory. What sort of strategy is this?"*

*"This is in* The Art of War *too," replied Han Hsin. "It is just that you have failed to notice it! Does it not say in* The Art of War: *'Drive them into a fatal position and they will come out alive; place them in a hopeless spot and they will survive'? Moreover, I did not have at my disposal troops that I had trained and led from past times, but was forced, as the saying goes, to round up men from the market place and use them to fight with. Under such circumstances, if I had not placed them in a desperate situation where each man was obliged to fight for his own life, but had allowed them to remain in a safe place, they would have all run away. Then what good would they have been to me?"*

*"Indeed!" his generals exclaimed in admiration. "We would never have thought of that."*

*RECORDS OF THE HISTORIAN,* SZUMA CHIEN, CIRCA 145 B.C.–CIRCA 86 B.C.

particularly good at it have had some success. But over two thousand years ago, the Chinese strategist Sun-tzu came to believe that listening to speeches, no matter how rousing, was too passive an experience to have an enduring effect. Instead Sun-tzu talked of a "death ground"—a place where an army is backed up against some geographical feature like a mountain, a river, or a forest and has no escape route. Without a way to retreat, Sun-tzu argued, an army fights with double or triple the spirit it would have on open terrain, because death is viscerally present. Sun-tzu advocated deliberately stationing soldiers on death ground to give them the desperate edge that makes men fight like the devil. That is what Cortés did in Mexico, and it is the only sure way to create a real fire in the belly. The world is ruled by necessity: People change their behavior only if they have to. They will feel urgency only if their lives depend on it.

Death ground is a psychological phenomenon that goes well beyond the battlefield: it is any set of circumstances in which you feel enclosed and without options. There is very real pressure at your back, and you cannot retreat. Time is running out. Failure—a form of psychic death—is staring you in the face. You must act or suffer the consequences.

Understand: we are creatures who are intimately tied to our environment—we respond viscerally to our circumstances and to the people around us. If our situation is easy and relaxed, if people are friendly and warm, our natural tension unwinds. We may even grow bored and tired; our environment is failing to challenge us, although we may not realize it. But put yourself in a high-stakes situation—a psychological death ground—and the dynamic changes. Your body responds to danger with a surge of energy; your mind focuses. Urgency is forced on you; you are compelled to waste no more time.

The trick is to use this effect deliberately from time to time, to practice it on yourself as a kind of wake-up call. The following five actions are designed to put you on a psychological death ground. Reading and thinking about them won't work; you must put them into effect. They are forms of pressure to apply to yourself. Depending on whether you want a low-intensity jolt for regular use or a real shock, you can turn the level up or down. The scale is up to you.

**Stake everything on a single throw.** In 1937 the twenty-eight-year-old Lyndon B. Johnson—at the time the Texas director of the National Youth Administration—faced a dilemma. The Texas congressman James Buchanan had suddenly died. Since loyal Texan voters tended to return incumbents to office, a Texan congressional seat generally came available only every ten or twenty years—and Johnson wanted to be in Congress by the time he was thirty; he did not have ten years to wait. But he was very young and was virtually unknown in Buchanan's old district, the tenth. He would be facing political heavyweights whom voters would heavily favor. Why try something that seemed doomed to failure? Not only would the race be a waste of money, but the humiliation, if Johnson lost badly, could derail his long-term ambitions.

Johnson considered all this—then decided to run. Over the next few weeks, he campaigned intensely, visiting the district's every backwater village and town, shaking the poorest farmer's hand, sitting in drugstores to meet people who had never come close to talking to a candidate before. He pulled every trick in the book—old-style rallies and barbecues, newfangled radio ads. He worked night and day—and hard. By the time the race was over, Johnson was in a hospital, being treated for exhaustion and appendicitis. But, in one of the great upsets in American political history, he had won.

By staking his future on one throw, Johnson put himself in a death-ground situation. His body and spirit responded with the energy he needed. Often we try too many things at one time, thinking that one of them will bring us success—but in these situations our minds are diffused, our efforts halfhearted. It is better to take on one daunting challenge, even one that others think foolish. Our future is at stake; we cannot afford to lose. So we don't.

*Unlimited possibilities are not suited to man; if they existed, his life would only dissolve in the boundless. To become strong, a man's life needs the limitations ordained by duty and voluntarily accepted. The individual attains significance as a free spirit only by surrounding himself with these limitations and by determining for himself what his duty is.*

*The I Ching*, China, circa eighth century B.C.

**Act before you are ready.** In 49 B.C. a group of Roman senators, allied with Pompey and fearing the growing power of Julius Caesar, ordered the great general to disband his army or be considered a traitor to the Republic. When Caesar received this decree, he was in southern Gaul (modern-day France) with only five thousand men; the rest of his legions were far to the north, where he had been campaigning. He had no intention of obeying the decree—that would have been suicide—but it would be weeks before the bulk of his army could join him. Unwilling to wait, Caesar told his captains, "Let the die be cast," and he and his five thousand men crossed the Rubicon, the river marking the border between Gaul and Italy. Leading troops onto Italian soil meant war with Rome. Now there was no turning back; it was fight or die. Caesar was compelled to concentrate his forces, to not waste a single man, to act with speed, and to be as creative as possible. He marched on Rome. By seizing the initiative, he frightened the senators, forcing Pompey to flee.

*Death is nothing, but to live defeated is to die every day.*

Napoleon Bonaparte, 1769–1821

We often wait too long to act, particularly when we face no outside pressure. It is sometimes better to act before you think you are ready—to force the issue and cross the Rubicon. Not only will you take your opponents by surprise, you will also have to make the most of your resources. You have committed yourself and cannot turn back. Under pressure your creativity will flourish. Do this often and you will develop your ability to think and act fast.

*When danger is greatest.—It is rare to break one's leg when in the course of life one is toiling upwards—it happens much more often when one starts to take things easy and to choose the easy paths.*

Friedrich Nietzsche, 1844–1900

**Enter new waters.** The Hollywood studio MGM had been good to Joan Crawford: it had discovered her, made her a star, crafted her image. By the early 1940s, though, Crawford had had enough. It was all too comfortable; MGM kept casting her in the same kinds of roles, none of them a challenge. So, in 1943, Crawford did the unthinkable and asked out of her contract.

The consequences for Crawford could have been terrible; to challenge the studio system was considered highly unwise. Indeed, when she then signed up with Warner Brothers, predictably enough she was offered the same mediocre sorts of scripts. She turned them down. On the verge of being fired, she finally found the part she had been looking for: the title role in *Mildred Pierce*, which, however, she was not offered. Setting to work on the director, Michael Curtiz, she managed to change his mind and land the role. She gave the performance of her life, won her only Best Actress Oscar, and resurrected her career.

In leaving MGM, Crawford was taking a big chance. If she failed to succeed at Warner Brothers, and quickly, her career would be over. But Crawford thrived on risk. When she was challenged, when she felt on edge, she burst with energy and was at her best. Like Crawford, you sometimes have to force yourself onto death ground—leaving stale relationships and comfortable situations behind, cutting your ties to the past. If you give yourself no way out, you will have to make your new endeavor work. Leaving the past for unknown terrain is like a death—and feeling this finality will snap you back to life.

*Be absolute for death; either death or life*
*Shall thereby be the sweeter. Reason thus with life:*
*If I do lose thee, I do lose a thing*
*That none but fools would keep: a breath thou art,*
*Servile to all the skyey influences,*
*That dost this habituation, where thou keep'st,*
*Hourly afflict: merely, thou art death's fool;*
*For him thou labour'st by thy flight to shun*
*And yet runn'st toward him still. Thou art not noble;*
*For all the accommodations that thou bear'st*
*Are nursed by baseness. Thou'rt by no means valiant;*
*For thou dost fear the soft and tender fork*
*Of a poor worm. Thy best of rest is sleep,*
*And that thou oft provokest; yet grossly fear'st Thy death, which is no more.*

*Measure for Measure*, William Shakespeare, 1564–1616

**Make it "you against the world."** Compared to sports like football, baseball is slow and has few outlets for aggression. This was a problem for the hitter Ted Williams, who played best when he was angry—when he felt that it was him against the world. Creating this mood on the field was difficult for Williams, but early on, he discovered a secret weapon: the press. He got into the habit of insulting sportswriters, whether just by refusing to cooperate with them or by verbally abusing them. The reporters returned the favor, writing scathing articles on his character, questioning his talent, trumpeting the slightest drop in his batting average. It was when Williams was hammered by the press, though, that he played best. He would go on a hitting tear, as if to prove them wrong. In 1957, when he carried on a yearlong feud with the papers, he played perhaps his greatest season and won the batting title at what for a baseball player is the advanced age of forty. As one journalist wrote, "Hate seems to activate his reflexes like adrenaline stimulates the heart. Animosity is his fuel!"

For Williams the animosity of the press and, with the press, of the public, was a kind of constant pressure that he could read, hear, and feel. They hated him, they doubted him, they wanted to see him fail; he would show them. And he did. A fighting spirit needs a little edge, some anger and hatred to fuel it. So do not sit back and wait for people to get aggressive; irritate and infuriate them deliberately. Feeling cornered by a multitude of people who dislike you, you will fight like hell. Hatred is a powerful emotion. Remember: in any battle you are putting your name and reputation on the line; your enemies will relish your failure. Use that pressure to make yourself fight harder.

*O gentlemen, the time of life is short!*
*To spend that shortness basely were too long,*
*If life did ride upon a dial's point,*
*Still ending at the arrival of an hour.*
*An if we live, we live to tread on kings;*
*If die, brave death, when princes die with us!*

*King Henry IV, Part I*, William Shakespeare, 1564–1616

**Keep yourself restless and unsatisfied.** Napoleon had many qualities that made him perhaps history's greatest general, but the one that raised him to the heights and kept him there was his boundless energy. During campaigns he worked eighteen to twenty-hour days. If necessary, he would go without sleep for several days, yet sleeplessness rarely reduced his capacities. He would work in the bath, at the theater, during a dinner party. Keeping his eye on every detail of the war, he would ride endless miles on horseback without tiring or complaining.

Certainly Napoleon had extraordinary endurance, but there was more to it than that: he never let himself rest, was never satisfied. In 1796, in his first real position of command, he led the French to a remarkable victory in Italy, then immediately went on another campaign, this time in Egypt. There, unhappy with the way the war was going and with a lack of political power that he felt was cutting into his control over military affairs, he returned to France and conspired to become first consul. This achieved, he immediately set out on his second Italian campaign. And on he went, immersing himself in new wars, new challenges, that required him to call on his limitless energy. If he did not meet the crisis, he would perish.

When we are tired, it is often because we are bored. When no real challenge faces us, a mental and physical lethargy sets in. "Sometimes death only comes from a lack of energy," Napoleon once said, and lack of energy comes from a lack of challenges, comes when we have taken on less than we are capable of. Take a risk and your body and mind will respond with a rush of energy. Make risk a constant practice; never let yourself settle down. Soon living on death ground will become a kind of addiction—you won't be able to do without it. When soldiers survive a brush with death, they often feel an exhilaration that they want to have again. Life has more meaning in the face of death. The risks you keep taking, the challenges you keep overcoming, are like symbolic deaths that sharpen your appreciation of life.

**Image:**
Fire. By itself it has no force; it depends on its environment. Give it air, dry timber, a wind to fan the flames, and it gains a terrifying momentum, growing hotter, feeding off itself, consuming everything in its path. Never leave such power to chance.

**Authority:** When you will survive if you fight quickly and perish if you do not, this is called [death] ground. . . . Put them in a spot where they have no place to go, and they will die before fleeing. If they are to die there, what can they not do? Warriors exert their full strength. When warriors are in great danger, then they have no fear. When there is nowhere to go, they are firm, when they are deeply involved, they stick to it. If they have no choice, they will fight.
—The Art of War, *Sun-tzu (fourth century B.C.)*

## REVERSAL

If the feeling of having nothing to lose can propel you forward, it can do the same for others. You must avoid any conflict with people in this position. Maybe they are living in terrible conditions or, for whatever reason, are suicidal; in any case they are desperate, and desperate people will risk everything in a fight. This gives them a huge advantage. Already defeated by circumstances, they have nothing to lose. You do. Leave them alone.

Conversely, attacking enemies when their morale is low gives you the advantage. Maybe they are fighting for a cause they know is unjust or for a leader they do not respect. Find a way to lower their spirits even further. Troops with low morale are discouraged by the slightest setback. A show of force will crush their fighting spirit.

Always try to lower the other side's sense of urgency. Make your enemies think they have all the time in the world; when you suddenly appear at their border, they are in a slumbering state, and you will easily overrun them. While you are sharpening your fighting spirit, always do what you can to blunt theirs.

PART

# II

# ORGANIZATIONAL (TEAM) WARFARE

You may have brilliant ideas, you may be able to invent unbeatable strategies—but if the group that you lead, and that you depend on to execute your plans, is unresponsive and uncreative, and if its members always put their personal agendas first, your ideas will mean nothing. You must learn the lesson of war: it is the structure of the army—the chain of command and the relationship of the parts to the whole—that will give your strategies force.

The primary goal in war is to build speed and mobility into the very structure of your army. That means having a single authority on top, avoiding the hesitancy and confusion of divided leadership. It means giving soldiers a sense of the overall goal to be accomplished and the latitude to take action to meet that goal; instead of reacting like automatons, they are able to respond to events in the field. Finally, it means motivating soldiers, creating an overall esprit de corps that gives them irresistible momentum. With forces organized in this manner, a general can adapt to circumstances faster than the enemy can, gaining a decided advantage.

This military model is extremely adaptable to any group. It has one simple requirement: before formulating a strategy or taking action, understand the structure of your group. You can always change it and redesign it to fit your purposes. The following three chapters will help you focus on this critical issue and give you strategic options—possible organizational models to follow, as well as disastrous mistakes to avoid.

# 5

# AVOID THE SNARES OF GROUPTHINK

## THE COMMAND-AND-CONTROL STRATEGY

*The problem in leading any group is that people inevitably have their own agendas. If you are too authoritarian, they will resent you and rebel in silent ways. If you are too easy-going, they will revert to their natural selfishness and you will lose control. You have to create a chain of command in which people do not feel constrained by your influence yet follow your lead. Put the right people in place—people who will enact the spirit of your ideas without being automatons. Make your commands clear and inspiring, focusing attention on the team, not the leader. Create a sense of participation, but do not fall into Groupthink—the irrationality of collective decision making. Make yourself look like a paragon of fairness, but never relinquish unity of command.*

*How very different is the cohesion between that of an army rallying around one flag carried into battle at the personal command of one general and that of an allied military force extending 50 or 100 leagues, or even on different sides of the theater! In the first case, cohesion is at its strongest and unity at its closest. In the second case, the unity is very remote, often consisting of no more than a shared political intention, and therefore only scanty and imperfect, while the cohesion of the parts is mostly weak and often no more than an illusion.*

*On War*, Carl von Clausewitz, 1780–1831

## THE BROKEN CHAIN

World War I began in August 1914, and by the end of that year, all along the Western Front, the British and French were caught in a deadly stalemate with the Germans. Meanwhile, though, on the Eastern Front, Germany was badly beating the Russians, allies of Britain and France. Britain's military leaders had to try a new strategy, and their plan, backed by First Lord of the Admiralty Winston Churchill and others, was to stage an attack on Gallipoli, a peninsula on Turkey's Dardanelles Strait. Turkey was an ally of Germany's, and the Dardanelles was the gateway to Constantinople, the Turkish capital (present-day Istanbul). If the Allies could take Gallipoli, Constantinople would follow, and Turkey would have to leave the war. In addition, using bases in Turkey and the Balkans, the Allies could attack Germany from the southeast, dividing its armies and weakening its ability to fight on the Western Front. They would also have a clear supply line to Russia. Victory at Gallipoli would change the course of the war.

The plan was approved, and in March 1915, General Sir Ian Hamilton was named to lead the campaign. Hamilton, at sixty-two, was an able strategist and an experienced commander. He and Churchill felt certain that their forces, including Australians and New Zealanders, would outmatch the Turks. Churchill's orders were simple: take Constantinople. He left the details to the general.

Hamilton's plan was to land at three points on the southwestern tip of the Gallipoli peninsula, secure the beaches, and sweep north. The landings took place on April 27. From the beginning almost everything went wrong: the army's maps were inaccurate, its troops landed in the wrong places, the beaches were much narrower than expected. Worst of all, the Turks fought back unexpectedly fiercely and well. At the end of the first day, most of the Allies' 70,000 men had landed, but they were unable to advance beyond the beaches, where the Turks would hold them pinned down for several weeks. It was another stalemate; Gallipoli had become a disaster.

All seemed lost, but in June, Churchill convinced the government to send more troops and Hamilton devised a new plan. He would land 20,000 men at Suvla Bay, some twenty miles to the north. Suvla was a vulnerable target: it had a large harbor, the terrain was low-lying and easy, and it was defended by only a handful of Turks. An invasion here would force the Turks to divide their forces, freeing up the Allied armies to the south. The stalemate would be broken, and Gallipoli would fall.

To command the Suvla operation Hamilton was forced to accept the most senior Englishman available for the job, Lieutenant General Sir Frederick Stopford. Under him, Major General Frederick Hammersley would lead the Eleventh Division. Neither of these men was Hamilton's first choice. Stopford, a sixty-one-year-old military teacher, had never led troops in war and saw artillery bombardment as the only way to win a battle; he was also in poor health. Hammersley, for his part, had suffered a nervous breakdown the previous year.

*In war it is not men, but the man, that counts.*

NAPOLEON BONAPARTE, 1769–1821

Hamilton's style was to tell his officers the purpose of an upcoming battle but leave it to them how to bring it about. He was a gentleman, never blunt or forceful. At one of their first meetings, for example, Stopford requested changes in the landing plans to reduce risk. Hamilton politely deferred to him.

Hamilton did have one request. Once the Turks knew of the landings at Suvla, they would rush in reinforcements. As soon as the Allies were ashore, then, Hamilton wanted them to advance immediately to a range of hills four miles inland, called Tekke Tepe, and to get there before the Turks. From Tekke Tepe the Allies would dominate the peninsula. The order was simple enough, but Hamilton, so as not to offend his subordinate, expressed it in the most general terms. Most crucially, he specified no time frame. He was sufficiently vague that Stopford completely misinterpreted him: instead of trying to reach Tekke Tepe "as soon as possible," Stopford thought he should advance to the hills "if possible." That was the order he gave Hammersley. And as Hammersley, nervous about the whole campaign, passed it down to his colonels, the order became less urgent and vaguer still.

Also, despite his deference to Stopford, Hamilton overruled the lieutenant general in one respect: he denied a request for more artillery bombardments to loosen up the Turks. Stopford's troops would outnumber the Turks at Suvla ten to one, Hamilton replied; more artillery was superfluous.

The attack began in the early morning of August 7. Once again much turned bad: Stopford's changes in the landing plans made a mess. As his officers came ashore, they began to argue, uncertain about their positions and objectives. They sent messengers to ask their next step: Advance? Consolidate? Hammersley had no answers. Stopford had stayed on a boat offshore, from which to control the battlefield—but on that boat he was impossible to reach quickly enough to get prompt orders from him. Hamilton was on an island still farther away. The day was frittered away in argument and the endless relaying of messages.

The next morning Hamilton began to sense that something had gone very wrong. From reconnaissance aircraft he knew that the flat land around Suvla was essentially empty and undefended; the way to Tekke Tepe was open—the troops had only to march—but they were staying where they were. Hamilton decided to visit the front himself. Reaching Stopford's boat late that afternoon, he found the general in a self-congratulatory mood: all 20,000 men had gotten ashore. No, he had not yet ordered the troops to advance to the hills; without artillery he was afraid the Turks might counterattack, and he needed the day to consolidate his positions and to land supplies. Hamilton strained to control himself: he had heard an hour earlier that Turkish reinforcements had been seen hurrying toward Suvla. The Allies would have to secure Tekke Tepe

*Any army is like a horse, in that it reflects the temper and the spirit of its rider. If there is an uneasiness and an uncertainty, it transmits itself through the reins, and the horse feels uneasy and uncertain.*

*LONE STAR PREACHER*, COLONEL JOHN W. THOMASON, JR., 1941

this evening, he said—but Stopford was against a night march. Too dangerous. Hamilton retained his cool and politely excused himself.

In near panic, Hamilton decided to visit Hammersley at Suvla. Much to his dismay, he found the army lounging on the beach as if it were a bank holiday. He finally located Hammersley—he was at the far end of the bay, busily supervising the building of his temporary headquarters. Asked why he had failed to secure the hills, Hammersley replied that he had sent several brigades for the purpose, but they had encountered Turkish artillery and his colonels had told him they could not advance without more instructions. Communications between Hammersley, Stopford, and the colonels in the field were taking forever, and when Stopford had finally been reached, he had sent the message back to Hammersley to proceed cautiously, rest his men, and wait to advance until the next day. Hamilton could control himself no longer: a handful of Turks with a few guns were holding up an army of 20,000 men from marching a mere four miles! Tomorrow morning would be too late; the Turkish reinforcements were on their way. Although it was already night, Hamilton ordered Hammersley to send a brigade immediately to Tekke Tepe. It would be a race to the finish.

Hamilton returned to a boat in the harbor to monitor the situation. At sunrise the next morning, he watched the battlefield through binoculars—and saw, to his horror, the Allied troops in headlong retreat to Suvla. A large Turkish force had arrived at Tekke Tepe thirty minutes before them. In the next few days, the Turks managed to regain the flats around Suvla and to pin Hamilton's army on the beach. Some four months later, the Allies gave up their attack on Gallipoli and evacuated their troops.

### Interpretation

In planning the invasion at Suvla, Hamilton thought of everything. He understood the need for surprise, deceiving the Turks about the landing site. He mastered the logistical details of a complex amphibious assault. Locating the key point—Tekke Tepe—from which the Allies could break the stalemate in Gallipoli, he crafted an excellent strategy to get there. He even tried to prepare for the kind of unexpected contingencies that can always happen in battle. But he ignored the one thing closest to him: the chain of command, and the circuit of communications by which orders, information, and decisions would circulate back and forth. He was dependent on that circuit to give him control of the situation and allow him to execute his strategy.

The first links in the chain of command were Stopford and Hammersley. Both men were terrified of risk, and Hamilton failed to adapt himself to their weakness: his order to reach Tekke Tepe was polite, civilized, and unforceful, and Stopford and Hammersley interpreted it according to their fears. They saw Tekke Tepe as a possible goal to aim for once the beaches were secured.

The next links in the chain were the colonels who were to lead the assault on Tekke Tepe. They had no contact with Hamilton on his island or with Stopford on his boat, and Hammersley was too overwhelmed to lead them. They themselves were terrified of acting on their own and maybe messing up a plan they had never understood; they hesitated at every step. Below the colonels were officers and soldiers who, without leadership, were left wandering on the beach like lost ants. Vagueness at the top turned into confusion and lethargy at the bottom. Success depended on the speed with which information could pass in both directions along the chain of command, so that Hamilton could understand what was happening and adapt faster than the enemy. The chain was broken, and Gallipoli was lost.

When a failure like this happens, when a golden opportunity slips through your fingers, you naturally look for a cause. Maybe you blame your incompetent officers, your faulty technology, your flawed intelligence. But that is to look at the world backward; it ensures more failure. The truth is that everything starts from the top. What determines your failure or success is your style of leadership and the chain of command that you design. If your orders are vague and halfhearted, by the time they reach the field they will be meaningless. Let people work unsupervised and they will revert to their natural selfishness: they will see in your orders what they want to see, and their behavior will promote their own interests.

Unless you adapt your leadership style to the weaknesses of the people in your group, you will almost certainly end up with a break in the chain of command. Information in the field will reach you too slowly. A proper chain of command, and the control it brings you, is not an accident; it is your creation, a work of art that requires constant attention and care. Ignore it at your peril.

*For what the leaders are, that, as a rule, will the men below them be.*
*—Xenophon (430?–355? B.C.)*

## REMOTE CONTROL

In the late 1930s, U.S. Brigadier General George C. Marshall (1880–1958) preached the need for major military reform. The army had too few soldiers, they were badly trained, current doctrine was ill suited to modern technology—the list of problems went on. In 1939, President Franklin D. Roosevelt had to select his next army chief of staff. The appointment was critical: World War II had begun in Europe, and Roosevelt believed that the United States was sure to get involved. He understood the need for military reform, so he bypassed generals with more seniority and experience and chose Marshall for the job.

The appointment was a curse in disguise, for the War Department was hopelessly dysfunctional. Many of its generals had monstrous egos

*What must be the result of an operation which is but partially understood by the commander, since it is not his own conception? I have undergone a pitiable experience as prompter at head-quarters, and no one has a better appreciation of the value of such services than myself; and it is particularly in a council of war that such a part is absurd. The greater the number and the higher the rank of the military officers who compose the council, the more difficult will it be to accomplish the triumph of truth and reason, however small be the amount of dissent. What would have been the action of a council of war to which Napoleon proposed the movement of Arcola, the crossing of the Saint-Bernard, the maneuver at Ulm, or that at Gera and Jena? The timid would have regarded them as rash, even to madness, others would have seen a thousand difficulties of execution, and all would have concurred in rejecting them; and if, on the contrary, they had been adopted, and had been executed by any one but Napoleon, would they not certainly have proved failures?*

BARON ANTOINE-HENRI DE JOMINI, 1779–1869

and the power to impose their way of doing things. Senior officers, instead of retiring, took jobs in the department, amassing power bases and fiefdoms that they did everything they could to protect. A place of feuds, waste, communication breakdowns, and overlapping jobs, the department was a mess. How could Marshall revamp the army for global war if he could not control it? How could he create order and efficiency?

Some ten years earlier, Marshall had served as the assistant commander of the Infantry School at Fort Benning, Georgia, where he had trained many officers. Throughout his time there, he had kept a notebook in which he recorded the names of promising young men. Soon after becoming chief of staff, Marshall began to retire the older officers in the War Department and replace them with these younger men whom he had personally trained. These officers were ambitious, they shared his desire for reform, and he encouraged them to speak their minds and show initiative. They included men like Omar Bradley and Mark Clark, who would be crucial in World War II, but no one was more important than the protégé Marshall spent the most time on: Dwight D. Eisenhower.

The relationship began a few days after the attack on Pearl Harbor, when Marshall asked Eisenhower, then a colonel, to prepare a report on what should be done in the Far East. The report showed Marshall that Eisenhower shared his ideas on how to run the war. For the next few months, he kept Eisenhower in the War Plans Division and watched him closely: the two men met every day, and in that time Eisenhower soaked up Marshall's style of leadership, his way of getting things done. Marshall tested Eisenhower's patience by indicating that he planned to keep him in Washington instead of giving him the field assignment that he desperately wanted. The colonel passed the test. Much like Marshall himself, he got along well with other officers yet was quietly forceful.

In July 1942, as the Americans prepared to enter the war by fighting alongside the British in North Africa, Marshall surprised one and all by naming Eisenhower commander in the European Theater of Operations. Eisenhower was by this time a lieutenant general but was still relatively unknown, and in his first few months in the job, as the Americans fared poorly in North Africa, the British clamored for a replacement. But Marshall stood by his man, offering him advice and encouragement. One key suggestion was for Eisenhower to develop a protégé, much as Marshall had with him—a kind of roving deputy who thought the way he did and would act as his go-between with subordinates. Marshall's suggestion for the post was Major General Bradley, a man he knew well; Eisenhower accepted the idea, essentially duplicating the staff structure that Marshall had created in the War Department. With Bradley in place, Marshall left Eisenhower alone.

Marshall positioned his protégés throughout the War Department, where they quietly spread his way of doing things. To make the task easier, he cut the waste in the department with utter ruthlessness, reducing from sixty to six the number of deputies who reported to him. Marshall hated excess; his reports to Roosevelt made him famous for his ability to summarize a complex situation in a few pages. The six men who reported to him found that any report that lasted a page too long simply went unread. He would listen to their oral presentations with rapt attention, but the minute they wandered from the topic or said something not thought through, he would look away, bored, uninterested. It was an expression they dreaded: without saying a word, he had made it known that they had displeased him and it was time for them to leave. Marshall's six deputies began to think like him and to demand from those who reported to them the efficiency and streamlined communications style he demanded of them. The speed of the information flow up and down the line was now quadrupled.

Marshall exuded authority but never yelled and never challenged men frontally. He had a knack for communicating his wishes indirectly —a skill that was all the more effective since it made his officers think about what he meant. Brigadier General Leslie R. Groves, the military director of the project to develop the atom bomb, once came to Marshall's office to get him to sign off on $100 million in expenditures. Finding the chief of staff engrossed in paperwork, he waited while Marshall diligently compared documents and made notes. Finally Marshall put down his pen, examined the $100 million request, signed it, and returned it to Groves without a word. The general thanked him and was turning to leave when Marshall finally spoke: "It may interest you to know what I was doing: I was writing the check for $3.52 for grass seed for my lawn."

The thousands who worked under Marshall, whether in the War Department or abroad in the field, did not have to see him personally to feel his presence. They felt it in the terse but insightful reports that reached them from his deputies, in the speed of the responses to their questions and requests, in the department's efficiency and team spirit. They felt it in the leadership style of men like Eisenhower, who had absorbed Marshall's diplomatic yet forceful way of doing things. In a few short years, Marshall transformed the War Department and the U.S. Army. Few really understood how he had done it.

*"Do you think every Greek here can be a king?*
*It's no good having a carload of commanders. We need*
*One commander, one king, the one to whom Zeus,*
*Son of Cronus the crooked, has given the staff*
*And the right to make decisions for his people."*
*And so Odysseus mastered the army. The men all*
*Streamed back from their ships and huts and assembled*
*With a roar.*

*The Iliad*, Homer, circa Ninth Century B.C.

## Interpretation

When Marshall became chief of staff, he knew that he would have to hold himself back. The temptation was to do combat with everyone in every problem area: the recalcitrance of the generals, the political feuds, the layers of waste. But Marshall was too smart to give in to that temptation. First, there were too many battles to fight, and they would exhaust him. He'd get frustrated, lose time, and probably give himself a heart attack. Second, by trying to micromanage the department, he would become embroiled in petty entanglements and lose sight of the larger picture. And finally he would come across as a bully. The only way to slay this many-headed monster, Marshall knew, was to step back. He

*Reports gathered and presented by the General Staff, on the one hand, and by the Statistical Bureau, on the other, thus constituted the most important sources of information at Napoleon's disposal. Climbing through the chain of command, however, such reports tend to become less and less specific; the more numerous the stages through which they pass and the more standardized the form in which they are presented, the greater the danger that they will become so heavily profiled (and possibly sugar-coated or merely distorted by the many summaries) as to become almost meaningless. To guard against this danger and to keep subordinates on their toes, a commander needs to have in addition a kind of directed telescope—the metaphor is an apt one—which he can direct, at will, at any part of the enemy's forces, the terrain, or his own army in order to bring in information that is not only less structured than that passed on by the normal channels but also tailored to meet his momentary (and specific) needs. Ideally, the regular reporting system should tell the commander which questions to ask, and the directed telescope should enable him to answer those questions. It was the two systems*

had to rule indirectly through others, controlling with such a light touch that no one would realize how thoroughly he dominated.

The key to Marshall's strategy was his selection, grooming, and placement of his protégés. He metaphorically cloned himself in these men, who enacted the spirit of his reforms on his behalf, saving him time and making him appear not as a manipulator but as a delegator. His cutting of waste was heavy-handed at first, but once he put his stamp on the department, it began to run efficiently on its own—fewer people to deal with, fewer irrelevant reports to read, less wasted time on every level. This streamlining achieved, Marshall could guide the machine with a lighter touch. The political types who were clogging the chain of command were either retired or joined in the team spirit he infused. His indirect style of communicating amused some of his staff, but it was actually a highly effective way of asserting his authority. An officer might go home chuckling about finding Marshall fussing over a gardening bill, but it would slowly dawn on him that if he wasted a penny, his boss would know.

Like the War Department that Marshall inherited, today's world is complex and chaotic. It is harder than ever to exercise control through a chain of command. You cannot supervise everything yourself; you cannot keep your eye on everyone. Being seen as a dictator will do you harm, but if you submit to complexity and let go of the chain of command, chaos will consume you.

The solution is to do as Marshall did: operate through a kind of remote control. Hire deputies who share your vision but can think on their own, acting as you would in their place. Instead of wasting time negotiating with every difficult person, work on spreading a spirit of camaraderie and efficiency that becomes self-policing. Streamline the organization, cutting out waste—in staff, in the irrelevant reports on your desk, in pointless meetings. The less attention you spend on petty details, the more time you will have for the larger picture, for asserting your authority generally and indirectly. People will follow your lead without feeling bullied. That is the ultimate in control.

*Madness is the exception in individuals but the rule in groups.*
—*Friedrich Nietzsche (1844–1900)*

## KEYS TO WARFARE

Now more than ever, effective leadership requires a deft and subtle touch. The reason is simple: we have grown more distrustful of authority. At the same time, almost all of us imagine ourselves as authorities in our own right—officers, not foot soldiers. Feeling the need to assert themselves, people today put their own interests before the team. Group unity is fragile and can easily crack.

These trends affect leaders in ways they barely know. The tendency is to give more power to the group: wanting to seem democratic, leaders poll the whole staff for opinions, let the group make decisions, give subordinates input into the crafting of an overall strategy. Without realizing it, these leaders are letting the politics of the day seduce them into violating one of the most important rules of warfare and leadership: unity of command. Before it is too late, learn the lessons of war: divided leadership is a recipe for disaster, the cause of the greatest military defeats in history.

*together, cutting across each other and wielded by Napoleon's masterful hand, which made the revolution in command possible.*

COMMAND IN WAR, MARTIN VAN CREVELD, 1985

Among the foremost of these defeats was the Battle of Cannae, in 216 B.C., between the Romans and the Carthaginians led by Hannibal. The Romans outnumbered the Carthaginians two to one but were virtually annihilated in a perfectly executed strategic envelopment. Hannibal, of course, was a military genius, but the Romans take much of the blame for their own defeat: they had a faulty command system, with two tribunes sharing leadership of the army. Disagreeing over how to fight Hannibal, these men fought each other as much as they fought him, and they made a mess of things.

Nearly two thousand years later, Frederick the Great, king of Prussia and leader of its army, outfought and outlasted the five great powers aligned against him in the Seven Years' War partly because he made decisions so much faster than the alliance generals, who had to consult each other in every move they made. In World War II, General Marshall was well aware of the dangers of divided leadership and insisted that one supreme commander should lead the Allied armies. Without his victory in this battle, Eisenhower could not have succeeded in Europe. In the Vietnam War, the unity of command enjoyed by the North Vietnamese general Vo Nguyen Giap gave him a tremendous advantage over the Americans, whose strategy was crafted by a crowd of politicians and generals.

Divided leadership is dangerous because people in groups often think and act in ways that are illogical and ineffective—call it Groupthink. People in groups are political: they say and do things that they think will help their image within the group. They aim to please others, to promote themselves, rather than to see things dispassionately. Where an individual can be bold and creative, a group is often afraid of risk. The need to find a compromise among all the different egos kills creativity. The group has a mind of its own, and that mind is cautious, slow to decide, unimaginative, and sometimes downright irrational.

This is the game you must play: Do whatever you can to preserve unity of command. Keep the strings to be pulled in your hands; the overarching strategic vision must come from you and you alone. At the same time, hide your tracks. Work behind the scenes; make the group feel involved in your decisions. Seek their advice, incorporating their good ideas, politely deflecting their bad ones. If necessary, make minor, cosmetic strategy changes to assuage the insecure political animals in the group, but ultimately trust your own vision. Remember the dangers of

*Tomorrow at dawn you depart [from St. Cloud] and travel to Worms, cross the Rhine there, and make sure that all preparations for the crossing of the river by my guard are being made there. You will then proceed to Kassel and make sure that the place is being put in a state of defense and provisioned. Taking due security precautions, you will visit the fortress of Hanau. Can it be secured by a coup de main? If necessary, you will visit the citadel of Marburg too. You will then travel on to Kassel and report to me by way of my chargé d'affaires at that place, making sure that he is in fact there. The voyage from Frankfurt to Kassel is not to take place by night, for you are to observe anything that might interest me. From Kassel you are to travel, also by day, by the shortest way to Köln. The land between Wesel, Mainz, Kassel, and Köln is to be reconnoitered. What roads and good communications exist there? Gather information about communications between Kassel and Paderborn. What is the significance of Kassel? Is the place armed and capable of resistance? Evaluate the forces of the Prince Elector in regard to their present state, their artillery, militia, strong places. From Köln you will travel to meet me at Mainz; you are to keep to the right bank*

group decision making. The first rule of effective leadership is never to relinquish your unity of command.

Control is an elusive phenomenon. Often, the harder you tug at people, the less control you have over them. Leadership is more than just barking out orders; it takes subtlety.

Early in his career, the great Swedish film director Ingmar Bergman was often overwhelmed with frustration. He had visions of the films he wanted to make, but the work of being a director was so taxing and the pressure so immense that he would lash out at his cast and crew, shouting orders and attacking them for not giving him what he wanted. Some would stew with resentment at his dictatorial ways, others became obedient automatons. With almost every new film, Bergman would have to start again with a new cast and crew, which only made things worse. But eventually he put together a team of the finest cinematographers, editors, art directors, and actors in Sweden, people who shared his high standards and whom he trusted. That let him loosen the reins of command; with actors like Max von Sydow, he could just suggest what he had in mind and watch as the great actor brought his ideas to life. Greater control could now come from letting go.

A critical step in creating an efficient chain of command is assembling a skilled team that shares your goals and values. That team gives you many advantages: spirited, motivated people who can think on their own; an image as a delegator, a fair and democratic leader; and a saving in your own valuable energy, which you can redirect toward the larger picture.

In creating this team, you are looking for people who make up for your deficiencies, who have the skills you lack. In the American Civil War, President Abraham Lincoln had a strategy for defeating the South, but he had no military background and was disdained by his generals. What good was a strategy if he could not realize it? But Lincoln soon found his teammate in General Ulysses S. Grant, who shared his belief in offensive warfare and who did not have an oversize ego. Once Lincoln discovered Grant, he latched on to him, put him in command, and let him run the war as he saw fit.

Be careful in assembling this team that you are not seduced by expertise and intelligence. Character, the ability to work under you and with the rest of the team, and the capacity to accept responsibility and think independently are equally key. That is why Marshall tested Eisenhower for so long. You may not have as much time to spare, but never choose a man merely by his glittering résumé. Look beyond his skills to his psychological makeup.

Rely on the team you have assembled, but do not be its prisoner or give it undue influence. Franklin D. Roosevelt had his infamous "brain trust," the advisers and cabinet members on whom he depended for their ideas and opinions, but he never let them in on the actual decision making, and he kept them from building up their own power base within the administration. He saw them simply as tools, extending his own abilities and saving him valuable time. He understood unity of command and was never seduced into violating it.

A key function of any chain of command is to supply information rapidly from the trenches, letting you adapt fast to circumstances. The shorter and more streamlined the chain of command, the better for the flow of information. Even so, information is often diluted as it passes up the chain: the telling details that reveal so much become standardized and general as they are filtered through formal channels. Some on the chain, too, will interpret the information for you, filtering what you hear. To get more direct knowledge, you might occasionally want to visit the field yourself. Marshall would sometimes drop in on an army base incognito to see with his own eyes how his reforms were taking effect; he would also read letters from soldiers. But in these days of increasing complexity, this can consume far too much of your time.

What you need is what the military historian Martin van Creveld calls "a directed telescope": people in various parts of the chain, and elsewhere, to give you instant information from the battlefield. These people—an informal network of friends, allies, and spies—let you bypass the slow-moving chain. The master of this game was Napoleon, who created a kind of shadow brigade of younger officers in all areas of the military, men chosen for their loyalty, energy, and intelligence. At a moment's notice, he would send one of these men to a far-off front or garrison, or even to enemy headquarters (ostensibly as a diplomatic envoy), with secret instructions to gather the kind of information he could not get fast enough through normal channels. In general, it is important to cultivate these directed telescopes and plant them throughout the group. They give you flexibility in the chain, room to maneuver in a generally rigid environment.

The single greatest risk to your chain of command comes from the political animals in the group. People like this are inescapable; they spring up like weeds in any organization. Not only are they out for themselves, but they build factions to further their own agendas and fracture the cohesion you have built. Interpreting your commands for their own purposes, finding loopholes in any ambiguity, they create invisible breaks in the chain.

Try to weed them out before they arrive. In hiring your team, look at the candidates' histories: Are they restless? Do they often move from place to place? That is a sign of the kind of ambition that will keep them from fitting in. When people seem to share your ideas exactly, be wary: they are probably mirroring them to charm you. The court of Queen Elizabeth I of England was full of political types. Elizabeth's solution was to keep her opinions quiet; on any issue, no one outside her inner circle knew where she stood. That made it hard for people to mirror her, to disguise their intentions behind a front of perfect agreement. Hers was a wise strategy.

*on the Rhine and submit a short appreciation of the country around Dusseldorf, Wesel, and Kassel. I shall be at Mainz on the 29th in order to receive your report. You can see for yourself how important it is for the beginning of the campaign and its progress that you should have the country well imprinted on your memory.*

NAPOLEON'S WRITTEN INSTRUCTIONS TO FIELD GENERAL, QUOTED IN *COMMAND IN WAR*, MARTIN VAN CREVELD, 1985

Another solution is to isolate the political moles—to give them no room to maneuver within the organization. Marshall accomplished this by infusing the group with his spirit of efficiency; disrupters of that spirit stood out and could quickly be isolated. In any event, do not be naïve. Once you identify the moles in the group, you must act fast to stop them from building a power base from which to destroy your authority.

Finally, pay attention to the orders themselves—their form as well as their substance. Vague orders are worthless. As they pass from person to person, they are hopelessly altered, and your staff comes to see them as symbolizing uncertainty and indecision. It is critical that you yourself be clear about what you want before issuing your orders. On the other hand, if your commands are too specific and too narrow, you will encourage people to behave like automatons and stop thinking for themselves—which they must do when the situation requires it. Erring in neither direction is an art.

Here, as in so much else, Napoleon was the master. His orders were full of juicy details, which gave his officers a feel for how his mind worked while also allowing them interpretive leeway. He would often spell out possible contingencies, suggesting ways the officer could adapt his instructions if necessary. Most important, he made his orders inspiring. His language communicated the spirit of his desires. A beautifully worded order has extra power; instead of feeling like a minion, there only to execute the wishes of a distant emperor, the recipient becomes a participant in a great cause. Bland, bureaucratic orders filter down into listless activity and imprecise execution. Clear, concise, inspiring orders make officers feel in control and fill troops with fighting spirit.

**Image:** The Reins. A horse with no bridle is useless, but equally bad is the horse whose reins you pull at every turn, in a vain effort at control. Control comes from almost letting go, holding the reins so lightly that the horse feels no tug but senses the slightest change in tension and responds as you desire. Not everyone can master such an art.

**Authority:** Better one bad general than two good ones. —*Napoleon Bonaparte (1769–1821)*

## REVERSAL

No good can ever come of divided leadership. If you are ever offered a position in which you will have to share command, turn it down, for the enterprise will fail and you will be held responsible. Better to take a lower position and let the other person have the job.

It is always wise, however, to take advantage of your opponent's faulty command structure. Never be intimidated by an alliance of forces against you: if they share leadership, if they are ruled by committee, your advantage is more than enough. In fact, do as Napoleon did and seek out enemies with that kind of command structure. You cannot fail to win.

# 6

# SEGMENT YOUR FORCES

## THE CONTROLLED-CHAOS STRATEGY

*The critical elements in war are speed and adaptability—the ability to move and make decisions faster than the enemy. But speed and adaptability are hard to achieve today. We have more information than ever before at our fingertips, making interpretation and decision making more difficult. We have more people to manage, those people are more widely spread, and we face more uncertainty. Learn from Napoleon, warfare's greatest master: speed and adaptability come from flexible organization. Break your forces into independent groups that can operate and make decisions on their own. Make your forces elusive and unstoppable by infusing them with the spirit of the campaign, giving them a mission to accomplish, and then letting them run.*

*Finally, a most important point to be considered is that the revolutionary system of command employed by Napoleon was the outcome not of any technological advances, as one might expect, but merely of superior organization and doctrine. The technical means at the emperor's disposal were not a whit more sophisticated than those of his opponents; he differed from them in that he possessed the daring and ingenuity needed to transcend the limits that technology had imposed on commanders for thousands of years. Whereas Napoleon's opponents sought to maintain control and minimize uncertainty by keeping their forces closely concentrated, Napoleon chose the opposite way, reorganizing and decentralizing his army in such a way as to enable its parts to operate independently for a limited period of time and consequently tolerate a higher degree of uncertainty. Rather than allowing the technological means at hand to dictate the method of strategy and the functioning of command, Napoleon made profitable use of the very limitations imposed by the technology.*

COMMAND IN WAR, MARTIN VAN CREVELD, 1985

## CALCULATED DISORDER

In 1800, by defeating Austria in the Battle of Marengo, Napoleon gained control of northern Italy and forced the Austrians to sign a treaty recognizing French territorial gains there and in Belgium. For the next five years, an uneasy peace held sway—but Napoleon crowned himself emperor of France, and many in Europe began to suspect that this Corsican upstart had limitless ambitions. Karl Mack, the Austrian quartermaster general and an older and influential member of the Austrian military, advocated a preemptive strike against France, with an army large enough to guarantee victory. He told his colleagues, "In war the object is to beat the enemy, not merely to avoid being beaten."

Mack and like-minded officers slowly gained influence, and in April 1805, Austria, England, and Russia signed a treaty of alliance to wage war on France and force her to return to her pre-Napoleonic borders. That summer they formulated their plan: 95,000 Austrian troops would attack the French in northern Italy, redressing the humiliating defeat of 1800. Another 23,000 troops would secure the Tyrol, between Italy and Austria. Mack would then lead a force of 70,000 men west along the Danube into Bavaria, preventing this strategically located country from allying itself with France. Once encamped in Bavaria, Mack and his army would await the arrival a few weeks later of 75,000 troops from Russia; the two armies would link up, and this unstoppable force would march west into France. Meanwhile the English would attack the French at sea. More troops would later be funneled into each war zone, making for an army totaling 500,000 men overall—the largest military force ever assembled in Europe up to that point. Not even Napoleon could withstand an army more than twice the size of his own, moving in on him from all sides.

In the middle of September, Mack began his phase of the campaign by advancing along the Danube to Ulm, in the heart of Bavaria. Having established his camp there, he felt hugely satisfied. Mack loathed disorder and uncertainty. He tried to think of everything in advance, to come up with a clear plan and make sure everyone stuck to it—"clockwork warfare," he called it. He thought his plan was perfect; nothing could go wrong. Napoleon was doomed.

Mack had once been captured and forced to spend three years in France, where he had studied Napoleon's style of war. A key Napoleonic strategy was to make the enemy divide his forces, but now the trick was reversed: with trouble in Italy, Napoleon could not afford to send more than 70,000 French troops across the Rhine into Germany and Bavaria. The moment he crossed the Rhine, the Austrians would know his intentions and would act to slow his march; his army would need at least two months to reach Ulm and the Danube. By then the Austrians would already have linked up with the Russians and swept through the Alsace and France. The strategy was as close to foolproof as any Mack had ever known. He savored the role he would play in destroying

Napoleon, for he hated the man and all he represented—undisciplined soldiers, the fomenting of revolution throughout Europe, the constant threat to the status quo. For Mack the Russians could not arrive in Ulm too soon.

Near the end of September, however, Mack began to sense something wrong. To the west of Ulm lay the Black Forest, between his own position and the French border. Suddenly scouts were telling him that a French army was passing through the forest in his direction. Mack was bewildered: it made the best sense for Napoleon to cross the Rhine into Germany farther to the north, where his passage east would be smoother and harder to stop. But now he was yet again doing the unexpected, funneling an army through a narrow opening in the Black Forest and sending it straight at Mack. Even if this move were just a feint, Mack had to defend his position, so he sent part of his army west into the Black Forest to stem the French advance long enough for the Russians to come to his aid.

A few days later, Mack began to feel horribly confused. The French were proceeding through the Black Forest, and some of their cavalry had come quite far. At the same time, though, word reached Mack of a large French army somewhere to the north of his position. The reports were contradictory: some said this army was at Stuttgart, sixty miles northwest of Ulm; others had it more to the east or even farther to the north or—quite close, near the Danube. Mack could get no hard information, since the French cavalry that had come through the Black Forest blocked access to the north for reconnaissance. The Austrian general now faced what he feared most—uncertainty—and it was clouding his ability to think straight. Finally he ordered all of his troops back to Ulm, where he would concentrate his forces. Perhaps Napoleon intended to do battle at Ulm. At least Mack would have equal numbers.

In early October, Austrian scouts were at last able to find out what was really going on, and it was a nightmare. A French army had crossed the Danube to the east of Ulm, blocking Mack's way back to Austria and cutting off the Russians. Another army lay to the south, blocking his route to Italy. How could 70,000 French soldiers appear in so many places at once? And move so fast? Gripped by panic, Mack sent probes in every direction. On October 11 his men discovered a weak point: only a small French force barred the way north and east. There he could push through and escape the French encirclement. He began to prepare for the march. But two days later, when he was on the point of ordering the retreat, his scouts reported that a large French force had appeared overnight, blocking the northeastern route as well.

On October 20, finding out that the Russians had decided not to come to his rescue, Mack surrendered. Over 60,000 Austrian soldiers were taken prisoner with hardly a shot fired. It was one of the most splendidly bloodless victories in history.

In the next few months, Napoleon's army turned east to deal with the

*We find our attention drawn repeatedly to what one might call "the organizational dimension of strategy." Military organizations, and the states that develop them, periodically assess their own ability to handle military threats. When they do so they tend to look at that which can be quantified: the number of troops, the quantities of ammunition, the readiness rates of key equipment, the amount of transport, and so on. Rarely, however, do they look at the adequacy of their organization as such, and particularly high level organization, to handle these challenges. Yet as Pearl Harbor and other cases suggest, it is in the deficiency of organizations that the embryo of misfortune develops.*

*MILITARY MISFORTUNES: THE ANATOMY OF FAILURE IN WAR*, ELIOT A. COHEN AND JOHN GOOCH, 1990

*The fact that, historically speaking, those armies have been most successful which did not turn their troops into automatons, did not attempt to control everything from the top, and allowed subordinate commanders considerable latitude has been abundantly demonstrated. The Roman centurions and military tribunes; Napoleon's marshals; Moltke's army commanders; Ludendorff's storm detachments . . .—all these are examples, each within its own stage of technological development, of the way things were done in some of the most successful military forces ever.*

COMMAND IN WAR, MARTIN VAN CREVELD, 1985

Russians and remaining Austrians, culminating in his spectacular victory at Austerlitz. Meanwhile Mack languished in an Austrian prison, sentenced to two years for his role in this humiliating defeat. There he racked his brains (losing his sanity in the process, some said): Where had his plan gone wrong? How had an army appeared out of nowhere to his east, so easily swallowing him up? He had never seen anything like it, and he was trying to figure it out to the end of his days.

## Interpretation

History should not judge General Mack too harshly, for the French armies he faced in the fall of 1805 represented one of the greatest revolutions in military history. For thousands of years, war had been fought in essentially the same way: the commander led his large and unified army into battle against an opponent of roughly equal size. He would never break up his army into smaller units, for that would violate the military principle of keeping one's forces concentrated; furthermore, scattering his forces would make them harder to monitor, and he would lose control over the battle.

Suddenly Napoleon changed all that. In the years of peace between 1800 and 1805, he reorganized the French military, bringing different forces together to form the Grande Armée, 210,000 men strong. He divided this army into several corps, each with its own cavalry, infantry, artillery, and general staff. Each was led by a marshal general, usually a young officer of proven strength in previous campaigns. Varying in size from 15,000 to 30,000 men, each corps was a miniature army headed by a miniature Napoleon.

*Patton's philosophy of command was: "Never tell people how to do things. Tell them what to do and they will surprise you with their ingenuity."*

PATTON: A GENIUS FOR WAR, CARLO D'ESTE, 1995

The key to the system was the speed with which the corps could move. Napoleon would give the marshals their mission, then let them accomplish it on their own. Little time was wasted with the passing of orders back and forth, and smaller armies, needing less baggage, could march with greater speed. Instead of a single army moving in a straight line, Napoleon could disperse and concentrate his corps in limitless patterns, which to the enemy seemed chaotic and unreadable.

This was the monster that Napoleon unleashed on Europe in September 1805. While a few corps were dispatched to northern Italy as a holding force against Austria's planned invasion there, seven corps moved east into Germany in a scattered array. A reserve force with much cavalry was sent through the Black Forest, drawing Mack to the west—and so making it harder for him to understand what was happening to the north and easier to entrap. (Napoleon understood Mack's simple psychology and how the appearance of disorder would paralyze him.) Meanwhile, with Stuttgart as a pivot, the seven corps wheeled south to the Danube and cut off Mack's various escape routes. One corps marshal, hearing that the northeastern route was weakly held, did not wait for Napoleon to send orders but simply sped and covered it on his own. Wherever Mack went, he would hit a corps large enough to hold him until the rest of the French army could tighten the circle. It was like a pack of coyotes against a rabbit.

Understand: the future belongs to groups that are fluid, fast, and nonlinear. Your natural tendency as a leader may be to want to control the group, to coordinate its every movement, but that will just tie you to the past and to the slow-moving armies of history. It takes strength of character to allow for a margin of chaos and uncertainty—to let go a little—but by decentralizing your army and segmenting it into teams, you will gain in mobility what you lose in complete control. And mobility is the greatest force multiplier of them all. It allows you to both disperse and concentrate your army, throwing it into patterns instead of advancing in straight lines. These patterns will confuse and paralyze your opponents. Give your different corps clear missions that fit your strategic goals, then let them accomplish them as they see fit. Smaller teams are faster, more creative, more adaptable; their officers and soldiers are more engaged, more motivated. In the end, fluidity will bring you far more power and control than petty domination.

*Separate to live, unite to fight.*
*—Napoleon Bonaparte (1769–1821)*

*Agamemnon smiled and moved on,*
*Coming next to the two captains*
*Who shared the name Ajax*
*As they were strapping on their helmets.*
*Behind them a cloud of infantry loomed . . .*
*Agamemnon*
*Was glad to see them, and his words flew out:*
*"Ajax, both of you, Achaean commanders,*
*I would be out of line if I issued you orders.*
*You push your men to fight hard on your own.*
*By Father Zeus, by Athena and Apollo,*
*If all of my men had your kind of heart,*
*King Priam's city would soon bow her head,*
*Taken and ravaged under our hands."*

*The Iliad*, Homer, circa Ninth Century B.C.

## KEYS TO WARFARE

The world is full of people looking for a secret formula for success and power. They do not want to think on their own; they just want a recipe to follow. They are attracted to the idea of strategy for that very reason. In their minds strategy is a series of steps to be followed toward a goal. They want these steps spelled out for them by an expert or a guru. Believing in the power of imitation, they want to know exactly what some great person has done before. Their maneuvers in life are as mechanical as their thinking.

To separate yourself from such a crowd, you need to get rid of a common misconception: the essence of strategy is not to carry out a brilliant plan that proceeds in steps; it is to put yourself in situations where you have more options than the enemy does. Instead of grasping at Option A as the single right answer, true strategy is positioning yourself to be able to do A, B, or C depending on the circumstances. That is strategic depth of thinking, as opposed to formulaic thinking.

Sun-tzu expressed this idea differently: what you aim for in strategy, he said, is *shih*, a position of potential force—the position of a boulder perched precariously on a hilltop, say, or of a bowstring stretched taut. A tap on the boulder, the release of the bowstring, and potential force is violently unleashed. The boulder or arrow can go in any direction; it is geared to the actions of the enemy. What matters is not following preordained steps but placing yourself in *shih* and giving yourself options.

Napoleon was probably unaware of Sun-tzu's concept of *shih*, yet he

*It was during this period of post-war introspection and evaluation that one of the fundamental military concepts of Scharnhorst and Gneisenau coalesced into a clearly defined doctrine understandable to and understood by all officers in the Army. This was the concept of* Auftragstaktik, *or mission tactics. Moltke himself inserted in the draft of a new tactical manual for senior commanders the following lines: "A favorable situation will never be exploited if commanders wait for orders. The highest commander and the youngest soldier must always be conscious of the fact that omission and inactivity are worse than resorting to the wrong expedient." . . . Nothing epitomized the outlook and performance of the German General Staff, and of the German Army which it coordinated, more than this concept of mission tactics: the responsibility of each German officer and noncommissioned officer . . . to do without question or doubt whatever the situation required, as he saw it. This meant that he should act without awaiting orders, if action seemed necessary. It also meant that he should act contrary to orders, if these did not seem to be consistent with the situation.*

had perhaps history's greatest understanding of it. Once he had positioned his seven corps in their seemingly chaotic pattern along the Rhine and his reserve forces in the Black Forest, he was in *shih.* Wherever Mack turned, whatever he did, the Austrians were doomed. Napoleon had endless options while Mack had only a few, and all of them bad.

Napoleon had always aimed at his version of *shih,* and he perfected it in the 1805 campaign. Obsessed with structure and organization, he developed the corps system, building flexibility into the very skeleton of his army. The lesson is simple: a rigid, centralized organization locks you into linear strategies; a fluid, segmented army gives you options, endless possibilities for reaching *shih.* Structure *is* strategy—perhaps the most important strategic choice you will make. Should you inherit a group, analyze its structure and alter it to suit your purposes. Pour your creative energy into its organization, making fluidity your goal. In doing so you will be following in the footsteps not only of Napoleon but of perhaps the greatest war machine in modern times, the Prussian (and later German) army.

Shortly after Napoleon's devastating defeat of the Prussians at the Battle of Jena in 1806 (see chapter 2), the Prussian leaders did some soul-searching. They saw they were stuck in the past; their way of doing things was too rigid. Suddenly the military reformers, including Carl von Clausewitz, were taken seriously and given power. And what they decided to do was unprecedented in history: they would institutionalize success by designing a superior army structure.

At the core of this revolution was the creation of a general staff, a cadre of officers specially trained and educated in strategy, tactics, and leadership. A king, a prime minister, or even a general might be incompetent at war, but a group of brilliant and well-trained officers on the army's staff could compensate for his failures. The structure of this body was unfixed: each new chief of staff could alter its size and function to suit his needs and the times. After each campaign or training exercise, the staff would rigorously examine itself and its performance. A whole section was created for the purpose of these examinations and for the study of military history. The general staff would learn from its mistakes and those of others. It was to be a work permanently in progress.

The most important reform was the development of the *Auftragstaktik* (mission-oriented command system). In German there are two words for "command": *Auftrag* and *Befehl.* A *Befehl* is an order to be obeyed to the letter. An *Auftrag* is much more general: it is a statement of overall mission, a directive to be followed in its spirit, not its letter. The *Auftragstaktik*—inspired by Prussia's archenemy Napoleon and the leeway he gave his marshals—permeated the general staff. Officers were first inculcated with the philosophy of German warfare: speed, the need to take the offensive, and so on. Then they were put through exercises to help them develop their ability to think on their own, to make decisions that met the overall philosophy but responded to the circumstances of the moment. Leading the equivalent of a corps in battle, officers were given missions to accomplish and then were let loose. They were judged by the results of their actions, not on how those results were achieved.

The general staff (with a few interruptions) was in place from 1808 to the end of World War II. During that period the Germans consistently outfought other armies in the field–including the Allies in World War I, despite the severe limitations of trench warfare. Their success culminated in the most devastating military victory in modern history: the 1940 blitzkrieg invasion of France and the Low Countries, when the German army ran rings around the rigid defenses of the French. It was the structure of their army, and their use of the *Auftragstaktik*, that gave them more options and greater potential force.

The German general staff should serve as the organizational model for any group that aims at mobility and strategic depth. First, the staff's structure was fluid, allowing its leaders to adapt it to their own needs. Second, it examined itself constantly and modified itself according to what it had learned. Third, it replicated its structure through the rest of the army: its officers trained the officers below them, and so on down the line. The smallest team was inculcated with the overall philosophy of the group. Finally, rather than issuing rigid orders, the staff embraced the mission command, the *Auftragstaktik*. By making officers and soldiers feel more creatively engaged, this tactic improved their performance and sped up the decision-making process. Mobility was written into the system.

The key to the *Auftragstaktik* is an overall group philosophy. This can be built around the cause you are fighting for or a belief in the evil of the enemy you face. It can also include the style of warfare—defensive, mobile, ruthlessly aggressive—that best suits it. You must bring the group together around this belief. Then, through training and creative exercises, you must deepen its hold on them, infuse it into their blood. Now, when you unleash your corps on their missions, you can trust their decisions and feel confident in your power to coordinate them.

The Mongol hordes led by Genghis Khan in the first half of the thirteenth century were perhaps the closest precursors to Napoleon's corps. Genghis, who preached a philosophy of Mongol superiority, was a master of mobility in warfare. His segmented forces could disperse and concentrate in complicated patterns; the armies that faced them were shocked at how chaotic they seemed, so impossible to figure out, yet they maneuvered with amazing coordination. Mongol soldiers knew what to do, and when, without being told. For their victims the only explanation was that they were possessed by the devil.

The sinister coordination of the Mongols, however, was actually the result of rigorous training. Every winter in peacetime, Genghis would run the Great Hunt, a three-month-long operation in which he would scatter the entire Mongol army along an eighty-mile line in the steppes of Central Asia and what is now Mongolia. A flag in the ground hundreds

*To make perfectly clear that action contrary to orders was not considered either as disobedience or lack of discipline, German commanders began to repeat one of Moltke's favorite stories, of an incident observed while visiting the headquarters of Prince Frederick Charles. A major, receiving a tongue-lashing from the Prince for a tactical blunder, offered the excuse that he had been obeying orders, and reminded the Prince that a Prussian officer was taught that an order from a superior was tantamount to an order from the King. Frederick Charles promptly responded: "His Majesty made you a major because he believed you would know when not to obey his orders." This simple story became guidance for all following generations of German officers.*

*A Genius for War: The German Army and General Staff, 1807–1945*,
Colonel T.N. Dupuy,
1977

*[Tom] Yawkey was thirty years old when he bought the Red Sox, a hopelessly bankrupt team that had won only forty-three games the previous season and averaged only 2,365 paying customers. The ball club became his toy. Because he loved his players, he spoiled them rotten. And because he spoiled them rotten, they praised him to the skies. . . . There is a well-publicized exchange in which Bobby Doerr asks Tommy Henrich why the Red Sox weren't able to beat the Yankees in big games. "Weren't we good enough?" Doerr asks. It wasn't that they weren't good enough, Henrich answers. "Your owner was too good to you. The Red Sox didn't have to get into the World Series to drive Cadillacs. The Yankees did." . . . [The Red Sox organization] was an amateur operation . . . pitted against the toughest, most professional operation of all time.*

HITTER: THE LIFE AND TURMOILS OF TED WILLIAMS, ED LINN, 1993

of miles away marked the hunt's endpoint. The line would advance, driving before it all the animals in its path. Slowly, in an intricately choreographed maneuver, the ends of the line would curve to form a circle, trapping the animals within. (The hunt's endpoint would form the center of the circle.) As the circle tightened, the animals were killed; the most dangerous of them, the tigers, were left till last. The Great Hunt exercised the Mongols' ability to communicate through signals at a distance, coordinate their movements with precision, know what to do in different circumstances, and act without waiting for orders. Even bravery became an exercise, when individual soldiers would have to take on a tiger. Through hunting and a form of play, Genghis could instill his philosophy, develop cohesion and trust among his men, and tighten his army's discipline.

In unifying your own hordes, find exercises to increase your troops' knowledge of and trust in each other. This will develop implicit communication skills between them and their intuitive sense of what to do next. Time will not then be wasted in the endless transmission of messages and orders or in constantly monitoring your troops in the field. If you can disguise these exercises as play, as in the Great Hunt, so much the better.

Throughout the 1940s and '50s, two great baseball organizations did battle: the Boston Red Sox, built around Ted Williams, and the New York Yankees, with their great hitter Joe DiMaggio. The owner of the Red Sox, Tom Yawkey, believed in pampering his players, creating a pleasant environment for them, developing friendships with them. A happy team would play well, he thought. For this purpose he went drinking with his men, played cards with them, checked them in to nice hotels on tour. He also meddled in managerial decisions, always with an eye toward making things better for his players and keeping them happy.

The Yankees' philosophy was very different, emphasizing discipline and victory at all costs. The organization's separate parts stayed out of one another's business—they understood the team ethos and knew they would be judged on results. The manager was left to make his own decisions. Yankee players felt an intense need to live up to the team's winning traditions; they were afraid of losing.

In those two decades, the Red Sox players fought among themselves, fell into factions, whined and complained at any perceived slight, and won just one pennant. The Yankees were cohesive and spirited; they won thirteen pennants and ten World Series. The lesson is simple: do not confuse a chummy, clublike atmosphere with team spirit and cohesion. Coddling your soldiers and acting as if everyone were equal will ruin discipline and promote the creation of factions. Victory will forge stronger bonds than superficial friendliness, and victory comes from discipline, training, and ruthlessly high standards.

Finally, you need to structure your group according to your soldiers' strengths and weaknesses, to their social circumstances. To do that you must be attuned to the human side of your troops; you must understand them, and the spirit of the times, inside and out.

*In a real sense, maximum disorder was our equilibrium.*

T. E. LAWRENCE, 1885–1935

During the American Civil War, the Union generals struggled with the ragtag nature of their army. Unlike the disciplined, well-trained troops of the Confederacy, many Northern soldiers had been forcibly conscripted at the last minute; they were pioneers, rugged frontiersmen, and they were fiercely independent. Some generals tried desperately to instill discipline, and mostly they failed. Others just paid attention to map strategy, while their armies continued to perform badly.

General William Tecumseh Sherman had a different solution: he changed his organization to suit the personalities of his men. He created a more democratic army, encouraged initiative in his officers, let them dress as they saw fit; he loosened outward discipline to foster morale and group spirit. Like frontiersmen generally, his soldiers were restless and nomadic, so he exploited their mobility and kept his army in perpetual motion, always marching faster than his enemies could. Of all the Union armies, Sherman's were the most feared and performed the best.

Like Sherman, do not struggle with your soldiers' idiosyncrasies, but rather turn them into a virtue, a way to increase your potential force. Be creative with the group's structure, keeping your mind as fluid and adaptable as the army you lead.

**Image:**
The Spider's Web.
Most animals attack along a straight line; the spider weaves a web, adapted to its location and spun in a pattern, whether simple or complex. Once the web is woven, the work is done. The spider has no need to hunt; it simply waits for the next fool to fall into the web's barely visible strands.

**Authority:** Thus the army . . . moves for advantage, and changes through segmenting and reuniting. Thus its speed is like the wind, its slowness like the forest; its invasion and plundering like a fire. . . . It is as difficult to know as the darkness; in movement it is like thunder.—The Art of War, *Sun-tzu, (fourth century* B.C.*)*

## REVERSAL

Since the structure of your army has to be suited to the people who compose it, the rule of decentralization is flexible: some people respond better to rigid authority. Even if you run a looser organization, there may be times when you will have to tighten it and give your officers less freedom. Wise generals set nothing in stone, always retaining the ability to reorganize their army to fit the times and their changing needs.

# 7

# TRANSFORM YOUR WAR INTO A CRUSADE

## MORALE STRATEGIES

*The secret to motivating people and maintaining their morale is to get them to think less about themselves and more about the group. Involve them in a cause, a crusade against a hated enemy. Make them see their survival as tied to the success of the army as a whole. In a group in which people have truly bonded, moods and emotions are so contagious that it becomes easy to infect your troops with enthusiasm. Lead from the front: let your soldiers see you in the trenches, making sacrifices for the cause. That will fill them with the desire to emulate and please you. Make both rewards and punishments rare but meaningful. Remember: a motivated army can work wonders, making up for any lack of material resources.*

*You can do nothing with an army that is an amalgam of a hundred people here, a hundred people there, and so on. What can be achieved with four thousand men, united and standing shoulder to shoulder, you cannot do with forty or even four hundred thousand men who are divided and pulled this way and that by internal conflicts. . . .*

*RULES OF WAR AND BRAVERY*, MUBARAKSHAH, PERSIA, THIRTEENTH CENTURY

## THE ART OF MAN MANAGEMENT

We humans are selfish by nature. Our first thoughts in any situation revolve around our own interests: How will this affect *me*? How will it help *me*? At the same time, by necessity, we try to disguise our selfishness, making our motives look altruistic or disinterested. Our inveterate selfishness and our ability to disguise it are problems for you as a leader. You may think that the people working for you are genuinely enthusiastic and concerned—that is what they say, that is what their actions suggest. Then slowly you see signs that this person or that is using his or her position in the group to advance purely personal interests. One day you wake up to find yourself leading an army of selfish, conniving individuals.

That is when you start thinking about morale—about finding a way to motivate your troops and forge them into a group. Perhaps you try artfully to praise people, to offer them the possibility of reward—only to find you have spoiled them, strengthening their selfishness. Perhaps you try punishments and discipline—only to make them resentful and defensive. Perhaps you try to fire them up with speeches and group activities—but people are cynical nowadays; they will see right through you.

The problem is not what you are doing but the fact that it comes late. You have begun to think about morale only after it has become an issue, not before. That is your mistake. Learn from history's great motivators and military leaders: the way to get soldiers to work together and maintain morale is to make them feel part of a group that is fighting for a worthy cause. That distracts them from their own interests and satisfies their human need to feel part of something bigger than they are. The more they think of the group, the less they think of themselves. They soon begin to link their own success to the group's; their own interests and the larger interests coincide. In this kind of army, people know that selfish behavior will disgrace them in the eyes of their companions. They become attuned to a kind of group conscience.

Morale is contagious: put people in a cohesive, animated group and they naturally catch that spirit. If they rebel or revert to selfish behavior, they are easily isolated. You must establish this dynamic the minute you become the group's leader; it can only come from the top—that is, from you.

The ability to create the right group dynamic, to maintain the collective spirit, is known in military language as "man management." History's great generals—Alexander the Great, Hannibal, Napoleon—were all masters of the art, which for military men is more than simply important: in battle it can be the deciding issue, a matter of life and death. In war, Napoleon once said, "The moral is to the physical as three to one." He meant that his troops' fighting spirit was crucial in the outcome of the battle: with motivated soldiers he could beat an army three times the size of his own.

To create the best group dynamic and prevent destructive morale problems, follow these eight crucial steps culled from the writings and experiences of the masters of the art. It is important to follow as many of the steps as possible; none is less important than any other.

*What stronger breast-plate than a heart untainted!*
*Thrice is he arm'd that hath his quarrel just,*
*And he but naked, though lock'd up in steel,*
*Whose conscience with injustice is corrupted.*

King Henry V,
William Shakespeare,
1564–1616

**Step 1: Unite your troops around a cause. Make them fight for an idea.** Now more than ever, people have a hunger to believe in something. They feel an emptiness, which, left alone, they might try to fill with drugs or spiritual fads, but you can take advantage of it by channeling it into a cause you can convince them is worth fighting for. Bring people together around a cause and you create a motivated force.

The cause can be anything you wish, but you should represent it as progressive: it fits the times, it is on the side of the future, so it is destined to succeed. If necessary, you can give it a veneer of spirituality. It is best to have some kind of enemy to hate—an enemy can help a group to define itself in opposition. Ignore this step and you are left with an army of mercenaries. You will deserve the fate that usually awaits such armies.

*There are always moments when the commander's place is not back with his staff but up with the troops. It is sheer nonsense to say that maintenance of the men's morale is the job of the battalion commander alone. The higher the rank, the greater the effect of the example. The men tend to feel no kind of contact with a commander who, they know, is sitting somewhere in headquarters. What they want is what might be termed a physical contact with him. In moments of panic, fatigue, or disorganization, or when something out of the ordinary has to be demanded from them, the personal example of the commander works wonders, especially if has had the wit to create some sort of legend around himself.*

Field Marshal
Erwin Rommel,
1891–1944

**Step 2: Keep their bellies full.** People cannot stay motivated if their material needs go unmet. If they feel exploited in any way, their natural selfishness will come to the surface and they will begin to peel off from the group. Use a cause—something abstract or spiritual—to bring them together, but meet their material needs. You do not have to spoil them by overpaying them; a paternalistic feeling that they are being taken care of, that you are thinking of their comfort, is more important. Attending to their physical needs will make it easier to ask more of them when the time comes.

**Step 3: Lead from the front.** The enthusiasm with which people join a cause inevitably wanes. One thing that speeds up its loss, and that produces discontent, is the feeling that the leaders do not practice what they preach. Right from the beginning, your troops must see you leading from the front, sharing their dangers and sacrifices—taking the cause as seriously as they do. Instead of trying to push them from behind, make them run to keep up with you.

**Step 4: Concentrate their *ch'i.*** There is a Chinese belief in an energy called ch'i, which dwells in all living things. All groups have their own level of ch'i, physical and psychological. A leader must understand this energy and know how to manipulate it.

Idleness has a terrible effect on ch'i. When soldiers are not working, their spirits lower. Doubts creep in, and selfish interests take over. Similarly, being on the defensive, always waiting and reacting to what the enemy dishes out, will also lower ch'i. So keep your soldiers busy, acting

*During the Spring and Autumn era, the state of Qi was invaded by the states of Jin and Yan. At first the invaders overcame the military forces of Qi. One of the eminent nobles of the court of Qi recommended the martialist Tian Rangju to the lord of Qi. To this man, later called Sima Rangju, is attributed the famous military handbook "Sima's Art of War." . . . The lord of Qi then summoned Rangju to discuss military matters with him. The lord was very pleased with what Rangju had to say, and he made him a general, appointing him to lead an army to resist the aggression of the forces of Yan and Jin. Rangju said, "I am lowly in social status, yet the lord has promoted me from the ranks and placed me above even the grandees. The soldiers are not yet loyal to me, and the common people are not familiar with me; as a man of little account, my authority is slight. I request one of your favorite ministers, someone honored by the state, to be overseer of the army." The lord acceded to this request and appointed a nobleman to be the overseer. Rangju took his leave, arranging to meet the nobleman at the military headquarters at noon the following day. Then Rangju hastened back to set up a sundial and a waterclock to await the new*

for a purpose, moving in a direction. Do not make them wait for the next attack; propelling them forward will excite them and make them hungry for battle. Aggressive action concentrates ch'i, and concentrated ch'i is full of latent force.

**Step 5: Play to their emotions.** The best way to motivate people is not through reason but through emotion. Humans, however, are naturally defensive, and if you begin with an appeal to their emotions—some histrionic harangue—they will see you as manipulative and will recoil. An emotional appeal needs a setup: lower their defenses, and make them bond as a group, by putting on a show, entertaining them, telling a story. Now they have less control over their emotions and you can approach them more directly, moving them easily from laughter to anger or hatred. Masters of man management have a sense of drama: they know when and how to hit their soldiers in the gut.

**Step 6: Mix harshness and kindness.** The key to man management is a balance of punishment and reward. Too many rewards will spoil your soldiers and make them take you for granted; too much punishment will destroy their morale. You need to hit the right balance. Make your kindness rare and even an occasional warm comment or generous act will be powerfully meaningful. Anger and punishment should be equally rare; instead your harshness should take the form of setting very high standards that few can reach. Make your soldiers compete to please you. Make them struggle to see less harshness and more kindness.

**Step 7: Build the group myth.** The armies with the highest morale are armies that have been tested in battle. Soldiers who have fought alongside one another through many campaigns forge a kind of group myth based on their past victories. Living up to the tradition and reputation of the group becomes a matter of pride; anyone who lets it down feels ashamed. To generate this myth, you must lead your troops into as many campaigns as you can. It is wise to start out with easy battles that they can win, building up their confidence. Success alone will help bring the group together. Create symbols and slogans that fit the myth. Your soldiers will want to belong.

**Step 8: Be ruthless with grumblers.** Allow grumblers and the chronically disaffected any leeway at all and they will spread disquiet and even panic throughout the group. As fast as you can, you must isolate them and get rid of them. All groups contain a core of people who are more motivated and disciplined than the rest—your best soldiers. Recognize them, cultivate their goodwill, and set them up as examples. These people will serve as natural ballasts against those who are disaffected and panicky.

*You know, I am sure, that not numbers or strength brings victory in war; but whichever army goes into battle stronger in soul, their enemies generally cannot withstand them.*
*—Xenophon (430?–355? B.C.)*

## HISTORICAL EXAMPLES

**1.** In the early 1630s, Oliver Cromwell (1599–1658), a provincial gentleman farmer in Cambridgeshire, England, fell victim to a depression and to constant thoughts of death. Deep in crisis, he converted to the Puritan religion, and suddenly his life took a new turn: he felt he had experienced a direct communion with God. Now he believed in providence, the idea that everything happens for a reason and according to God's will. Whereas before he had been despondent and indecisive, now he was filled with purpose: he thought himself among God's elect.

Eventually Cromwell became a member of Parliament and a vocal defender of the common people in their grievances against the aristocracy. Yet he felt marked by providence for something larger than politics: he had visions of a great crusade. In 1642, Parliament, in a bitter struggle with Charles I, voted to cut off the king's funds until he agreed to limits on royal power. When Charles refused, civil war broke out between the Cavaliers (supporters of the king, who wore their hair long) and the Roundheads (the rebels, so called since they cropped their hair short). Parliament's most fervent supporters were Puritans like Cromwell, who saw the war against the king as his chance—more than his chance, his calling.

Although Cromwell had no military background, he hurriedly formed a troop of sixty horsemen from his native Cambridgeshire. His aim was to incorporate them in a larger regiment, gain military experience by fighting under another commander, and slowly prove his worth. He was confident of ultimate victory, for he saw his side as unbeatable: after all, God was on their side, and all his men were believers in the cause of creating a more pious England.

Despite his lack of experience, Cromwell was something of a military visionary: he imagined a new kind of warfare spearheaded by a faster, more mobile cavalry, and in the war's first few months he proved a brave and effective leader. He was given more troops to command but soon realized that he had grossly overestimated the fighting spirit of those on his side: time and again he led cavalry charges that pierced enemy lines, only to watch in disgust as his soldiers broke order to plunder the enemy camp. Sometimes he tried to hold part of his force in reserve to act as reinforcements later in the battle, but the only command they listened to was to advance, and in retreat they were hopelessly disordered. Representing themselves as crusaders, Cromwell's men were re-

*overseer. Now this new overseer was a proud and haughty aristocrat, and he imagined that as overseer he was leading his own army. Because of his pride and arrogance, he did not see any need to hurry, in spite of his promise with Rangju the martial master. His relatives and close associates gave him a farewell party, and he stayed to drink with them. At noon the next day, the new overseer had not arrived at headquarters. Rangju took down the sundial and emptied the water-clock. He assembled the troops and informed them of the agreement with the new overseer. That evening the nobleman finally arrived. Rangju said to him, "Why are you late?" He said, "My relatives, who are grandees, gave me a farewell party, so I stayed for that." Rangju said, "On the day a military leader receives his orders, he forgets about his home; when a promise is made in the face of battle, one forgets his family; when the war drums sound, one forgets his own body. Now hostile states have invaded our territory; the state is in an uproar; the soldiers are exposed at the borders; the lord cannot rest or enjoy his food; the lives of the common people all depend on you—how can you talk about farewell parties?"*

*Rangju then summoned the officer in charge of military discipline and asked him, "According to military law, what happens to someone who arrives later than an appointed time?" The officer replied, "He is supposed to be decapitated." Terrified, the aristocrat had a messenger rush back to report this to the lord and beseech him for help. But the haughty nobleman was executed before the messenger even returned, and his execution was announced to the army. The soldiers all shook with fear. Eventually the lord sent an emissary with a letter pardoning the nobleman, who was, after all, the new overseer of the army. The emissary galloped right into camp on horseback with the lord's message. Rangju said, "When a general is in the field, there are orders he doesn't take from the ruler."*

*He also said to the disciplinary officer, "It is a rule that there shall be no galloping through camp, yet now the emissary has done just that. What should be done with him?" The officer said, "He should be executed." The emissary was petrified, but Rangju said, "It is not proper to kill an emissary of the lord," and had two of the emissary's attendants executed in his stead. This too was announced to the army. Rangju sent the emissary back to report to*

vealed by battle as mercenaries, fighting for pay and adventure. They were useless.

In 1643, when Cromwell was made a colonel at the head of his own regiment, he decided to break with the past. From now on, he would recruit only soldiers of a certain kind: men who, like himself, had experienced religious visions and revelations. He sounded out the aspirants, tested them for the depth of their faith. Departing from a long tradition, he appointed commoners, not aristocrats, as officers; as he wrote to a friend, "I had rather have a plain russet-coated captain that knows what he fights for, and loves what he knows, than that which you call a gentleman and is nothing else." Cromwell made his recruits sing psalms and pray together. In a stern check on bad discipline, he taught them to see all their actions as part of God's plan. And he looked after them in an unusual way for the times, making sure they were well fed, well clothed, and promptly paid.

When Cromwell's army went into battle, it was now a force to reckon with. The men rode in tight formation, loudly singing psalms. As they neared the king's forces, they would break into a "pretty round trot," not the headlong and disorderly charge of other troops. Even in contact with the enemy, they kept their order, and they retreated with as much discipline as when they advanced. Since they believed that God was with them, they had no fear of death: they could march straight up a hill into enemy fire without breaking step. Having gained control over his cavalry, Cromwell could maneuver them with infinite flexibility. His troops won battle after battle.

In 1645, Cromwell was named lieutenant general of the cavalry in the New Model Army. That year, at the Battle of Naseby, his disciplined regiment was crucial in the Roundheads' victory. A few days later, his cavalry finished off the Royalist forces at Langport, effectively putting an end to the first stage of the Civil War.

## Interpretation

That Cromwell is generally considered one of history's great military leaders is all the more remarkable given that he learned soldiery on the job. During the second stage of the Civil War, he became head of the Roundhead armies, and later, after defeating King Charles and having him executed, he became Lord Protector of England. Although he was ahead of his times with his visions of mobile warfare, Cromwell was not a brilliant strategist or field tactician; his success lay in the morale and discipline of his cavalry, and the secret to those was the quality of the men he recruited—true believers in his cause. Such men were naturally open to his influence and accepting of his discipline. With each new victory, they grew more committed to him and more cohesive. He could ask the most of them.

Above all else, then, pay attention to your staff, to those you recruit to your cause. Many will pretend to share your beliefs, but your first battle will show that all they wanted was a job. Soldiers like these are mercenaries and will get you nowhere. True believers are what you want; expertise and impressive résumés matter less than character and the capacity for sacrifice. Recruits of character will give you a staff already open to your influence, making morale and discipline infinitely easier to attain. This core personnel will spread the gospel for you, keeping the rest of the army in line. As far as possible in this secular world, make battle a religious experience, an ecstatic involvement in something transcending the present.

*the lord, and then he set out with the army. When the soldiers made camp, Rangju personally oversaw the digging of wells, construction of stoves, preparation of food and drink, and care of the sick. He shared all of the supplies of the leadership with the soldiers, personally eating the same rations as they. He was especially kind to the weary and weakened. After three days, Rangju called the troops to order. Even those who were ill wanted to go along, eager to go into battle for Rangju. When the armies of Jin and Yan heard about this, they withdrew from the state of Qi. Now Rangju led his troops to chase them down and strike them. Eventually he recovered lost territory and returned with the army victorious.*

MASTERING THE ART OF WAR: *ZHUGE LIANG'S AND LIU JI'S COMMENTARIES ON THE CLASSIC BY SUN-TZU*, TRANSLATED BY THOMAS CLEARY, 1989

**2.** In 1931 the twenty-three-year-old Lyndon Baines Johnson was offered the kind of job he had been dreaming of: secretary to Richard Kleberg, newly elected congressman from Texas's Fourteenth Congressional District. Johnson was a high-school debating teacher at the time, but he had worked on several political campaigns and was clearly a young man of ambition. His students at Sam Houston High—in Houston, Texas—assumed that he would quickly forget about them, but, to the surprise of two of his best debaters, L. E. Jones and Gene Latimer, he not only kept in touch, he wrote to them regularly from Washington. Six months later came a bigger surprise still: Johnson invited Jones and Latimer to Washington to work as his assistants. With the Depression at its height, jobs were scarce—particularly jobs with this kind of potential. The two teenagers grabbed the opportunity. Little did they know what they were in for.

The pay was ridiculously low, and it soon became clear that Johnson intended to work the two men to their human limit. They put in eighteen- or twenty-hour days, mostly answering constituents' mail. "The chief has a knack, or, better said, a genius for getting the most out of those around him," Latimer later wrote. "He'd say, 'Gene, it seems L.E.'s a little faster than you today.' And I'd work faster. 'L.E., he's catching up with you.' And pretty soon, we'd both be pounding [the typewriter] for hours without stopping, just as fast as we could."

Jones didn't usually take orders too well, but he found himself working harder and harder for Johnson. His boss seemed destined for something great: that Johnson would scale the heights of power was written all over his face—and he would bring the ambitious Jones along with him. Johnson could also turn everything into a cause, making even the most trivial issue a crusade for Kleberg's constituents, and Jones felt part of that crusade—part of history.

The most important reason for both Jones's and Latimer's willingness to work so hard, though, was that Johnson worked still harder. When Jones trudged into the office at five in the morning, the lights would already be on, and Johnson would be hard at work. He was also the last to leave. He never asked his employees to do anything he wouldn't do himself. His energy was intense, boundless, and contagious. How could you let such a man down by working less hard than he did?

Not only was Johnson relentlessly demanding, but his criticisms were often cruel. Occasionally, though, he would do Jones and Latimer some unexpected favor or praise them for something they hadn't realized he had noticed. At moments like this, the two young men quickly forgot the many bitter moments in their work. For Johnson, they felt, they would go to the ends of the earth.

And indeed Johnson rose through the ranks, first winning influence within Kleberg's office, then gaining the attention of President Franklin D. Roosevelt himself. In 1935, Roosevelt named Johnson Texas state director for the recently built National Youth Administration. Now Johnson began to build a larger team around the core of his two devoted assistants; he also built loyalties in a scattering of others for whom he found jobs in Washington. The dynamic he had created with Jones and Latimer now repeated itself on a larger scale: assistants competed for his attention, tried to please him, to meet his standards, to be worthy of him and of his causes.

In 1937, when Congressman James Buchanan suddenly died, the seat for Texas's Tenth District unexpectedly fell empty. Despite the incredible odds against him—he was still relatively unknown and way too young—Johnson decided to run and called in his chips: his carefully cultivated acolytes poured into Texas, becoming chauffeurs, canvassers, speechwriters, barbecue cooks, crowd entertainers, nurses—whatever the campaign needed. In the six short weeks of the race, Johnson's foot soldiers covered the length and breadth of the Tenth District. And in front of them at every step was Johnson himself, campaigning as if his life depended on it. One by one, he and his team won over voters in every corner of the district, and finally, in one of the greatest upsets in any American political race, Johnson won the election. His later career, first as a senator, then as U.S. president, obscured the foundation of his first great success: the army of devoted and tireless followers that he had carefully built up over the previous five years.

THE WOLVES AND THE DOGS AT WAR

*One day, enmity broke out between the dogs and the wolves. The dogs elected a Greek to be their general. But he was in no hurry to engage in battle, despite the violent intimidation of the wolves. "Understand," he said to them, "why I deliberately put off engagement. It is because one must always take counsel before acting. The wolves, on the one hand, are all of the same race, all of the same color. But our soldiers have very varied habits, and each one is proud of his own country. Even their colors are not uniform: some are black, some russet, and others white or ash-grey. How can I lead into battle those who are not in harmony and who are all dissimilar?'* In all armies it is unity of will and purpose which assures victory over the enemy.

*FABLES*, AESOP, SIXTH CENTURY B.C.

## Interpretation

Lyndon Johnson was an intensely ambitious young man. He had neither money nor connections but had something more valuable: an understanding of human psychology. To command influence in the world, you need a power base, and here human beings—a devoted army of followers—are more valuable than money. They will do things for you that money cannot buy.

That army is tricky to build. People are contradictory and defensive: push them too hard and they resent you; treat them well and they take you for granted. Johnson avoided those traps by making his staff want his approval. To do that he led from the front. He worked harder than any of his staff, and his men saw him do it; failing to match him would have made them feel guilty and selfish. A leader who works that hard stirs competitive instincts in his men, who do all they can to prove themselves worthier than their teammates. By showing how much of his own time and effort he was willing to sacrifice, Johnson earned their respect. Once he had that respect, criticism, even when harsh, became an effective motivator, making his followers feel they were disappointing him. At the same time, some kind act out of the blue would break down any ability to resist him.

Understand: morale is contagious, and you, as leader, set the tone. Ask for sacrifices you won't make yourself (doing everything through assistants) and your troops grow lethargic and resentful; act too nice, show too much concern for their well-being, and you drain the tension from their souls and create spoiled children who whine at the slightest pressure or request for more work. Personal example is the best way to set the proper tone and build morale. When your people see your devotion to the cause, they ingest your spirit of energy and self-sacrifice. A few timely criticisms here and there and they will only try harder to please you, to live up to your high standards. Instead of having to push and pull your army, you will find them chasing after you.

*Hannibal was the greatest general of antiquity by reason of his admirable comprehension of the morale of combat, of the morale of the soldier, whether his own or the enemy's. He shows his greatness in this respect in all the different incidents of war, of campaign, of action. His men were not better than the Roman soldiers. They were not as well-armed, one-half less in number. Yet he was always the conqueror. He understood the value of morale. He had the absolute confidence of his people. In addition, he had the art, in commanding an army, of always securing the advantage of morale.*

COLONEL CHARLES ARDANT DU PICQ, 1821–70

**3.** In May of 218 B.C., the great general Hannibal, of Carthage in modern Tunisia, embarked on a bold plan: he would lead an army through Spain, Gaul, and across the Alps into northern Italy. His goal was to defeat Rome's legions on their own soil, finally putting an end to Rome's expansionist policies.

The Alps were a tremendous obstacle to military advance—in fact, the march of an army across the high mountains was unprecedented. Yet in December of that year, after much hardship, Hannibal reached northern Italy, catching the Romans completely off guard and the region undefended. There was a price to pay, however: of Hannibal's original 102,000 soldiers, a mere 26,000 survived, and they were exhausted, hungry, and demoralized. Worse, there was no time to rest: a Roman army was on its way and had already crossed the Po River, only a few miles from the Carthaginian camp.

On the eve of his army's first battle with the fearsome Roman legions, Hannibal somehow had to bring his worn-out men alive. He decided to put on a show: gathering his army together, he brought in a group of prisoners and told them that if they fought one another to the death in a gladiatorial contest, the victors would win freedom and a place in the Carthaginian army. The prisoners agreed, and Hannibal's soldiers were treated to hours of bloody entertainment, a great distraction from their troubles.

When the fighting was over, Hannibal addressed his men. The contest had been so enjoyable, he said, because the prisoners had fought so intensely. That was partly because the weakest man grows fierce when losing means death, but there was another reason as well: they had the chance to join the Carthaginian army, to go from being abject prisoners to free soldiers fighting for a great cause, the defeat of the hated Romans. You soldiers, said Hannibal, are in exactly the same position. You face a much stronger enemy. You are many miles from home, on hostile territory, and you have nowhere to go—in a way you are prisoners, too. It is either freedom or slavery, victory or death. But fight as these men fought today and you will prevail.

*Four brave men who do not know each other will not dare to attack a lion. Four less brave, but knowing each other well, sure of their reliability and consequently of mutual aid, will attack resolutely. There is the science of the organization of armies in a nutshell.*

COLONEL CHARLES ARDANT DU PICQ, 1821–70

The contest and speech got hold of Hannibal's soldiers, and the next day they fought with deadly ferocity and defeated the Romans. A series of victories against much larger Roman legions followed.

Nearly two years later, the two sides met at Cannae. Before the battle, with the armies arrayed within sight of each other, the Carthaginians could see that they were hopelessly outnumbered, and fear passed through the ranks. Everyone went quiet. A Carthaginian officer called Gisgo rode out in front of the men, taking in the Roman lines; stopping before Hannibal, he remarked, with a quaver in his voice, on the disparity in numbers. "There is one thing, Gisgo, that you have not noticed," Hannibal replied: "In all that great number of men opposite, there is not a single one whose name is Gisgo."

Gisgo burst out laughing, so did those within hearing, and the joke passed through the ranks, breaking the tension. No, the Romans had no Gisgo. Only the Carthaginians had Gisgo, and only the Carthaginians had Hannibal. A leader who could joke at a moment like this had to feel supremely confident—and if the leader were Hannibal, that feeling was probably justified.

*The Greeks met the Trojans without a tremor.*
*Agamemnon ranged among them, commanding:*
*"Be men, my friends. Fight with valor*
*And with a sense of shame before your comrades.*
*You're less likely to be killed with a sense of shame.*
*Running away never won glory or a fight."*

*THE ILIAD*, HOMER, CIRCA NINTH CENTURY B.C.

Just as the troops had been swept with anxiety, now they were infected with self-assurance. At Cannae that day, in one of the most devastating victories in history, the Carthaginians crushed the Roman army.

## Interpretation

Hannibal was a master motivator of a rare kind. Where others would harangue their soldiers with speeches, he knew that to depend on words was to be in a sorry state: words only hit the surface of a soldier, and a leader must grab his men's hearts, make their blood boil, get into their minds, alter their moods. Hannibal reached his soldiers' emotions indirectly, by relaxing them, calming them, taking them outside their problems and getting them to bond. Only then did he hit them with a speech that brought home their precarious reality and swayed their emotions.

At Cannae a one-line joke had the same effect: instead of trying to persuade the troops of his confidence, Hannibal showed it to them. Even as they laughed at the joke about Gisgo, they bonded over it and understood its inner meaning. No need for a speech. Hannibal knew that subtle changes in his men's mood could spell the difference between victory and defeat.

Like Hannibal, you must aim indirectly at people's emotions: get them to laugh or cry over something that seems unrelated to you or to the issue at hand. Emotions are contagious—they bring people together and make them bond. Then you can play them like a piano, moving them from one emotion to the other. Oratory and eloquent pleas only irritate and insult us; we see right through them. Motivation is subtler than that. By advancing indirectly, setting up your emotional appeal, you will get inside instead of just scratching the surface.

*He suddenly lost concern for himself, and forgot to look at a menacing fate. He became not a man but a member. He felt that something of which he was a part—a regiment, an army, a cause, or a country—was in a crisis. He was welded into a common personality which was dominated by a single desire. For some moments he could not flee, no more than a little finger can commit a revolution from a hand. . . .*
*There was a consciousness always of the presence of his comrades about him. He felt the subtle battle brotherhood more potent even than the cause for which they were fighting. It was a mysterious fraternity born of the smoke and danger of death.*

*The Red Badge of Courage*, Stephen Crane, 1871–1900

**4.** In the 1930s and '40s, the Green Bay Packers were one of the most successful teams in professional football, but by the late '50s they were the worst. What went wrong? The team had many talented players, like the former All-American Paul Hornung. The owners cared about it deeply and kept hiring new coaches, new players—but nothing could slow the fall. The players tried; they hated losing. And, really, they weren't that bad—they came close to winning many of the games they lost. So what could they do about it?

The Packers hit bottom in 1958. For the 1959 season, they tried the usual trick, bringing in a new coach and general manager: Vince Lombardi. The players mostly didn't know much about the man, except that he had been an assistant coach for the New York Giants.

As the players convened to meet the new coach, they expected the typical speech: this is the year to turn things around; I'm going to get tough with you; no more business as usual. Lombardi did not disappoint them: in a quiet, forceful tone, he explained a new set of rules and code of conduct. But a few players noticed something different about Lombardi: he oozed confidence—no shouts, no demands. His tone and manner suggested that the Packers were already a winning team; they just had to live up to it. Was he an idiot or some kind of visionary?

Then came the practices, and once again the difference was not so much how they were conducted as the spirit behind them—they *felt* different. They were shorter but more physically demanding, almost to the point of torture. And they were intense, with the same simple plays endlessly repeated. Unlike other coaches, Lombardi explained what he was doing: installing a simpler system, based not on novelty and surprise but on efficient execution. The players had to concentrate intensely—the slightest mistake and they were doing extra laps or making the whole team do extra laps. And Lombardi changed the drills constantly: the players were never bored and could never relax their mental focus.

Earlier coaches had always treated a few players differently: the stars. They had a bit of an attitude, and they took off early and stayed up late. The other men had come to accept this as part of the pecking order, but deep down they resented it. Lombardi, though, had no favorites; for him there were no stars. "Coach Lombardi is very fair," said defensive tackle Henry Jordan. "He treats us all the same—like dogs." The players liked that. They enjoyed seeing Hornung yelled at and disciplined just as much as the others.

Lombardi's criticisms were relentless and got under his players' skins. He seemed to know their weak points, their insecurities. How did he know, for instance, that Jordan hated to be criticized in front of the others? Lombardi exploited his fear of public lashings to make him try harder. "We were always trying to show [Lombardi] he was wrong," commented one player. "That was his psych."

*Once more unto the breach, dear friends, once more;*
*Or close the wall up with our English dead.*
*In peace there's nothing so becomes a man*
*As modest stillness and humility:*
*But when the blast of war blows in our ears,*
*Then imitate the action of the tiger;*
*Stiffen the sinews, summon up the blood,*
*Disguise fair nature with hard-favour'd rage;*
*Then lend the eye a terrible aspect;*
*Let it pry through the portage of the head*
*Like the brass cannon; let the brow o'erwhelm it*
*As fearfully as doth a galled rock*
*O'erhang and jutty his confounded base,*
*Swill'd with the wild and wasteful ocean.*
*Now set the teeth and stretch the nostril wide,*
*Hold hard to the breath and bend up every spirit*
*To his full height. On, on, you noblest English,*
*Whose blood is fet from fathers of war-proof!*
*Fathers that, like so many Alexanders,*
*Have in these parts from morn till even fought*
*And sheathed their swords for lack of argument:*
*Dishonour not your mothers; now attest*
*That those whom you call'd fathers did beget you.*

The practices grew more intense still; the players had never worked this hard in their lives. Yet they found themselves showing up earlier and staying later. By the season's first game, Lombardi had prepared them for every contingency. Sick of training, they were grateful to be playing in a real game at last—and, to their surprise, all that work made the game a lot easier. They were more prepared than the other team and less tired in the fourth quarter. They won their first three games. With this sudden success, their morale and confidence soared.

The Packers finished the year with a 7–5 record, a remarkable turnaround from 1958's 1-10-1. After one season under Lombardi, they had become the most tight-knit team in professional sports. No one wanted to leave the Packers. In 1960 they reached the championship game, and in 1961 they won it, with many more to follow. Over the years various of Lombardi's Packers would try to explain how he had transformed them, but none of them could really say how he had pulled it off.

### Interpretation

When Vince Lombardi took over the Packers, he recognized the problem right away: the team was infected with adolescent defeatism. Teenagers will often strike a pose that is simultaneously rebellious and lackadaisical. It's a way of staying in place: trying harder brings more risk of failure, which they cannot handle, so they lower their expectations, finding nobility in slacking off and mediocrity. Losing hurts less when they embrace it.

Groups can get infected with this spirit without realizing it. All they need is a few setbacks, a few adolescent-minded individuals, and slowly expectations lower and defeatism sets in. The leader who tries to change the group's spirit directly—yelling, demanding, disciplining—actually plays into the teenage dynamic and reinforces the desire to rebel.

Lombardi was a motivational genius who saw everything in psychological terms. To him the National Football League teams were virtually equal in talent. The differences lay in attitude and morale: reversing the Packers' defeatism would translate into wins, which would lift their morale, which in turn would bring more wins. Lombardi knew he had to approach his players indirectly—had to trick them into changing. He began with a show of confidence, talking as if he assumed they were winners who had fallen on bad times. That got under their skins, far more than they realized. Then, in his practices, Lombardi didn't make demands—a defensive, whiny approach that betrays insecurity. Instead he changed the practices' spirit, making them quiet, intense, focused, workmanlike. He knew that willpower is tied to what you believe possible; expand that belief and you try harder. Lombardi created a better team—which won its first game—by making its players see possibilities. Defeat was no longer comfortable.

Understand: a group has a collective personality that hardens over time, and sometimes that personality is dysfunctional or adolescent. Changing it is difficult; people prefer what they know, even if it doesn't work. If you lead this kind of group, do not play into its negative dynamic. Announcing intentions and making demands will leave people defensive and feeling like children. Like Lombardi, play the wily parent. Ask more of them. Expect them to work like adults. Quietly alter the spirit with which things are done. Emphasize efficiency: anybody can be efficient (it isn't a question of talent), efficiency breeds success, and success raises morale. Once the spirit and personality of the group start to shift, everything else will fall into place.

*Be copy now to men of grosser blood,*
*And teach them how to war. And you, good yeomen,*
*Whose limbs were made in England, show us here*
*The mettle of your pasture; let us swear*
*That you are worth your breeding; which I doubt not;*
*For there is none of you so mean and base,*
*That hath not noble lustre in your eyes.*
*I see you stand like greyhounds in the slips,*
*Straining upon the start. The game's afoot:*
*Follow your spirit, and upon this charge*
*Cry "God for Harry, England, and Saint George!"*

*King Henry V*,
William Shakespeare, 1564–1616

**5.** In April 1796 the twenty-six-year-old Napoleon Bonaparte was named commander of the French forces fighting the Austrians in Italy. For many officers his appointment was something of a joke: they saw their new leader as too short, too young, too inexperienced, and even too badly groomed to play the part of "general." His soldiers, too, were underpaid, underfed, and increasingly disillusioned with the cause they were fighting for, the French Revolution. In the first few weeks of the campaign, Napoleon did what he could to make them fight harder, but they were largely resistant to him.

On May 10, Napoleon and his weary forces came to the Bridge of Lodi, over the river Adda. Despite his uphill struggle with his troops, he had the Austrians in retreat, but the bridge was a natural place to take a stand, and they had manned it with soldiers on either side and with well-placed artillery. Taking the bridge would be costly—but suddenly the French soldiers saw Napoleon riding up in front of them, in a position of extreme personal risk, directing the attack. He delivered a stirring speech, then launched his grenadiers at the Austrian lines to cries of "*Vive la République!*" Caught up in the spirit, his senior officers led the charge.

The French took the bridge, and now, after this relatively minor operation, Napoleon's troops suddenly saw him as a different man. In fond recognition of his courage, they gave him a nickname: "Le Petit Caporal." The story of Napoleon facing the enemy at the Bridge of Lodi passed through the ranks. As the campaign wore on, and Napoleon won victory after victory, a bond developed between the soldiers and their general that went beyond mere affection.

Between battles Napoleon would sometimes wander among the soldiers' campfires, mingling with them. He himself had risen through the ranks—he had once been an ordinary gunner—and he could talk to the men as no other general could. He knew their names, their histories, even in what battles they'd been wounded. With some men he would pinch an earlobe between his finger and thumb and give it a friendly tweak.

Napoleon's soldiers did not see him often, but when they did, it was as if an electrical charge passed through them. It was not just his personal presence; he knew exactly when to show up—before a big battle or when morale had slipped for some reason. At these moments he would tell them they were making history together. If a squad were about to lead a charge or seemed in trouble, he would ride over and yell, "Thirty-eighth: I know you! Take me that village—at the charge!" His soldiers felt they weren't just obeying orders, they were living out a great drama.

*Mercenary and auxiliary arms are useless and dangerous; and if one keeps his state founded on mercenary arms, one will never be firm or secure; for they are disunited, ambitious, without discipline, unfaithful; bold among friends, among enemies cowardly; no fear of God, no faith with men; ruin is postponed only as long as attack is postponed; and in peace you are despoiled by them, in war by the enemy. The cause of this is that they have no love nor cause to keep them in the field other than a small stipend, which is not sufficient to make them want to die for you.*

*The Prince*, Niccolò Machiavelli, 1513

Napoleon rarely showed anger, but when he did, his men felt worse than just guilty or upset. Late in the first Italian campaign, Austrian troops had forced some of his troops into a humiliating retreat for which there was no excuse. Napoleon visited their camp personally. "Soldiers, I am not satisfied with you," he told them, his large gray eyes seemingly on fire. "You have shown neither bravery, discipline, nor perseverance. . . . You have allowed yourselves to be driven from positions where a handful of men could have stopped an army. Soldiers of the Thirty-ninth and Eighty-fifth, you are not French soldiers. General, chief of staff, let it be inscribed on their colors: 'They no longer form part of the Army of Italy!' " The soldiers were astounded. Some cried; others begged for another chance. They repented their weakness and turned completely around: the Thirty-ninth and Eighty-fifth would go on to distinguish themselves for strengths they had never shown previously.

Some years later, during a difficult campaign against the Austrians in Bavaria, the French won a hard-fought victory. The next morning Napoleon reviewed the Thirteenth Regiment of Light Infantry, which had played a key role in the battle, and asked the colonel to name its bravest man. The colonel thought for a moment: "Sir, it is the drum major." Napoleon immediately asked to see the young bandsman, who appeared, quaking in his boots. Then Napoleon announced loudly for everyone to hear, "They say that you are the bravest man in this regiment. I appoint you a knight of the Legion of Honor, baron of the Empire, and award you a pension of four thousand francs." The soldiers gasped. Napoleon was famous for his well-timed promotions and for promoting soldiers on merit, making even the lowliest private feel that if he proved himself, he could someday be a marshal. But a drum major becoming a baron overnight? That was entirely beyond their experience. Word of it spread rapidly through the troops and had an electrifying effect—particularly on the newest conscripts, the ones who were most homesick and depressed.

Throughout his long, very bloody campaigns and even his heart-wrenching defeats—the bitter winter in Russia, the eventual exile to Elba, the final act at Waterloo—Napoleon's men would go to the ends of the earth for Le Petit Caporal and for no one else.

Interpretation

Napoleon was the greatest man manager in history: he took millions of unruly, undisciplined, unsoldierly young men, recently liberated by the French Revolution, and molded them into one of the most successful fighting forces ever known. Their high morale was all the more remarkable for the ordeals he put them through. Napoleon used every trick in the book to build his army. He united them around a cause, spreading first the ideas of the French Revolution, later the glory of France as a growing empire. He treated them well but never spoiled them. He appealed not to their greed but to their thirst for glory and recognition. He led from the front, proving his bravery again and again. He kept his men moving—there was always a new campaign for glory. Having bonded with them, he skillfully played on their emotions. More than soldiers fighting in an army, his men felt themselves part of a myth, united under the emperor's legendary eagle standards.

*If you wish to be loved by your soldiers, husband their blood and do not lead them to slaughter.*

FREDERICK THE GREAT, 1712–86

Of all Napoleon's techniques, none was more effective than his use of punishments and rewards, all staged for the greatest dramatic impact. His personal rebukes were rare, but when he was angry, when he punished, the effect was devastating: the target felt disowned, outcast. As if exiled from the warmth of his family, he would struggle to win back the general's favor and then never to give him a reason to be angry again. Promotions, rewards, and public praise were equally rare, and when they came, they were always for merit, never for some political calculation. Caught between the poles of wanting never to displease Napoleon and yearning for his recognition, his men were pulled into his sway, following him devotedly but never quite catching up.

Learn from the master: the way to manage people is to keep them in suspense. First create a bond between your soldiers and yourself. They respect you, admire you, even fear you a little. To make the bond stronger, hold yourself back, create a little space around yourself; you are warm yet with a touch of distance. Once the bond is forged, appear less often. Make both your punishments and your praises rare and unexpected, whether for mistakes or for successes that may seem minor at the time but have symbolic meaning. Understand: once people know what pleases you and what angers you, they turn into trained poodles, working to charm you with apparent good behavior. Keep them in suspense: make them think of you constantly and want to please you but never know just how to do it. Once they are in the trap, you will have a magnetic pull over them. Motivation will become automatic.

**Image:**
The Ocean's Tide. It ebbs and flows so powerfully that no one in its path can escape its pull or move against it. Like the moon, you are the force that sets the tide, which carries everything along in its wake.

**Authority:** The Way means inducing the people to have the same aim as the leadership, so that they will share death and share life, without fear of danger. —*Sun-tzu (fourth century* B.C.*)*

## REVERSAL

If morale is contagious, so is its opposite: fear and discontent can spread through your troops like wildfire. The only way to deal with them is to cut them off before they turn into panic and rebellion.

In 58 B.C., when Rome was fighting the Gallic War, Julius Caesar was preparing for battle against the Germanic leader Ariovistus. Rumors about the ferocity and size of the German forces were flying, and his army was panicky and mutinous. Caesar acted fast: first he had the rumormongers arrested. Next he addressed his soldiers personally, reminding them of their brave ancestors who had fought and defeated the Germans. He would not lead their weaker descendants into battle; since the Tenth Legion alone seemed immune to the growing panic, he would take them alone. As Caesar prepared to march with the valiant Tenth Legion, the rest of the army, ashamed, begged him to forgive them and let them fight. With a show of reluctance, he did so, and these once frightened men fought fiercely.

In such cases you must act like Caesar, turning back the tide of panic. Waste no time, and deal with the whole group. People who spread panic or mutiny experience a kind of madness in which they gradually lose contact with reality. Appeal to their pride and dignity, make them feel ashamed of their moment of weakness and madness. Remind them of what they have accomplished in the past, and show them how they are falling short of the ideal. This social shaming will wake them up and reverse the dynamic.

PART

# DEFENSIVE WARFARE

To fight in a defensive manner is not a sign of weakness; it is the height of strategic wisdom, a powerful style of waging war. Its requirements are simple: First, you must make the most of your resources, fighting with perfect economy and engaging only in battles that are necessary. Second, you must know how and when to retreat, luring an aggressive enemy into an imprudent attack. Then, waiting patiently for his moment of exhaustion, launch a vicious counterattack.

In a world that frowns on displays of overt aggression, the ability to fight defensively—to let others make the first move and then wait for their own mistakes to destroy them—will bring you untold power. Because you waste neither energy nor time, you are always ready for the next inevitable battle. Your career will be long and fruitful.

To fight this way, you must master the arts of deception. By seeming weaker than you are, you can draw the enemy into an ill-advised attack; by seeming stronger than you are—perhaps through an occasional act that is reckless and bold—you can deter the enemy from attacking you. In defensive warfare you are essentially leveraging your weaknesses and limitations into power and victory.

The following four chapters will instruct you in the basic arts of defensive warfare: economy of means, counterattack, intimidation and deterrence, and how to retreat skillfully and lie low when under aggressive attack.

# 8

# PICK YOUR BATTLES CAREFULLY

## THE PERFECT-ECONOMY STRATEGY

*We all have limitations—our energies and skills will take us only so far. Danger comes from trying to surpass our limits. Seduced by some glittering prize into overextending ourselves, we end up exhausted and vulnerable. You must know your limits and pick your battles carefully. Consider the hidden costs of a war: time lost, political goodwill squandered, an embittered enemy bent on revenge. Sometimes it is better to wait, to undermine your enemies covertly rather than hitting them straight on. If battle cannot be avoided, get them to fight on your terms. Aim at their weaknesses; make the war expensive for them and cheap for you. Fighting with perfect economy, you can outlast even the most powerful foe.*

*In the utilization of a theater of war, as in everything else, strategy calls for economy of strength. The less one can manage with, the better; but manage one must, and here, as in commerce, there is more to it than mere stinginess.*

CARL VON CLAUSEWITZ, 1780–1831

## THE SPIRAL EFFECT

In 281 B.C. war broke out between Rome and the city of Tarentum, on Italy's east coast. Tarentum had begun as a colony of the Greek city of Sparta; its citizens still spoke Greek, considered themselves cultured Spartans, and thought other Italian cities barbaric. Rome meanwhile was an emerging power, locked in a series of wars with neighboring cities.

The prudent Romans were reluctant to take on Tarentum. It was Italy's wealthiest city at the time, rich enough to finance its allies in a war against Rome; it was also too far away, off in the southeast, to pose an immediate threat. But the Tarentines had sunk some Roman ships that had wandered into their harbor, killing the fleet's admiral, and when Rome had tried to negotiate a settlement, its ambassadors had been insulted. Roman honor was at stake, and it readied itself for war.

Tarentum had a problem: it was wealthy but had no real army. Its citizens had gotten used to easy living. The solution was to call in a Greek army to fight on its behalf. The Spartans were otherwise occupied, so the Tarentines called on King Pyrrhus of Epirus (319–272 B.C.), the greatest Greek warrior king since Alexander the Great.

Epirus was a small kingdom in west-central Greece. It was a poor land, sparsely populated, with meager resources, but Pyrrhus—raised on stories of Achilles, from whom his family claimed to be descended, and of Alexander the Great, a distant cousin—was determined to follow in the footsteps of his illustrious ancestors and relatives, expanding Epirus and carving out his own empire. As a young man, he had served in the armies of other great military men, including Ptolemy, a general of Alexander's who now ruled Egypt. Pyrrhus had quickly proved his value as a warrior and leader. In battle he had become known for leading dangerous charges, earning himself the nickname "The Eagle." Back in Epirus he had built up his small army and trained it well, even managing to defeat the much larger Macedonian army in several battles.

Pyrrhus's reputation was on the rise, but it was hard for a small country like his to gain ascendancy over more powerful Greek neighbors like the Macedonians, the Spartans, and the Athenians. And the Tarentines' offer was tempting: First, they promised him money and a large army raised from allied states. Second, by defeating the Romans, he could make himself master of Italy, and from Italy he could take first Sicily, then Carthage in North Africa. Alexander had moved east to create his empire; Pyrrhus could move west and dominate the Mediterranean. He accepted the offer.

In the spring of 280 B.C., Pyrrhus set sail with the largest Greek army ever to cross into Italy: 20,000 foot soldiers, 3,000 horsemen, 2,000 bowmen, and twenty elephants. Once in Tarentum, though, he realized he had been tricked: not only did the Tarentines have no army, they had made no effort to assemble one, leaving Pyrrhus to do it himself. Pyrrhus wasted no time: he declared a military dictatorship in the city and began to build and train an army from among the Tarentines as fast as possible.

Pyrrhus's arrival in Tarentum worried the Romans, who knew his reputation as a strategist and fighter. Deciding to give him no time to prepare, they quickly sent out an army, forcing Pyrrhus to make do with what he had and he set off to face them. The two armies met near the town of Heraclea. Pyrrhus and his troops were outnumbered and at one point were on the verge of defeat, when he unleashed his secret weapon: his elephants, with their massive weight, loud, fearsome trumpeting, and soldiers on top, firing arrows down at will. The Romans had never faced elephants in battle before, and panic spread among them, turning the tide of the fight. Soon the disciplined Roman legions were in headlong retreat.

"The Eagle" had won a great victory. His fame spread across the Italian peninsula; he was indeed the reincarnation of Alexander the Great. Now other cities sent him reinforcements, more than making up for his losses at Heraclea. But Pyrrhus was worried. He had lost many veterans in the battle, including key generals. More important, the strength and discipline of the Roman legions had impressed him—they were like no other troops he had faced. He decided to try to negotiate a peaceful settlement with the Romans, offering to share the peninsula with them. At the same time, though, he marched on Rome, to give the negotiations urgency and to make it clear that unless the Romans sued for peace, they would face him again.

Meanwhile the defeat at Heraclea had had a powerful effect on the Romans, who were not easily intimidated and did not take defeat lightly. Immediately after the battle, a call went out for recruits, and young men responded in droves. The Romans proudly rejected the offer of a settlement; they would never share Italy.

The two armies met again near the town of Asculum, not far from Rome, in the spring of 279 B.C. This time their numbers were about equal. The first day of battle was fierce, and once again the Romans seemed to have the edge, but on the second day Pyrrhus, a strategic master, managed to lure the Roman legions onto terrain better suited to his own style of maneuvering, and he gained the advantage. As was his wont, near the end of the day he personally led a violent charge at the heart of the Roman legions, elephants in front. The Romans scattered, and Pyrrhus was once again victorious.

King Pyrrhus had now scaled the heights, yet he felt only gloom and foreboding. His losses had been terrible; the ranks of the generals he depended on were decimated, and he himself had been badly wounded. At the same time, the Romans seemed inexhaustible, undaunted by their defeat. When congratulated on his victory at Asculum, he replied, "If we defeat the Romans in one more such battle, we shall be totally ruined."

Pyrrhus, however, was already ruined. His losses at Asculum were too large to be quickly replaced, and his remaining forces were too few to fight the Romans again. His Italian campaign was over.

### Interpretation

From the story of King Pyrrhus and his famous lament after the Battle of Asculum comes the expression "Pyrrhic victory," signifying a triumph that is as good as a defeat, for it comes at too great a cost. The victor is too exhausted to exploit his win, too vulnerable to face the next battle. And indeed, after the "victory" at Asculum, Pyrrhus staggered from one disaster to the next, his army never quite strong enough to defeat his growing hosts of enemies. This culminated in his untimely death in battle, ending Epirus's hopes to become a power in Greece.

Pyrrhus could have avoided this downward spiral. Advance intelligence would have told him about both the disciplined ferocity of the Romans and the decadence and treachery of the Tarentines, and, knowing this, he could have taken more time to build an army or canceled the expedition altogether. Once he saw that he had been tricked, he could have turned back; after Heraclea there was still time to retrench, consolidate, quit while he was ahead. Had he done any of this, his story might have had a different ending. But Pyrrhus could not stop himself—the dream was too alluring. Why worry about the costs? He could recover later. One more battle, one more victory, would seal the deal.

Pyrrhic victories are much more common than you might think. Excitement about a venture's prospects is natural before it begins, and if the goal is enticing, we unconsciously see what we want to see—more of the possible gains, fewer of the possible difficulties. The further we go, the harder it becomes to pull back and rationally reassess the situation. In such circumstances the costs tend not just to mount—they spiral out of control. If things go badly, we get exhausted, which leads us to make mistakes, which lead to new, unforeseen problems, which in turn lead to new costs. Any victories we might have along the way are meaningless.

Understand: the more you want the prize, the more you must compensate by examining what getting it will take. Look beyond the obvious costs and think about the intangible ones: the goodwill you may squander by waging war, the fury of the loser if you win, the time that winning may take, your debt to your allies. You can always wait for a better time; you can always try something more in line with your resources. Remember: history is littered with the corpses of people who ignored the costs. Save yourself unnecessary battles and live to fight another day.

*When the weapons have grown dull and spirits depressed, when our strength has been expended and resources consumed, then others will take advantage of our exhaustion to arise. Then even if you have wise generals you cannot make things turn out well in the end.*

—The Art of War, *Sun-tzu (fourth century B.C.)*

## STRENGTHS AND WEAKNESSES

When Queen Elizabeth I (1533–1603) ascended the throne of England in 1558, she inherited a second-rate power: the country had been racked by civil war, and its finances were in a mess. Elizabeth dreamed of creating a long period of peace in which she could slowly rebuild England's foundations and particularly its economy: a government with money was a government with options. England, a small island with limited resources, could not hope to compete in war with France and Spain, the great powers of Europe. Instead it would gain strength through trade and economic stability.

Year by year for twenty years, Elizabeth made progress. Then, in the late 1570s, her situation suddenly seemed dire: an imminent war with Spain threatened to cancel all the gains of the previous two decades. The Spanish king, Philip II, was a devout Catholic who considered it his personal mission to reverse the spread of Protestantism. The Low Countries (now Holland and Belgium) were properties of Spain at the time, but a growing Protestant rebellion was threatening its rule, and Philip went to war with the rebels, determined to crush them. Meanwhile his most cherished dream was to restore Catholicism to England. His short-term strategy was a plot to have Elizabeth assassinated and then to place her half sister, the Catholic Mary Queen of Scots, on the British throne. In case this plan failed, his long-term strategy was to build an immense armada of ships and invade England.

*He whom the ancients called an expert in battle gained victory where victory was easily gained. Thus the battle of the expert is never an exceptional victory, nor does it win him reputation for wisdom or credit for courage. His victories in battle are unerring. Unerring means that he acts where victory is certain, and conquers an enemy that has already lost.*

*The Art of War*,
Sun-tzu,
Fourth Century B.C.

Philip did not keep his intentions well hidden, and Elizabeth's ministers saw war as inevitable. They advised her to send an army to the Low Countries, forcing Philip to put his resources there instead of into an attack on England—but Elizabeth balked at that idea; she would send small forces there to help the Protestant rebels avert a military disaster, but she would not commit to anything more. Elizabeth dreaded war; maintaining an army was a huge expense, and all sorts of other hidden costs were sure to emerge, threatening the stability she had built up. If war with Spain really was inevitable, Elizabeth wanted to fight on her own terms; she wanted a war that would ruin Spain financially and leave England safe.

Defying her ministers, Elizabeth did what she could to keep the peace with Spain, refusing to provoke Philip. That bought her time to put aside funds for building up the British navy. Meanwhile she worked in secret to damage the Spanish economy, which she saw as its only weak spot. Spain's enormous, expanding empire in the New World made it powerful, but that empire was far away. To maintain it and profit from it, Philip was entirely dependent on shipping, a vast fleet that he paid for with enormous loans from Italian bankers. His credit with these banks depended on the safe passage of his ships bringing gold from the New World. The power of Spain rested on a weak foundation.

And so Queen Elizabeth unleashed her greatest captain, Sir Francis Drake, on the Spanish treasure ships. He was to appear to be operating on his own, a pirate out for his own profit. No one was to know of the connection between him and the queen. With each ship that he captured, the interest rate on Philip's loans crept upward, until eventually the Italian bankers were raising the rate more because of the threat of Drake than because of any specific loss. Philip had hoped to launch his armada against England by 1582; short of money, he had to delay. Elizabeth had bought herself more time.

*Achilles now routed the Trojans and pursued them towards the city, but his course, too, was run. Poseidon and Apollo, pledged to avenge the deaths of Cycnus and Troilus, and to punish certain insolent boasts that Achilles had uttered over Hector's corpse, took counsel together. Veiled with cloud and standing by the Scaean gate, Apollo sought out Paris in the thick of battle, turned his bow and guided the fatal shaft. It struck the one vulnerable part of Achilles's body, the right heel, and he died in agony.*

*THE GREEK MYTHS, VOL. 2,* ROBERT GRAVES, 1955

Meanwhile, much to the chagrin of Philip's finance ministers, the king refused to scale back the size of the invading armada. Building it might take longer, but he would just borrow more money. Seeing his fight with England as a religious crusade, he would not be deterred by mere matters of finance.

While working to ruin Philip's credit, Elizabeth put an important part of her meager resources into building up England's spy network—in fact, she made it the most sophisticated intelligence agency in Europe. With agents throughout Spain, she was kept informed of Philip's every move. She knew exactly how large the armada was to be and when it was to be launched. That allowed her to postpone calling up her army and reserves until the very last moment, saving the government money.

Finally, in the summer of 1588, the Spanish Armada was ready. It comprised 128 ships, including twenty large galleons, and a vast number of sailors and soldiers. Equal in size to England's entire navy, it had cost a fortune. The Armada set sail from Lisbon in the second week of July. But Elizabeth's spies had fully informed her of Spain's plans, and she was able to send a fleet of smaller, more mobile English ships to harass the Armada on its way up the French coast, sinking its supply ships and generally creating chaos. As the commander of the English fleet, Lord Howard of Effingham, reported, "Their force is wonderful great and strong; and yet we pluck their feathers little by little."

Finally the Armada came to anchor in the port of Calais, where it was to link up with the Spanish armies stationed in the Low Countries. Determined to prevent it from picking up these reinforcements, the English gathered eight large ships, loaded them with flammable substances, and set them on course for the Spanish fleet, which was anchored in tight formation. As the British ships approached the harbor under full sail, their crews set them on fire and evacuated. The result was havoc, with dozens of Spanish ships in flames. Others scrambled for safe water, often colliding with one another. In their haste to put to sea, all order broke down.

The loss of ships and supplies at Calais devastated Spanish discipline and morale, and the invasion was called off. To avoid further attacks on the return to Spain, the remaining ships headed not south but north, planning to sail home around Scotland and Ireland. The English did not even bother with pursuit; they knew that the rough weather in those waters would do the damage for them. By the time the shattered Armada returned to Spain, forty-four of its ships had been lost and most of the rest were too damaged to be seaworthy. Almost two-thirds of its sailors and soldiers had perished at sea. Meanwhile England had lost not a single ship, and barely a hundred men had died in action.

It was a great triumph, but Elizabeth wasted no time on gloating. To save money, she immediately decommissioned the navy. She also refused to listen to advisers who urged her to follow up her victory by attacking the Spanish in the Low Countries. Her goals were limited: to exhaust Philip's resources and finances, forcing him to abandon his dreams of Catholic dominance and instituting a delicate balance of power in Europe. And this, indeed, was ultimately her greatest triumph, for Spain never recovered financially from the disaster of the Armada and soon gave up its designs on England altogether.

Interpretation

The defeat of the Spanish Armada has to be considered one of the most cost-effective in military history: a second-rate power that barely maintained a standing army was able to face down the greatest empire of its time. What made the victory possible was the application of a basic military axiom: attack their weaknesses with your strengths. England's strengths were its small, mobile navy and its elaborate intelligence network; its weaknesses were its limited resources in men, weaponry, and money. Spain's strengths were its vast wealth and its huge army and fleet; its weaknesses were the precarious structure of its finances, despite their magnitude, and the lumbering size and slowness of its ships.

Elizabeth refused to fight on Spain's terms, keeping her army out of the fray. Instead she attacked Spain's weaknesses with her strengths: plaguing the Spanish galleons with her smaller ships, wreaking havoc on the country's finances, using special ops to grind its war machine to a halt. She was able to control the situation by keeping England's costs down while making the war effort more and more expensive for Spain. Eventually a time came when Philip could only fail: if the Armada sank, he would be ruined for years to come, and even if the Armada triumphed, victory would come so dear that he would ruin himself trying to exploit it on English soil.

Understand: no person or group is completely either weak or strong. Every army, no matter how invincible it seems, has a weak point, a place left unprotected or undeveloped. Size itself can be a weakness in the end. Meanwhile even the weakest group has something it can build on, some hidden strength. Your goal in war is not simply to amass a stockpile of weapons, to increase your firepower so you can blast your enemy away. That is wasteful, expensive to build up, and leaves you vulnerable to guerrilla-style attacks. Going at your enemies blow by blow, strength against strength, is equally unstrategic. Instead you must first assess their weak points: internal political problems, low morale, shaky finances, overly centralized control, their leader's megalomania. While carefully keeping your own weaknesses out of the fray and preserving your strength for the long haul, hit their Achilles' heel again and again. Having their weaknesses exposed and preyed upon will demoralize them, and, as they tire, new weaknesses will open up. By carefully calibrating strengths and weaknesses, you can bring down your Goliath with a slingshot.

*Limitations are troublesome, but they are effective. If we live economically in normal times, we are prepared for times of want. To be sparing saves us from humiliation. Limitations are also indispensable in the regulation of world conditions. In nature there are fixed limits for summer and winter, day and night, and these limits give the year its meaning. In the same way, economy, by setting fixed limits upon expenditures, acts to preserve property and prevent injury to the people.*

*THE I CHING*, CHINA, CIRCA EIGHTH CENTURY B.C.

*Abundance makes me poor.*
*—Ovid (43 B.C.–A.D. 17)*

## KEYS TO WARFARE

Reality can be defined by a sharp series of limitations on every living thing, the final boundary being death. We have only so much energy to expend before we tire; only so much in the way of food and resources is available to us; our skills and capacities can go only so far. An animal lives within those limits: it does not try to fly higher or run faster or expend endless energy amassing a pile of food, for that would exhaust it and leave it vulnerable to attack. It simply tries to make the most of what it has. A cat, for instance, instinctively practices an economy of motion and gesture, never wasting effort. People who live in poverty, similarly, are acutely aware of their limits: forced to make the most of what they have, they are endlessly inventive. Necessity has a powerful effect on their creativity.

*In all this—in selection of nutriment, of place and climate, of recreation—there commands an instinct of self-preservation which manifests itself most unambiguously as an instinct for self-defense. Not to see many things, not to hear them, not to let them approach one—first piece of ingenuity, first proof that one is no accident but a necessity. The customary word for this self-defensive instinct is taste. Its imperative commands, not only to say No when Yes would be a piece of "selflessness," but also to say No as little as possible. To separate oneself, to depart from that to which No would be required again and again. The rationale is that defensive expenditures, be they never so small, become a rule, a habit, lead to an extraordinary and perfectly superfluous impoverishment. Our largest expenditures are our most frequent small ones. Warding off, not letting come close, is an expenditure—one should not deceive oneself over this—a strength squandered on negative objectives. One can merely through the constant need to ward off become too weak any longer to defend oneself. . . . Another form of sagacity and self-defense consists in reacting as seldom as possible and withdrawing from*

The problem faced by those of us who live in societies of abundance is that we lose a sense of limit. We are carefully shielded from death and can pass months, even years, without contemplating it. We imagine endless time at our disposal and slowly drift further from reality; we imagine endless energy to draw on, thinking we can get what we want simply by trying harder. We start to see everything as limitless—the goodwill of friends, the possibility of wealth and fame. A few more classes and books and we can extend our talents and skills to the point where we become different people. Technology can make anything achievable.

Abundance makes us rich in dreams, for in dreams there are no limits. But it makes us poor in reality. It makes us soft and decadent, bored with what we have and in need of constant shocks to remind us that we are alive. In life you must be a warrior, and war requires realism. While others may find beauty in endless dreams, warriors find it in reality, in awareness of limits, in making the most of what they have. Like the cat, they look for the perfect economy of motion and gesture—the way to give their blows the greatest force with the least expenditure of effort. Their awareness that their days are numbered—that they could die at any time—grounds them in reality. There are things they can never do, talents they will never have, lofty goals they will never reach; that hardly bothers them. Warriors focus on what they *do* have, the strengths that they *do* possess and that they must use creatively. Knowing when to slow down, to renew, to retrench, they outlast their opponents. They play for the long term.

Through the final years of French colonial rule in Vietnam and on through the Vietnam War, the military leader of the Vietnamese insurgents was General Vo Nguyen Giap. In first the French and then the Americans, he faced an enemy with vastly superior resources, firepower, and training. His own army was a ragtag collection of peasants; they had morale, a deep sense of purpose, but little else. Giap had no trucks to carry supplies, and his communications were nineteenth century. Another general would have tried to catch up, and Giap had the opportunity—he had the offer of trucks, radios, weapons, and training from China—but he saw them as a trap. It wasn't only that he didn't want to spend his limited funds on such things; in the long run, he believed, all they would do was turn the North Vietnamese into a weaker version of their enemy. Instead he chose to make the most of what he had, turning his army's weaknesses into virtues.

*situations and relationships in which one would be condemned as it were to suspend one's freedom, one's initiative, and become a mere reagent.*

*Ecce Homo*,
Friedrich Nietzsche,
1888

Trucks could be spotted from the air, and the Americans could bomb them. But the Americans could not bomb supply lines they could not see. Exploiting his resources, then, Giap used a vast network of peasant coolies to carry supplies on their backs. When they came to a river, they would cross it on rope bridges hung just below the surface of the water. Right up to the end of the war, the Americans were still trying to figure out how North Vietnam supplied its armies in the field.

Meanwhile Giap developed hit-and-run guerrilla tactics that gave him enormous potential to disrupt American supply lines. To fight, move troops, and ferry supplies, the Americans used helicopters, which gave them tremendous mobility. But the war ultimately had to be fought on the ground, and Giap was endlessly inventive in using the jungle to neutralize American air power, disorient American foot soldiers, and camouflage his own troops. He could not hope to win a pitched battle against superior U.S. weaponry, so he put his effort into spectacular, symbolic, demoralizing attacks that would drive home the futility of the war when they appeared on American TV. With the minimum that he had, he created the maximum effect.

Armies that seem to have the edge in money, resources, and firepower tend to be predictable. Relying on their equipment instead of on knowledge and strategy, they grow mentally lazy. When problems arise, their solution is to amass more of what they already have. But it's not what you have that brings you victory, it's how you use it. When you have less, you are naturally more inventive. Creativity gives you an edge over enemies dependent on technology; you will learn more, be more adaptable, and you will outsmart them. Unable to waste your limited resources, you will use them well. Time will be your ally.

If you have less than your enemy, do not despair. You can always turn the situation around by practicing perfect economy. If you and your enemy are equals, getting hold of more weaponry matters less than making better use of what you have. If you have more than your enemy, fighting economically is as important as ever. As Pablo Picasso said, Even if you are wealthy, act poor. The poor are more inventive, and often have more fun, because they value what they have and know their limits. Sometimes in strategy you have to ignore your greater strength and force yourself to get the maximum out of the minimum. Even if you have the technology, fight the peasant's war.

This does not mean that you disarm or fail to exploit what advantages you may have in matériel. In Operation Desert Storm, the U.S. campaign against Iraq in 1991, American military strategists made full use of their superior technology, particularly in the air, but they did not depend on this for victory. They had learned the lesson of their debacle twenty years earlier in Vietnam, and their maneuvers showed the kind of deceptive feints and use of mobility associated with smaller, guerrilla-like forces. This combination of advanced technology and creative flair proved devastating.

War is a balance of ends and means: a general might have the best plan to achieve a certain end, but unless he has the means to accomplish it, his plan is worthless. Wise generals through the ages, then, have learned to begin by examining the means they have at hand and then to develop their strategy out of those tools. That is what made Hannibal a brilliant strategist: he would always think first of the givens—the makeup of his own army and of the enemy's, their respective proportions of cavalry and infantry, the terrain, his troops' morale, the weather. That would give him the foundation not only for his plan of attack but for the ends he wanted to achieve in this particular encounter. Instead of being locked in to a way of fighting, like so many generals, he constantly adjusted his ends to his means. That was the strategic advantage he used again and again.

The next time you launch a campaign, try an experiment: do not think about either your solid goals or your wishful dreams, and do not plan out your strategy on paper. Instead think deeply about what you have—the tools and materials you will be working with. Ground yourself not in dreams and plans but in reality: think of your own skills, any political advantage you might have, the morale of your troops, how creatively you can use the means at your disposal. Then, out of that process, let your plans and goals blossom. Not only will your strategies be more realistic, they will be more inventive and forceful. Dreaming first of what you want and then trying to find the means to reach it is a recipe for exhaustion, waste, and defeat.

Do not mistake cheapness for perfect economy—armies have failed by spending too little as often as by spending too much. When the British attacked Turkey during World War I, hoping to knock it out of the war and then attack Germany from the east, they began by sending a fleet to break through the Dardanelles Strait and head for the Turkish capital of Constantinople. The fleet made good progress, but even so, after several weeks some ships had been sunk, more lives than expected had been lost, and the venture in general was proving costly. So the British called off the naval campaign, deciding instead to land an army on the peninsula of Gallipoli and fight through by land. That route seemed safer and cheaper—but it turned into a months-long fiasco that cost thousands of lives and in the end led nowhere, for the Allies eventually gave up and pulled out their troops. Years later, Turkish documents were uncovered that revealed that the British fleet had been on the verge of success: in another day or two, it would have broken through and Constantinople would probably have fallen. The whole course of the war might have been changed. But the British had overeconomized; at the last moment, they had pulled their punches, worrying about cost. In the end the cost of trying to win on the cheap wound up punitively expensive.

Perfect economy, then, does not mean hoarding your resources. That is not economy but stinginess—deadly in war. Perfect economy means finding a golden mean, a level at which your blows count but do not wear you out. Overeconomizing will wear you out more, for the war will drag on, its costs growing, without your ever being able to deliver a knockout punch.

Several tactics lend themselves to economy in fighting. First is the use of deception, which costs relatively little but can yield powerful results. During World War II the Allies used a complicated series of deceptions to make the Germans expect an attack from many different directions, forcing them to spread themselves thin. Hitler's Russian campaign was much weakened by the need to keep troops in France and the Balkans, to defend from attacks there—attacks that never came. Deception can be a great equalizer for the weaker side. Its arts include the gathering of intelligence, the spreading of misinformation, and the use of propaganda to make the war more unpopular within the enemy camp.

Second, look for opponents you can beat. Avoid enemies who have nothing to lose—they will work to bring you down whatever it costs. In the nineteenth century, Otto von Bismarck built up Prussia's military power on the backs of weaker opponents such as the Danes. Easy victories enhance morale, develop your reputation, give you momentum, and, most important, do not cost you much.

There will be times when your calculations misfire; what had seemed to be an easy campaign turns out hard. Not everything can be foreseen. Not only is it important to pick your battles carefully, then, but you must also know when to accept your losses and quit. In 1971 the boxers Muhammad Ali and Joe Frazier, both at the heights of their careers, met for the world heavyweight championship. It was a grueling match, one of the most exciting in history; Frazier won by a decision after nearly knocking out Ali in the fifteenth round. But both men suffered horribly in the fight; both threw a lot of good punches. Wanting revenge, Ali gained a rematch in 1974—another grueling fifteen-round affair—and won by a decision. Neither boxer was happy, both wanted a more conclusive result, so they met again in 1975, in the famous "Thrilla in Manila." This time Ali won in the fourteenth round, but neither man was ever the same again: these three fights had taken too much out of them, shortening their careers. Pride and anger had overtaken their powers of reason. Do not fall into such a trap; know when to stop. Do not soldier on out of frustration or pride. Too much is at stake.

Finally, nothing in human affairs stays the same. Over time either your efforts will tend to slow down—a kind of friction will build up, whether from unexpected exterior events or from your own actions—or momentum will help to move you forward. Wasting what you have will create friction, lowering your energy and morale. You are essentially

*Every limitation has its value, but a limitation that requires persistent effort entails a cost of too much energy. When, however, the limitation is a natural one (as, for example, the limitation by which water flows only downhill), it necessarily leads to success, for then it means a saving of energy. The energy that otherwise would be consumed in a vain struggle with the object is applied wholly to the benefit of the matter in hand, and success is assured.*

*THE I CHING*, CHINA, CIRCA EIGHTH CENTURY B.C.

slowing yourself down. Fighting economically, on the other hand, will build momentum. Think of it as finding your level—a perfect balance between what you are capable of and the task at hand. When the job you are doing is neither above nor below your talents but at your level, you are neither exhausted nor bored and depressed. You suddenly have new energy and creativity. Fighting with perfect economy is like hitting that level—less resistance in your path, greater energy unleashed. Oddly enough, knowing your limits will expand your limits; getting the most out of what you have will let you have more.

**Image:** The Swimmer. The water offers resistance; you can move only so fast. Some swimmers pound at the water, trying to use force to generate speed—but they only make waves, creating resistance in their path. Others are too delicate, kicking so lightly they barely move. Consummate swimmers hit the surface with perfect economy, keeping the water in front of them smooth and level. They move as fast as the water will let them and cover great distances at a steady pace.

**Authority:** The value of a thing sometimes lies not in what one attains with it but in what one pays for it—what it costs us. *—Friedrich Nietzsche (1844–1900)*

## REVERSAL

There can never be any value in fighting uneconomically, but it is always a wise course to make your opponent waste as much of his resources as possible. This can be done through hit-and-run tactics, forcing him to expend energy chasing after you. Lure him into thinking that one big offensive will ruin you; then bog that offensive down in a protracted war in which he loses valuable time and resources. A frustrated opponent exhausting energy on punches he cannot land will soon make mistakes and open himself up to a vicious counterattack.

# 9

# TURN THE TABLES

# THE COUNTERATTACK STRATEGY

*Moving first—initiating the attack—will often put you at a disadvantage: you are exposing your strategy and limiting your options. Instead discover the power of holding back and letting the other side move first, giving you the flexibility to counterattack from any angle. If your opponents are aggressive, bait them into a rash attack that will leave them in a weak position. Learn to use their impatience, their eagerness to get at you, as a way to throw them off balance and bring them down. In difficult moments do not despair or retreat: any situation can be turned around. If you learn how to hold back, waiting for the right moment to launch an unexpected counterattack, weakness can become strength.*

## DISGUISED AGGRESSION

In September 1805, Napoleon Bonaparte faced the greatest crisis until that moment in his career: Austria and Russia had joined in an alliance against him. To the south, Austrian troops were attacking the French soldiers occupying northern Italy; to the east, the Austrian general Karl Mack was leading a large force into Bavaria. A sizable Russian army under General Mikhail Kutusov was on its way to join Mack's army, and this allied force, once merged and expanded, would head for France. East of Vienna, more Russian and Austrian troops were waiting to be deployed wherever needed. Napoleon's armies were outnumbered two to one.

Napoleon's plan was to try to defeat each of the alliance's armies one by one, using his smaller but more mobile corps to fight them before they could join forces. While committing enough troops to produce a stalemate in Italy, he moved into Bavaria before Kutusov could reach it and forced Mack's ignominious surrender at Ulm, with hardly a shot being fired (see chapter 6). This bloodless victory was a masterpiece, but to exploit it to its fullest, Napoleon needed to catch Kutusov before the Russian general could himself be reinforced by more Russian or Austrian troops. To that end, Napoleon sent the bulk of his army east, toward Vienna, hoping to trap the retreating Russian forces. But the pursuit bogged down: the weather was bad, the French troops were tired, their marshals made mistakes, and, most important, the wily Kutusov was cleverer in retreat than in attack. Managing to elude the French, he reached the town of Olmütz, northeast of Vienna, where the remaining Austro-Russian forces were stationed.

Now the situation reversed: suddenly it was Napoleon who was in grave danger. The strength of his corps was their mobility; relatively small, they were vulnerable individually and worked best when operating close enough to one another to come fast to one another's support. Now they were dispersed in a long line from Munich to Vienna, which Napoleon had taken after his victory over Mack at Ulm. The men were hungry, tired, and short of supplies. The Austrians fighting the French in northern Italy had given up the battle there and were in retreat—but that put them heading northeast, posing a threat to Napoleon's southern flank. To the north, the Prussians, seeing that Napoleon was in trouble, were considering joining the alliance. If that happened, they could wreak havoc on Napoleon's extended lines of communication and supply—and the two armies moving in from north and south could squeeze him to death.

Napoleon's options were abysmal. To continue the pursuit of Kutusov would further extend his lines. Besides, the Russians and Austrians were now 90,000 strong and in an excellent position at Olmütz. To stay put, on the other hand, was to risk being slowly swallowed by armies on all sides. Retreat seemed the only solution, and it was what his generals

*The technique of "according with" the enemy's expectations and desires requires first determining what they believe and want, then apparently conforming to them until the situation can be exploited: Definition: When the enemy wants to take something and you yield it, it is termed "according with." . . . In general, when going contrary to something merely solidifies it, it is better to accord with it in order to lead them to flaws. If the enemy wants to advance, be completely flexible and display weakness in order to induce an advance. If the enemy wants to withdraw, disperse and open an escape route for their retreat. If the enemy is relying upon a strong front, establish your own front lines far off, solidly assuming a defensive posture in order to observe their arrogance. If the enemy relies upon their awesomeness, be emptily respectful but substantially plan while awaiting their laxness. Draw them forward and cover them, release and capture them. Exploit their arrogance, capitalize on their laxity.*

SEVENTEENTH-CENTURY MING DYNASTY TEXT, QUOTED IN *THE TAO OF SPYCRAFT*, RALPH D. SAWYER advised, but with the weather deteriorating (it was mid-November) and the enemy sure to harass him, that would be costly, too. And retreat would mean that his victory at Ulm had been wasted—a tremendous blow to the morale of his men. That would virtually invite the Prussians to join the war, and his enemies the English, seeing him vulnerable, might go so far as to invade France. Whatever path he chose seemed to lead to disaster. For several days he went into deep thought, ignoring his advisers and poring over maps.

*A rapid, powerful transition to the attack—the glinting sword of vengeance—is the most brilliant moment of the defense.*

CARL VON CLAUSEWITZ, 1780–1831

Meanwhile, at Olmütz, the Austrian and Russian leaders—among them the Austrian Emperor Francis I and the young czar Alexander I—watched Napoleon's moves with intense curiosity and excitement. They had him where they wanted him; surely they would be able to recoup the disaster at Ulm and then some.

On November 25, alliance scouts reported that Napoleon had moved a large part of his army to Austerlitz, halfway between Vienna and Olmütz. There it looked as if his forces were occupying the Pratzen Heights, a position that would indicate preparation for battle. But Napoleon had only some 50,000 men with him; he was outnumbered nearly two to one. How could he hope to face the allies? Even so, on November 27, Francis I offered him an armistice. Napoleon was formidable, and even at those odds, fighting him was a risk. In truth, Francis was also trying to buy enough time to envelop the French army completely, but none of the alliance generals thought Napoleon would fall for that trick.

To their surprise, however, Napoleon seemed eager to come to terms. Suddenly the czar and his generals had a new thought: he was panicking, grasping at straws. That suspicion seemed borne out almost immediately, when, on November 29, Napoleon abandoned the Pratzen Heights almost as soon as he had taken them, assuming a position to their west and repeatedly repositioning his cavalry. He appeared utterly confused. The next day he asked for a meeting with the czar himself. Instead the czar sent an emissary, who reported back that Napoleon had been unable to disguise his fear and doubt. He had seemed on edge, emotional, even distraught. The emissary's conditions for armistice had been harsh, and although Napoleon had not agreed to them, he had listened quietly, seeming chastened, even intimidated. This was music to the ears of the young czar, who was burning for his first engagement with Napoleon. He was tired of waiting.

By abandoning the Pratzen Heights, Napoleon seemed to have put himself in a vulnerable position: his southern lines were weak, and his route of retreat, southwest toward Vienna, was exposed. An allied army could take the Pratzen Heights, pivot south to break through that weak point in his lines and cut off his retreat, then move back north to surround his army and destroy him. Why wait? A better chance would never come. Czar Alexander and his younger generals prevailed over the hesitant Austrian emperor and launched the attack.

It began early on the morning of December 2. While two smaller divisions faced off against the French from the north, pinning them down, a stream of Russian and Austrian soldiers moved toward the Pratzen Heights, took them, then wheeled to the south, aiming at the French weak point. Although they met resistance from the outnumbered enemy, they quickly broke through and were soon able to take the key positions that would allow them to turn north and surround Napoleon. But at 9:00 A.M., as the last alliance troops (some 60,000 men in all) made their way to the heights and headed south, word reached the allied commanders that something unexpected was afoot: a large French force, invisible to them beyond the Pratzen Heights, was suddenly heading due east, straight for the town of Pratzen itself and the center of the allied lines.

Kutusov saw the danger: the allies had advanced so many men into the gap in the French lines that they had left their own center exposed. He tried to turn back the last troops heading south, but it was too late. By 11:00 A.M. the French had retaken the heights. Worse, French troops had come up from the southwest to reinforce the southern position and prevent the allies from surrounding the French. Everything had turned around. Through the town of Pratzen, the French were now pouring through the allied center and were swiftly moving to cut off the retreat of the allied troops to their south.

Each part of the allied army—north, center, and south—was now effectively isolated from the others. The Russians in the southernmost position tried to retreat farther to the south, but thousands of them lost their lives in the frozen lakes and marshes in their path. By 5:00 P.M. the rout was complete, and a truce was called. The Austro-Russian army had suffered terrible casualties, far more than the French. The defeat was so great that the alliance collapsed; the campaign was over. Somehow Napoleon had snatched victory from defeat. Austerlitz was the greatest triumph of his career.

*A sudden inspiration then came to William [at the Battle of Hastings, A.D. 1066], suggested by the disaster which had befallen the English right in the first conflict. He determined to try the expedient of a feigned flight, a stratagem not unknown to Bretons and Normans of earlier ages. By his orders a considerable portion of the assailants suddenly wheeled about and retired in seeming disorder. The English thought, with more excuse on this occasion than on the last, that the enemy was indeed routed, and for the second time a great body of them broke the line and rushed after the retreating squadrons. When they were well on their way down the slope, William repeated his former procedure. The intact portion of his host fell upon the flanks of the pursuers, while those who had simulated flight faced about and attacked them in front. The result was again a foregone conclusion: the disordered men of the fyrd were hewn to pieces, and few or none of them escaped back to their comrades on the height.*

*HISTORY OF THE ART OF WAR IN THE MIDDLE AGES*, SIR CHARLES OMAN, 1898

## Interpretation

In the crisis leading up to the Battle of Austerlitz, Napoleon's advisers and marshals had thought only of retreat. Sometimes it is better, they believed, to accept a setback willingly and go on the defensive. On the other side stood the czar and his allies, who had Napoleon weak. Whether they waited to envelop him or attacked right away, they were on the offensive.

In the middle was Napoleon, who, as a strategist, stood far above both his own advisers and marshals, on the one hand, and the czar and alliance generals on the other. His superiority lay in the fluidity of his thinking: he did not conceive war in mutually exclusive terms of defense and offense. In his mind they were inextricably linked: a defensive position was the perfect way to disguise an offensive maneuver, a counterattack; an offensive maneuver was often the best way to defend a weak position. What Napoleon orchestrated at Austerlitz was neither retreat nor attack but something far more subtle and creative: he fused defense and offense to set up the perfect trap.

First, having taken Vienna, Napoleon advanced to Austerlitz, apparently taking the offensive. That startled the Austrians and Russians, even though they still heavily outnumbered him. Next he backed off and took a defensive position; then he seemed to switch between offense and defense, giving every appearance of confusion. In his meeting with the czar's emissary, he seemed confused personally as well as strategically. It was all high drama, staged by Napoleon to make him look weak and vulnerable, inviting attack.

These maneuvers fooled the allies into giving up prudence, striking out at Napoleon with total abandon and exposing themselves in the process. Their defensive position at Olmütz was so strong and dominant that only leaving it would ruin it, and that was precisely what Napoleon lured them into doing. Then, instead of defending himself against their rash attack, he suddenly switched to the offensive himself, the counterattack. In doing so he altered the dynamic of the battle not only physically but psychologically: when an attacking army suddenly has to go on the defensive, its spirit crumbles. And indeed the alliance troops panicked, retreating to the frozen lakes that Napoleon had intended as their graveyard all along.

Most of us only know how to play either offensively or defensively. Either we go into attack mode, charging our targets in a desperate push to get what we want, or we try frantically to avoid conflict and, if it is forced on us, to ward off our enemies as best we can. Neither approach works when it excludes the other. Making offense our rule, we create enemies and risk acting rashly and losing control of our own behavior, but constant defensiveness backs us into a corner, becomes a bad habit. In either case we are predictable.

Instead consider a third option, the Napoleonic way. At times you seem vulnerable and defensive, getting your opponents to disregard you as a threat, to lower their guard. When the moment is right and you sense an opening, you switch to the attack. Make your aggression controlled and your weakness a ploy to disguise your intentions. In a dangerous moment, when those around you see only doom and the need to retreat, that is when you smell an opportunity. By playing weak you can seduce your aggressive enemies to come at you full throttle. Then catch them off guard by switching to the offense when they least expect it. Mixing offense and defense in this fluid fashion, you will stay one step ahead of your inflexible opponents. The best blows are the ones they never see coming.

*When the enemy finds itself in a predicament and wants to engage us in a decisive battle, wait; when it is advantageous for the enemy but not for us to fight, wait; when it is expedient to remain still and whoever moves first will fall into danger, wait; when two enemies are engaged in a fight that will result in defeat or injury, wait; when the enemy forces, though numerous, suffer from mistrust and tend to plot against one another, wait; when the enemy commander, though wise, is handicapped by some of his cohorts, wait.*

*The Wiles of War: 36 Military Strategies from Ancient China*, translated by Sun Haichen, 1991

*However desperate the situation and circumstances, don't despair. When there is everything to fear, be unafraid. When surrounded by dangers, fear none of them. When without resources, depend on resourcefulness. When surprised, take the enemy itself by surprise.*
*—Sun-tzu,* The Art of War *(fourth century* B.C.*)*

*These two main principles of application are specifically related to the tactical value assigned to the personality of the opponent in combat. According to the unilateral principle of application, the personality of the opponent was considered the primary target of an attack or counterattack, for the purpose of either total or partial subjugation. According to the bilateral principle of application, on the other hand, the opponent's personality was viewed not merely as a target, but also (and by certain bujutsu masters, primarily) as an instrument—that is, as the unwilling but nevertheless useful vector of his own subjugation. . . . . . . It is the principle of bilateral application which seems to represent a tactical differentiation between Japanese bujutsu and the martial arts of the West. Lafcadio Hearn, for example, considered this principle "a uniquely Oriental idea," asking, "What Western brain could have elaborated this strange teaching: never to oppose force to force, but only to direct and utilize the power of attack; to overthrow the enemy solely by his own strength—to vanquish him solely by his own efforts?" (Smith, 128) . . . Takuan, writing about the art of swordsmanship in particular, refers to the strategic*

## JUJITSU

In 1920 the Democratic Party nominated Ohio governor James Cox as its candidate to succeed the retiring President Woodrow Wilson. At the same time, it named thirty-eight-year-old Franklin Delano Roosevelt as its vice presidential nominee. Roosevelt had served as the assistant secretary of the navy under Wilson; more important, he was the cousin of Theodore Roosevelt, still very popular after his presidency in the first decade of the century.

The Republican nominee was Warren G. Harding, and the campaign was a grueling affair. The Republicans had a lot of money; they avoided talking about the issues and played up Harding's folksy image. Cox and Roosevelt responded to the Republicans by going on a vigorous offensive, basing their campaign on a single issue of Wilson's: American participation in the League of Nations, which they hoped would bring peace and prosperity. Roosevelt campaigned all over the country, delivering speech after speech—the idea was to counter the Republicans' money with sheer effort. But the race was a disaster: Harding won the presidency in one of the biggest landslides in American electoral history.

The following year, Roosevelt was stricken with polio and lost the use of his legs. Coming just after the disastrous 1920 campaign, his illness marked a turning point in his life: suddenly made aware of his physical fragility and mortality, he retreated into himself and reassessed. The world of politics was vicious and violent. To win an election, people would do anything, stooping to all kinds of personal attacks. The public official moving in this world was under pressure to be as unscrupulous as everyone else and survive as best he could—but that approach did not suit Roosevelt personally and took too much out of him physically. He decided to craft a different political style, one that would separate him from the crowd and give him a constant advantage.

In 1932, after a stint as governor of New York, Roosevelt ran as the Democratic presidential nominee against the Republican incumbent, Herbert Hoover. The country was in the midst of the Depression, and Hoover seemed incapable of dealing with it. Given the weakness of his record, a defensive hand was a difficult one for him to play, and, like the Democrats in 1920, he went vigorously on the offensive, attacking Roosevelt as a socialist. Roosevelt in turn traveled the country, speaking on his ideas for getting America out of the Depression. He didn't give many specifics, nor did he respond to Hoover's attacks directly—but he radiated confidence and ability. Hoover meanwhile seemed shrill and aggressive. The Depression would probably have doomed him to defeat whatever he did, but he lost far bigger than expected: the size of Roosevelt's victory—nearly an electoral sweep—surprised one and all.

In the weeks following the election, Roosevelt essentially hid from public view. Slowly his enemies on the right began to use his absence to attack him, circulating speculation that he was unprepared for the challenge of the job. The criticisms became pointed and aggressive. At his inauguration, however, Roosevelt gave a rousing speech, and in his first months in office, now known as the "Hundred Days," he switched from the appearance of inactivity to a powerful offensive, hurrying through legislation that made the country feel as if something were finally being done. The sniping died.

Over the next few years, this pattern repeatedly recurred. Roosevelt would face resistance: The Supreme Court, say, would overturn his programs, and enemies on all sides (Senator Huey Long and labor leader John L. Lewis on the left, Father Charles Coughlin and wealthy businessmen on the right) would launch hostile campaigns in the press. Roosevelt would retreat, ceding the spotlight. In his absence the attacks would seem to pick up steam, and his advisers would panic—but Roosevelt was just biding his time. Eventually, he knew, people would tire of these endless attacks and accusations, particularly because, by refusing to reply to them, he made them inevitably one-sided. Then—usually a month or two before election time—he would go on the offensive, defending his record and attacking his opponents suddenly and vigorously enough to catch them all off guard. The timing would also jolt the public, winning him their attention.

In the periods when Roosevelt was silent, his opponents' attacks would grow, and grow more shrill—but that only gave him material he could use later, taking advantage of their hysteria to make them ridiculous. The most famous example of this came in 1944, when that year's Republican presidential nominee, Thomas Dewey, launched a series of personal attacks on Roosevelt, questioning the activities of his wife, his sons, and even his dog, the Scotch terrier Fala, whom Dewey accused of being pampered at the taxpayers' expense. Roosevelt countered in a campaign speech,

> The Republican leaders have not been content to make personal attacks upon me—or my sons—they now include my little dog, Fala. Unlike the members of my family, Fala resents this. When he learned that the Republican fiction writers had concocted a story that I left him behind on an Aleutian island and had sent a destroyer back to find him—at a cost to the taxpayer of 2 or 3, or 8 or 20 million dollars—his Scotch soul was furious. He has not been the same dog since. I am accustomed to hearing malicious falsehoods about myself, but I think I have the right to object to libelous statements about my dog.

*value of the bilateral principle in the strategy of counterattack against an opponent, when he advised his pupil to "make use of his attack by turning it on to himself. Then, his sword meant to kill you becomes your own and the weapon will fall on the opponent himself. In Zen this is known as 'seizing the enemy's spear and using it as the weapon to kill him' " (Suzuki, 96)*

*The ancient schools of jujutsu were very empathetic on this subject. . . . Ju-Jutsu (literally "soft art"), as its name implies, is based upon the principle of opposing softness or elasticity to hardness or stiffness. Its secret lies in keeping one's body full of* ki, *with elasticity in one's limbs, and in being ever on the alert to turn the strength of one's foe to one's own advantage with the minimum employment of one's own muscular force.*

*Secrets of the Samurai*, Oscar Ratti and Adele Westbrook, 1973

Devastatingly funny, the speech was also ruthlessly effective. And how could his opponents reply to it when it quoted their own words right back at them? Year after year Roosevelt's opponents exhausted themselves attacking him, scoring points at moments when it didn't matter and losing one landslide election after another to him.

*To undertake the military operations, the army must prefer stillness to movement. It reveals no shape when still but exposes its shape in movement. When a rash movement leads to exposure of the shape of the army, it will fall victim to the enemy. But for movement, the tiger and leopard will not fall into trap, the deer will not run into snare, the birds will not be stuck by net, and the fish and turtles will not be caught by hooks. All these animals become prey to man because of their movement. Therefore the wise man treasures stillness. By keeping still, he can dispel temerity and cope with the temerarious enemy. When the enemy exposes a vulnerable shape, seize the chance to subdue it.* The Book of Master Weiliao *observes, "The army achieves victory by stillness." Indeed, the army should not move without careful thought, much less take reckless action.*

The Wiles of War: 36 Military Strategies from Ancient China, translated by Sun Haichen, 1991

Interpretation

Roosevelt could not bear to feel cornered, to have no options. This was partly because of his flexible nature; he preferred to bend to circumstances, changing direction effortlessly as needed. It also came out of his physical limitations—he hated to feel hemmed in and helpless. Early on, when Roosevelt campaigned in the usual aggressive way of American politics, arguing his case and attacking his opponents, he felt hopelessly constricted. Through experiment he learned the power of holding back. Now he let his opponents make the first move: whether by attacking him or by detailing their own positions, they would expose themselves, giving him openings to use their own words against them later on. By staying silent under their attacks, he would goad them into going too far (nothing is more infuriating than engaging with someone and getting no response) and ending up shrill and irrational, which played badly with the public. Once their own aggression had made them vulnerable, Roosevelt would come in for the kill.

Roosevelt's style can be likened to jujitsu, the Japanese art of self-defense. In jujitsu a fighter baits opponents by staying calm and patient, getting them to make the first aggressive move. As they come at the fighter and either strike at him or grab hold of him—either push or pull—the fighter moves with them, using their strength against them. As he deftly steps forward or back at the right moment, the force of their own momentum throws them off balance: often they actually fall, and even if they don't, they leave themselves vulnerable to a counterblow. Their aggression becomes their weakness, for it commits them to an obvious attack, exposing their strategy and making it hard for them to stop.

In politics, jujitsu style yields endless benefits. It gives you the ability to fight without seeming aggressive. It saves energy, for your opponents tire while you stay above the fray. And it widens your options, allowing you to build on what they give you.

Aggression is deceptive: it inherently hides weakness. Aggressors cannot control their emotions. They cannot wait for the right moment, cannot try different approaches, cannot stop to think about how to take their enemies by surprise. In that first wave of aggression, they seem strong, but the longer their attack goes on, the clearer their underlying weakness and insecurity become. It is easy to give in to impatience and make the first move, but there is more strength in holding back, patiently letting the other person make the play. That inner strength will almost always prevail over outward aggression.

Time is on your side. Make your counterattacks swift and sudden—like the cat who creeps on padded paws to suddenly pounce on its prey. Make jujitsu your style in almost everything you do: it is your way of responding to aggression in everyday life, your way of facing circumstances. Let events come to you, saving valuable time and energy for those brief moments when you blaze with the counterattack.

*The soundest strategy in war is to postpone operations until the moral disintegration of the enemy renders the delivery of the mortal blow both possible and easy.*
—*Vladimir Lenin (1870–1924)*

## KEYS TO WARFARE

Thousands of years ago, at the dawn of military history, various strategists in different cultures noticed a peculiar phenomenon: in battle, the side that was on the defensive often won in the end. There seemed to be several reasons for this. First, once the aggressor went on the attack, he had no more surprises in store—the defender could clearly see his strategy and take protective action. Second, if the defender could somehow turn back this initial attack, the aggressor would be left in a weak position; his army was disorganized and exhausted. (It requires more energy to take land than to hold it.) If the defenders could take advantage of this weakness to deliver a counterblow, they could often force the aggressor to retreat.

Based on these observations, the art of the counterattack was developed. Its basic tenets were to let the enemy make the first move, actively baiting him into an aggressive attack that would expend his energy and unbalance his lines, then taking advantage of his weakness and disorganization. This art was refined by theorists such as Sun-tzu and practiced to perfection by leaders like Philip of Macedon.

The counterattack is, in fact, the origin of modern strategy. The first real example of an indirect approach to war, it represents a major breakthrough in thinking: instead of being brutal and direct, the counterattack is subtle and deceptive, using the enemy's energy and aggression to bring about his downfall. Although it is one of the oldest and most basic strategies in warfare, it remains in many ways the most effective and has proven highly adaptable to modern conditions. It was the strategy of choice of Napoleon Bonaparte, T. E. Lawrence, Erwin Rommel, and Mao Tse-tung.

The counterattack principle is infinitely applicable to any competitive environment or form of conflict, since it is based on certain truths of human nature. We are inherently impatient creatures. We find it hard to wait; we want our desires to be satisfied as quickly as possible. This is a tremendous weakness, for it means that in any given situation we often

THE HEFFALUMP TRAP

*Piglet and Pooh have fallen into a Hole in the Floor of the Forest. They have Agreed that it is Really a Heffalump Trap, which makes Piglet Nervous. He imagines that a Heffalump has Landed Close By:*
*Heffalump (gloatingly):* "Ho-ho!"
*Piglet ( carelessly):* "Tra-la-la, tra-la-la."
*Heffalump (surprised, and not quite so sure of himself): "Ho-ho!"*
*Piglet (more carelessly still): "Tiddle-um-tum, tiddle-um-tum."*
*Heffalump (beginning to say Ho-ho and turning it awkwardly into a cough): "H'r'm! What's all this?"*
*Piglet (surprised): "Hullo! This is a trap I've made, and I'm waiting for a Heffalump to fall into it."*
*Heffalump (greatly disappointed): "Oh!" (after a long silence): "Are you sure?"*
*Piglet: "Yes."*
*Heffalump: "Oh!" (nervously): "I—I thought it was a trap* I'd *made to catch Piglets."*
*Piglet (surprised): "Oh, no!"*
*Heffalump: "Oh!" (apologetically): "I—I must have got it wrong, then."*
*Piglet: "I'm afraid so." (politely): "I'm sorry." (he goes on humming.)*
*Heffalump: "Well—well—I—well. I suppose I'd better be getting back?"*
*Piglet (looking up carelessly): "Must you? Well, if you see Christopher Robin*

*anywhere, you might tell him I want him."*
*Heffalump (eager to please): "Certainly! Certainly!" (he hurries off.)*
*Pooh (who wasn't going to be there, but we find we can't do without him): "Oh Piglet, how brave and clever you are!"*
*Piglet (modestly): "Not at all, Pooh." (And then, when Christopher Robin comes, Pooh can tell him all about it.)*

THE HOUSE AT POOH CORNER, A. A. MILNE, 1928

commit ourselves without enough thought. In charging ahead we limit our options and get ourselves into trouble. Patience, on the other hand, particularly in war, pays unlimited dividends: it allows us to sniff out opportunities, to time a counterblow that will catch the enemy by surprise. A person who can lie back and wait for the right moment to take action will almost always have an advantage over those who give in to their natural impatience.

The first step in mastering the counterattack is to master yourself, and particularly the tendency to grow emotional in conflict. When the great baseball player Ted Williams made the major leagues with the Boston Red Sox, he took a look around. He was now a member of an elite—the best hitters in the country. They all had sharp vision, quick reflexes, and strong arms, but relatively few of them could control their impatience at the plate—and pitchers preyed on that weakness, getting them to swing on losing pitches. Williams separated himself out, and made himself perhaps the greatest pure hitter in baseball history, by developing his patience and a kind of hitter's counterattack: he would wait, and keep waiting, for the best pitch to swing at. Good pitchers are masters at making a hitter feel frustrated and emotional, but Williams would not be baited: whatever they did, he would wait for the pitch that was right for him. In fact, he turned the situation around: given his ability to wait, it was the pitcher, not Williams, who would end up impatient and throwing the wrong pitch as a result.

*The notion of "catching"* (utsuraseru) *applies to many things: yawning and sleepiness, for example. Time can also be "catching." In a large-scale battle, when the enemy is restless and trying to bring a quick conclusion to the battle, pay no attention. Instead, try to pretend that you are calm and quiet with no urgent need to end the battle. The enemy will then be affected by your calm and easy attitude and become less alert. When this "catching" occurs, quickly execute a strong attack to defeat the enemy. . . . There is also a concept called "making one drunk," which is similar to the notion of "catching." You can make your opponent feel bored, carefree, or feeble spirited. You should study these matters well.*

THE BOOK OF FIVE RINGS, MIYAMOTO MUSASHI, 1584–1645

Once you learn patience, your options suddenly expand. Instead of wearing yourself out in little wars, you can save your energy for the right moment, take advantage of other people's mistakes, and think clearly in difficult situations. You will see opportunities for counterattack where others see only surrender or retreat.

The key to the successful counterattack is staying calm while your opponent gets frustrated and irritable. In sixteenth-century Japan, there emerged a novel way of fighting called Shinkage: the swordsman would begin the fight by mirroring his opponent's every move, copying his every footstep, every blink, every gesture, every twitch. This would drive the enemy crazy, for he would be unable to read the Shinkage samurai's moves or get any sense of what he was up to. At some point he would lose patience and strike out, lowering his guard. The Shinkage samurai would inevitably parry this attack and follow up with a fatal counterblow.

Shinkage samurai believed that the advantage in a life-and-death swordfight lay not in aggression but in passivity. By mirroring their enemy's moves, they could understand his strategy and thinking. By being calm and observant—patient—they could detect when their opponent had decided to attack; the moment would register in his eyes or in a slight movement of his hands. The more irritated he became and the harder he tried to hit the Shinkage fighter, the greater his imbalance and vulnerability. Shinkage samurai were virtually unbeatable.

Mirroring people—giving back to them just what they give you—is a powerful method of counterattack. In daily life, mirroring and passivity can charm people, flattering them into lowering their defenses and opening themselves to attack. It can also irritate and discomfit them. Their thoughts become yours; you are feeding off them like a vampire, your passive front disguising the control you are exercising over their minds. Meanwhile you are giving them nothing of yourself; they cannot see through you. Your counterattack will come as a complete surprise to them.

The counterattack is a particularly effective strategy against what might be called "the barbarian"—the man or woman who is especially aggressive by nature. Do not be intimidated by these types; they are in fact weak and are easily swayed and deceived. The trick is to goad them by playing weak or stupid while dangling in front of them the prospect of easy gains.

During the era of the Warring States in ancient China, the state of Qi found itself threatened by the powerful armies of the state of Wei. The Qi general consulted the famous strategist Sun Pin (a descendant of Sun-tzu himself), who told him that the Wei general looked down on the armies of Qi, believing that their soldiers were cowards. That, said Sun Pin, was the key to victory. He proposed a plan: Enter Wei territory with a large army and make thousands of campfires. The next day make half that number of campfires, and the day after that, half that number again. Putting his trust in Sun Pin, the Qi general did as he was told.

The Wei general, of course, was carefully monitoring the invasion, and he noted the dwindling campfires. Given his predisposition to see the Qi soldiers as cowards, what could this mean but that they were defecting? He would advance with his cavalry and crush this weak army; his infantry would follow, and they would march into Qi itself. Sun Pin, hearing of the approaching Wei cavalry and calculating how fast they were moving, retreated and stationed the Qi army in a narrow pass in the mountains. He had a large tree cut down and stripped of its bark, then wrote on the bare log, "The general of Wei will die at this tree." He set the log in the path of the pursuing Wei army, then hid archers on both sides of the pass. In the middle of the night, the Wei general, at the head of his cavalry, reached the place where the log blocked the road. Something was written on it; he ordered a torch lit to read it. The torchlight was the signal and the lure: the Qi archers rained arrows on the trapped Wei horsemen. The Wei general, realizing he had been tricked, killed himself.

Sun Pin based his baiting of the Wei general on his knowledge of the man's personality, which was arrogant and violent. By turning these qualities to his advantage, encouraging his enemy's greed and aggression, Sun Pin could control the man's mind. You, too, should look for the emotion that your enemies are least able to manage, then bring it to the surface. With a little work on your part, they will lay themselves open to your counterattack.

*The other improvement was his father's inspiration. Lyndon Johnson was very dejected as he sat, on the day the* Express *poll appeared, in his parents' home in Johnson City after hours of campaigning, talking to his parents, his brother, his Uncle Tom, his cousin Ava Johnson Cox, and Ava's eight-year-old son, William, known as "Corky." The leaders were almost all against him, he said; he had several large rallies scheduled, and he had not been able to persuade a single prominent individual to introduce him. So, Ava recalls—in a recollection echoed by Lyndon's brother—"his Daddy said, 'If you can't use that route, why don't you go the other route?' " "What other route?" Lyndon asked—and his Daddy mapped it out for him. There was a tactic, Sam Johnson said, that could make the leaders' opposition work for him, instead of against him. The same tactic, Sam said, could make the adverse newspaper polls work for him, instead of against him. It could even make the youth issue work for him. If the leaders were against him, he told his son, stop trying to conceal that fact; emphasize it—in a dramatic fashion. If he was behind in the race, emphasize that—in a dramatic fashion. If he was younger than the other candidates, emphasize that.*

*Lyndon asked his father what he meant, and his father told him. If no leader would introduce Lyndon, Sam said, he should stop searching for mediocre adults as substitutes, but instead should be introduced by an outstanding young child. And the child should introduce him not as an adult would introduce him, but with a poem, a very special poem. . . . And when Lyndon asked who the child should be, Sam smiled, and pointed to Ava's son. In an area in which horsemanship was one of the most esteemed talents, Corky Cox was, at the age of eight, already well known for the feats of riding and calf-roping with which he had swept the children's events in recent rodeos; the best young cowboy in the Hill Country, people were calling him. "Corky can do it," Sam said. All the next day, Sam trained him. "He wanted Corky to really shout out 'thousands,' " Ava recalls. "He wanted him to smack down his hand every time he said that word. I can still see Uncle Sam smacking down his hand on the kitchen table to show Corky how." And that night, at a rally in Henly, in Hays County, Lyndon Johnson told the audience, "They say I'm a young candidate. Well, I've got a young campaign manager, too," and he called Corky to the podium,*

In our own time, the family therapist Jay Haley has observed that for many difficult people acting out is a strategy—a method of control. They give themselves the license to be impossible and neurotic. If you react by getting angry and trying to make them stop, you are doing just what they want: they are engaging your emotions and dominating your attention. If, on the other hand, you simply let them run amok, you put them still more in control. But Haley discovered that if you encourage their difficult behavior, agree with their paranoid ideas, and push them to go further, you turn the dynamic around. This is not what they want or expect; now they're doing what *you* want, which takes the fun out of it. It is the jujitsu strategy: you are using their energy against them. In general, encouraging people to follow their natural direction, to give in to their greed or neuroses, will give you more control over them than active resistance will. Either they get themselves into terrible trouble or they become hopelessly confused, all of which plays into your hands.

Whenever you find yourself on the defensive and in trouble, the greatest danger is the impulse to overreact. You will often exaggerate your enemy's strength, seeing yourself as weaker than is actually the case. A key principle of counterattack is never to see a situation as hopeless. No matter how strong your enemies seem, they have vulnerabilities you can prey upon and use to develop a counterattack. Your own weakness can become a strength if you play it right; with a little clever manipulation, you can always turn things around. That is how you must look at every apparent problem and difficulty.

An enemy seems powerful because he has a particular strength or advantage. Maybe it's money and resources; maybe it's the size of his army or of his territory; maybe, more subtly, it's his moral standing and reputation. Whatever his strength might be, it is actually a potential weakness, simply because he relies on it: neutralize it and he is vulnerable. Your task is to put him in a situation in which he cannot use his advantage.

In 480 B.C., when the Persian king Xerxes invaded Greece, he had a huge advantage in the size of his army and particularly his navy. But the Athenian general Themistocles was able to turn that strength into weakness: he lured the Persian fleet into the narrow straits off the island of Salamis. In these choppy, difficult waters, the very size of the fleet, its apparent strength, became a nightmare: it was completely unable to maneuver. The Greeks counterattacked and destroyed it, ending the invasion.

If your opponent's advantage comes from a superior style of fighting, the best way to neutralize it is to learn from it, adapting it to your own purposes. In the nineteenth century, the Apaches of the American Southwest were for many years able to torment U.S. troops through guerrilla-style tactics that were perfectly suited to the terrain. Nothing seemed to work until General George Crook hired disaffected Apaches to teach him their way of fighting and serve as scouts. Adapting their style of warfare, Crook neutralized the Apaches' strengths and finally defeated them.

As you neutralize your enemy's strengths, you must similarly reverse your own weaknesses. If your forces are small, for example, they are also mobile; use that mobility to counterattack. Perhaps your reputation is lower than your opponent's; that just means you have less to lose. Sling mud—some of it will stick, and gradually your enemy will sink to your level. Always find ways to turn your weakness to advantage.

Difficulties with other people are inevitable; you must be willing to defend yourself and sometimes to take the offensive. The modern dilemma is that taking the offensive is unacceptable today—attack and your reputation will suffer, you will find yourself politically isolated, and you will create enemies and resistance. The counterattack is the answer. Let your enemy make the first move, then play the victim. Without overt manipulation on your part, you can control your opponents' minds. Bait them into a rash attack; when it ends up in disaster, they will have only themselves to blame, and everyone around them will blame them, too. You win both the battle of appearances and the battle on the field. Very few strategies offer such flexibility and power.

*and Corky, smacking down his hand, recited a stanza of Edgar A. Guest's "It Couldn't Be Done":*

There are thousands to tell you it cannot be done,
There are thousands to prophesy failure;
There are thousands to point out to you one by one,
The dangers that wait to assail you.
But just buckle in with a bit of a grin,
Just take off your coat and go to it;
Just start in to sing as you tackle the thing
That "cannot be done," and you'll do it.

*The Path to Power: The Years of Lyndon Johnson*, vol. 1, Robert A. Caro, 1990

**Image:** The Bull. It is large, its stare is intimidating, and its horns can pierce your flesh. Attacking it and trying to escape it are equally fatal. Instead stand your ground and let the bull charge your cape, giving it nothing to hit, making its horns useless. Get it angry and irritated—the harder and more furiously it charges, the faster it wears itself down. A point will come when you can turn the game around and go to work, carving up the once fearsome beast.

**Authority:** The whole art of war consists in a well-reasoned and extremely circumspect defensive, followed by a rapid and audacious attack. —*Napoleon Bonaparte (1769–1821)*

## REVERSAL

The counterattack strategy cannot be applied in every situation: there will always be times when it is better to initiate the attack yourself, gaining control by putting your opponents on the defensive before they have time to think. Look at the details of the situation. If the enemy is too smart to lose patience and attack you, or if you have too much to lose by waiting, go on the offensive. It is also usually best to vary your methods, always having more than one strategy to draw on. If your enemies think you always wait to counterattack, you have the perfect setup for moving first and surprising them. So mix things up. Watch the situation and make it impossible for your opponents to predict what you will do.

*Conditions are such that the hostile forces favored by the time are advancing. In this case retreat is the right course, and it is through retreat that success is achieved. But success consists in being able to carry out the retreat correctly. Retreat is not to be confused with flight. Flight means saving oneself under any circumstances, whereas retreat is a sign of strength. We must be careful not to miss the right moment while we are in full possession of power and position. Then we shall be able to interpret the signs of the time before it is too late and to prepare for provisional retreat instead of being drawn into a desperate life-and-death struggle. Thus we do not simply abandon the field to the opponent; we make it difficult for him to advance by showing perseverance in single acts of resistance. In this way we prepare, while retreating, for the counter-movement. Understanding the laws of a constructive retreat of this sort is not easy. The meaning that lies hidden in such a time is important.*

*THE I CHING*, CHINA, CIRCA EIGHTH CENTURY B.C.

# 10

# CREATE A THREATENING PRESENCE

## DETERRENCE STRATEGIES

*The best way to fight off aggressors is to keep them from attacking you in the first place. To accomplish this you must create the impression of being more powerful than you are. Build up a reputation: You're a little crazy. Fighting you is not worth it. You take your enemies with you when you lose. Create this reputation and make it credible with a few impressive—impressively violent—acts. Uncertainty is sometimes better than overt threat: if your opponents are never sure what messing with you will cost, they will not want to find out. Play on people's natural fears and anxieties to make them think twice.*

*If your organization is small in numbers, then do what Gideon did: conceal the members in the dark but raise a din and clamor that will make the listener believe that your organization numbers many more than it does. . . . Always remember the first rule of power tactics:* Power is not only what you have but what the enemy thinks you have.

*RULES FOR RADICALS*, SAUL D. ALINSKY, 1972

## REVERSE INTIMIDATION

Inevitably in life you will find yourself facing people who are more aggressive than you are—crafty, ruthless people who are determined to get what they want. Fighting them head-on is generally foolish; fighting is what they are good at, and they are unscrupulous to boot. You will probably lose. Trying to fend them off by giving them part of what they are after, or otherwise pleasing or appeasing them, is a recipe for disaster: you are only showing your weakness, inviting more threats and attacks. But giving in completely, surrendering without a fight, hands them the easy victory they crave and makes you resentful and bitter. It can also become a bad habit, the path of least resistance in dealing with difficult situations.

Instead of trying to avoid conflict or whining about the injustice of it all, consider an option developed over the centuries by military leaders and strategists to deal with violent and acquisitive neighbors: reverse intimidation. This art of deterrence rests on three basic facts about war and human nature: First, people are more likely to attack you if they see you as weak or vulnerable. Second, they cannot know for sure that you're weak; they depend on the signs you give out, through your behavior both present and past. Third, they are after easy victories, quick and bloodless. That is why they prey on the vulnerable and weak.

Deterrence is simply a matter of turning this dynamic around, altering any perception of yourself as weak and naïve and sending the message that battle with you will not be as easy as they had thought. This is generally done by taking some visible action that will confuse aggressors and make them think they have misread you: you may indeed be vulnerable, but they are not sure. You're disguising your weakness and distracting them. Action has much more credibility than mere threatening or fiery words; hitting back, for instance, even in some small, symbolic way, will show that you mean what you say. With so many other people around who are timid and easy prey, the aggressor will most likely back off and move on to someone else.

This form of defensive warfare is infinitely applicable to the battles of daily life. Appeasing people can be as debilitating as fighting them; deterring them, scaring them out of attacking you or getting in your way, will save you valuable energy and resources. To deter aggressors you must become adept at deception, manipulating appearances and their perceptions of you—valuable skills that can be applied to all aspects of daily warfare. And finally, by practicing the art as needed, you will build for yourself a reputation as someone tough, someone worthy of respect and a little fear. The passive-aggressive obstructionists who try to undermine you covertly will also think twice about taking you on.

The following are five basic methods of deterrence and reverse intimidation. You can use them all in offensive warfare, but they are particularly effective in defense, for moments when you find yourself vulnerable and under attack. They are culled from the experiences and writings of the greatest masters of the art.

**Surprise with a bold maneuver.** The best way to hide your weakness and to bluff your enemies into giving up their attack is to take some unexpected, bold, risky action. Perhaps they had thought you were vulnerable, and now you are acting as someone who is fearless and confident. This will have two positive effects: First, they will tend to think your move is backed up by something real—they will not imagine you could be foolish enough to do something audacious just for effect. Second, they will start to see strengths and threats in you that they had not imagined.

**Reverse the threat.** If your enemies see you as someone to be pushed around, turn the tables with a sudden move, however small, designed to scare *them*. Threaten something they value. Hit them where you sense they may be vulnerable, and make it hurt. If that infuriates them and makes them attack you, back off a moment and then hit them again when they're not expecting it. Show them you are not afraid of them and that you are capable of a ruthlessness they had not seen in you. You needn't go too far; just inflict a little pain. Send a short, threatening message to indicate that you are capable of a lot worse.

**Seem unpredictable and irrational.** In this instance you do something suggesting a slightly suicidal streak, as if you felt you had nothing to lose. You show that you are ready to take your enemies down with you, destroying their reputations in the process. (This is particularly effective with people who have a lot to lose themselves—powerful people with sterling reputations.) To defeat you will be costly and perhaps self-destructive. This will make fighting you very unattractive. You are not acting out emotionally; that is a sign of weakness. You are simply hinting that you are a little irrational and that your next move could be almost anything. Crazy opponents are terrifying—no one likes fighting people who are unpredictable and have nothing to lose.

**Play on people's natural paranoia.** Instead of threatening your opponents openly, you take action that is indirect and designed to make them think. This might mean using a go-between to send them a message—to tell some disturbing story about what you are capable of. Or maybe you "inadvertently" let them spy on you, only to hear something that should give them cause for concern. Making your enemies think they have found out you are plotting a countermove is more effective than telling them so yourself; make a threat and you may have to live up to it, but making them think you are working treacherously against them is another story. The more veiled menace and uncertainty you generate, the more their imaginations will run away with them and the more dangerous an attack on you will seem.

**Establish a frightening reputation.** This reputation can be for any number of things: being difficult, stubborn, violent, ruthlessly efficient.

*A certain person said the following. There are two kinds of dispositions, inward and outward, and a person who is lacking in one or the other is worthless. It is, for example, like the blade of a sword, which one should sharpen well and then put in its scabbard, periodically taking it out and knitting one's eyebrows as in an attack, wiping off the blade, and then placing it in its scabbard again. If a person has his sword out all the time, he is habitually swinging a naked blade; people will not approach him and he will have no allies. If a sword is always sheathed, it will become rusty, the blade will dull, and people will think as much of its owner.*

*HAGAKURE: THE BOOK OF THE SAMURAI*, YAMAMOTO TSUNETOMO, 1659–1720

*Brinkmanship is . . . the deliberate creation of a recognizable risk, a risk that one does not completely control. It is the tactic of deliberately letting the situation get somewhat out of hand, just because its being out of hand may be intolerable to the other party and force his accommodation. It means harassing and intimidating an adversary by exposing him to a shared risk, or deterring him by showing that if he makes a contrary move he may disturb us so that we slip over the brink whether we want to or not, carrying him with us.*

*THINKING STRATEGICALLY*, AVINASH K. DIXIT AND BARRY J. NALEBUFF, 1991

Build up that image over the years and people will back off from you, treating you with respect and a little fear. Why obstruct or pick an argument with someone who has shown he will fight to the bitter end? Someone strategic yet ruthless? To create this image, you may every now and then have to play a bit rough, but eventually it will become enough of a deterrent to make those occasions rare. It will be an offensive weapon, scaring people into submission before they even meet you. In any event, you must build your reputation carefully, allowing no inconsistencies. Any holes in this kind of image will make it worthless.

*Injuring all of a man's ten fingers is not as effective as chopping off one.*
—*Mao Tse-tung (1893–1976)*

## DETERRENCE AND REVERSE INTIMIDATION IN PRACTICE

**1.** In March 1862, less than a year after the start of the American Civil War, the Confederates' situation looked bleak: they had lost a series of important battles, their generals were squabbling, morale was low, and recruits were hard to find. Sensing the South's great weakness, a large Union army under Major General George B. McClellan headed toward the Virginia coast, planning to march from there west to Richmond, the capital of the South. There were enough Confederate troops in the area to hold off McClellan's army for a month or two, but Southern spies reported that Union troops stationed near Washington were about to be transferred to the march on Richmond. If these troops reached McClellan—and they were promised by Abraham Lincoln himself—Richmond would be doomed; and if Richmond fell, the South would have to surrender.

The Confederate general Stonewall Jackson was based in Virginia's Shenandoah Valley at the head of 3,600 men, a ragtag group of rebels he had recruited and trained. His job was merely to defend the fertile valley against a Union army in the area, but as he pondered the developing campaign against Richmond, he saw the possibility of something much greater. Jackson had been a classmate of McClellan's at West Point and knew that underneath his brash, talkative exterior he was basically timid, overly anxious about his career and making any mistakes. McClellan had 90,000 men ready for the march on Richmond, almost double the available Confederate forces, but Jackson knew that this cautious man would wait to fight until his army was overwhelming; he wanted the extra troops that Lincoln had promised him. Lincoln, however, would not release those forces if he saw danger elsewhere. The Shenandoah Valley was to the southwest of Washington. If Jackson could possibly create enough confusion as to what was happening there, he could disrupt the Union plans and perhaps save the South from disaster.

On March 22, Jackson's spies reported that two-thirds of the Union army stationed in the Shenandoah Valley, under General Nathaniel Banks, was heading east to join McClellan. Soon an army near Washington, led by General Irvin McDowell, would move toward Richmond as well. Jackson wasted no time: he marched his men fast to the north to attack the Union soldiers still in the valley, near Kernstown. The battle was fierce, and at the end of the day Jackson's soldiers were forced to retreat. To them the engagement seemed to have been a defeat, even a disaster: outnumbered nearly two to one, they had suffered terrible casualties. But Jackson, always a hard man to figure out, seemed oddly satisfied.

A few days later, Jackson received the news he had been waiting for: Lincoln had ordered Banks's army to return to the valley and McDowell's army to stay where it was. The battle at Kernstown had gotten his attention and made him worry—only a little, but enough. Lincoln did not know what Jackson was up to or how large his army was, but he wanted the Shenandoah Valley pacified no matter what. Only then would he release Banks and McDowell. McClellan was forced to agree with that logic, and although he had the men to march on Richmond right away, he wanted to wait for the reinforcements who would make the attack a sure thing.

After Kernstown, Jackson retreated south, away from Banks, and lay low for a few weeks. In early May, thinking that the Shenandoah Valley had been secured, Lincoln sent McDowell toward Richmond, and Banks prepared to join him. Again Jackson was ready: he marched his army in a completely bizarre fashion, first to the east, toward McDowell, then back west into the Valley. Not even his own soldiers knew what he was doing. Mystified by these strange maneuvers, Lincoln imagined—but wasn't sure—that Jackson was marching to fight McDowell. Once again he halted McDowell's march south, kept half of Banks's army in the valley, and sent the other half to help McDowell defend himself against Jackson.

Suddenly the Union's plans, which had seemed so perfect, were in disarray, its troops too scattered to support each other. Now Jackson went in for the kill: he linked up with other Confederate divisions in the area and, on May 24, marched on the Union army—now divided and dangerously diminished—that remained in the valley. Jackson maneuvered onto its flank and sent it in headlong retreat north to the Potomac River. His pursuit of this army sent a wave of panic through Washington: this now dreaded general, commanding forces that seemed to have doubled in size overnight, was heading straight for the capital.

Secretary of War Edwin Stanton telegraphed Northern governors to alert them to the threat and to muster troops for the city's defense. Reinforcements quickly arrived to halt the Confederate advance. Meanwhile Lincoln, determined to eliminate Jackson once and for all, ordered half of McDowell's army west to join in the fight to destroy this pest and the other half to return to Washington to secure the capital. McClellan could only agree.

*One classic response to a particularly vicious beanball was exemplified by a play Jackie Robinson made in the summer of 1953. Sal Maglie of the New York Giants was "Sal the Barber," mostly because his high inside fast balls "shaved" hitters' chins. Maglie was candid and friendly when he wasn't pitching. "You have to make the batter afraid of the ball or, anyway, aware that he can get hurt," Maglie told me matter-of-factly one afternoon over drinks at his apartment in Riverdale. "A lot of pitchers think they do that by throwing at a hitter when the count is two strikes and no balls. The trouble there is that the knockdown is expected. You don't scare a guy by knocking him down when he knows he's going to be knocked down." "Then when, Sal?" I asked. "A good time is when the count is two and two. He's looking to swing. You knock him down then and he gets up shaking. Now curve him and you have your out. Of course, to do that you have to be able to get your curve over the plate on a three-and-two count. Not every pitcher can." Maglie could break three different curves over the plate, three and two. He had particular success against such free-swinging sluggers as Roy Campanella and Gil Hodges. But it is simplistic to say Maglie intimidated Campanella and Hodges. Rather, his unpredictable patterns disrupted their timing and concentration. He had less success with Pee Wee Reese and Jackie Robinson, and one day in Ebbets Field, by throwing a shoulder-high fast ball behind Robinson, Maglie brought matters to detonation. The knockdowns thrown at [Cookie] Lavagetto, the fatal pitch thrown at Ray Chapman, roared toward the temple. A batter gets away from that pitch by ducking backward. (Chapman's freeze reaction, though not unknown, is rare.) Angered or frustrated by Robinson that afternoon in Brooklyn, Maglie threw his best fast ball behind the hitter, shoulder high. That was and is dangerous and inexcusable. As a batter strides forward, he loses height. Reflex makes him duck backward. A batter's head moves directly into the path of the fast ball thrown behind him shoulder high. Robinson started to duck into Maglie's pitch and then his phenomenal reflexes enabled him to stop, as it were, in mid-duck. The ball sailed just behind the back of Robinson's neck. Robinson glared but did not lose his poise. Maglie threw an outside curve, and Robinson bunted toward Whitey Lockman, the Giant's first baseman. By making Lockman field the bunt, Robinson was forcing Maglie to leave the pitcher's mound*

Once again Jackson retreated, but by now his plan had worked to perfection. In three months, with only 3,600 men, he had diverted well over 60,000 Northern troops, bought the South enough time to coordinate the defense of Richmond, and completely altered the course of the war.

## Interpretation

The story of Stonewall Jackson in the Shenandoah Valley illustrates a simple truth: what matters in war, as in life generally, is not necessarily how many men you have or how well supplied you are but how your enemies see you. If they think you are weak and vulnerable, they act aggressively, which in and of itself can put you in trouble. If they suddenly think you are strong, or unpredictable, or have hidden resources, they back off and reassess. Getting them to change their plans and treat you more carefully can by itself alter the war. In any struggle, some things will be outside your control; you may not be able to put together a large army or defend all your weak points, but you can always affect people's perceptions of you.

Jackson altered Union perceptions first by his bold attack on Kernstown, which made Lincoln and McClellan think he had more troops than he did—they could not imagine that anyone would be so stupid as to send only 3,600 men against a Union stronghold. If Jackson was stronger than they had imagined, that meant they needed more men in the Shenandoah Valley, which cut into the troops available for the march on Richmond. Next Jackson began behaving unpredictably, creating the impression of having not only a large army but also some strange and worrying plan. Lincoln's and McClellan's inability to figure out this plan stopped them in their tracks, making them divide their forces to take care of the possible dangers. Finally Jackson attacked boldly one more time. He did not have nearly enough men to threaten Washington, but Lincoln could not be sure of that. Like a conjuror, Jackson created a bogeyman out of an army that in essence was laughably small.

You must take control over people's perceptions of you by playing with appearances, mystifying and misleading them. Like Jackson, it is best to mix audacity with unpredictability and unorthodoxy and act boldly in moments of weakness or danger. That will distract people from any holes in your armor, and they'll be afraid there may be more to you than meets the eye. Then, if you make your behavior hard to read, you'll only seem more powerful, since actions that elude interpretation attract attention, worry, and a bit of awe. In this way you will throw people off balance and onto their heels. Kept at a distance, they will be unable to tell how far you are bluffing them. Aggressors will back off. Appearance and perception—you are not someone to mess with—will become reality.

**2.** King Edward I of England was a fierce thirteenth-century warrior-king who was determined to conquer all of the British Isles. First he battered the Welsh into submission; then he set his sights on Scotland, laying siege to towns and castles and razing to the ground the communities that dared to resist him. He was even more brutal with the Scots who fought back, including the famous Sir William Wallace: he hunted them down and had them publicly tortured and executed.

Only one Scottish lord eluded Edward: Robert the Bruce, Earl of Carrick (1274–1329), who had somehow escaped to the remote fastness of northern Scotland. So Edward captured the rebel's family and friends, killing the men and imprisoning the women in cages. Bruce remained defiant. In 1306 he had himself crowned Scotland's king; whatever it took, he vowed to revenge himself on Edward and throw the English out of Scotland. Hearing this, Edward became even more determined to capture this final piece in his Scottish wars, but in 1307 he died, before the job was done.

Edward's son, now Edward II, did not share his father's lust for war. Edward I had left the island secure. The new king did not have to worry about Scotland; England was far wealthier, and its armies were well equipped, well fed, well paid, and experienced. In fact, their recent wars had made them the most-feared fighters in Europe. At any moment Edward II could field a great army against the Scots, whose weapons and armor were primitive. He felt confident that he could handle Robert the Bruce.

A few months into the reign of Edward II, Bruce managed to take some Scottish castles held by the English and burn them to the ground. When Edward sent forces against him, Bruce refused to fight and fled with his small army into the forest. Edward sent more men to secure his remaining strongholds in Scotland and exact revenge on Bruce, but now Scots soldiers suddenly began to raid England. Highly mobile, these pirates on horseback devastated the northern English countryside, destroying crops and livestock. The English campaign in Scotland had become too costly, so it was called off—but a few years later Edward tried again.

This time an English army penetrated farther into Scotland, but again, in response, Scottish raiders rode south into England, wreaking still more havoc on farms and property. And in Scotland itself Bruce's army burned their own countrymen's crops, leaving the English invaders nothing to eat. As before, the English wore themselves out chasing Bruce, but to no avail—the Scots refused battle. Bivouacked in their camps, the English soldiers would hear bagpipes and horns out in the dark at night, making it impossible to sleep. Hungry, tired, and irritated to no end, they soon retreated back to northern England, only to find their own land barren of crops and cattle. Morale sank. No one wanted to fight in Scotland anymore. Slowly one castle after another fell back into Scottish hands.

In 1314 the Scots finally engaged in direct combat with the English, at the Battle of Bannockburn, and defeated them. It was a most humiliating loss for Edward II, who swore to avenge it. In 1322 he decided to

*and cover first. There he would be in Robinson's path, and Jack, going at full and full-muscled tilt, intended to run over Maglie, signing his name in spikes on the pitcher's spine. Saturnine, Faustian, brooding Sal Maglie refused to leave the mound. At a critical moment, the Barber lost his nerve. Davey Williams, the Giants' second baseman, rushed over, and as he was reaching for Lockman's throw, Robinson crashed into him, a knee catching Williams in the lower back. Robinson's knee was so swollen a day later that he could not play. Williams never really recovered. He dropped out of the major leagues two seasons later, at twenty-eight. . . . "Actually," Robinson himself said a few days later, "I'm sorry that Williams got hurt. But when Maglie threw behind me, he was starting a really dangerous business, and I was going to put a stop to it before he hit Gil or Campy or Pee Wee in the head. . . ."*
*After that I saw Maglie start eight games against the Dodgers, but I never saw him throw another fast ball behind a hitter. The grim, intimidating beanballer had been intimidated himself, and by a bunt.*

*The Head Game*, Roger Kahn, 2000

finish Bruce off once and for good with a vigorous campaign worthy of his father. Organizing and personally leading the largest army yet to fight the rebellious Scots, Edward got as far as Edinburgh Castle. At one point he sent foragers out to look for food in the countryside; they returned with a single decrepit bull and an empty wagon. Dysentery swept the English troops. Edward was forced to retreat, and when he reached northern England, he saw that the Scots had once again razed the fields there, and more thoroughly than ever. Hunger and disease finished off the remnants of his army. The campaign was such a disaster that a rebellion broke out among Edward's lords: he fled but in 1327 was captured and killed.

The following year Edward's son, Edward III, negotiated a peace with the Scots, granting Scotland its independence and recognizing Robert the Bruce as its rightful king.

*Another anecdote explaining* iwao-no-mi *concerns an accomplished warrior who had reached the highest stage of the art of sword fighting. Having been enlightened as to the true meaning of the art of sword fighting, which should be based on the promotion of well-being of people rather than the destruction or killing of others, this great master was not interested in fighting any longer. His ability in the art of sword fighting was absolutely unquestionable; he was respected and feared by everyone. He walked the streets with a cane like a bored old man and yet wherever he went people looked at him with intense fear and respect. People were careful not to anger him and the old man was nonchalant. This is akin to having a huge rock hanging above a mountain path. People are afraid of the rock, which they believe may come down at any moment, and so they walk quietly and carefully under the rock. But the rock is actually very stable, being planted in the ground so deeply that it will never fall down. But people do not know it, and they continue to fear that it will fall down if they make any kind of loud noise as they walk under it. The rock just sits there completely indifferent to its surroundings and people's fear and awe.*

*A Way to Victory: The Annotated Book of Five Rings*, translated and commentary by Hidy Ochiai, 2001

### Interpretation

The English thought they could move on Scotland with impunity anytime they wanted. The Scots were poorly equipped, and their leadership was bitterly divided: seeing such weakness, what could prevent English conquest? Trying to stop what seemed inevitable, Robert the Bruce evolved a novel strategy. When the English attacked, he did not take them on directly; he would have lost. Instead he hit them indirectly but where it hurt, doing exactly to the English what they were doing to him: ruining his country. He continued to play tit for tat until the English understood that every time they attacked Scotland, they would get a bloody nose in exchange: they would lose valuable farmland, be harassed, fight in abysmal conditions. They slowly lost their hunger for the fight, then finally gave up.

The essence of this deterrence strategy is the following: when someone attacks you or threatens you, you make it clear that he will suffer in return. He—or she—may be stronger, he may be able to win battles, but you will make him pay for each victory. Instead of taking him on directly, you hurt something he values, something close to home. You make him understand that every time he bothers you he can expect damage, even if on a smaller scale. The only way to make you stop attacking him in your irritating fashion is for him to stop attacking you. You are like a wasp on his skin: most people leave wasps alone.

**3.** One morning in 1474, King Louis XI (1423–83)—France's infamous "Spider King," so named because he always wove the most intricate and well-conceived plots against his enemies—went into a vehement rant against the Duke of Milan. The courtiers present that January day listened in amazement as the normally composed and careful king spun out his suspicions: although the duke's father had been a friend, the son could not be trusted; he was working against France, breaking the treaty between the two countries. On and on the king went: perhaps he would have to take action against the duke. Suddenly, to the courtiers' dismay, a man slipped quietly out of the room. It was Christopher da Bollate, the Milanese ambassador to France. Bollate had been received graciously by the king earlier that morning but then had retreated into the background; Louis must have forgotten he was there. The king's diatribe could cause quite a diplomatic mess.

Later that day Louis invited Bollate to his private rooms and, lounging on his bed, began an apparently casual conversation. Drifting into politics, he described himself as a supporter of the Duke of Milan's: he would do anything, he said, to help the duke expand his power. Then he asked, "Tell me, Christopher, has it been reported to you what I said this morning in council? Tell me the truth—was it not some courtier who told you?" Bollate confessed that he had actually been in the room during the king's tirade and had heard the king's words himself. He also protested that the Duke of Milan was a loyal friend of France. Louis replied that he had his doubts about the duke and had cause to be angry—but then he immediately changed the subject to something pleasant, and Bollate eventually left.

The next day the king sent three councilors to visit Bollate. Was he comfortable in his lodgings? Was he happy with his treatment from the king? Was there anything they could do to improve his stay at the French court? They also wanted to know if he was going to pass on the king's words to the duke. The king, they said, considered Bollate a friend, a confidant; he had merely been venting his emotions. It meant nothing. Bollate should forget the whole thing.

Of course, none of these men—the councilors, the courtiers, Bollate—knew that the king had done all this deliberately. Louis was certain that the perfidious ambassador—whom he hardly considered a friend, let alone a confidant—would report what he had said in detail to the duke. He knew that the duke was treacherous, and this was precisely how Louis wanted to send him a warning. And it seemed the message got through: for the next several years, the duke was an obedient ally.

*Once, when a group of five or six pages were traveling to the capital together in the same boat, it happened that their boat struck a regular ship late at night. Five or six seamen from the ship leapt aboard and loudly demanded that the pages give up their boat's anchor, in accord with the seaman's code. Hearing this, the pages ran forward yelling, "The seaman's code is something for people like you! Do you think that we samurai are going to let you take equipment from a boat carrying warriors? We will cut you down and throw you into the sea to the last man!" With that, all the seamen fled back to their own ship. At such a time, one must act like a samurai. For trifling occasions it is better to accomplish things simply by yelling. By making something more significant than it really is and missing one's chance, an affair will not be brought to a close and there will be no accomplishment at all.*

*HAGAKURE: THE BOOK OF THE SAMURAI*, YAMAMOTO TSUNETOMO, 1659–1720

## Interpretation

The Spider King was a man who always plotted several moves in advance. In this case he knew that if he spoke politely and diplomatically to the ambassador of his worries about the duke, his words would carry no weight—they would seem like whining. If he vented his anger directly to the ambassador, on the other hand, he would look out of control. A direct thrust is also easily parried: the duke would just mouth reassurances, and the treachery would go on. By transmitting his threat indirectly, however, Louis made it stick. That the duke was not meant to know he was angry made his anger truly ominous: it meant he was planning something and wanted to keep the duke from suspecting it and knowing his true feelings. He delivered his threat insidiously to make the duke ponder his intentions and to instill an uneasy fear.

*It was thus that, during the 1930s, the diplomacy of Mussolini's Italy was greatly enhanced by a stance of restless bellicosity and by a mirage of great military strength: an army of "eight million bayonets," whose parades were dashing affairs of* bersaglieri *on the run and roaring motorized columns; and an air force greatly respected, not least for its spectacular long-range flights to the North Pole and South America; and a navy that could acquire many impressive ships because so little of its funding was wasted on gunnery trials and navigation. By a military policy in which stage management dominated over the sordid needs of war preparation, Mussolini sacrificed real strength for the sake of hugely magnified images of what little strength there was—but the results of suasion that those images evoked were very real: Britain and France were both successfully dissuaded from interfering with Italy's conquest of Ethiopia, its intervention in Spain, and the subjection of Albania; and none dared oppose Italy's claim to be accepted as a Great Power, whose interests had to be accommodated sometimes in tangible ways such as the licenses obtained by Italian banks in Bulgaria, Hungary,*

When we are under attack, the temptation is to get emotional, to tell the aggressors to stop, to make threats as to what we'll do if they keep going. That puts us in a weak position: we've revealed both our fears and our plans, and words rarely deter aggressors. Sending them a message through a third party or revealing it indirectly through action is much more effective. That way you signal that you are already maneuvering against them. Keep the threat veiled: if they can only glimpse what you are up to, they will have to imagine the rest. Making them see you as calculating and strategic will have a chilling effect on their desires to harm or attack you. It is not worth the risk to find out what you may be up to.

**4.** In the early 1950s, John Boyd (1927–97) served with distinction as a fighter pilot in the Korean War. By the middle of that decade, he was the most respected flight instructor at Nellis Air Force Base in Nevada; he was virtually unbeatable in practice dogfights, so good that he was asked to rewrite the manual on fighter-pilot tactics. He had developed a style that would demoralize and terrorize, get inside the opponent's head, disrupt his ability to react. Boyd was clever and fearless. But none of his training and skill, none of his brushes with death as a pilot, prepared him for the bloodless backstabbing, political maneuvering, and indirect warfare of the Pentagon, where he was assigned in 1966 to help design lightweight jet fighters.

As Major Boyd quickly discovered, Pentagon bureaucrats were more concerned with their careers than with national defense. They were less interested in developing the best new fighter than in satisfying contractors, often buying their new technological gear regardless of its suitability. Boyd, as a pilot, had trained himself to see every situation as a kind of strategic combat, and in this instance he decided to transfer his skills and style of warfare to the jungles of the Pentagon. He would intimidate, discourage, and outsmart his opponents.

Boyd believed that a streamlined jet fighter of the kind he was designing could outperform any plane in the world. But contractors hated his design, because it was inexpensive—it did not highlight the technology they were trying to peddle. Meanwhile Boyd's colleagues in the Pentagon had their own pet projects. Competing for the same pot of money, they did everything they could to sabotage or transform his design.

Boyd developed a defense: Outwardly he looked a little dumb. He wore shabby suits, smoked a nasty cigar, kept a wild look in his eye. He seemed to be just another emotional fighter pilot, promoted too fast and too soon. But behind the scenes he mastered every detail. He made sure he knew more than his opponents: he could quote statistics, studies, and engineering theories to support his own project and poke holes through theirs. Contractors would show up in meetings with glossy presentations delivered by their top engineers; they would make fantastic claims to dazzle the generals. Boyd would listen politely, seem impressed, and then suddenly, without warning, he would go on the offensive—deflating their optimistic claims, showing in detail that the numbers did not add up, revealing the hype and the fakery. The more they protested, the more vicious Boyd got, bit by bit tearing their project to shreds.

Blindsided by a man they had grossly underestimated, time and again the contractors would leave these meetings vowing revenge. But what could they do? He had already shot down their numbers and turned their proposals to mush. Caught in the act of oversell, they had lost all credibility. They would have to accept their defeat. Soon they learned to avoid Boyd: instead of trying to sabotage him, they hoped he would fail on his own.

In 1974, Boyd and his team had finished the design of a jet they had been working on, and it seemed certain to be approved. But part of Boyd's strategy had been to build up a network of allies in different parts of the Pentagon, and these men told him that there was a group of three-star generals who hated the project and were planning his defeat. They would let him brief the various officials in the chain of command, all of whom would give him their go-ahead; then there would be a final meeting with the generals, who would scuttle the project as they had planned to all along. Having gotten that far, though, the project would look as if it had been given a fair hearing.

In addition to his network of allies, Boyd always tried to make sure he had at least one powerful supporter. This was usually easy to find: in a political environment like the Pentagon, there was always some general or other powerful official who was disgusted with the system and was happy to be Boyd's secret protector. Now Boyd called on his most powerful ally, Secretary of Defense James Schlesinger, and won Schlesinger's personal approval for the project. Then, at the meeting with the generals, whom he could tell were inwardly gloating that they finally had him, Boyd announced, "Gentlemen, I am authorized by the secretary of defense to inform you that this is not a decision brief. This briefing is for information purposes only." The project, he said, had already been approved. He went on to deliver his presentation, making it as long as possible—twisting the knife in their backs. He wanted them to feel humiliated and wary of messing with him again.

As a fighter pilot, Boyd had trained himself to think several moves ahead of his opponents, always aiming to surprise them with some terrifying maneuver. He incorporated this strategy into his bureaucratic battles. When a general gave him some order that was clearly designed to ruin the plans for his lightweight jet, he would smile, nod, and say, "Sir, I'll be happy to follow that order. But I want you to put it in writing." Generals liked to issue commands verbally rather than putting them on paper as a way to cover themselves in case things went bad. Caught off guard, the general would either have to drop the order or deny the request to put it in writing—which, if publicized, would make him look terrible. Either way he was trapped.

*Romania, and Yugoslavia). Only Mussolini's last-minute decision to enter the war in June 1940—when his own considerable prudence was overcome by the irresistible temptation of sharing in the spoils of the French collapse—brought years of successful deception (and self-deception) to an end.*

*Strategy: The Logic of War and Peace*, Edward N. Luttwak, 1987

After several years of dealing with Boyd, generals and their minions learned to avoid him—and his foul cigars, his verbal abuse, his knife-twisting tactics—like the plague. Given this wide berth, he was able to push his designs for the F-15 and F-16 through the Pentagon's almost impossible process, leaving an enduring imprint on the air force by creating two of its most famous and effective jet fighters.

### Interpretation

Boyd realized early on that his project was unpopular at the Pentagon and that he would meet opposition and obstruction up and down the line. If he tried to fight everyone, to take on every contractor and general, he would exhaust himself and go down in flames. Boyd was a strategist of the highest order—his thinking would later have a major influence on Operation Desert Storm—and a strategist never hits strength against strength; instead he probes the enemy's weaknesses. And a bureaucracy like the Pentagon inevitably has weaknesses, which Boyd knew how to locate.

The people in Boyd's Pentagon wanted to fit in and be liked. They were political people, careful about their reputations; they were also very busy and had little time to waste. Boyd's strategy was simple: over the years he would establish a reputation for being difficult, even nasty. To get involved with Boyd could mean an ugly public fight that would sully your reputation, waste your time, and hurt you politically. In essence Boyd transformed himself into a kind of porcupine. No animal wants to take on a creature that can do so much damage, no matter how small it is; even tigers will leave it alone. And being left alone gave Boyd staying power, allowing him to survive long enough to shepherd the F-15 and F-16 through.

Reputation, Boyd knew, is key. Your own reputation may not be intimidating; after all, we all have to fit in, play politics, seem nice and accommodating. Most often this works fine, but in moments of danger and difficulty being seen as so nice will work against you: it says that you can be pushed around, discouraged, and obstructed. If you have never been willing to fight back before, no threatening gesture you make will be credible. Understand: there is great value in letting people know that when necessary you can let go of your niceness and be downright difficult and nasty. A few clear, violent demonstrations will suffice. Once people see you as a fighter, they will approach you with a little fear in their hearts. And as Machiavelli said, it is more useful to be feared than to be loved.

**Image:**
The Porcupine. It seems rather stupid and slow, easy prey, but when it is threatened or attacked, its quills stand erect. If touched, they come out easily in your flesh, and trying to extract them makes their hooked ends go deeper and deeper, causing still more damage. Those who have fought with a porcupine learn never to repeat the experience. Even without fighting it, most people know to avoid it and leave it in peace.

**Authority:** When opponents are unwilling to fight with you, it is because they think it is contrary to their interests, or because you have misled them into thinking so. —*Sun-tzu (fourth century* B.C.*)*

## REVERSAL

The purpose of strategies of deterrence is to discourage attack, and a threatening presence or action will usually do the job. In some situations, though, you can more safely achieve the same thing by doing the opposite: play dumb and unassuming. Seem inoffensive, or already defeated, and people may leave you alone. A harmless front can buy you time: that is how Claudius survived the violent, treacherous world of Roman politics on his way to becoming emperor—he seemed too innocuous to bother with. This strategy needs patience, though, and is not without risk: you are deliberately making yourself the lamb among the wolves.

In general, you have to keep your attempts at intimidation under control. Be careful not to become intoxicated by the power fear brings: use it as a defense in times of danger, not as your offense of choice. In the long run, frightening people creates enemies, and if you fail to back up your tough reputation with victories, you will lose credibility. If your opponent gets angry enough to decide to play the same game back at you, you may also escalate a squabble into a retaliatory war. Use this strategy with caution.

# 11

# TRADE SPACE FOR TIME

## THE NONENGAGEMENT STRATEGY

*Retreat in the face of a strong enemy is a sign not of weakness but of strength. By resisting the temptation to respond to an aggressor, you buy yourself valuable time—time to recover, to think, to gain perspective. Let your enemies advance; time is more important than space. By refusing to fight, you infuriate them and feed their arrogance. They will soon overextend themselves and start making mistakes. Time will reveal them as rash and you as wise. Sometimes you can accomplish most by doing nothing.*

## RETREAT TO ADVANCE

In the early 1930s, Mao Tse-tung (1893–1976) was a rising star in the Chinese Communist Party. A civil war had broken out between the Communists and the Nationalists; Mao led campaigns against the Nationalists, using guerrilla tactics to beat them time and again, despite being greatly outnumbered. He also served as the chairman of the fledgling Chinese Communist government, and his provocative essays on strategy and philosophy were widely read.

Then a power struggle broke out among the Communists: a group of Soviet-educated intellectuals known as the 28 Bolsheviks tried to gain control of the party. They despised Mao, seeing his taste for guerrilla warfare as a sign of timidity and weakness and his advocacy of a peasant revolution backward. Instead they advocated frontal warfare, fighting the Nationalists directly for control of key cities and regions, as the Communists had done in Russia. Slowly the 28B isolated Mao and stripped him of both political and military power. In 1934 they put him under virtual house arrest on a farm in Hunan.

Mao's friends and comrades felt he had suffered a dizzying fall from grace. But more troubling than the fall itself was his apparent acceptance of it: he did not rally supporters to fight back, he stopped publishing, he effectively disappeared. Perhaps the 28B had been right: Mao was a coward.

That same year the Nationalists—led by General Chiang Kai-shek—launched a new campaign to destroy the Communists. Their plan was to encircle the Red Army in its strongholds and kill every last soldier, and this time they seemed likely to succeed. The 28B fought back bravely, battling to hold on to the few cities and regions under Communist control, but the Nationalists outnumbered them, were better equipped, and had German military advisers to help them. The Nationalists took city after city and slowly surrounded the Communists.

Thousands deserted the Red Army, but finally its remaining soldiers—around 100,000 of them—managed to break out of the Nationalist encirclement and head northwest. Mao joined them in their flight. Only now did he begin to speak up and question the 28B strategy. They were retreating in a straight line, he complained, making it easier for the Nationalists to chase them, and they were moving too slowly, carrying too many documents, file cabinets, and other trappings from their old offices. They were acting as if the whole army were merely moving camp and planning to keep fighting the Nationalists in the same way, fighting over cities and land. Mao argued that this new march should not be a momentary retreat to safer ground, but something larger. The whole concept of the party needed rethinking: instead of copying the Bolsheviks, they should create a distinctly Chinese revolution based on the peasantry, China's single largest population group. To accomplish this they needed time and freedom from attack. They should head southwest, to the farthest reaches of China, where the enemy could not reach them.

Red Army officers began to listen to Mao: his guerrilla tactics had been successful before, and the 28B strategy was clearly failing. They slowly adopted his ideas. They traveled more lightly; they moved only at night; they feinted this way and that to throw the Nationalists off their scent; wherever they went, they conducted rallies to recruit peasants to their cause. Somehow Mao had become the army's de facto leader. Although outnumbered a hundred to one, under his leadership the Red Army managed to escape the Nationalists and, in October 1935, to arrive at the remote reaches of Shan-hsi Province, where it would finally be safe.

After crossing twenty-four rivers and eighteen mountain ranges and having many near misses with disaster, the army came to the end of its "Long March." It was radically reduced—it now numbered only 6,000—but a new kind of party had been forged, the kind Mao had wanted all along: a hard-core group of devoted followers who believed in a peasant revolution and embraced guerrilla warfare. Safe from attack in Shan-hsi, this purified party was slowly able first to recover, then to spread its gospel. In 1949 the Communists finally defeated the Nationalists for good and exiled them from mainland China.

*Six in the fourth place means:*
*The army retreats. No blame.*
*In face of a superior enemy, with whom it would be hopeless to engage in battle, an orderly retreat is the only correct procedure, because it will save the army from defeat and disintegration. It is by no means a sign of courage or strength to insist upon engaging in a hopeless struggle regardless of circumstances.*

*THE I CHING*, CHINA, CIRCA EIGHTH CENTURY B.C.

### Interpretation

Mao was born and raised on a farm, and Chinese farm life could be harsh. A farmer had to be patient, bending with the seasons and the capricious climate. Thousands of years earlier, the Taoist religion had emerged from this hard life. A key concept in Taoism is that of *wei wu*—the idea of action through inaction, of controlling a situation by not trying to control it, of ruling by abdicating rule. *Wei wu* involves the belief that by reacting and fighting against circumstances, by constantly struggling in life, you actually move backward, creating more turbulence in your path and difficulties for yourself. Sometimes it is best to lie low, to do nothing but let the winter pass. In such moments you can collect yourself and strengthen your identity.

Growing up on a farm, Mao had internalized these ideas and applied them constantly in politics and war. In moments of danger, when his enemies were stronger, he was not afraid to retreat, although he knew that many would see this as a sign of weakness. Time, he knew, would show up the holes in his enemies' strategy, and he would use that time to reflect on himself and gain perspective on the whole situation. He made his period of retreat in Hunan not a negative humiliation but a positive strategy. Similarly, he used the Long March to forge a new identity for the Communist Party, creating a new kind of believer. Once his winter had passed, he reemerged—his enemies succumbing to their own weaknesses, himself strengthened by a period of retreat.

War is deceptive: you may think that you are strong and that you are making advances against an enemy, but time may show that you were actually marching into great danger. You can never really know, since our immersion in the present deprives us of true perspective. The best you can do is to rid yourself of lazy, conventional patterns of thinking. Advancing is not always good; retreating is not always weak. In fact, in moments of danger or trouble, refusing to fight is often the best strategy: by disengaging from the enemy, you lose nothing that is valuable in the long run and gain time to turn inward, rethink your ideas, separate the true believers from the hangers-on. Time becomes your ally. By doing nothing outwardly, you gain inner strength, which will translate into tremendous power later, when it is time to act.

*Space I can recover. Time, never.*
*—Napoleon Bonaparte (1769–1821)*

## KEYS TO WARFARE

The problem we all face in strategy, and in life, is that each of us is unique and has a unique personality. Our circumstances are also unique; no situation ever really repeats itself. But most often we are barely aware of what makes us different—in other words, of who we really are. Our ideas come from books, teachers, all kinds of unseen influences. We respond to events routinely and mechanically instead of trying to understand their differences. In our dealings with other people, too, we are easily infected by their tempo and mood. All this creates a kind of fog. We fail to see events for what they are; we do not know ourselves.

Your task as a strategist is simple: to see the differences between yourself and other people, to understand yourself, your side, and the enemy as well as you can, to get more perspective on events, to know things for what they are. In the hubbub of daily life, this is not easy—in fact, the power to do it can come only from knowing when and how to retreat. If you are always advancing, always attacking, always responding to people emotionally, you have no time to gain perspective. Your strategies will be weak and mechanical, based on things that happened in the past or to someone else. Like a monkey, you will imitate instead of create. Retreating is something you must do every now and then, to find yourself and detach yourself from infecting influences. And the best time to do this is in moments of difficulty and danger.

Symbolically the retreat is religious, or mythological. It was only by escaping into the desert that Moses and the Jews were able to solidify their identity and reemerge as a social and political force. Jesus spent his forty days in the wilderness, and Mohammed, too, fled Mecca at a time of great peril for a period of retreat. He and just a handful of his most devoted supporters used this period to deepen their bonds, to understand who they were and what they stood for, to let time work its good. Then this little band of believers reemerged to conquer Mecca and the Arabian Peninsula and later, after Mohammed's death, to defeat the Byzantines and the Persian empire, spreading Islam over vast territories.

Around the world every mythology has a hero who retreats, even to Hades itself in the case of Odysseus, to find himself.

If Moses had stayed and fought in Egypt, the Jews would be a footnote in history. If Mohammed had taken on his enemies in Mecca, he would have been crushed and forgotten. When you fight someone more powerful than you are, you lose more than your possessions and position; you lose your ability to think straight, to keep yourself separate and distinct. You become infected with the emotions and violence of the aggressor in ways you cannot imagine. Better to flee and use the time your flight buys to turn inward. Let the enemy take land and advance; you will recover and turn the tables when the time comes. The decision to retreat shows not weakness but strength. It is the height of strategic wisdom.

The essence of retreat is the refusal to engage the enemy in any way, whether psychologically or physically. You may do this defensively, to protect yourself, but it can also be a positive strategy: by refusing to fight aggressive enemies, you can effectively infuriate and unbalance them.

*Opportunities are changing ceaselessly. Those who get there too early have gone too far, while those who get there too late cannot catch up. As the sun and moon go through their courses, time does not go along with people. Therefore, sages do not value huge jewels as much as they value a little time. Time is hard to find and easy to lose.*

*Huainanzi*, China, second century B.C.

During World War I, England and Germany fought a side war in East Africa, where each of them had a colony. In 1915 the English commander, Lieutenant General Jan Smuts, moved against the much smaller German army in German East Africa, led by Colonel Paul von Lettow-Vorbeck. Smuts was hoping for a quick win; as soon as he had finished off the Germans, his troops could move to more important theaters of war. But von Lettow-Vorbeck refused to engage him and retreated south. Smuts marched in pursuit.

Time and again Smuts thought he had von Lettow-Vorbeck cornered, only to find that the German officer had moved on just hours earlier. As if a drawn by a magnet, Smuts followed von Lettow-Vorbeck across rivers, mountains, and forests. Their supply lines extended over hundreds of miles, his soldiers were now vulnerable to small, harassing actions from the Germans, which destroyed their morale. Bogged down in pestilential jungles, as time went by, Smuts's army was decimated by hunger and disease, all without ever fighting a real battle. By the end of the war, von Lettow-Vorbeck had managed to lead his enemy on a four-year cat-and-mouse chase that had completely tied up valuable English forces and yielded them nothing in return.

Smuts was a persistent, thorough, aggressive leader who liked to defeat his opponents through maneuver in the field. Von Lettow-Vorbeck played on this taste: he refused to engage Smuts in frontal battle but stayed enticingly close, just beyond reach, holding out the possibility of engagement so as to keep the Englishmen pushing forward into the wilderness. Infuriated to no end, Smuts continued the chase. Von Lettow-Vorbeck used Africa's vast spaces and inhospitable climate to destroy the English.

Most people respond to aggression by in some way getting involved with it. It is almost impossible to hold back. By disengaging completely and retreating, you show great power and restraint. Your enemies are desperate for you to react; retreat infuriates and provokes them into further attack. So keep retreating, exchanging space for time. Stay calm and balanced. Let them take the land they want; like the Germans, lure them into a void of nonaction. They will start to overextend themselves and make mistakes. Time is on your side, for you are not wasting any of it in useless battles.

War is notoriously full of surprises, of unforeseen events that can slow down and ruin even the best-laid plan. Carl von Clausewitz called this "friction." War is a constant illustration of Murphy's Law: if anything can go wrong, it will. But when you retreat, when you exchange space for time, you are making Murphy's Law work for you. So it was with von Lettow-Vorbeck: he set up Smuts as the victim of Murphy's Law, giving him enough time to make the worst come to pass.

During the Seven Years' War (1756–63), Frederick the Great of Prussia was faced with Austrian, French, and Russian armies on every side, all determined to carve him up. A strategist who usually favored aggressive attack, Frederick this time went on the defensive, crafting his maneuvers to buy himself time and slip the net his enemies were trying to catch him in. Year after year he managed to avoid disaster, though barely. Then, suddenly, Czarina Elizabeth of Russia died. She had hated Frederick bitterly, but her nephew and successor to the throne, Czar Peter III, was a perverse young boy who had not liked his aunt and who greatly admired Frederick the Great. He not only pulled Russia out of the war, he allied himself with the Prussians. The Seven Years' War was over; the miracle Frederick needed had come to pass. Had he surrendered at his worst point or tried to fight his way out, he would have lost everything. Instead he maneuvered to create time for Murphy's Law to do its work on his enemies.

War is a physical affair, which takes place somewhere specific: generals depend on maps and plan strategies to be realized in particular locations. But time is just as important as space in strategic thought, and knowing how to use time will make you a superior strategist, giving an added dimension to your attacks and defense. To do this you must stop thinking of time as an abstraction: in reality, beginning the minute you are born, time is all you have. It is your only true commodity. People can take away your possessions, but—short of murder—not even the most powerful aggressors can take time away from you unless you let them. Even in prison your time is your own, if you use it for your own purposes. To waste your time in battles not of your choosing is more than just a mistake, it is stupidity of the highest order. Time lost can never be regained.

**Image:** The Desert Sands. In the desert there is nothing to feed on and nothing to use for war: just sand and empty space. Retreat to the desert occasionally, to think and see with clarity. Time moves slowly there, which is what you need. When under attack, fall back into the desert, luring your enemies into a place where they lose all sense of time and space and fall under your control.

**Authority:** To remain disciplined and calm while waiting for disorder to appear amongst the enemy is the art of self-possession. *—Sun-tzu (fourth century B.C.)*

## REVERSAL

When enemies attack you in overwhelming force, instead of retreating you may sometimes decide to engage them directly. You are inviting martyrdom, perhaps even hoping for it, but martyrdom, too, is a strategy, and one of ancient standing: martyrdom makes you a symbol, a rallying point for the future. The strategy will succeed if you are important enough—if your defeat has symbolic meaning—but the circumstances must work to highlight the rightness of your cause and the ugliness of the enemy's. Your sacrifice must also be unique; too many martyrs, spread over too much time, will spoil the effect. In cases of extreme weakness, when facing an impossibly large enemy, martyrdom can be used to show that your side's fighting spirit has not been extinguished, a useful way to keep up morale. But, in general, martyrdom is a dangerous weapon and can backfire, for you may no longer be there to see it through, and its effects are too strong to be controlled. It can also take centuries to work. Even when it may prove symbolically successful, a good strategist avoids it. Retreat is always the better strategy.

Retreat must never be an end in itself; at some point you have to turn around and fight. If you don't, retreat is more accurately called surrender: the enemy wins. Combat is in the long run unavoidable. Retreat can only be temporary.

PART

# IV

# OFFENSIVE WARFARE

The greatest dangers in war, and in life, come from the unexpected: people do not respond the way you had thought they would, events mess up your plans and produce confusion, circumstances are overwhelming. In strategy this discrepancy between what you want to happen and what does happen is called "friction." The idea behind conventional offensive warfare is simple: by attacking the other side first, hitting its points of vulnerability, and seizing the initiative and never letting it go, you create your own circumstances. Before any friction can creep in and undermine your plans, you move to the offensive, and your relentless maneuvers force so much friction on the enemy that he collapses.

This is the form of warfare practiced by the most successful captains in history, and the secret to their success is a perfect blend of strategic cleverness and audacity. The strategic element comes in the planning: setting an overall goal, crafting ways to reach it, and thinking the whole plan through in intense detail. This means thinking in terms of a campaign, not individual battles. It also means knowing the strengths and weaknesses of the other side, so that you can calibrate your strikes to its vulnerabilities. The more detailed your planning, the more confident you will feel as you go into battle, and the easier it will be to stay on course once the inevitable problems arise. In the attack itself, though, you must strike with such spirit and audacity that you put your enemies on their heels, giving irresistible momentum to your offensive.

The following eleven chapters will initiate you into this supreme form of warfare. They will help you to put your desires and goals into a larger framework known as "grand strategy." They will show you how to look at your enemies and uncover their secrets. They will describe how a solid base of planning will give you fluid options for attack and how specific maneuvers (the flanking maneuver, the envelopment) and styles of attack (hitting centers of gravity, forcing the enemy into positions of great weakness) that work brilliantly in war can be applied in life. Finally, they will show you how to finish off your campaign. Without a vigorous conclusion that meets your overall goals, everything you have done will be worthless. Mastering the various components of offensive warfare will give all of your attacks in life much greater force.

# 12

# LOSE BATTLES BUT WIN THE WAR

## GRAND STRATEGY

*Everyone around you is a strategist angling for power, all trying to promote their own interests, often at your expense. Your daily battles with them make you lose sight of the only thing that really matters: victory in the end, the achievement of greater goals, lasting power. Grand strategy is the art of looking beyond the battle and calculating ahead. It requires that you focus on your ultimate goal and plot to reach it. In grand strategy you consider the political ramifications and long-term consequences of what you do. Instead of reacting emotionally to people, you take control, and make your actions more dimensional, subtle, and effective. Let others get caught up in the twists and turns of the battle, relishing their little victories. Grand strategy will bring you the ultimate reward: the last laugh.*

*Readiness is everything. Resolution is indissolubly bound up with caution. If an individual is careful and keeps his wits about him, he need not become excited or alarmed. If he is watchful at all times, even before danger is present, he is armed when danger approaches and need not be afraid. The superior man is on his guard against what is not yet in sight and on the alert for what is not yet within hearing; therefore he dwells in the midst of difficulties as though they did not exist. . . . If reason triumphs, the passions withdraw of themselves.*

THE I CHING, CHINA, CIRCA EIGHTH CENTURY B.C.

## THE GREAT CAMPAIGN

Growing up at the Macedonian court, Alexander (356–322 B.C.) was considered a rather strange young man. He enjoyed the usual boyish pursuits, such as horses and warfare; having fought alongside his father, King Philip II, in several battles, he had proved his bravery. But he also loved philosophy and literature. His tutor was the great thinker Aristotle, under whose influence he loved to argue about politics and science, looking at the world as dispassionately as possible. Then there was his mother, Olympias: a mystical, superstitious woman, she had had visions at Alexander's birth that he would one day rule the known world. She told him about them and filled him with stories of Achilles, from whom her family claimed descent. Alexander adored his mother (while hating his father) and took her prophecies most seriously. From early on in life, he carried himself as if he were more than the son of a king.

Alexander was raised to be Philip's successor, and the state he was to inherit had grown considerably during his father's reign. Over the years the king had managed to build up the Macedonian army into the supreme force in all Greece. He had defeated Thebes and Athens and had united all the Greek city-states (except Sparta) into a Hellenic league under his leadership. He was a crafty, intimidating ruler. Then, in 336 B.C., a disgruntled nobleman assassinated him. Suddenly seeing Macedonia as vulnerable, Athens declared its independence from the league. The other city-states followed suit. Tribes from the north now threatened to invade. Almost overnight Philip's small empire was unraveling.

When Alexander came to the throne, he was only twenty, and many considered him unready. It was a bad time for learning on the job; the Macedonian generals and political leaders would have to take him under their wing. They advised him to go slowly, to consolidate his position in both the army and Macedonia and then gradually reform the league through force and guile. That was what Philip would have done. But Alexander would not listen; he had another plan, or so it seemed. Without giving his enemies in and beyond Macedonia time to organize against him, he led the army south and reconquered Thebes in a series of lightning maneuvers. Next he marched on the Athenians, who, fearing his retribution, begged forgiveness and pleaded to be readmitted to the league. Alexander granted their wish.

The eccentric young prince had shown himself to be a bold and unpredictable king—attacking when he was not meant to, yet showing Athens unexpected mercy. He was hard to read, but his first maneuvers as king had won him many admirers. His next move, however, was still stranger and more audacious: instead of working to consolidate his gains and strengthen the fragile league, he proposed to launch a crusade against the Persian Empire, the Greeks' great enemy. Some 150 years earlier, the Persians had tried to invade Greece. They had almost succeeded, and it remained their dream to try it again and get it right. With

Persia a constant threat, the Greeks could never rest easy, and their maritime trade was cramped by the power of the Persian navy.

In 334 B.C., Alexander led a united army of 35,000 Greeks across the Dardanelle Straits and into Asia Minor, the westernmost part of the Persian Empire. In their first encounter with the enemy, at the Battle of the Granicus, the Greeks routed the Persians. Alexander's generals could only admire his boldness: he seemed poised to conquer Persia, fulfilling his mother's prophecy in record time. He succeeded through speed and by seizing the initiative. Now soldiers and generals alike expected him to head straight east into Persia to finish off the enemy army, which seemed surprisingly weak.

Once again Alexander confounded expectations, suddenly deciding to do what he had never done before: take his time. That would have seemed wise when he first came to power, but now it seemed likely to give the Persians the one thing they needed: time to recover and replenish. Yet Alexander led his army not east but south, down the coast of Asia Minor, freeing local towns from Persian rule. Next he zigzagged east and then south again, through Phoenicia and into Egypt, quickly defeating the weak Persian garrison there. The Egyptians hated their Persian rulers and welcomed Alexander as their liberator. Now Alexander could use Egypt's vast stores of grain to feed the Greek army and help keep the Greek economy stable, while depriving Persia of valuable resources.

As the Greeks advanced farther from home, the Persian navy, which could land an army almost anywhere in the Mediterranean to attack them from the rear or flank, was a worrying threat. Before Alexander set out on his expedition, many had advised him to build up the Greek navy and take the battle to the Persians by sea as well as land. Alexander had ignored them. Instead, as he passed through Asia Minor and then along the coast of Phoenicia, he simply captured Persia's principal ports, rendering their navy useless.

These small victories, then, had a greater strategic purpose. Even so, they would have meant little had the Greeks been unable to defeat the Persians in battle—and Alexander seemed to be making that victory more difficult. The Persian king, Darius, was concentrating his forces east of the Tigris River; he had numbers and his choice of location and could wait in ease for Alexander to cross the river. Had Alexander lost his taste for battle? Had Persian and Egyptian culture softened him? It seemed so: he had begun to wear Persian clothes and to adopt Persian customs. He was even seen worshipping Persian gods.

As the Persian army retreated east of the Tigris, large areas of the Persian empire had come under Greek control. Now Alexander spent much of his time not on warfare but on politics, trying to see how best to govern these regions. He decided to build on the Persian system already in place, keeping the same titles for jobs in the governmental bureaucracy, collecting the same tribute that Darius had done. He changed only

THE FOX AND THE MONKEY ELECTED KING

*The monkey, having danced in an assembly of the animals and earned their approval, was elected by them to be king. The fox was jealous. So, seeing a piece of meat one day in a snare, he led the monkey to it, saying that he had found a treasure. But rather than take it for himself, he had kept guard over it, as its possession was surely a prerogative of royalty. The fox then urged him to take it.*

*The monkey approached it, taking no care, and was caught in the trap. When he accused the fox of luring him into a trap, the fox replied: "Monkey, you want to reign over all the animals, but look what a fool you are!"*
It is thus that those who throw themselves into an enterprise without sufficient thought not only fail, but even become a laughing stock.

*FABLES*, AESOP, SIXTH CENTURY B.C.

the harsh, unpopular aspects of Persian rule. Word quickly spread of his generosity and gentleness toward his new subjects. Town after town surrendered to the Greeks without a fight, only too glad to be part of Alexander's growing empire, which transcended Greece and Persia. He was the unifying factor, the benevolent overseeing god.

Finally, in 331 B.C., Alexander marched on the main Persian force at Arbela. What his generals had not understood was that, deprived of the use of its navy, its rich lands in Egypt, and the support and tribute of almost all of its subjects, the Persian Empire had already crumbled. Alexander's victory at Arbela merely confirmed militarily what he had already achieved months earlier: he was now the ruler of the once mighty Persian Empire. Fulfilling his mother's prophecy, he controlled almost all of the known world.

*Epistemologically speaking, the source of all erroneous views on war lies in* idealist *and* mechanistic *tendencies. . . . People with such tendencies are subjective and one-sided in their approach to problems. They indulge in groundless and purely subjective talk, basing themselves upon a single aspect or temporary manifestation* [*and*] *magnify it with similar subjectivity into the whole of the problem. . . . Only by opposing idealistic and mechanistic tendencies and taking an objective all-sided view in making a study of war can we draw correct conclusions on the question of war.*

*Selected Military Writings*, Mao Tse-tung, 1893–1976

## Interpretation

Alexander the Great's maneuvers bewildered his staff: they seemed to have no logic, no consistency. Only later could the Greeks look back and really see his magnificent achievement. The reason they could not understand him was that Alexander had invented a whole new way of thinking and acting in the world: the art of grand strategy.

In grand strategy you look beyond the moment, beyond your immediate battles and concerns. You concentrate instead on what you want to achieve down the line. Controlling the temptation to react to events as they happen, you determine each of your actions according to your ultimate goals. You think in terms not of individual battles but of a campaign.

Alexander owed his novel style of strategizing to his mother and to Aristotle. His mother had given him a sense of destiny and a goal: to rule the known world. From the age of three, he could see in his mind's eye the role he would play when he was thirty. From Aristotle he learned the power of controlling his emotions, seeing things dispassionately, thinking ahead to the consequences of his actions.

Trace the zigzags of Alexander's maneuvers and you will see their grand-strategic consistency. His quick actions against first Thebes, then Persia, worked psychically on his soldiers and on his critics. Nothing quiets an army faster than battle; Alexander's sudden crusade against the hated Persians was the perfect way to unite the Greeks. Once he was in Persia, though, speed was the wrong tactic. Had Alexander advanced, he would have found himself controlling too much land too quickly; running it would have exhausted his resources, and in the ensuing power vacuum, enemies would have sprung up everywhere. Better to proceed slowly, to build on what was there, to win hearts and minds. Instead of wasting money on building a navy, better simply to make the Persian navy unusable. To pay for the kind of extended campaign that would bring long-term success, first seize the rich lands of Egypt. None of Alexander's actions were wasted. Those who saw his plans bear fruit, in ways they themselves had been entirely unable to predict, thought him a kind of god—and certainly his control over events deep in the future seemed more godlike than human.

To become a grand strategist in life, you must follow the path of Alexander. First, clarify your life—decipher your own personal riddle—by determining what it is you are destined to achieve, the direction in which your skills and talents seem to push you. Visualize yourself fulfilling this destiny in glorious detail. As Aristotle advised, work to master your emotions and train yourself to think ahead: "This action will advance me toward my goal, this one will lead me nowhere." Guided by these standards, you will be able to stay on course.

Ignore the conventional wisdom about what you should or should not be doing. It may make sense for some, but that does not mean it bears any relation to your own goals and destiny. You need to be patient enough to plot several steps ahead—to wage a campaign instead of fighting battles. The path to your goal may be indirect, your actions may be strange to other people, but so much the better: the less they understand you, the easier they are to deceive, manipulate, and seduce. Following this path, you will gain the calm, Olympian perspective that will separate you from other mortals, whether dreamers who get nothing done or prosaic, practical people who accomplish only small things.

*What I particularly admire in Alexander is, not so much his campaigns . . . but his political sense. He possessed the art of winning the affection of the people.*

—*Napoleon Bonaparte (1769–1821)*

*There is, however, much difference between the East and the West in cultural heritages, in values, and in ways of thinking. In the Eastern way of thinking, one starts with the whole, takes everything as a whole and proceeds with a comprehensive and intuitive synthesization [combinaton]. In the Western way of thinking, however, one starts with the parts, takes [divides] a complex matter into component parts and then deals with them one by one, with an emphasis on logical analysis. Accordingly, Western traditional military thought advocates a direct military approach with a stress on the use of armed forces.*

THE STRATEGIC ADVANTAGE: SUN ZI & WESTERN APPROACHES TO WAR, CAO SHAN, ED., 1997

## TOTAL WARFARE

In 1967 the leaders of the American war effort in Vietnam thought they were finally making progress. They had launched a series of operations to search out and destroy the Vietcong—North Vietnamese soldiers who had infiltrated South Vietnam and had come to control much of its countryside. These guerrilla fighters were elusive, but the Americans had inflicted heavy losses on them in the few battles they had managed to force on them that year. The new South Vietnamese government, supported by the Americans, seemed relatively stable, which could help to win it approval among the Vietnamese people. To the north, bombing raids had knocked out many of North Vietnam's airfields and heavily damaged its air force. Although massive antiwar demonstrations had broken out in the United States, polls showed that most Americans supported the war and believed that the end was in sight.

Since the Vietcong and the North Vietnamese army had proved rather ineffective in head-to-head battle against the might of American firepower and technology, the strategy was to somehow lure them into a major engagement. That would be the turning point of the war. And by the end of 1967, intelligence indicated that the North Vietnamese were about to fall into just such a trap: their commander, General Vo Nguyen Giap, was planning a major offensive against the U.S. marine outpost at Khe Sanh. Apparently he wanted to repeat his greatest success, the battle at Dien Bien Phu in 1954, in which he had defeated the French army, driving the French out of Vietnam for good.

Khe Sanh was a key strategic outpost. It was located a mere fourteen miles from the demilitarized zone that separated North from South Vietnam. It was also six miles from the border of Laos, site of a stretch of the famous Ho Chi Minh Trail, the North Vietnamese supply route to the Vietcong in the South. General William C. Westmoreland, the overall U.S. commander, was using Khe Sanh to monitor enemy activity to the north and west. Dien Bien Phu had served a similar role for the French, and Giap had been able to isolate and destroy it. Westmoreland would not allow Giap to repeat that feat. He built well-protected airstrips around Khe Sanh, ensuring full use of his helicopters and control of the air. He called up substantial numbers of troops from the south to the Khe Sanh area, just in case he needed them. He also ordered 6,000 additional marines to reinforce the outpost. But a major attack on Khe Sanh was nothing he wanted to discourage: in frontal battle the enemy would finally expose itself to severe defeat.

In the first few weeks of 1968, all eyes were on Khe Sanh. The White House and the U.S. media were certain that the decisive battle of the war was about to begin. Finally, at dawn on January 21, 1968, the North Vietnamese army launched a vicious assault. As both sides dug in, the battle turned into a siege.

Soon after the engagement began, the Vietnamese were to celebrate their lunar New Year, the holiday called Tet. It was a period of revelry, and in time of war it was also a traditional moment to declare a truce. This year was no different; both sides agreed to halt the fighting during Tet. Early on the morning of January 31, however, the first day of the New Year, reports began to trickle in from all over South Vietnam: virtually every major town and city, as well as the most important American bases, had come under Vietcong attack. An army general, tracking the assault pattern on a map, said it "resembled a pinball machine, lighting up with each raid."

Parts of Saigon itself had been overrun by enemy soldiers, some of whom had managed to blow their way through the wall of the U.S. embassy, the very symbol of the American presence in Vietnam. Marines regained control of the embassy in a bloody fight, which was widely seen on American television. The Vietcong also attacked the city's radio station, the presidential palace, and Westmoreland's own compound at the Tan Son Nhut air base. The city quickly descended into street fighting and chaos.

Outside Saigon, provincial cities, too, came under siege. Most prominent was the North Vietnamese capture of Hue, the ancient Vietnamese capital and a city revered by Buddhists. Insurgents managed to take control of virtually the whole city.

Meanwhile the attacks on Khe Sanh continued in waves. It was hard for Westmoreland to tell what the main target was: were the battles to the south merely a means of drawing forces away from Khe Sanh, or was it the other way around? Within a few weeks, in all parts of South Vietnam, the Americans regained the upper hand, retaking control of Saigon and securing their air bases. The sieges at Hue and Khe Sanh took longer, but massive artillery and air bombardments eventually doomed the insurgents, as well as leveling entire sections of Hue.

After what later became known as the Tet Offensive was over, Westmoreland likened it to the Battle of the Bulge, near the end of World War II. There the Germans had managed to surprise the Allies by staging a bold incursion into eastern France. In the first few days, they had advanced rapidly, creating panic, but once the Allies recovered, they had managed to push the Germans back—and eventually it became apparent that the battle was the German military's death knell, their last shot. So it was, Westmoreland argued, with the North Vietnamese army at Khe Sanh and the Vietcong throughout the South: they had suffered terrible casualties, far more than the Americans had—in fact, the entire Vietcong infrastructure had been wiped out. They would never recover; at long last the enemy had revealed itself and had been badly mauled.

The Americans thought Tet had been a tactical disaster for the North. But another viewpoint began to trickle in from home: the drama at the U.S. embassy, the siege of Hue, and the attacks on air bases had kept millions of Americans glued to their television sets. Until then the Vietcong had operated mostly in the countryside, barely visible to the American public. Now, for the first time, they were apparent in major cities, wreaking havoc and destruction. Americans had been told the war was winding down and winnable; these images said otherwise. Suddenly the war's purpose seemed less clear. How could South Vietnam remain stable in the face of this ubiquitous enemy? How could the Americans ever claim a clear victory? There was really no end in sight.

American opinion polls tracked a sharp turn against the war. Antiwar demonstrations broke out all over the country. President Lyndon Johnson's military advisers, who had been telling him that South Vietnam was coming under control, now confessed that they were no longer so optimistic. In the New Hampshire Democratic primary that March, Johnson was stunned by his defeat by Senator Eugene McCarthy, who had galvanized the growing antiwar sentiment. Shortly thereafter Johnson announced that he would not run for reelection in the upcoming presidential race and that he would slowly disengage American forces from Vietnam.

The Tet Offensive was indeed the turning point in the Vietnam War, but not in the direction that Westmoreland and his staff had foreseen.

*When dark inertia increases,*
*obscurity and inactivity, negligence*
*and delusion, arise.*
*When lucidity prevails,*
*the self whose body dies*
*enters the untainted worlds*
*of those who know reality.*
*When he dies in passion,*
*he is born among lovers of action;*
*so when he dies in dark inertia,*
*he is born into wombs of folly.*
*The fruit of good conduct*
*is pure and untainted, they say,*
*but suffering is the fruit of passion,*
*ignorance the fruit of dark inertia.*
*From lucidity knowledge is born;*
*from passion comes greed;*
*from dark inertia come negligence,*
*delusion, and ignorance.*
*Men who are lucid go upward;*
*men of passion stay in between;*
*men of dark inertia,*
*caught in vile ways,*
*sink low.*

*The Bhagavad Gita: Krishna's Counsel in Time of War,* circa First Century A.D.

## Interpretation

For the American strategists, the success of the war depended mostly on the military. By using their army and superior weaponry to kill as many Vietcong as possible and gain control of the countryside, they would ensure the stability of the South Vietnamese government. Once the South was strong enough, North Vietnam would give up the fight.

The North Vietnamese saw the war very differently. By nature and practice, they viewed conflict in much broader terms. They looked at the political situation in the South, where American search-and-destroy missions were alienating South Vietnamese peasants. The North Vietnamese, meanwhile, did everything they could to win the peasants over and earned for themselves an army of millions of silent sympathizers. How could the South be secure when the Americans had failed to capture the hearts and minds of the Vietnamese farmers? The North Vietnamese also looked to the American political scene, where, in 1968, there was to be a presidential election. And they looked at American culture, where support for the war was wide but not deep. The Vietnam War was the first televised war in history; the military was trying to control information about the war, but the images on television spoke for themselves.

*At this the grey-eyed goddess Athena smiled, and gave him a caress, her looks being changed now, so she seemed a woman, tall and beautiful and no doubt skilled at weaving splendid things. She answered briskly: "Whoever gets around you [Odysseus] must be sharp and guileful as a snake; even a god might bow to you in ways of dissimulation. You! You chameleon! Bottomless bag of tricks! Here in your own country would you not give your stratagems a rest or stop spellbinding for an instant? . . . Two of a kind, we are, contrivers, both. Of all men now alive you are the best in plots and story telling. My own fame is for wisdom among the gods—deceptions, too.*

THE ODYSSEY, HOMER, CIRCA NINTH CENTURY B.C.

On and on the North Vietnamese went, continually broadening their outlook and analyzing the war's global context. And out of this study they crafted their most brilliant strategy: the Tet Offensive. Using their army of peasant sympathizers in the South, they were able to infiltrate every part of the country, smuggling in arms and supplies under the cover of the Tet holiday. The targets they hit were not only military but televisual: their attacks in Saigon, base of most of the American media (including the CBS newsman Walter Cronkite, visiting at the time) were spectacular; Hue and Khe Sanh were also places heavily covered by American reporters. They also struck symbolic locations—embassies, palaces, air bases—that would suck in media attention. On television all this would create the dramatic (and deceptive) impression that the Vietcong were everywhere while American bombing raids and pacification programs had gotten nowhere. In effect, the goal of the Tet Offensive was not a military target but the American public in front of its televisions. Once Americans lost faith—and in an election year—the war was doomed. The North Vietnamese did not have to win a single pitched battle on the field, and in fact they never did. But by extending their vision beyond the battlefield to politics and culture, they won the war.

We always tend to look at what is most immediate to us, taking the most direct route toward our goals and trying to win the war by winning as many battles as we can. We think in small, microlevel terms and react to present events—but this is petty strategy. Nothing in life happens in isolation; everything is related to everything else and has a broader context. That context includes people outside your immediate circle whom your actions affect, the public at large, the whole world; it includes politics, for every choice in modern life has political ramifications; it includes culture, the media, the way the public sees you. Your task as a grand strategist is to extend your vision in all directions—not only looking further into the future but also seeing more of the world around you, more than your enemy does. Your strategies will become insidious and impossible to thwart. You will be able to harness the relationships between events, one battle setting up the next, a cultural coup setting up a political coup. You will bring the war to arenas your enemies have ignored, catching them by surprise. Only grand strategy can yield grand results.

*War is the continuation of politics by other means.*
*—Carl von Clausewitz (1780–1831)*

## KEYS TO WARFARE

Thousands of years ago, we humans elevated ourselves above the animal world and never looked back. Figuratively speaking, the key to this evolutionary advance was our powers of vision: language, and the ability to reason that it gave us, let us see more of the world around us. To protect itself from a predator, an animal depended on its senses and instincts; it could not see around the corner or to the other end of the forest. We humans, on the other hand, could map the entire forest, study the habits of dangerous animals and even nature itself, gaining deeper, wider knowledge of our environment. We could see dangers coming before they were here. This expanded vision was abstract: where an animal is locked in the present, we could see into the past and glimpse as far as our reason would take us into the future. Our sight expanded further and further into time and space, and we came to dominate the world.

Somewhere along the line, however, we stopped evolving as rational creatures. Despite our progress there is always a part of us that remains animal, and that animal part can respond only to what is most immediate in our environment—it is incapable of thinking beyond the moment. The dilemma affects us still: the two sides of our character, rational and animal, are constantly at war, making almost all of our actions awkward. We reason and plan to achieve a goal, but in the heat of action we become emotional and lose perspective. We use cleverness and strategy to grab for what we want, but we do not stop to think about whether what we want is necessary, or what the consequences of getting it will be. The extended vision that rationality brings us is often eclipsed by the reactive, emotional animal within—the stronger side of our nature.

More than we are today, the ancient Greeks were close to the passage of the human race from animal to rational. To them our dual nature made us tragic, and the source of tragedy was limited vision. In classical Greek tragedies such as *Oedipus Rex*, the protagonist may think he knows the truth and knows enough about the world to act in it, but his vision is

*Then he saw Odysseus and asked:*
*"Now tell me about this one, dear child,*
*Shorter than Agamemnon by a head*
*But broader in the shoulders and chest.*
*His armor is lying on the ground*
*And he's roaming the ranks like a ram,*
*That's it, just like a thick-fleeced ram*
*Striding through a flock of silvery sheep."*
*And Helen, Zeus' child:*
*"That is Laertes' son,*
*The master strategist Odysseus, born and bred*
*In the rocky hills of Ithaca. He knows*
*Every trick there is, and his mind runs deep."*
*Antenor turned to her and observed astutely:*
*"Your words are not off the mark there, madam.*
*Odysseus came here once before, on an embassy*
*For your sake along with Menelaus.*
*I entertained them courteously in the great hall*
*And learned each man's character and depth of mind.*
*Standing in a crowd of Trojans, Menelaus,*
*With his wide shoulders, was more prominent,*
*But when both were seated Odysseus was lordlier.*
*When it came time for each to speak in public*
*And weave a spell of wisdom with their words,*
*Menelaus spoke fluently enough, to the point*

limited by his emotions and desires. He has only a partial perspective on life and on his own actions and identity, so he acts imprudently and causes suffering. When Oedipus finally understands his own role in all his misfortunes, he tears out his eyes—symbols of his tragic limitation. He can see out into the world but not inward into himself.

The Greeks, however, also recognized the potential for a higher human possibility. Far above the sphere of mortals were the gods on Mount Olympus, who had perfect vision of the world and of both the past and the future; and the human race shared something with them as well as with animals—we were not only part animal but part divine. Furthermore, those able to see further than others, to control their animal nature and think before they acted, were humans of the most deeply human kind—the ones best able to use the reasoning powers that separate us from animals. As opposed to human stupidity (limited vision), the Greeks imagined an ideal human prudence. Its symbol was Odysseus, who always thought before he acted. Having visited Hades, the land of the dead, he was in touch with ancestral history and the past; and he was also always curious, eager for knowledge, and able to view human actions, his own and other people's, with a dispassionate eye, considering their long-term consequences. In other words, like the gods, if to a lesser extent, he had the skill of looking into the future. The consummate realist, the man of vision, Odysseus was a character in the epic poetry of Homer, but there were also historical versions of the ideal: the political figure and military leader Themistocles, for example, and Alexander the Great, raised to heights of combined intellect and action by Aristotle.

The prudent man might seem cold, his rationality sucking pleasure out of life. Not so. Like the pleasure-loving gods on Mount Olympus, he has the perspective, the calm detachment, the ability to laugh, that come with true vision, which gives everything he does a quality of lightness—these traits comprising what Nietzsche calls the "Apollonian ideal." (Only people who can't see past their noses make things heavy.) Alexander, the great strategist and man of action, was also famous for revelry and festivity. Odysseus loved adventure; no one was better at the experience of pleasure. He was simply more reasonable, more balanced, less vulnerable to his own emotions and moods, and he left less tragedy and turmoil in his wake.

This calm, detached, rational, far-seeing creature, called "prudent" by the Greeks, is what we shall call the "grand strategist."

We are all of us to some extent strategists: we naturally want control over our lives, and we plot for power, consciously or unconsciously angling to get what we want. We use strategies, in other words, but they tend to be linear and reactive and are often fractured and struck off course by emotional responses. Clever strategists can go far, but all but a few make mistakes. If they are successful, they get carried away and overreach; if they face setbacks—and setbacks are inevitable over a lifetime—they are easily overwhelmed. What sets grand strategists apart is the ability to look more deeply into both themselves and others, to understand and learn from the past and to have a clear sense of the future, to the extent that it can be predicted. Simply, they see more, and their extended vision lets them carry out plans over sometimes-long periods of time—so long that those around them may not even realize that they have a plan in mind. They strike at the roots of a problem, not at its symptoms, and hit their mark cleanly. In moving toward becoming a grand strategist, you follow in the path of Odysseus and rise toward the condition of the gods. It is not so much that your strategies are more clever or manipulative as that they exist on a higher plane. You have made a qualitative leap.

In a world where people are increasingly incapable of thinking consequentially, more animal than ever, the practice of grand strategy will instantly elevate you above others.

To become a grand strategist does not involve years of study or a total transformation of your personality. It simply means more effective use of what you have—your mind, your rationality, your vision. Having evolved as a solution to the problems of warfare, grand strategy is a military concept. And an examination of its historical development will reveal the key to making it work for you in daily life.

In the early history of warfare, a ruler or general who understood strategy and maneuver could exercise power. He could win battles, carve out an empire, or at the very least defend his own city or state. But problems came with strategy on this level. More than any other human activity, war plays havoc with emotion, stirs the animal within. In plotting war a king would depend on things like his knowledge of the terrain and his understanding of both the enemy's forces and his own; his success would depend on his ability to see these things clearly. But this vision was likely to be clouded. He had emotions to respond to, desires to realize; he could not think his goals through. Wanting to win, he would underestimate the enemy's strength or overestimate his own. When Xerxes of Persia invaded Greece in 480 B.C., he thought he had a perfectly rational plan. There was much he had not taken into account, and disaster followed.

Other rulers actually won their battles only to grow drunk on victory and not know when to stop, stirring up implacable hatred, distrust, and the desire for revenge all around them, culminating in war on several fronts and total defeat—as in the destruction of the warlike Assyrian Empire, its capital of Nineveh eternally buried in the sand. In cases like that, victory in battle brought only danger, exposing the conqueror to ruinous cycles of attack and counterattack.

In ancient times, strategists and historians from Sun-tzu to Thucydides became conscious of this recurring self-destructive pattern in warfare and began to work out more rational ways to fight. The first step was to think beyond the immediate battle. Supposing you won victory, where would it leave you—better off or worse? To answer that question, the

*And very clearly, but briefly, since he is not*
*A man of many words. Being older, he spoke first.*
*Then Odysseus, the master strategist, rose quickly,*
*But just stood there, his eyes fixed on the ground.*
*He did not move his staff forward or backward*
*But held it steady. You would have thought him*
*A dull, surly lout without any wit. But when he*
*Opened his mouth and projected his voice*
*The words fell down like snowflakes in a blizzard.*
*No mortal could have vied with Odysseus then,*
*And we no longer held his looks against him."*

*The Iliad*, Homer, circa Ninth century B.C.

logical step was to think ahead, to the third and fourth battles on, which connected like links in a chain. The result was the concept of the campaign, in which the strategist sets a realistic goal and plots several steps ahead to get there. Individual battles matter only in the way they set up the next ones down the line; an army can even deliberately lose a battle as part of a long-term plan. The victory that matters is that of the overall campaign, and everything is subordinated to that goal.

Forgetting our objectives.—*During the journey we commonly forget its goal. Almost every profession is chosen and commenced as a means to an end but continued as an end in itself. Forgetting our objectives is the most frequent of all acts of stupidity.*

FRIEDRICH NIETZSCHE, 1844–1900

This kind of strategy represented a qualitative advance. Think of chess, where the grand master, instead of focusing only on the move at hand and making it solely in reaction to what the other player has just done, must visualize the entire chessboard deep into the future, crafting an overall strategy, using the moves of the pawns now to set up those of the more powerful pieces later on. Thinking in terms of the campaign gave strategy a new depth. The strategist used more and more of the map.

War on this level required that the strategist think deeply in all directions before launching the campaign. He had to know the world. The enemy was just one part of the picture; the strategist also had to anticipate the reactions of allies and neighboring states—any missteps with them and the entire plan could unravel. He had to imagine the peace after the war. He had to know what his army was capable of over time and ask no more of it than that. He had to be realistic. His mind had to expand to meet the complexities of the task—and all this before a single blow was exchanged.

Yet strategic thinking on this level yielded limitless benefits. A victory on the battlefield would not seduce the leader into an unconsidered move that might ultimately set the campaign back, nor would a defeat unnerve him. When something unexpected happened—and the unexpected is to be expected in war—the solution he improvised to meet it would have to suit goals far on the horizon. His subordination of his emotions to strategic thought would give him more control during the course of the campaign. He would keep his perspective in the heat of battle. He would not get caught up in the reactive and self-destructive pattern that had destroyed so many armies and states.

This principle of campaigning was only relatively recently christened "grand strategy," but it has existed in various forms since ancient times. It is clearly visible in Alexander's conquest of Persia, in the Roman and Byzantine empires' control of vast territories with small armies, in the disciplined campaigns of the Mongols, in Queen Elizabeth I's defeat of the Spanish Armada, in the Duke of Marlborough's brilliantly conceived campaigns against the Hapsburgs. In modern times North Vietnam's defeat first of the French, then of the United States—in the latter case without winning a single major battle—must be considered a consummate use of the art.

Military history shows that the key to grand strategy—the thing that separates it from simple, garden-variety strategy—is its particular quality of forethought. Grand strategists think and plan further into the future before taking action. Nor is their planning simply a matter of accumulating knowledge and information; it involves looking at the world with a dispassionate eye, thinking in terms of the campaign, planning indirect, subtle steps along the way whose purpose may only gradually become visible to others. Not only does this kind of planning fool and disorient the enemy; for the strategist it has the psychological effects of calm, a sense of perspective, flexibility to change in the moment while keeping the ultimate goal in mind. Emotions are easier to control; vision is far-seeing and clear. Grand strategy is the apex of rationality.

Grand strategy has four main principles, distilled below from case histories of the most successful practitioners of the art. The more you can incorporate these principles into your plans, the better the results.

**Focus on your greater goal, your destiny.** The first step toward becoming a grand strategist—the step that will make everything else fall into place—is to begin with a clear, detailed, purposeful goal in mind, one rooted in reality. We often imagine that we generally operate by some kind of plan, that we have goals we are trying to reach. But we're usually fooling ourselves; what we have are not goals but wishes. Our emotions infect us with hazy desire: we want fame, success, security—something large and abstract. This haziness unbalances our plans from the beginning and sets them on a chaotic course. What have distinguished all history's grand strategists and can distinguish you, too, are specific, detailed, focused goals. Contemplate them day in and day out, and imagine how it will feel to reach them and what reaching them will look like. By a psychological law peculiar to humans, clearly visualizing them this way will turn into a self-fulfilling prophecy.

Having clear objectives was crucial to Napoleon. He visualized his goals in intense detail—at the beginning of a campaign, he could see its last battle clearly in his mind. Examining a map with his aides, he would point to the exact spot where it would end—a ridiculous prediction, it might seem, since not only is war in any period subject to chance and to whatever the enemy comes up with to surprise you, but the maps of Napoleon's era were notoriously unreliable. Yet time and again his predictions would prove uncannily correct. He would also visualize the campaign's aftermath: the signing of the treaty, its conditions, how the defeated Russian czar or Austrian emperor would look, and exactly how the achievement of this particular goal would position Napoleon for his next campaign.

As a young man, Lyndon B. Johnson, despite his limited education, was determined to become president one day. Dream turned into obsession: he could picture himself as president, strutting the world's stage. As he advanced in his career, he never did anything without one eye on this ultimate objective. In 1957, Johnson, by that time a Texas senator, supported a civil rights bill. That damaged him in Texas but elevated him

*Plot against the difficult while it remains easy,*
*Act against the great while it is still minute.*
*Difficult affairs throughout the realm invariably commence with the easy,*
*Great affairs throughout the realm inevitably commence with the small.*
*For this reason the Sage never acts against the great and is thus able to complete greatness.*
*What is tranquil remains easily grasped,*
*What has not yet betrayed signs is easy to plot against.*
*The brittle is easily split,*
*The minute is easily scattered.*
*Act upon them before they attain being,*
*Control them before they become chaotic.*
*Trees that require both arms to embrace*
*Are born from insignificant saplings.*
*A nine-story tower commences with a little accumulated earth,*
*A journey of a thousand kilometers begins beneath one's feet.*

*Tao Te Ching*, Lao-tzu, circa 551–479 B.C.

THE WILD BOAR AND THE FOX

*A wild boar was sharpening his tusks on a tree trunk one day. A fox asked him why he did this when there was neither huntsman nor danger threatening him. "I do so for a good reason," he replied. "For if I am suddenly surprised by danger I wouldn't have the time to sharpen my tusks. But now I will find them ready to do their duty."*

The fable shows that it is no good waiting until danger comes to be ready.

*FABLES*, AESOP, SIXTH CENTURY B.C.

nationally: apparently a senator from the South had stuck his neck out, risking his job. Johnson's vote caught the attention of John F. Kennedy, who, in the campaign of 1960, nominated him for vice president—the job that was ultimately Johnson's stepping-stone to the presidency.

Clear long-term objectives give direction to all of your actions, large and small. Important decisions become easier to make. If some glittering prospect threatens to seduce you from your goal, you will know to resist it. You can tell when to sacrifice a pawn, even lose a battle, if it serves your eventual purpose. Your eyes are focused on winning the campaign and nothing else.

Your goals must be rooted in reality. If they are simply beyond your means, essentially impossible for you to realize, you will grow discouraged, and discouragement can quickly escalate into a defeatist attitude. On the other hand, if your goals lack a certain dimension and grandeur, it can be hard to stay motivated. Do not be afraid to be bold. In the large sense, you are working out for yourself what Alexander experienced as his destiny and what Friedrich Nietzsche called your "life's task"—the thing toward which your natural leanings and aptitudes, talents and desires, seem to point you. Assigning yourself a life task will inspire and guide you.

The goal's nature is critical: some objectives, if realized, will hurt you in the long run. The objectives of grand strategy in the true sense are to build a solid foundation for future expansion, to make you more secure, to increase your power. When Israel seized the Sinai Desert during the Six-Day War in 1967, what seemed to make sense was creating a kind of buffer zone between itself and Egypt. In fact, this just meant more territory to patrol and control, and it created a cause to motivate enduring hostility in the Egyptian populace. The Sinai was also vulnerable to surprise attack, which is what ended up happening in the Yom Kippur War of 1973. Since holding on to the desert, though seductive, ultimately disserved the needs of security, in the terms of grand strategy it was probably a mistake. It is sometimes hard to know what the long-term effects of achieving a goal will be, but the more seriously and realistically you examine the possibilities downwind, the less likely you are to miscalculate.

**Widen your perspective.** Grand strategy is a function of vision, of seeing further in time and space than the enemy does. The process of foresight is unnatural: we can only ever live in the present, which is the ground for our consciousness, and our subjective experiences and desires narrow the scope of our vision—they are like a prison we inhabit. Your task as a grand strategist is to force yourself to widen your view, to take in more of the world around you, to see things for what they are and for how they may play out in the future, not for how you wish them to be. Every event has a reason, a causal chain of relationships that made it happen; you have to dig deep into that reality, instead of seeing only the surfaces of things. The closer you get to objectivity, the better your strategies and the easier the path to your goals.

You can take a step in this direction by always trying to look at the world through the eyes of other people—including, most definitely, your enemy—before engaging in war. Your own cultural preconceptions are a major hindrance to seeing the world objectively. Looking through other people's eyes is not a question of political correctness or of some soft, hazy sensitivity; it makes your strategies more effective. During the Vietnam War, the North Vietnamese intensely studied the American cultural scene. They looked for shifts in public opinion and strained to understand the U.S. political system and the social effects of television. American strategists, on the other hand, revealed an absolutely minimal understanding of the alien cultures of Vietnam—whether the South Vietnamese culture they were supporting or the North Vietnamese culture they were trying to fight. Blinded by their obsession with stopping the spread of communism, they failed to note the far deeper influences of culture and religion on the North Vietnamese way of fighting. Theirs was a grand-strategic blunder of the highest order.

Grand strategists keep sensitive antennae attuned to the politics of any situation. Politics is the art of promoting and protecting your own interests. You might think it was largely a question of parties and factions, but every individual is, among other things, a political creature seeking to secure his or her own position. Your behavior in the world always has political consequences, in that the people around you will analyze it in terms of whether it helps or harms them. To win the battle at the cost of alienating potential allies or creating intractable enemies is never wise.

Taking politics into account, you must figure out your grand strategy with a mind to gaining support from other people—to creating and strengthening a base. In the Roman Civil War in 49 B.C., Julius Caesar faced off against Pompey, who was then the more experienced military man. Caesar gained the edge by planning his maneuvers with an eye to their effect on public opinion in Rome. Lacking support in the Senate, he built support among the general public. Caesar was a brilliant political animal, and what made him so was his grasp of the public psyche: he understood their self-interest and shaped his strategies accordingly. Being political means understanding people—seeing through their eyes.

**Sever the roots.** In a society dominated by appearances, the real source of a problem is sometimes hard to grasp. To work out a grand strategy against an enemy, you have to know what motivates him or is the source of his power. Too many wars and battles drag on because neither side knows how to strike at the other's roots. As a grand strategist, you must expand your vision not only far and wide but under. Think hard, dig deep, do not take appearances for reality. Uncover the roots of the trouble and you can strategize to sever them, ending the war or problem with finality.

When the Carthaginian general Hannibal invaded Italy in 218 B.C., various Roman generals strained to defeat him, but none was effective.

*This is as it should be. No major proposal required for war can be worked out in ignorance of political factors; and when people talk, as they often do, about harmful political influence on the management of war, they are not really saying what they mean. Their quarrel should be with the policy itself, not with its influence. If the policy is right—that is, successful—any intentional effect it has on the conduct of war can only be to the good. If it has the opposite effect the policy itself is wrong. Only if statesmen look to certain military moves and actions to produce effects that are foreign to their nature do political decisions influence operations for the worse. In the same way as a man who has not fully mastered a foreign language sometimes fails to express himself correctly, so statesmen often issue orders that defeat the purpose they are meant to serve. Time and again that has happened, which demonstrates that a certain grasp of military affairs is vital for those in charge of general policy. Before continuing, we must guard against a likely misinterpretation. We are far from believing that a minister of war immersed in his files, an erudite engineer or even an experienced soldier would, simply on the*

The Roman general later called Scipio Africanus looked at the situation differently: the problem wasn't Hannibal himself, or his base in Spain, or his ability to restock his supplies by sea from Carthage; the problem was Carthage itself. This was a country with an intractable hatred of Rome, and a long power struggle had endured between the two. Instead of taking on Hannibal, a brilliant military man, in Italy, then, Scipio invaded Carthage, forcing Hannibal to leave Italy to defend his homeland. The attack on Carthage was more than a mere feint to draw Hannibal away; it was a sizable invasion. Scipio's grand strategy worked to perfection: not only did he defeat Hannibal in battle, he destroyed Carthage as a rival power, permanently ending its ability to stand up to Rome.

A part of grand strategy related to severing the roots is seeing dangers as they start to sprout, then cutting them down before they get too big to handle. A grand strategist knows the value of preemptive action.

**Take the indirect route to your goal.** The greatest danger you face in strategy is losing the initiative and finding yourself constantly reacting to what the other side does. The solution, of course, is to plan ahead but also to plan subtly—to take the indirect route. Preventing your opponent from seeing the purpose of your actions gives you an enormous advantage.

So make your first move merely a setup, designed to extract a response from your opponent that opens him up to what comes next. Hit him directly and he reacts, taking a defensive pose that may allow him to parry your next blow; but if he can't see the point of your strike, or if it misleads him as to where the next one will come from, he is defenseless and blind. The key is to maintain control of your emotions and plot your moves in advance, seeing the entire chessboard.

The film director Alfred Hitchcock made this strategy a life principle. His every action a setup designed to yield results down the road, he calmly thought ahead and moved step by step. His goal was to make a film that matched his original vision, uncorrupted by the influence of the actors, producers, and other staff who necessarily came along later. By controlling every detail of the film's screenplay, he made it almost impossible for the producer to interfere. Should the producer try to meddle during the actual shooting, Hitchcock would have a camera ready on set with no film in it. He could pretend to take the extra shots that the producer wanted, letting the producer feel powerful without risk to the end result. Hitchcock did the same with actors: instead of telling them directly what to do, he would infect them with the emotion he wanted—fear, anger, desire—by the way he treated them on set. Every step on the campaign trail fit perfectly into the next one.

In working on the level not of the battle but of the campaign, your first step is crucial. It should usually be deceptively soft and indirect, making it harder to read. The Japanese bombing of Pearl Harbor during World War II was a devastating surprise, but as the first move of a campaign it was a disaster. The Japanese showed their hand too quickly; rallying American public opinion to an intense level of anger, they ensured that the Americans would prosecute the war to the bitter end—and it was the Americans who had the greater military resources. Always pay attention to the first step of the campaign. It sets the tempo, determines the enemy's mind-set, and launches you in a direction that had better be the right one.

The Prussian military theoretician Carl von Clausewitz famously argued that war is the continuation of politics by other means. He meant that every nation has goals—security, well-being, prosperity—that it ordinarily pursues through politics, but when another nation or internal force thwarts their achievement through politics, war is the natural result. War is never merely about victory on the battlefield or the simple conquest of land; it is about the pursuit of a policy that cannot be realized in any other way than through force.

When a war is lost, however, all fingers usually point at the military. We may sometimes go over the generals' heads, to the politicians who declared war in the first place; during and after the Vietnam War, for example, some blamed the loss on the government's failure to commit to the war with full effort. More often, though, the postgame analysis is military—we pore over the war's battles, critiquing the officers' moves. And of course it is the military that has planned and fought the war, but even so, the real problem is a problem of grand strategy. According to von Clausewitz, failure in war is a failure of policy. The goals of the war, and the policies that drove it, were unrealistic, inappropriate, blind to other factors.

This idea is the philosophy of the grand strategist. Whenever anything goes wrong, it is human nature to blame this person or that. Let other people engage in such stupidity, led around by their noses, seeing only what is immediately visible to the eye. You see things differently. When an action goes wrong—in business, in politics, in life—trace it back to the policy that inspired it in the first place. The goal was misguided.

This means that you yourself are largely the agent of anything bad that happens to you. With more prudence, wiser policies, and greater vision, you could have avoided the danger. So when something goes wrong, look deep into yourself—not in an emotional way, to blame yourself or indulge your feelings of guilt, but to make sure that you start your next campaign with a firmer step and greater vision.

*basis of their particular experience, make the best director of policy—always assuming that the prince himself is not in control. Far from it. What is needed in the post is distinguished intellect and strength of character. He can always get the necessary military information somehow or other. The military and political affairs of France were never in worse hands than when the brothers Belle-Isle and the Duc de Choiseul were responsible—good soldiers though they all were.*

*On War*,
Carl von Clausewitz,
1780–1831

**Image:**
The Mountaintop.
Down on the battle-
field, everything is
smoke and confusion. It is
hard to tell friend from foe,
to see who is winning, to foresee
the enemy's next move. The general
must climb high above the fray, to the
mountaintop, where everything becomes
clearer and more in focus. There he can see beyond
the battlefield—to the movements of reserves, to
the enemy camp, to the battle's future shape. Only
from the mountaintop can the general direct the war.

**Authority:** It is a common mistake in going to war to begin at the wrong end, to act first and to wait for disaster to discuss the matter. *—Thucydides (between 460 and 455 B.C.–circa 400 B.C.)*

REVERSAL

Grand strategy involves two dangers that you must consider and combat. First, the successes it brings you in your first campaigns may have the same effect on you that easy victory on the battlefield gives a general: drunk on triumph, you may lose the sense of realism and proportion on which your future moves depend. Even such supreme grand strategists as Julius Caesar and Napoleon eventually fell victim to this dynamic: losing their sense of reality, they began to believe that their instincts were infallible. The greater the victory, the greater the danger. As you get older, as you move to your next campaign, you must retrench, strain doubly hard to rein in your emotions, and maintain a sense of realism.

Second, the detachment necessary to grand strategy can bring you to a point where you find it hard to act. Understanding the world too well, you see too many options and become as indecisive as Hamlet. No matter how far we progress, we remain part animal, and it is the animal in us that fires our strategies, gives them life, animates us to fight. Without the desire to fight, without a capacity for the violence war churns up, we cannot deal with danger.

The prudent Odysseus types are comfortable with both sides of their nature. They plan ahead as best they can, see far and wide, but when it comes time to move ahead, they move. Knowing how to control your emotions means not repressing them completely but using them to their best effect.

# 13

# KNOW YOUR ENEMY
# THE INTELLIGENCE STRATEGY

*The target of your strategies should be less the army you face than the mind of the man or woman who runs it. If you understand how that mind works, you have the key to deceiving and controlling it. Train yourself to read people, picking up the signals they unconsciously send about their innermost thoughts and intentions. A friendly front will let you watch them closely and mine them for information. Beware of projecting your own emotions and mental habits onto them; try to think as they think. By finding your opponents' psychological weaknesses, you can work to unhinge their minds.*

*He who knows the enemy and himself Will never in a hundred battles be at risk.*

SUN-TZU, FOURTH CENTURY B.C.

## THE MIRRORED ENEMY

In June 1838, Lord Auckland, the British governor-general of India, called a meeting of his top officials to discuss a proposed invasion of Afghanistan. Auckland and other British ministers had become increasingly concerned at Russia's growing influence in the area. The Russians had already made an ally of Persia; they were now trying to do the same with Afghanistan, and if they were successful, the British in India would find themselves potentially cut off by land to the west and vulnerable to more incursions by the Russians. Instead of trying to outdo the Russians and negotiate an alliance with the Afghan ruler, Dost Mahomed, Auckland proposed what he thought was a surer solution: invade Afghanistan and install a new ruler—Shah Soojah, a former Afghan leader forced out of power twenty-five years earlier—who would then be indebted to the English.

*[As to the second case], that of being drawn into one [a trap or ambush] . . . you must be shrewd about not believing easily things not in accord with reason. For example, if the enemy puts some booty before you, you ought to believe that within it there is a hook and that it conceals some trick. If many of the enemy are put to flight by your few, if a few of the enemy assail your many, if the enemy turn in sudden flight, . . . you ought to fear a trick. And you should never believe that the enemy does not know how to carry on his affairs; rather, if you hope to be less deceived . . . and . . . run less risk, in proportion as your enemy is weaker, in proportion as he is less cautious, you should the more respect him.*

*THE ART OF WAR*, NICCOLÒ MACHIAVELLI, 1521

Among the men listening to Auckland that day was William Macnaghten, the forty-five-year-old chief secretary of the Calcutta government. Macnaghten thought the invasion a brilliant idea: a friendly Afghanistan would secure British interests in the area and even help to spread British influence. And the invasion could hardly fail. The British army would have no trouble sweeping away the primitive Afghan tribesmen; they would present themselves as liberators, freeing the Afghans from Russian tyranny and bringing to the country the support and civilizing influence of England. As soon as Shah Soojah was in power, the army would leave, so that British influence over the grateful shah, although powerful, would be invisible to the Afghan public. When it came time for Macnaghten to give his opinion on the proposed invasion, his support of it was so sound and enthusiastic that Lord Auckland not only decided to go ahead, he named Macnaghten the queen's envoy to Kabul, the Afghan capital—the top British representative in Afghanistan.

Meeting little resistance along the way, in August 1839 the British army reached Kabul. Dost Mahomed fled to the mountains, and the shah reentered the city. To the local inhabitants, this was a strange sight: Shah Soojah, whom many could barely remember, looked old and submissive alongside Macnaghten, who rode into Kabul wearing a bright-colored uniform topped by a cocked hat fringed with ostrich feathers. Why had these people come? What were they doing here?

With the shah back in power, Macnaghten had to reassess the situation. Reports came in informing him that Dost Mahomed was building an army in the mountains to the north. Meanwhile, to the south, it seemed that in invading the country the British had insulted some local chieftains by plundering their lands for food. These chiefs were now stirring up trouble. It was also clear that the shah was unpopular with his former subjects, so unpopular that Macnaghten could not leave him and other British interests in the country unprotected. Reluctantly Macnaghten ordered most of the British army to remain in Afghanistan until the situation was stabilized.

Time went by, and eventually Macnaghten decided to allow the officers and soldiers of this increasingly long-standing occupying force to send for their families, so that life would be less harsh for them. Soon the wives and children came, along with their Indian servants. But where Macnaghten had imagined that the arrival of the soldiers' families would have a humanizing, civilizing effect, it only alarmed the Afghans. Were the British planning a permanent occupation? Everywhere the local people looked, there were representatives of British interests, talking loudly in the streets, drinking wine, attending theaters and horse races—strange imported pleasures that they had introduced to the country. Now their families were making themselves at home. A hatred of everything English began to take root.

There were those who warned Macnaghten about this, and to all of them he had the same answer: everything would be forgotten and forgiven when the army left Afghanistan. The Afghans were childlike, emotional people; once they felt the benefits of English civilization, they would be more than grateful. One matter, however, did worry the envoy: the British government was unhappy about the increasing expense of the occupation. Macnaghten would have to do something to cut costs, and he knew just where to begin.

Most of the mountain passes through which Afghanistan's main trade routes ran were held by the Ghilzye tribes, who for many years, over the lives of many different rulers of the country, had been paid a stipend to keep the passes open. Macnaghten decided to halve this stipend. The Ghilzyes responded by blocking the passes, and elsewhere in the country tribes sympathetic to the Ghilzyes rebelled. Macnaghten, caught off guard, tried to put these rebellions down, but he did not take them too seriously, and worried officers who told him to respond more vigorously were rebuked for overreacting. Now the British army would have to stay indefinitely.

The situation deteriorated quickly. In October 1841 a mob attacked the home of a British official and killed him. In Kabul local chiefs began to conspire to expel their British overlords. Shah Soojah panicked. For months he had begged Macnaghten to let him capture and kill his main rivals, an Afghan ruler's traditional method of securing his position. Macnaghten had told him that a civilized country did not use murder to solve its political problems. The shah knew that the Afghans respected strength and authority, not "civilized" values; to them his failure to deal with his enemies made him look weak and unrulerlike and left him surrounded by enemies. Macnaghten would not listen.

The rebellion spread, and Macnaghten now had to confront the fact that he did not have the manpower to put down a general uprising. But why should he panic? The Afghans and their leaders were naïve; he would regain the upper hand through intrigue and cleverness. To that end, Macnaghten publicly negotiated an agreement whereby British troops and citizens would leave Afghanistan, in exchange for which the

THE AGING LION AND THE FOX

*A lion who was getting old and could no longer obtain his food by force decided that he must resort to trickery instead. So he retired to a cave and lay down pretending to be ill. Thus, whenever any animals came to his cave to visit him, he ate them all as they appeared.*
*When many animals had disappeared, a fox figured out what was happening. He went to see the lion but stood at a safe distance outside the cave and asked him how he was.*
*"Oh, not very well," said the lion. "But why don't you come in?"*
*But the fox said: "I would come inside if I hadn't seen that a lot of footprints are pointing inwards towards your cave but none are pointing out."*
Wise men note the indications of dangers and thus avoid them.

*FABLES,* AESOP, SIXTH CENTURY B.C.

*Bait.—"Everyone has his price"—this is not true. But there surely exists for everyone a bait he cannot help taking. Thus to win many people over to a cause one needs only to put on it a gloss of philanthropy, nobility, charitableness, self-sacrifice—and on to what cause can one not put it?—: these are the sweetmeats and dainties for* their *soul; others have others.*

HUMAN, ALL TOO HUMAN, FRIEDRICH NIETZSCHE, 1886

Afghans would supply the retreating British with food. Privately, though, Macnaghten made it known to a few key chiefs that he was willing to make one of them the country's vizier—and load with him money—in exchange for putting down the rebellion and allowing the English to stay.

The chief of the eastern Ghilzyes, Akbar Khan, responded to this offer, and on December 23, 1841, Macnaghten rode out for a private meeting with him to seal the bargain. After exchanging greetings Akbar asked Macnaghten if he wanted to go ahead with the treachery they were planning. Thrilled to have turned the situation around, Macnaghten cheerily answered that he did. Without a word of explanation, Akbar signaled his men to grab Macnaghten and throw him in prison—he had no intention of betraying the other chiefs. Along the way a mob developed, caught hold of the unfortunate envoy, and with a fury built up over years of humiliation literally tore him to pieces. His limbs and head were paraded through the streets of Kabul, and his torso was hung from a meat hook in the bazaar.

In a matter of days, everything unraveled. The remaining British troops—some 4,500 of them, along with 12,000 camp followers—were forced to agree to an immediate retreat from Afghanistan, despite the bitter winter weather. The Afghans were to keep the retreating army supplied but did not do so. Certain that the British would never leave unless forced to, they harassed them relentlessly in their retreat. Civilians and soldiers alike quickly perished in the snow.

On January 13, British forces at the fort in Jalalabad saw a single horse struggling toward the gates. Its half-dead rider, Dr. William Brydon, was the sole survivor of the British army's doomed invasion of Afghanistan.

### Interpretation

The knowledge that would have averted the catastrophe was at Macnaghten's fingertips long before he launched the expedition. Englishmen and Indians who had lived in Afghanistan could have told him that the Afghan people were among the proudest and most independent on the planet. To them the image of foreign troops marching into Kabul would constitute an unforgivable humiliation. On top of that, they were not a people yearning for peace, prosperity, and reconciliation. In fact, they saw strife and confrontation as a healthy way of life.

Macnaghten had the information but refused to see it. Instead he projected onto the Afghans the values of an Englishman, which he mistakenly assumed were universal. Blinded by narcissism, he misread every signal along the way. As a result his strategic moves—leaving the British army occupying Kabul, halving the Ghilzyes' stipend, trying not to overplay his hand in putting down the rebellions—were exactly the opposite of what was needed. And on that fateful day when he literally lost his head, he made the ultimate miscalculation, imagining that money and an appeal to self-interest would buy loyalty among the very people he had so humiliated.

Blindness and narcissism like this are not so rare; we find them every day. Our natural tendency is to see other people as mere reflections of our own desires and values. Failing to understand the ways they are not like us, we are surprised when they do not respond as we had imagined. We unintentionally offend and alienate people, then blame them, not our inability to understand them, for the damage done.

Understand: if you let narcissism act as a screen between you and other people, you will misread them and your strategies will misfire. You must be aware of this and struggle to see others dispassionately. Every individual is like an alien culture. You must get inside his or her way of thinking, not as an exercise in sensitivity but out of strategic necessity. Only by knowing your enemies can you ever hope to vanquish them.

*Be submissive so that he will trust you and you will thereby learn about his true situation. Accept his ideas and respond to his affairs as if you were twins. Once you have learned everything, subtly gather in his power. Thus when the ultimate day arrives, it will seem as if Heaven itself destroyed him.*

—*Tai Kung,* Six Secret Teachings *(circa fourth century* B.C.*)*

*Confucius's evaluation of Yang Hu, a man who had been forced to flee from one state to another because he proved greedy and disloyal each time he acquired power, provides a simple example of projecting behavior on the basis of constancy. Based upon this repeated behavioral pattern, Confucius accurately predicted that Yang Hu would certainly suffer an ignominious end. More generally, Mencius subsequently stated: "A man who ceases his efforts where he should not will abandon them anywhere. A man who is parsimonious with those with whom he should be generous will be parsimonious everywhere." Granting that people generally acquire fixed habits early in life, a man's end may therefore be foreseen by midlife: "Someone who is still disliked at forty years of age will end by being so."*

Ralph D. Sawyer, *The Tao of Spycraft*, 1998

## THE CLOSE EMBRACE

In 1805, Napoleon Bonaparte humiliated the Austrians in the battles of Ulm and Austerlitz. In the subsequent treaty, he carved up the Austrian Empire, taking over its lands in Italy and Germany. For Napoleon all this was one part of a chess game. His ultimate goal was to make Austria an ally—a weak and subordinate ally, but one that would lend him weight in the courts of Europe, since Austria had been a central force in European politics. As part of this overall strategy, Napoleon requested a new Austrian ambassador to France: Prince Klemens von Metternich, at the time the Austrian ambassador to the Prussian court in Berlin.

Metternich, then thirty-two, came from one of Europe's most illustrious families. A speaker of impeccable French, a staunch conservative in politics, he was a paragon of breeding and elegance and an inveterate ladies' man. The presence of this polished aristocrat would add a sheen to the imperial court that Napoleon was creating. More important, winning over a man of such power—and Napoleon could be quite seductive in private meetings—would help in his grand strategy of making Austria a weak satellite. And Metternich's weakness for women would give Napoleon a way in.

The two men met for the first time in August 1806, when Metternich presented his credentials. Napoleon acted coolly. He dressed well for the occasion but kept his hat on, which in the mores of the time was rather rude. After Metternich's speech—short and ceremonious—Napoleon began to pace the room and talk politics in a way that made it clear he

*Coordination is less of a problem when political leaders themselves play an active part in the intelligence effort. When he was Senate majority leader, Lyndon Johnson cultivated an extensive intelligence system with sources all over Washington. At one point in the 1950s, Johnson complained to a reporter that he was focusing on internal Democratic problems while failing to cover divisions in the Senate GOP. To make his point, he pulled out a memorandum on a recent private meeting at which the reporter and several of his colleagues had gotten a briefing on GOP factionalism from Senator Thurston Morton (R-KY). Rowland Evans and Robert Novak recalled: "The Intelligence System was a marvel of efficiency. It was also rather frightening." Even in the White House, Johnson believed in firsthand political intelligence. According to his aide Harry McPherson, "I guess he called a lot of people, but I could usually count on it in the late afternoon, as he woke up from his nap, that I would get a call which would usually say, 'What do you know?' " McPherson would then pass along the latest news that he picked up from reporters and political figures.*

THE ART OF POLITICAL WARFARE, JOHN J. PITNEY, JR., 2000

was in command. (He liked to stand up to talk to people while they remained seated.) He made a show of speaking pointedly and concisely; he was not some Corsican rube for the sophisticated Metternich to play with. In the end he was sure he had made the impression he wanted.

Over the months to come, Napoleon and Metternich had many more such meetings. It was the emperor's plan to charm the prince, but the charm ran inescapably the other way: Metternich had a way of listening attentively, making apt comments, even complimenting Napoleon on his strategic insights. At those moments Napoleon would beam inside: here was a man who could truly appreciate his genius. He began to crave Metternich's presence, and their discussions of European politics became more and more frank. The two became friends of sorts.

Hoping to take advantage of Metternich's weakness for women, Napoleon set up his sister, Caroline Murat, to have an affair with the prince. He learned from her a few pieces of diplomatic gossip, and she told him that Metternich had come to respect him. In turn she also told Metternich that Napoleon was unhappy with his wife, Empress Josephine, who could not bear children; he was considering divorce. Napoleon did not seem upset that Metternich knew such things about his personal life.

In 1809, seeking revenge for its ignominious defeat at Austerlitz, Austria declared war on France. Napoleon only welcomed this event, which gave him a chance to beat the Austrians still more soundly than before. The war was hard fought, but the French prevailed, and Napoleon imposed a humiliating settlement, annexing whole sections of the Austrian Empire. Austria's military was dismantled, its government was overhauled, and Napoleon's friend Metternich was named foreign minister—exactly where Napoleon wanted him.

Several months later something happened that caught Napoleon slightly off guard but delighted him: the Austrian emperor offered him his eldest daughter, the Archduchess Marie Louise, in marriage. Napoleon knew that the Austrian aristocracy hated him; this had to be Metternich's work. Alliance by marriage with Austria would be a strategic tour de force, and Napoleon happily accepted the offer, first divorcing Josephine, then marrying Marie Louise in 1810.

Metternich accompanied the archduchess to Paris for the wedding, and now his relationship with Napoleon grew still warmer. Napoleon's marriage made him a member of one of Europe's greatest families, and to a Corsican, family was everything; he had won a dynastic legitimacy he had long craved. In conversation with the prince, he opened up even more than before. He was also delighted with his new empress, who revealed a keen political mind. He let her in on his plans for empire in Europe.

In 1812, Napoleon invaded Russia. Now Metternich came to him with a request: the formation of an army of 30,000 Austrian soldiers at Napoleon's disposal. In return Napoleon would let Austria rebuild its military. Napoleon saw no harm in this step; he was allied with Austria by marriage, and rearmament there would help him in the end.

Months later the Russian invasion had turned into a disaster, and Napoleon was forced to retreat, his army decimated. Now Metternich offered his services as a mediator between France and the other European powers. Centrally placed as it is, Austria had performed that task in the past, and anyway Napoleon had little choice: he needed time to recoup. Even if Austria's role as a mediator allowed it to reassert its independence, he had little to fear from his in-laws.

By the spring of 1813, negotiations had broken down and a new war was about to break out between the badly damaged France and a powerful alliance of Russia, Prussia, England, and Sweden. By this time the Austrian army had grown considerably; somehow Napoleon had to get his hands on it—but his spies reported that Metternich had entered into secret agreement with the Allies. Surely this had to be some sort of ploy: how could the Austrian emperor fight his son-in-law? Yet in a few weeks, it became official: unless France negotiated a peace, Austria would drop its mediating position and join the Allies.

Napoleon could not believe what he was hearing. He traveled to Dresden for a meeting with Metternich, which took place on June 26. The moment he saw the prince, he felt a shock: the friendly, nonchalant air was gone. In a rather cold tone, Metternich informed him that France must accept a settlement that would reduce it to its natural boundaries. Austria was obligated to defend its interests and the stability of Europe. Suddenly it occurred to the emperor: Metternich had been playing him all along, the family ties merely a ploy to blind him to Austrian rearmament and independence. "So I have perpetrated a very stupid piece of folly in marrying an archduchess of Austria?" Napoleon blurted out. "Since Your Majesty desires to know my opinion," Metternich replied, "I will candidly say that Napoleon, the conqueror, has made a mistake."

Napoleon refused to accept Metternich's dictated peace. In return Austria dropped its neutrality and joined the Allies, becoming their de facto military leader. And with Austria leading the way, they finally defeated Napoleon in April 1814 and exiled him to the Mediterranean island of Elba.

*In all the martial arts, in all the performing arts and still more in all the forms of human behavior, a man's postures or moves are based on the movements of his [invisible] mind. . . . In the Kage Style of swordsmanship a swordsman reads his opponent's mind through his postures or moves. . . .*
*What mind can penetrate his opponent's mind? It is a mind that has been trained and cultivated to the point of detachment with perfect freedom. It is as clear as a mirror that can reflect the motions within his opponent's mind. . . . When one stands face to face with his opponents, his mind must not be revealed in the form of moves. Instead his mind should reflect his opponent's mind like water reflecting the moon.*

LIVES OF MASTER SWORDSMEN, MAKOTO SUGAWARA, 1988

## Interpretation

Napoleon prided himself on his ability to gauge people's psychology and use it against them, but in this case he was outwitted by a man far superior at such a game. Metternich's modus operandi was the following: he would quietly study his enemies from behind his smiling, elegant exterior, his own apparent relaxation inviting them to open up. In his very first meeting with Napoleon, he saw a man straining to impress: he noticed that the bantam Napoleon walked on his toes, to look taller, and struggled to suppress his Corsican accent. Later meetings only confirmed Metternich's impression of a man who craved acceptance as the social equal of Europe's aristocracy. The emperor was insecure.

This insight won, Metternich used it to craft the perfect counter-

*When Munenori was granted an audience with the shogun, he sat down, put his hands on the tatami floor, as retainers always did to show their respect to their master. Suddenly, Iemitsu thrust a spear at the "unsuspecting" Munenori—and was surprised to find himself lying flat on his back! Munenori had sensed the shogun's intention before a move had been made, and swept Iemitsu's legs out from under him at the instant of the thrust.*

LIVES OF MASTER SWORDSMEN, MAKOTO SUGAWARA, 1988

strategy: the offer of marriage into the Austrian dynasty. To a Corsican, that would mean everything, and it would blind Napoleon to a simple reality: for aristocrats like Metternich and the Austrian emperor, family ties meant nothing compared to the survival of the dynasty itself.

Metternich's genius was to recognize the appropriate target for his strategy: not Napoleon's armies, which Austria could not hope to defeat—Napoleon was a general for the ages—but Napoleon's mind. The prince understood that even the most powerful of men remains human and has human weaknesses. By entering Napoleon's private life, being deferential and subordinate, Metternich could find his weaknesses and hurt him as no army could. By getting closer to him emotionally—through the emperor's sister Caroline, through the Archduchess Marie Louise, through their convivial meetings—he could choke him in a friendly embrace.

Understand: your real enemy is your opponent's mind. His armies, his resources, his intelligence, can all be overcome if you can fathom his weakness, the emotional blind spot through which you can deceive, distract, and manipulate him. The most powerful army in the world can be beaten by unhinging the mind of its leader.

And the best way to find the leader's weaknesses is not through spies but through the close embrace. Behind a friendly, even subservient front, you can observe your enemies, get them to open up and reveal themselves. Get inside their skin; think as they think. Once you discover their vulnerability—an uncontrollable temper, a weakness for the opposite sex, a gnawing insecurity—you have the material to destroy them.

*War is not an act of the will aimed at inanimate matter, as it is in the mechanical arts. . . . Rather, [it] is an act of the will aimed at a living entity that* reacts.

*—Carl von Clausewitz (1780–1831)*

## KEYS TO WARFARE

The greatest power you could have in life would come neither from limitless resources nor even consummate skill in strategy. It would come from clear knowledge of those around you—the ability to read people like a book. Given that knowledge, you could distinguish friend from foe, smoking out snakes in the grass. You could anticipate your enemies' malice, pierce their strategies, and take defensive action. Their transparency would reveal to you the emotions they could least control. Armed with that knowledge, you could make them tumble into traps and destroy them.

This kind of knowledge has been a military goal since the dawn of history. That is why the arts of intelligence gathering and spying were invented. But spies are unreliable; they filter information through their own preconceptions and prejudices, and since their trade places them precisely between one side and the other and forces them to be independent operators, they are notoriously hard to control and can turn against you. Then, too, the nuances that give people away—the tone in a speaker's voice, the look in his or her eyes—are inevitably missing from their reports. In the end the spy's information means nothing unless you are adept at interpreting human behavior and psychology. Without that skill you will see in it what you want to see, confirming your own prejudices.

The leaders who have made best use of intelligence—Hannibal, Julius Caesar, Prince Metternich, Winston Churchill, Lyndon Johnson during his career in the U.S. Senate—were all first and foremost great students of human nature and superior readers of men. They honed their skills through personal observation of people. Only with that foundation could the use of spies extend their powers of vision.

The first step in the process is to get over the idea that people are impenetrable mysteries and that only some trick will let you peek into their souls. If they seem mysterious, it is because almost all of us learn to disguise our true feelings and intentions from an early age. If we went around showing just how we felt and telling people what we planned to do, we would make ourselves vulnerable to malice, and if we always spoke our minds, we would offend a lot of people unnecessarily. So as we grow up, concealing much of what we are thinking becomes second nature.

This deliberate opacity makes the intelligence game difficult but not impossible. For even as people consciously struggle to conceal what is going on in their minds, they unconsciously want to reveal themselves. Hiding how we feel in social situations is exhausting; being able to show ourselves is a relief. We secretly want people to know us, even including our dark side. Even while we consciously struggle to control this hidden yearning, unconsciously we are always sending out signals that reveal a part of what is going on inside—slips of the tongue, tones of voice, styles of dress, nervous twitches, sudden irrational actions, a look in the eye that contradicts our words, the things we say after a drink.

Understand: day in and day out, people emit signals that reveal their intentions and deepest desires. If we do not pick them up, it is because we are not paying attention. The reason for this is simple: we are usually locked up in our own worlds, listening to our internal monologues, obsessed with ourselves and with satisfying our own egos. Like William Macnaghten, we tend to see other people merely as reflections of ourselves. To the extent that you can drop your self-interest and see people for who they are, divorced from your desires, you will become more sensitive to their signals.

The ability to read people was a critical survival skill for Japanese samurai and was particularly emphasized by the Shinkage school of swordsmanship. One of the school's earliest masters was the seventeenth-century samurai Yagyu Munenori. One spring afternoon in his later

*In my opinion, there are two kinds of eyes: one kind simply looks at things and the other sees through things to perceive their inner nature. The former should not be tense [so as to observe as much as possible]; the latter should be strong [so as to discern the workings of the opponent's mind clearly]. Sometimes a man can read another's mind with his eyes. In fencing, it is all right to allow your own eyes to express your will but never let them reveal your mind. This matter should be considered carefully and studied diligently.*

MIYAMOTO MUSASHI, 1584–1645

Anger as spy.—*Anger empties out the soul and brings even its dregs to light. That is why, if we know no other way of discovering the truth of the matter, we must know how to put our acquaintances, our adherents and opponents, into a rage, so as to learn all that is really being thought and undertaken against us.*

*HUMAN, ALL TOO HUMAN*, FRIEDRICH NIETZSCHE, 1886

*Then David fled from Nai'oth in Ramah, and came and said before Jonathan, "What have I done? What is my guilt? And what is my sin before your father, that he seeks my life?" And he said to him, "Far from it! You shall not die. Behold, my father does nothing either great or small without disclosing it to me; and why should my father hide this from me? It is not so." But David replied, "Your father knows well that I have found favor in your eyes; and he thinks, 'Let not Jonathan know this, lest he be grieved.' But truly, as the Lord lives and as your soul lives, there is but a step between me and death." Then said Jonathan to David, "Whatever you say, I will do for you." David said to Jonathan, "Behold, tomorrow is the new moon, and I should not fail to sit at table with the king; but let me go, that I may hide myself in the field till the third day at evening. If your father misses me at all, then say, 'David earnestly asked leave of me to run to Bethlehem his city; for there is a yearly sacrifice there for all the family.' If he says, 'Good!' it will be well with your servant; but if he is angry, then know that evil is determined by him." . . . And Jonathan said to David, "Come, let us go out into the field." So they both went out into the field. . . . So David hid himself in the field; and when the new moon came, the king sat down to eat food. The king*

years, Munenori was taking a peaceful walk through his gardens, admiring the cherry blossoms. He was accompanied by a page/protector, who walked behind him, sword raised, as was the custom. Suddenly Munenori stopped in his tracks. He had a feeling of danger. Looking around, he saw nothing to warrant this feeling, but even so he was so troubled that he returned to his house and sat with his back against a post to prevent a surprise attack.

After Munenori had sat there for a while, his page asked him what was the matter. The samurai confessed that while looking at the cherry blossoms he had had an intimation of imminent danger, of an enemy on the attack. What troubled him now was that the danger had apparently been imaginary—he must have hallucinated it. A samurai depended on his keen instincts to anticipate attack. If Munenori had lost that power, his life as a warrior was over.

Suddenly the page threw himself to the ground and confessed: as Munenori walked in the garden, the thought had come to the page that if he were to strike at his master while the samurai was lost in admiration of the cherry blossoms, not even this gifted swordsman could have parried his attack. Munenori had not lost his skill at all; quite the contrary—his incomparable sensitivity to other people's emotions and thoughts had allowed him to pick up sensations from someone behind his back, rather as a horse senses the energy of its rider or a dog the movements of its owner. An animal has that sensitivity because it pays complete attention. Similarly, the Shinkage school taught warriors to empty their minds, centering themselves in the moment as animals did and keeping themselves from getting derailed by any particular thought. This would allow the Shinkage warrior to read in his opponent's elbow or hand the slight tension that signaled an attack; he could look through his opponent's eyes and sense the coming blow or notice the nervous shuffle of the feet that indicated fear or confusion. A master like Munenori could virtually read someone's thoughts when the other person wasn't even visible.

The power taught by the Shinkage school—the same power possessed by Prince Metternich—was the ability to let go of one's ego, to submerge oneself temporarily in the other person's mind. You will be amazed at how much you can pick up about people if you can shut off your incessant interior monologue, empty your thoughts, and anchor yourself in the moment. The details you now see give you unfiltered information from which you can put together an accurate picture of people's weaknesses and desires. Be particularly attentive to their eyes: it takes a lot of effort to hide the eyes' message about a person's state of mind.

According to the baseball pitcher Bob Lemon, the great player Ted Williams "was the only hitter who you felt saw through you." In the struggle between pitcher and batter, the pitcher has the advantage of knowing what pitch he will throw. The hitter can only guess at that, which is why even the best of them usually connect only one out of every three or four times. Somehow Williams changed those odds.

Williams's method wasn't magic, or even intuition; it was simple enough. He made baseball pitchers his study, watching their patterns over the course of a game, a season, a career. He would ask the pitchers on his own team endless questions about their process, trying to get a feel for how they thought. At the plate he would empty his mind of everything but the pitcher, noticing the slightest hitch in his windup or change in his grip—anything that would signal his intentions. The end result seemed uncanny: at bat, Williams was able to think himself into the pitcher's mind and anticipate the pitch that was coming. Sometimes he would even see himself as another person—a pitcher trying to outwit the great hitter Ted Williams. As Williams demonstrates, the ability to mimic and get inside your enemies' thought patterns depends on collecting as much information on them as you can, analyzing their past behavior for its habitual patterns, and being alert to the signs they give off in the present.

It is of course critical that people be unaware you are watching them so closely. A friendly front, like Prince Metternich's to Napoleon, will help disguise what you're doing. Do not ask too many questions; the trick is to get people to relax and open up without prodding, shadowing them so quietly that they never guess what you're really up to.

Information is useless unless you know how to interpret it, how to use it to tell appearance from reality. You must learn how to recognize a range of psychological types. Be alert, for instance, to the phenomenon of the masked opposite: when someone strikingly manifests a particular personality trait, that trait may well be a cover-up. The oily character who is ingratiatingly effusive with flattery may be hiding hostility and ill will; the aggressive bully may be hiding insecurity; the moralizer may be making a show of purity to hide nefarious desires. Whether they're throwing dust in your eyes or their own—they may be trying to convince themselves that they're not what they're afraid they are—the opposite trait lurks below the surface.

In general, it is easier to observe people in action, particularly in moments of crisis. Those are the times when they either reveal their weakness or struggle so hard to disguise it that you can see through the mask. You can actively probe them by doing things that seem harmless but get a response—maybe say something bold or provocative, then see how they react. Making people emotional, pushing their buttons, will touch some deep part of their nature. Either they will let slip some truth about themselves or they will put on a mask that you, in the laboratory situation you have created, will be able to peer behind.

A critical part of understanding people is gauging their powers of resistance. Without that knowledge you will either over- or underestimate them, depending on your own levels of fear and confidence. You need to know how much fight people have in them. Someone hiding his cowardice and lack of resolve can be made to surrender with a single violent push; someone desperate who has little to lose will fight to the bitter end.

*sat upon his seat, as at other times, upon the seat by the wall; Jonathan sat opposite, and Abner sat by Saul's side, but David's place was empty. Yet Saul did not say anything that day; for he thought, "Something has befallen him; he is not clean, surely he is not clean." But on the second day, the morrow after the new moon, David's place was empty. And Saul said to Jonathan his son, "Why has not the son of Jesse come to the meal, either yesterday or today?" Jonathan answered Saul, "David earnestly asked leave of me to go to Bethlehem; he said, 'Let me go; for our family holds a sacrifice in the city, and my brother has commanded me to be there. So now, if I have found favor in your eyes, let me get away, and see my brothers.' For this reason he has not come to the king's table." Then Saul's anger was kindled against Jonathan, and he said to him, "You son of a perverse, rebellious woman, do I not know that you have chosen the son of Jesse to your own shame, and to the shame of your mother's nakedness? For as long as the son of Jesse lives upon the earth, neither you nor your kingdom shall be established. Therefore send and fetch him to me, for he shall surely die." Then Jonathan answered Saul his father, "Why should he be put to death? What has he done?" But Saul cast his spear at him to smite him; so Jonathan knew that his father was determined to put David to death. And Jonathan rose from the table in fierce anger and ate no food the second day of the month, for he was grieved for David, because his father had disgraced him.*

1 SAMUEL 20:1–11, 24–34

The Mongols used to begin their campaigns with a battle whose only purpose was to test their opponent's strength and resolve. They would never deal with an enemy until they had gauged his morale. This set-up battle also had the benefit of revealing something of his strategy and thought.

The quality of the information you gather on your enemies is more important than the quantity. A single but crucial nugget can be the key to their destruction. When the Carthaginian general Hannibal saw that the Roman general he was facing was arrogant and hot-tempered, he would deliberately play weak, luring the man into a rash attack. Once Churchill saw that Hitler had a paranoid streak, becoming irrational at the merest hint of vulnerability, the British prime minister knew how to unhinge the German führer: by feigning to attack some marginal area like the Balkans, he could make him see threats on all sides and spread out his defenses, a critical military mistake.

Motive of attack.—*One attacks someone not only so as to harm him or to overpower him but perhaps only so as to learn how strong he is.*

FRIEDRICH NIETZSCHE, 1844–1900

In 1988, Lee Atwater was a political strategist on the team of the senior George Bush, who was then in the race to become that year's Republican presidential nominee. Discovering that Bush's main rival, Senator Robert Dole, had a terrible temper that his aides had to struggle to control, Atwater devised endless stratagems to push Dole's buttons. Not only did an upset Dole look unpresidential to the American public, but an emotional and angry man rarely thinks straight. A disturbed mind is one you can control and unbalance at will.

There are, of course, limits to how much intelligence gathering you can achieve by firsthand observation. A network of spies will extend your vision, particularly as you learn to interpret the information they bring you. An informal network is the best—a group of allies recruited over time to be your eyes and ears. Try to make friends with people at or near the source of information on your rival; one well-placed friend will yield far more than will a handful of paid spies. In Napoleon's time his intelligence network was second to none, but his best information came from friends whom he had carefully positioned in diplomatic circles around Europe.

*Last year at a great conference there was a certain man who explained his dissenting opinion and said he was resolved to kill the conference leader if it was not accepted. His motion was passed. After the procedures were over, the man said, "Their assent came quickly. I think that they are too weak and unreliable to be counselors to the master."*

*HAGAKURE: THE BOOK OF THE SAMURAI*, YAMAMOTO TSUNETOMO, 1659–1720

Always look for internal spies, people in the enemy camp who are dissatisfied and have an ax to grind. Turn them to your purposes and they will give you better information than any infiltrator you sneak in from outside. Hire people the enemy has fired—they will tell you how the enemy thinks. President Bill Clinton got his best intelligence on the Republicans from his adviser Dick Morris, who had worked for them for years and knew their weaknesses, both personal and organizational. A warning: never rely on one spy, one source of information, no matter how good. You risk being played or getting slanted, one-sided information.

Many people leave a paper trail of writings, interviews, and so on that is as revealing as anything you can learn from a spy. Well before World War II, Adolf Hitler's book *Mein Kampf* supplied a blueprint of his thinking and intentions, not to mention endless clues to his psychology.

His generals Erwin Rommel and Heinz Guderian also wrote about the new kind of blitzkrieg warfare they were preparing. People reveal a lot about themselves in their writing, partly intentionally—they are out to explain themselves, after all—and partly helplessly to the skilled reader-between-the-lines.

Finally, the enemy you are dealing with is not an inanimate object that will simply respond in an expected manner to your strategies. Your enemies are constantly changing and adapting to what you are doing. Innovating and inventing on their own, they try to learn from their mistakes and from your successes. So your knowledge of the enemy cannot be static. Keep your intelligence up to date, and do not rely on the enemy's responding the same way twice. Defeat is a stern teacher, and your beaten opponent today may be wiser tomorrow. Your strategies must take this possibility into account; your knowledge of the enemy must be not just deep but timely.

*Colonel John Cremony commented on their adeptness at seeming to "disappear" when he wrote "an Apache can conceal his swart body amidst the green grass, behind brown shrubs, or gray rocks, with so much address and judgement that any but the experienced would pass him by without detection at the distance of three or four yard" and noted that "they will watch for days, scanning your every movement, observing your every act; taking exact note of your party and all its belongings. Let no one suppose that these assaults are made upon the spur of the moment by bands accidentally encountered. Far from it; they are almost invariably the results of long watching—patient waiting—careful and rigorous observation, and anxious counsel."*

WARRIORS: WARFARE AND THE NATIVE AMERICAN INDIAN, NORMAN BANCROFT-HUNT, 1995

**Image:** The Shadow. Everyone has a shadow, a secret self, a dark side. This shadow comprises everything people try to hide from the world—their weaknesses, secret desires, selfish intentions. This shadow is invisible from a distance; to see it you must get up close, physically and most of all psychologically. Then it will come into relief. Follow close in your target's footsteps and he will not notice how much of his shadow he has revealed.

**Authority:** Thus the reason the farsighted ruler and his superior commander conquer the enemy at every move, and achieve successes far beyond the reach of the common crowd, is advance knowledge. Such knowledge cannot be had from ghosts and spirits, educed by comparison with past events, or verified by astrological calculations. It must come from people—people who know the enemy's situation. *—Sun-tzu (fourth century B.C.)*

*In principle, I should lay it down that the existence of secret agents should not be tolerated, as tending to augment the positive dangers of the evil against which they are used. That the spy will fabricate his information is a commonplace. But in the sphere of political and revolutionary action, relying partly on violence, the professional spy has every facility to fabricate the very facts themselves, and will spread the double evil of emulation in one direction, and of panic, hasty legislation, unreflecting hate, on the other.*

THE SECRET AGENT, JOSEPH CONRAD, 1857–1924

## REVERSAL

Even as you work to know your enemies, you must make yourself as formless and difficult to read as possible. Since people really only have appearances to go on, they can be readily deceived. Act unpredictably now and then. Throw them some golden nugget of your inner self—something fabricated that has nothing to do with who you really are. Be aware that they are scrutinizing you, and either give them nothing or feed them misinformation. Keeping yourself formless and inscrutable will make it impossible for people to defend themselves against you and render the intelligence they gather on you useless.

# 14

# OVERWHELM RESISTANCE WITH SPEED AND SUDDENNESS

## THE BLITZKRIEG STRATEGY

*In a world in which many people are indecisive and overly cautious, the use of speed will bring you untold power. Striking first, before your opponents have time to think or prepare, will make them emotional, unbalanced, and prone to error. When you follow with another swift and sudden maneuver, you will induce further panic and confusion. This strategy works best with a setup, a lull—your unexpected action catches your enemy off guard. When you strike, hit with unrelenting force. Acting with speed and decisiveness will garner you respect, awe, and irresistible momentum.*

*War is such that the supreme consideration is speed. This is to take advantage of what is beyond the reach of the enemy, to go by way of routes where he least expects you, and to attack where he has made no preparations.*

SUN-TZU, FOURTH CENTURY B.C.

## SLOW-SLOW-QUICK-QUICK

In 1218, Muhammad II, the shah of Khwarizm, received a visit from three ambassadors on behalf of Genghis Khan, the leader of the Mongol Empire to the east. The visitors bore magnificent gifts and, more important, the offer of a treaty between the two powers that would allow the reopening of the lucrative Silk Road connecting China and Europe. The shah's empire was immense, incorporating present-day Iran and much of Afghanistan. His capital, Samarkand, was fabulously wealthy, a symbol of his power, and increased trade along the route would only add to these riches. Since the Mongols made it clear they considered him the superior partner in the deal, the shah decided to sign the treaty.

A few months later, a Mongol caravan arrived in the city of Otrar, in the northeastern corner of the shah's empire, with the mission to buy luxury items for the Mongol court. The governor of Otrar suspected the men in the caravan of being spies. He had them killed, and he seized the goods they had brought to barter. Hearing of this outrage, Genghis Khan dispatched an ambassador, escorted by two soldiers, to the shah, calling for an apology. The demand—which presumed to put the two empires on equal footing—incensed the shah. He had the ambassador's head cut off and sent back to Genghis Khan. This, of course, meant war.

The shah was not afraid: his army, anchored by its well-trained Turkish cavalry, numbered over 400,000, at least twice the enemy's size. By defeating the Mongols in battle, the shah could finally take over their land. He assumed the Mongols would attack Transoxiana, the easternmost part of the shah's empire. Bordered to the east by the five-hundred-mile-long Syr Dar'ya River, to the north by the Kizil Kum Desert, and to the west by the Amu Dar'ya River, Transoxiana's interior was also home to two of the most important cities of the empire, Samarkand and Bukhara. The shah decided to establish a cordon of soldiers along the Syr Dar'ya, which the Mongols would have to cross to enter his empire. They could not cross from the north—the desert was impassable—and to go to the south would be too great a detour. Keeping the bulk of his army in the interior of Transoxiana, he would be able to place reinforcements where needed. He had an impregnable defensive position and superiority in numbers. Let the Mongols come. He would crush them.

In the summer of 1219, scouts reported that the Mongols were approaching the southern end of the Syr Dar'ya, through the Fergana Valley. The shah sent a large force, under the leadership of his son Jalal ad-Din, to destroy the enemy. After a fierce battle, the Mongols retreated. Jalal ad-Din reported back to his father that the Mongol army was not nearly as fearsome as their reputation. The men looked haggard, their horses emaciated, and they seemed none too eager to sustain a fight. The shah, believing the Mongols no match for his army, put more troops at the southern end of the cordon and waited.

A few months later, a Mongol battalion appeared without warning in the north, attacking the city of Otrar and capturing its governor—the same man responsible for the outrage against the Mongol traders. The Mongols killed him by pouring molten silver into his eyes and ears. Stunned by how quickly they had managed to reach Otrar, and from an unexpected direction, the shah decided to shift more troops to the north. These barbarians might move swiftly, he reasoned, but they could not overcome an entrenched army with such great numbers.

Next, however, two Mongol armies swept south from Otrar, running parallel to the Syr Dar'ya. One, under General Jochi, began to attack key towns along the river, while the other, under General Jebe, disappeared to the south. Like locusts, Jochi's army swarmed through the hills and lowlands near the river. The shah moved a good portion of his army to the river, keeping some reserves in Samarkand. Jochi's force was relatively small, 20,000 at most; these mobile units hit one position after another, without warning, burning forts and wreaking havoc.

Reports from the front lines began to give the shah a picture of these strange warriors from the east. Their army was all cavalry. Each Mongol not only rode a horse but was trailed by several more riderless horses, all mares, and when his own horse tired, he would mount a fresh one. These horses were light and fast. The Mongols were unencumbered by supply wagons; they carried their food with them, drank the mares' milk and blood, and killed and ate the horses that had become weak. They could travel twice as fast as their enemy. Their marksmanship was extraordinary—facing forward or in retreat, they could fire arrows with remarkable dexterity, making their attacks far deadlier than anything the shah's army had ever seen. Their divisions communicated over great distances with flags and torches; their maneuvers were precisely coordinated and nearly impossible to anticipate.

Dealing with this constant harassment exhausted the shah's forces. Now, suddenly, the army under General Jebe that had disappeared to the south reappeared heading northwest into Transoxiana at remarkable speed. The shah hurriedly sent south his last reserves, an army of 50,000, to do battle with Jebe. He still was not worried—his men had proved their superiority in direct combat, at the battle in the Fergana Valley.

This time, however, it was different. The Mongols unleashed strange weapons: their arrows were dipped in burning tar, which created smoke screens behind which their lightning-quick horsemen advanced, opening breaches in the lines of the shah's army through which more heavily armed cavalry would advance. Chariots darted back and forth behind the Mongol lines, bringing up constant supplies. The Mongols filled the sky with arrows, creating relentless pressure. They wore shirts of heavy silk. An arrow that managed to pierce the shirt would rarely reach the flesh and could be extracted easily by pulling at the shirt, all of this done while moving at great speed. Jebe's army annihilated the shah's forces.

The shah had one option left: retreat to the west, retrench, and slowly rebuild his army. As he began preparations, however, something beyond belief occurred: an army under Genghis Khan himself suddenly stood outside the gates of the city of Bukhara, to the west of Samarkand. Where had they come from? They could not have crossed the Kizil Kum Desert to the north. Their appearance seemed all but impossible, as if the devil himself had conjured them up. Bukhara soon fell, and within days Samarkand followed. Soldiers deserted, generals panicked. The shah, fearing for his life, fled with a handful of soldiers. The Mongols pursued him relentlessly. Months later, on a small island in the Caspian Sea, abandoned by one and all, wearing rags and begging for food, the former ruler of the wealthiest empire in the East finally died of hunger.

Interpretation

When Genghis Khan became the leader of the Mongol nation, he inherited perhaps the fastest army on the planet, but their swiftness had translated into limited military success. The Mongols might have perfected the art of fighting on horseback, but they were too undisciplined to exploit any advantage they gained this way or to coordinate for a large-scale attack. The genius of Genghis Khan was to transform the chaotic Mongol speed into something organized, disciplined, and strategic. He achieved this by adapting the ancient Chinese strategy of slow-slow-quick-quick.

The first step, a "slow," was to meticulously prepare before any campaign, which the Mongols always did to the highest degree. (In planning for the attack on the shah, the Mongols learned of a guide who knew of a chain of oases across the Kizil Kum Desert. This man was captured and later led Genghis Khan's army across the forbidding territory.) The second "slow" was a setup, which involved getting the enemy to lower its guard, lulling it into complacency. The Mongols, for example, deliberately lost the first battle in the Fergana Valley to feed the shah's arrogance. Then came the first "quick": fixing the enemy's attention forward with a swift frontal attack (Jebe's raids along the river). The final "quick" was a doubly swift blow from an unexpected direction. (Genghis's sudden appearance before the gates of Bukhara is considered by many the greatest military surprise in history.) A master of psychological warfare, Genghis Khan understood that men are most terrified by the unknown and unpredictable. The suddenness of his attacks made the speed of them doubly effective, leading to confusion and panic.

We live in a world in which speed is prized above almost all else, and acting faster than the other side has itself become the primary goal. But most often people are merely in a hurry, acting and reacting frantically to events, all of which makes them prone to error and wasting time in the long run. In order to separate yourself from the pack, to harness a speed that has devastating force, you must be organized and strategic. First, you prepare yourself before any action, scanning your enemy for weaknesses. Then you find a way to get your opponents to underestimate you, to lower their guard. When you strike unexpectedly, they will freeze up. When you hit again, it is from the side and out of nowhere. It is the unanticipated blow that makes the biggest impact.

*The less a thing is foreseen, the more . . . fright does it cause. This is nowhere seen better than in war, where every surprise strikes terror even to those who are much the stronger.*
*—Xenophon (430?–355? B.C.)*

## KEYS TO WARFARE

In May 1940 the German army invaded France and the Low Countries using a new form of warfare: the blitzkrieg. Advancing with incredible speed, the Germans coordinated tanks and airplanes in an attack that culminated in one of the quickest and most devastating victories in military history. The success of the blitzkrieg was largely due to the Allies' static, rigid defense—similar to the shah's defense against the Mongols. When the Germans breached this defense, the Allies could not adjust or react in time. The Germans advanced faster than their enemy could process what was happening. By the time the Allies decided upon a counterstrategy it was too late—conditions had changed. They were always a step behind.

Now more than ever, we find ourselves dealing with people who are defensive and cautious, who begin any action from a static position. The reason is simple: the pace of modern life is increasingly growing faster, full of distractions, annoyances, and interruptions. The natural response for many is to retreat inward, to erect psychological walls against the harsh realities of modern life. People hate the feeling of being rushed and are terrified of making a mistake. They unconsciously try to slow things down—by taking longer to make decisions, being noncommittal, defensive, and cautious.

Blitzkrieg warfare, adapted for daily combat, is the perfect strategy for these times. While those around you remain defensive and immobile, you surprise them with sudden and decisive action, forcing them to act before they are ready. They cannot respond, as they usually do, by being elusive or cautious. They will most likely become emotional and react imprudently. You have breached their defenses, and if you keep up the pressure and hit them again with something unexpected, you will send them into a kind of downward psychological spiral—pushing them into mistakes, which further deepens their confusion, and so the cycle goes on.

Many of those who practiced a form of blitzkrieg warfare on the battlefield used it to great effect in daily life. Julius Caesar—master of speed and surprise—was a great example of this. From out of the blue, he might form an alliance with a senator's bitterest enemy, forcing the senator either to change his opposition to Caesar or to risk a dangerous confrontation. Equally, he might unexpectedly pardon a man who had

CHEN/THE AROUSING (SHOCK, THUNDER)

*The hexagram Chen represents the eldest son, who seizes rule with energy and power. A yang line develops below two yin lines and presses upward forcibly. This movement is so violent that it arouses terror. It is symbolized by thunder, which bursts forth from the earth and by its shock causes fear and trembling.*

*THE JUDGEMENT*
*SHOCK brings success.*
*Shock comes—oh, oh!*
*Laughing words—*
*ha, ha!*
*The shock terrifies for a hundred miles. . . .*

*THE I CHING*, CHINA, CIRCA EIGHTH CENTURY B.C.

fought against him. Caught off guard, the man would become a loyal ally. Caesar's reputation for doing the unpredictable made people all the more cautious in his presence, further enhancing his ability to catch those around him unawares.

This strategy works wonders on those who are particularly hesitant and afraid of making any kind of mistake. In similar fashion, if you are facing an enemy that has divided leadership or internal cracks, a sudden and swift attack will make the cracks larger and cause internal collapse. Half of the success of Napoleon Bonaparte's form of blitzkrieg warfare was that he used it against armies of allies in which several bickering generals were in charge of strategy. Once his army broke through these armies' defenses, dissension would break out and they would fall apart from within.

The blitzkrieg strategy can be effective in diplomacy, too, as Henry Kissinger demonstrated. The former U.S. secretary of state would often take his time when beginning diplomatic negotiations, lulling the other side with bland banter. Then, with the deadline for the end of the talks approaching, he would suddenly hit them with a list of demands. Without enough time to process what was happening, they became prone to giving in or to becoming emotional and making mistakes. This was Kissinger's version of slow-slow-quick-quick.

For their initial thrust into France during World War II, the Germans chose to attack through the Ardennes Forest in southern Belgium. The forest, considered impenetrable by tank, was lightly guarded. Pushing through this weak point, the Germans were able to build up speed and momentum. In launching a blitzkrieg, you must begin by finding your enemy's weak point. Initiating the action where there will be less resistance will allow you to develop crucial momentum.

The success of this strategy depends on three things: a group that is mobile (often, the smaller the better), superior coordination between the parts, and the ability to send orders quickly up and down the chain of command. Do not depend on technology to accomplish this. During the Vietnam War, the U.S. military might in fact have been hindered by its superior communications—too much information to be processed made for slower response times. The North Vietnamese, who depended on a well-coordinated network of spies and informers, not gadgetry, made decisions more quickly and as a result were more nimble on the ground.

Shortly after being elected president in 1932, Franklin D. Roosevelt seemed to disappear from the public stage. The Depression was at its height, and for many Americans this was not very reassuring. Then, with his inauguration, Roosevelt changed tempo, giving a rousing speech that showed he had in fact been meditating deeply on the issues facing the country. In the weeks that followed, he came at Congress fast and furiously, with a series of bold legislative proposals. The intensity of this new direction was felt all the more because of the slow setup. More than mere drama, the momentum built by this strategy helped Roosevelt to

*But the genius in Ali takes his limitations and makes them virtues. Let us go step by step. I can't think of a past heavyweight champion whose punching power wasn't superior to Ali's. Yet in his first twenty fights, Ali, then Cassius Clay, won every one of them, scoring seventeen knockouts. . . . So what is Ali's mystery? Why does a man who every expert agrees has no punching power knock most of his opponents out, including a one-punch K.O. over Sonny Liston on Ali's first defense of his title? The answer is in speed and timing. Clay then, and Ali now, has the ability to let punches go with extreme quickness, but most important, at the right moment, just before the man in front of him is able to put his boxer's sense of anticipation to work. When that happens, the man getting hit doesn't see the punch. As a result, this man's brain can't prepare him to receive the impact of the blow. The eyes couldn't send the message back to the part of the body which would take the shock. So we arrive at one knockout of a conclusion: the punch that puts you to sleep is not so much the hard punch as the punch that you don't see coming.*

STING LIKE A BEE, JOSÉ TORRES AND BERT RANDOLPH SUGAR, 1971

convince the public that he meant business and was leading the country in the right direction. This momentum translated into support for his policies, which in turn helped spur confidence and turn the economy around.

Speed, then, is not only a powerful tool to use against an enemy, but it can also have a bracing, positive influence on those on your side. Frederick the Great noted that an army that moves quickly has higher morale. Velocity creates a sense of vitality. Moving with speed means there is less time for you and your army to make mistakes. It also creates a bandwagon effect: more and more people admiring your boldness, will decide to join forces with you. Like Roosevelt, make such decisive action as dramatic as possible: a moment of quiet and suspense on the stage before you make your startling entrance.

*Veni, vidi, vici* (I came, I saw, I conquered).

JULIUS CAESAR, 100–44 B.C.

**Image:** The Storm. The sky becomes still and calm, and a lull sets in, peaceful and soothing. Then, out of nowhere, lightning strikes, the wind picks up . . . and the sky explodes. It is the suddenness of the storm that is so terrifying.

**Authority:** You must be slow in deliberation and swift in execution. *—Napoleon Bonaparte (1769–1821)*

## REVERSAL

Slowness can have great value, particularly as a setup. To appear slow and deliberate, even a little foolish, will lull your enemies, infecting them with a sleepy attitude. Once their guard is down, an unexpected blow from the side will knock them out. Your use of slowness and speed, then, should be deliberate and controlled, never a natural tempo that you fall into.

In general, when facing a fast enemy, the only true defense is to be as fast or faster. Only speed can neutralize speed. Setting up a rigid defense, as the shah did against the Mongols, only plays into the hands of the swift and mobile.

# 15

# CONTROL THE DYNAMIC

## FORCING STRATEGIES

*People are constantly struggling to control you—getting you to act in their interests, keeping the dynamic on their terms. The only way to get the upper hand is to make your play for control more intelligent and insidious. Instead of trying to dominate the other side's every move, work to define the nature of the relationship itself. Shift the conflict to terrain of your choice, altering the pace and stakes to suit you. Maneuver to control your opponents' minds, pushing their emotional buttons, and compelling them to make mistakes. If necessary, let them feel they are in control in order to get them to lower their guard. If you control the overall direction and framing of the battle, anything they do will play into your hands.*

*"Pressing down the pillow" refers to one's efforts not to let the head of one's opponent rise. In battles based on martial strategy, it is taboo to let your opponent take the initiative, thus putting yourself on the defensive. You must try at all costs to lead your opponent by taking complete control of him. During combat, your opponent intends to dominate you as much as you want to dominate him, so it is vital that you pick up on your opponent's intentions and tactics so as to control him. . . . According to the principle of martial strategy, you must be able to control your opponent(s) at all times. Study this point well.*

THE BOOK OF FIVE RINGS, MIYAMOTO MUSASHI, 1584–1645

## THE ART OF ULTIMATE CONTROL

Control is an issue in all relationships. It is human nature to abhor feelings of helplessness and to strive for power. Whenever two people or groups interact, there is a constant maneuvering between them to define the relationship, to determine who has control over this and that. This battle of wills is inevitable. Your task as a strategist is twofold: First, recognize the struggle for control in all aspects of life, and never be taken in by those who claim they are not interested in control. Such types are often the most manipulative of all. Second, you must master the art of moving the other side like pieces on a chessboard, with purpose and direction. This art was cultivated by the most creative generals and military strategists throughout the ages.

War is above all else a struggle over who can control the actions of the other side to a greater extent. Military geniuses such as Hannibal, Napoleon, and Erwin Rommel discovered that the best way to attain control is to determine the overall pace, direction, and shape of the war itself. This means getting enemies to fight according to your tempo, luring them onto terrain that is unfamiliar to them and suited to you, playing to your strengths. And, most important of all, it means gaining influence over the frame of mind of your opponents, adapting your maneuvers to their psychological weaknesses.

The superior strategist understands that it is impossible to control exactly how an enemy will respond to this move or that. To attempt to do so will only lead to frustration and exhaustion. There is too much in war and in life that is unpredictable. But if the strategist can control the mood and mind-set of his enemies, it does not matter exactly how they respond to his maneuvers. If he can make them frightened, panicky, overly aggressive, and angry, he controls the wider scope of their actions and can trap them mentally before cornering them physically.

Control can be aggressive or passive. It can be an immediate push on the enemy, making him back up and lose the initiative. It can be playing possum, getting the enemy to lower his guard, or baiting him into a rash attack. The artist of control weaves both of these into a devastating pattern—hitting, backing off, baiting, overwhelming.

This art is infinitely applicable to the battles of everyday life. Many people tend to play unconscious games of domination or get caught up in trying to control someone else's every move. In trying to manage and determine too much, they exhaust themselves, make mistakes, push people away, and in the end lose control of the situation. If you understand and master the art, you will instantly become more creative in your approach to influencing and controlling the other side. By determining people's moods, the pace at which they must move, the stakes involved, you will find that almost anything people do in response to your maneuvers will fit into the overall dynamic you have shaped. They may know they are being controlled but be helpless to fight it, or they may

move in the direction you desire without realizing it. That is ultimate control.
The following are the four basic principles of the art.

*In short, I think like Frederick [the Great], one should always be the first to attack.*

NAPOLEON BONAPARTE, 1769–1821

**Keep them on their heels.** Before the enemy makes a move, before the element of chance or the unexpected actions of your opponents can ruin your plans, you make an aggressive move to seize the initiative. You then keep up a relentless pressure, exploiting this momentary advantage to the fullest. You do not wait for opportunities to open up; you make them yourself. If you are the weaker side, this will often more than level the playing field. Keeping your enemies on the defensive and in react mode will have a demoralizing effect on them.

**Shift the battlefield.** An enemy naturally wants to fight you on familiar terrain. Terrain in this sense means all of the details of the battle—the time and place, exactly what is being fought over, who is involved in the struggle, and so on. By subtly shifting your enemies into places and situations that are not familiar to them, you control the dynamic. Without realizing what is happening, your opponents find themselves fighting on your terms.

**Compel mistakes.** Your enemies depend on executing a strategy that plays to their advantages, that has worked in the past. Your task is twofold: to fight the battle in such a way that they cannot bring their strength or strategy into play and to create such a level of frustration that they make mistakes in the process. You do not give them enough time to do anything; you play to their emotional weaknesses, making them as irritable as possible; you bait them into deadly traps. It is less your action than their missteps that give you control.

**Assume passive control.** The ultimate form of domination is to make those on the other side think they are the ones in control. Believing they are in command, they are less likely to resist you or become defensive. You create this impression by moving with the energy of the other side, giving ground but slowly and subtly diverting them in the direction you desire. It is often the best way to control the overly aggressive and the passive-aggressive.

*One who excels at warfare compels men and is not compelled by others.*
*—Sun-tzu (fourth century B.C.)*

## HISTORICAL EXAMPLES

**1.** By the end of 1940, British forces in the Middle East had been able to secure their position in Egypt and take back a good part of Libya that the

Italians (an ally of Germany) had seized early in World War II. Having captured the important port town of Benghazi, the British were poised to advance farther west, all the way to Tripoli, allowing them to push the Italians out of the country for good. Then, unexpectedly, a halt was called in their advance. General Archibald Wavell, commander in chief of British forces in the Middle East, was waging battle on too many fronts. Since the Italians had proved themselves to be rather inept in desert warfare, the British felt they could afford to create a defensive line in Libya, build up their forces in Egypt, and launch a major offensive against the Italians by April of the following year.

News that a German armored brigade under the leadership of General Erwin Rommel had arrived in Tripoli in February 1941 did not alter the British plans. Rommel had been a superb commander during the blitzkrieg in France the previous year. But here he was under Italian command, dependent on the incompetent Italians for supplies, and his force was too small to make the British nervous. In addition, intelligence reports revealed that Hitler had sent him there with orders to do no more than block the British from advancing to Tripoli.

Then, without warning, at the end of March 1941, Rommel's tanks swept eastward. Rommel had broken up his small force into columns, and he hurled them in so many directions against the British defensive line that it was hard to fathom his intentions. These mechanized columns moved with incredible speed; advancing at night with lights dimmed, time and again they caught their enemy by surprise, suddenly appearing to their flank or rear. As their line was breached in multiple places, the British were compelled to retreat farther and farther east. To Wavell, who was following these events from Cairo, this was downright shocking and humiliating: Rommel was causing chaos with a disproportionately small number of tanks and severe supply limitations. Within a few weeks, the Germans had advanced to the border of Egypt.

What was most devastating about this offensive was the novel way in which Rommel fought. He used the desert as if it were an ocean. Despite supply problems and the difficult terrain, he kept his tanks in perpetual motion. The British could not let up their guard for a moment, and this mentally exhausted them. But his movements, though seemingly random, were always for a purpose. If he wanted to take a particular city, he would head in the opposite direction, then circle and attack from an unexpected side. He brought along an armada of trucks to kick up enough dust so that the British could not see where he was headed and to give the impression of a much larger force than was actually on the attack.

Rommel would ride with the front line, risking death so that he could make rapid judgments on the move, sending his columns here and there before the British had time to figure out the game. And he used his tanks in the opposite way of the British, to deadly effect. Instead of pushing them forward to punch holes in the enemy lines, he would send out his weakest tanks, then have them retreat at first contact; the British tanks would invariably swallow the bait and go in pursuit, kicking up so much of their own dust in the process that they would not see they were running straight into a line of German antitank guns. Once a sufficient number of British tanks had been taken out, Rommel would advance again, wreaking havoc behind the British lines.

*Given the same amount of intelligence, timidity will do a thousand times more damage in war than audacity.*

CARL VON CLAUSEWITZ, 1780–1831

Kept constantly on their heels, forced to come to rapid decisions in response to Rommel's moves, the British made endless mistakes. Not knowing where he might show up next, or from what direction, they spread their forces over dangerously vast areas. Before long, at the mere mention that a German column was approaching, Rommel at its head, the British would abandon their positions, even though they greatly outnumbered him. In the end the only thing that stopped him was Hitler's obsession with Russia, which bled Rommel of the supplies and reinforcements he needed to conquer Egypt.

### Interpretation

This is how Rommel analyzed the situation first confronting him: The enemy had a strong position to the east, which would only get stronger as more supplies and men came from Egypt. Rommel had a much smaller force, and the longer he waited, the more useless it would become. And so he decided to disobey Hitler's orders, risking his career on a truth he had learned in the blitzkrieg in France: making the first hit against the enemy completely alters the dynamic. If the enemy is the stronger side, it is upsetting and discouraging to be suddenly put on the defensive. Being larger and unprepared makes it harder to organize an orderly retreat.

To get his strategy to work, Rommel had to create maximum disorder in the enemy. In the ensuing confusion, the Germans would seem more formidable than they were. Speed, mobility, and surprise—as agents of such chaos—became ends in themselves. Once the enemy was on its heels, a deceptive maneuver—heading one way, then attacking from another—had double the effect. An enemy that is in retreat and without time to think will make endless mistakes if you keep up the pressure. Ultimately, the key to Rommel's success was to seize the initiative with one bold maneuver, then exploit this momentary advantage to the fullest.

Everything in this world conspires to put you on the defensive. At work, your superiors may want the glory for themselves and will discourage you from taking the initiative. People are constantly pushing and attacking you, keeping you in react mode. You are continually reminded of your limitations and what you cannot hope to accomplish. You are made to feel guilty for this and that. Such defensiveness on your part can become a self-fulfilling prophecy. Before anything, you need to liberate yourself from this feeling. By acting boldly, before others are ready, by moving to seize the initiative, you create your own circumstances rather than simply waiting for what life brings you. Your initial push alters the situation, on your terms. People are made to react to you, making you seem larger and more powerful than may be the case. The respect and fear you inspire will translate into offensive power, a reputation that precedes you. Like Rommel, you must also have a touch of madness: ready to disorient and confuse for its own sake, to keep advancing no matter the circumstance. It is up to you—be constantly defensive or make others feel it instead.

**2.** In 1932, Paramount Pictures, following a craze for gangster films, began production on *Night After Night.* The film was to star George Raft, who had recently made a name for himself in *Scarface.* Raft was cast to type as a typical gangster. But *Night After Night,* in a twist, was to have a comic edge to it. The producer, William Le Baron, was afraid that there was no one in the cast who had a light enough touch to pull this off. Raft, hearing of his concern, suggested he hire Mae West.

West was a celebrity in vaudeville and on Broadway, starring in plays she had written. She had made her reputation as a saucy, aggressive blonde with a devastating wit. Hollywood producers had thought of her before, but she was too bawdy for film. And by 1932 she was thirty-nine years old, on the plump side, and considered too old to be making a film debut. Nevertheless, Le Baron was willing to take a risk to liven up the picture. She would create a splash, provide an angle for promotion, then be sent back to Broadway, where she belonged. Paramount offered her a two-month contract at five thousand dollars a week, a generous deal for the times. West happily accepted.

At first West was a little difficult. She had been told to lose some weight, but she hated dieting and quickly gave up. Instead she had her hair dyed a rather indecent platinum blond. She hated the script—the dialog was flat and her character unimportant. The part needed to be rewritten, and West offered her services as a writer. Hollywood people were used to dealing with difficult actresses and had a panoply of tactics for taming them, particularly those who wanted their parts rewritten. What was unusual was an actress who offered to rewrite her own lines. Baffled by the request, even from someone who had written for Broadway, the studio executives came back with a firm refusal. Giving her that privilege would set a terrible precedent. West countered by refusing to continue with the film until they let her rewrite the dialog.

Paramount boss Adolph Zukor had seen West's screen test and liked her look and manner. The picture needed her. Zukor had a studio executive take her out to dinner on her birthday to try to cajole her; the goal was to calm her down enough so that they could begin shooting. Once cameras were rolling, he thought, they would find a way to get West to behave. But that night at dinner, West pulled out a check from her handbag and handed it to the executive. It was for twenty thousand dollars,

*When they came to the ford of Xanthus,*
*The eddying river that Zeus begot,*
*Achilles split the Trojans. Half he chased*
*Toward the city, across the plain where yesterday*
*The Greeks had fled from Hector's shining rage.*
*Hera, to slow this stampede of Trojans,*
*Spread a curtain of fog between them.*

*The others swerved—*
*And found themselves herded into the river.*
*They crashed down into the deep, silver water*
*As it tumbled and roared through its banks.*
*You could hear their screams as they floundered*
*And were whirled around in the eddies.*

Fire will sometimes cause a swarm of locusts
To rise in the air and fly to a river. The fire
Keeps coming, burning them instantly,
And the insects shrink down into the water.

*Just so Achilles. And Xanthus' noisy channel*
*Was clogged with chariots, horses, and men.*
*Achilles wasted no time. Leaving his spear*
*Propped against a tamarisk*
*And holding only his sword, he leapt from the bank*
*Like a spirit from hell bent on slaughter.*
*He struck over and over, in a widening spiral.*

the amount she had earned to that point. She was giving the money back to the studio and, thanking Paramount for the opportunity, told the executive she was leaving for New York the next morning.

Zukor, immediately apprised of this news, was caught completely off balance. West seemed willing to lose substantial money, risk a lawsuit for breach of contract, and guarantee that she would never work in Hollywood again. Zukor took another look at the script—perhaps she was right and the dialog was lousy. She would rather give up money and a career than be in an inferior picture! He decided to offer her a compromise: she could write her own dialog, and they would shoot two versions of the movie, one hers, one the studio's. That would cost a little more, but they would get West on film. If her version was better, which Zukor thought unlikely, that would only improve the picture; if not, they would go with the original version. Paramount couldn't lose.

West accepted the compromise and shooting began. One person, however, was not amused: the director, Archie L. Mayo, a man with an extensive résumé. Not only had West changed the script to suit her wisecracking style, she insisted on altering the blocking and camera setups to make the most of her lines. They fought and fought, until one day West refused to go on. She had demanded a shot of her disappearing up some stairs after delivering one of her patented wisecracks. This would give the audience time to laugh. Mayo thought it unnecessary and refused to shoot it. West walked off the set, and production came to a halt. Studio executives agreed that West's lines had lightened up the picture; let her have her way with the direction and shoot the shot, they told Mayo. They would edit it out later.

Production resumed. The other actress in her scenes, Alison Skipworth, had the distinct impression that West was determining the pace of the lines, getting the camera to focus on her, stealing the scene. Protesting that West was taking over the direction of the movie, Skipworth, too, was told not to worry—it would all be fixed in the editing.

When it came time to cut the film, however, West had so altered the mood and pace of her scenes that no editing could bring them back to the original; more important, her sense of timing and direction were solid. She had indeed improved the entire picture.

The film debuted in October 1932. The reviews were mixed, but almost all agreed that a new star was born. West's aggressive sexual style and wit fascinated the men in the audience. Though she appeared in just a few scenes, she was the only part of the film anyone seemed to remember. Lines she had written—"I'm a girl who lost her reputation and never missed it"—were quoted endlessly. As Raft later admitted, "Mae West stole everything but the cameras."

Audiences were soon clamoring for more Mae West—and Paramount, in financial trouble at the time, could not ignore them. At the age of forty, as plump as ever, West was signed to a long-term contract at the highest salary of any star in the studio. For her next film, *Diamond Lil*,

*Hideous groans rose from the wounded,*
*And the river water turned crimson with blood.*

Fish fleeing a dolphin's huge maw
Hide by the hundreds in the harbor's crannies,
But the dolphin devours whatever it catches.

*Likewise the Trojans beneath the riverbanks.*

*The Iliad*, Homer, circa Ninth century B.C.

she would have complete creative control. No other actress—or actor, for that matter—had ever pulled off such a coup and in so short a time.

Interpretation

When Mae West set foot in Hollywood, everything was stacked against her. She was old and passé. The director and an army of studio executives had just one goal: to use her in a scene or two to animate a dull picture, then ship her back to New York. She had no real power, and if she had chosen to fight on their battlefield—one in which actresses were a dime a dozen and exploited to the fullest—she would have gotten nowhere. West's genius, her form of warfare, was slowly but surely to shift the battlefield to terrain of her choice.

She began her war by playing the part of the blond bombshell, charming and seducing the Paramount men. Her screen test hooked them—she was difficult, but what actress wasn't? Next she asked to rewrite her lines and, receiving the expected refusal, ratcheted up the stakes by not budging. Returning the money she had been paid was her campaign's key moment: it subtly shifted the focus from a battle with an actress to the script itself. By showing herself ready to give up so much, she made Zukor begin to wonder more about the dialog than about her. After his compromise, West made her next maneuver, fighting over the blocking, the camera angles, the pacing of the shots. Her writing had become an accepted part of the scenery; the battle now was over her directing. Another compromise, which translated into another victory. Instead of fighting the studio executives on their terms, West had subtly shifted the battle to a field unfamiliar to them—fighting with an actress over the writing and directing of a film. On such terrain, against a smart and seductive woman, the army of Paramount men was lost and helpless.

Your enemies will naturally choose to fight on terrain that is to their liking, that allows them to use their power to best advantage. Cede them such power and you end up fighting on their terms. Your goal is to subtly shift the conflict to terrain of your choice. You accept the battle but alter its nature. If it is about money, shift it to something moral. If your opponents want to fight over a particular issue, reframe the battle to encompass something larger and more difficult for them to handle. If they like a slow pace, find a way to quicken it. You are not allowing your enemies to get comfortable or fight in their usual way. And an enemy who is lured onto unfamiliar terrain is one who has lost control of the dynamic. Once such control slips from his hands, he will compromise, retreat, make mistakes, and effect his own destruction.

**3.** By early 1864 the American Civil War had settled into a stalemate. Robert E. Lee's Army of Northern Virginia had managed to keep the Union forces away from Richmond, capital of the Confederacy. To the west the Confederates had established an impregnable defensive position at the town of Dalton, Georgia, blocking any Union advance on

Atlanta, the key industrial city of the South. President Abraham Lincoln, facing reelection that year and gravely worried about his chances if the stalemate continued, decided to name Ulysses S. Grant overall commander of the Union forces. Here was a man who would go on the offensive.

Grant's first move was to appoint his chief lieutenant, General William Tecumseh Sherman, to command the Union forces in Georgia. When Sherman arrived on the scene, he realized that any attempt to take Dalton was doomed from the start. The Confederate commander, General John Johnston, was a master at defensive warfare. With mountains to his rear and a solid position to his front, Johnston could simply stay put. A siege would take too long, and a frontal attack would be too costly. The situation seemed hopeless.

Sherman decided, then, that if he could not seize Dalton, he would take hold of Johnston's mind, striking fear in a man notorious for being conservative and cautious. In May 1864, Sherman sent three-fourths of his army into a direct attack on Dalton. With Johnston's attention held by this attack, Sherman then sneaked the Army of the Tennessee around the mountains to the town of Resaca, fifteen miles to the south of Dalton, blocking Johnston's only real route of retreat and only supply line. Terrified to find himself suddenly surrounded, Johnston had no choice but to give up his position at Dalton. He would not, however, play into Sherman's hands: he simply retreated to another defensive position that gave him maximum security, again inviting Sherman to attack him straight on. This quickly turned into a dance: Sherman would feint going one way, then would somehow divert a part of his army to the south of Johnston, who kept backing up . . . all the way to Atlanta.

The Confederate president, Jefferson Davis, disgusted by Johnston's refusal to fight, replaced him with General John Hood. Sherman knew that Hood was an aggressive commander, often even reckless. He also knew that neither the time nor the men were available to lay siege to Atlanta—Lincoln needed a quick victory. His solution was to send detachments to threaten Atlanta's defenses, but he made these forces temptingly small and weak. Hood could not resist the temptation to leave his stronghold in the city and move to the attack, only to find himself rushing into an ambush. This happened several times, and with each defeat, Hood's army became smaller and the morale of his men quickly deteriorated.

Now, with Hood's army tired and expecting disaster, Sherman played yet another trick. At the end of August, he marched his army southeast, past Atlanta, abandoning his supply lines. To Hood this could only mean that Sherman had given up the fight for Atlanta. Wild celebration broke out throughout the city. But Sherman had cunningly timed this march to coincide with the ripening of the corn, and with his men well fed and Hood unsuspecting, he cut off the final railway line still open to Atlanta and wheeled back to attack the unguarded city. Hood

*Let us admit that boldness in war even has its own prerogatives. It must be granted a certain power over and above successful calculations involving space, time, and magnitude of forces, for wherever it is superior, it will take advantage of its opponent's weakness. In other words, it is a genuinely creative force. This fact is not difficult to prove even scientifically. Whenever boldness encounters timidity, it is likely to be the winner, because timidity in itself implies a loss of equilibrium. Boldness will be at a disadvantage only in an encounter with deliberate caution, which may be considered bold in its own right, and is certainly just as powerful and effective; but such cases are rare. Timidity is the root of prudence in the majority of men. . . . The higher up the chain of command, the greater is the need for boldness to be supported by a reflective mind, so that boldness does not degenerate into purposeless bursts of blind passion.*

*On War*,
Carl von Clausewitz,
1780–1831

*The Olympians could now join battle with the giants. Heracles let loose his first arrow against Alcyoneus, the enemy's leader. He fell to the ground, but sprang up again revived, because this was his native soil of Phlegra. "Quick, noble Heracles!" cried Athene. "Drag him away to another country!" Heracles caught Alcyoneus up on his shoulders, and dragged him over the Thracian border, where he despatched him with a club.*

THE GREEK MYTHS, VOL. 1, ROBERT GRAVES, 1955

was forced to abandon Atlanta. This was the great victory that would ensure Lincoln's reelection.

Next came Sherman's strangest maneuver of all. He divided his army into four columns and, completely cutting himself loose from his supply lines, began a march east from Atlanta to Savannah and the sea. His men lived off the land, destroying everything in their path. Unencumbered by supply wagons, they moved with incredible speed. The four parallel columns were far enough apart that the southern forces could not tell where they were headed. The southern column seemed headed for Macon, the northern for Augusta. Confederate forces scrambled to cover both places, leaving the center open—which was exactly where Sherman planned to advance. Keeping the South on what he called "the horns of a dilemma," off balance and mystified as to his intentions, Sherman marched all the way to Savannah with hardly a battle.

The effect of this march was devastating. For the Confederate soldiers still fighting in Virginia, the ruin of Georgia—where many had left behind homes—was a terrible blow to their morale. Sherman's march cast a mood of deep gloom over the entire South. Slowly but surely it was losing its will to keep up the fight, Sherman's goal all along.

## Interpretation

In any conflict it is often the weaker side that in fact controls the dynamic. In this case the South was in control in both the strategic and the grand-strategic sense. In their immediate, local strategy, the Confederates had entrenched themselves in powerful defensive positions in Georgia and Virginia. The temptation for the North was to fight on the enemy's terms, to hurl division after division against these positions, at tremendous loss of life and with little chance of advancing. In the South's grand strategy, the longer this stalemate prevailed, the more likely Lincoln was to be thrown out of office. Then the war would end through negotiation. The South set the tempo for the battle (slow and grinding) and controlled the stakes.

As Sherman saw it, his goal was not to capture a city or to defeat the Confederates in battle. In his view the only way to win the war was to regain control of the dynamic. Instead of brutal, frontal attacks against Dalton or Atlanta, which would play into the South's hands, he operated indirectly. He frightened the timid Johnston into abandoning his stronghold and goaded the rash Hood into senseless attacks, in both instances playing upon the psychology of the opponent to force the issue. By constantly putting his enemy on the horns of a dilemma, where both staying put and moving were equally dangerous, he took control of the situation without having to waste men in battle. Most important, by demonstrating to the South with his destructive march that the longer the war dragged on, the worse it would be for them, he regained grand-strategic control of the war. For the Confederates, to keep fighting was slow suicide.

The worst dynamic in war, and in life, is the stalemate. It seems that whatever you do only feeds the stagnation. Once this happens, a kind of mental paralysis overcomes you. You lose the ability to think or respond in different ways. At such a point, all is lost. If you find yourself falling into such a dynamic—dealing with a defensive, entrenched opponent or trapped in a reactive relationship—you must become as creative as General Sherman. Deliberately shake up the pace of the slow waltz by doing something seemingly irrational. Operate outside the experience of the enemy, as Sherman did when he cut himself loose from his supplies. Move fast here and slow there. One major jolt given to the stale dynamic will shake it up, force the enemy to do something different. With the slightest change, you have room for greater change and taking control. Injecting novelty and mobility is often enough to unbalance the minds of your rigid and defensive opponents.

**4.** In 1833, Mr. Thomas Auld, the slaveholding owner of a plantation on Maryland's Eastern Shore, summoned back his slave Frederick Douglass, fifteen years old at the time, from Baltimore, where Douglass had just spent seven years serving Auld's brother. Now he was needed to work the fields of the plantation. But life in the city had changed Douglass in many ways, and to his chagrin he found it quite hard to disguise this from Auld. In Baltimore he had secretly managed to teach himself how to read and write, something no slave was allowed to do, for that would stimulate dangerous thoughts. On the plantation Douglass tried to teach as many slaves as possible to read; these efforts were quickly squashed. But what was worse for him was that he had developed a rather defiant attitude, what the slaveholder called impudence. He talked back to Auld, questioned some of his orders, and played all kinds of tricks to get more food. (Auld was notorious for keeping his slaves near starvation.)

One day Auld informed Douglass that he was hiring him out for a year to Mr. Edward Covey, a nearby farm renter who had earned a reputation as a consummate "breaker of young Negroes." Slaveholders would send him their most difficult slaves, and in exchange for their free labor Covey would beat every last ounce of rebellion out of them. Covey worked Douglass especially hard, and after a few months he was broken in body and spirit. He no longer desired to read books or engage in discussions with his fellow slaves. On his days off, he would crawl under the shade of a tree and sleep off his exhaustion and despair.

One especially hot day in August 1834, Douglass became ill and fainted. The next thing he knew, Covey was hovering over him, hickory slab in hand, ordering him back to work. But Douglass was too weak. Covey hit him on the head, opening a deep wound. He kicked him a few times, but Douglass could not move. Covey finally left, intending to deal with him later.

Douglass managed to get to his feet, staggered to the woods, and somehow made his way back to Auld's plantation. There he begged with

*Well, my dear reader, this battle with Mr. Covey,—undignified as it was, and as I fear my narration of it is—was the turning point in my* "life as a slave." *It rekindled in my breast the smouldering embers of liberty; it brought up my Baltimore dreams, and revived a sense of my own manhood. I was a changed being after that fight. I was* nothing *before; I WAS A MAN NOW. It recalled to life my crushed self-respect and my self-confidence, and inspired me with a renewed determination to be A FREEMAN. A man, without force, is without the essential dignity of humanity. Human nature is so constituted, that it cannot* honor *a helpless man, although it can pity him; and even this it cannot do long, if the signs of power do not arise. He only can understand the effect of this combat on my spirit, who has himself incurred something, hazarded something, in repelling the unjust and cruel aggressions of a tyrant. Covey was a tyrant, and a cowardly one, withal. After resisting him, I felt as I had never felt before. It was a resurrection from the dark and pestiferous tomb of slavery, to the heaven of comparative freedom. I was no longer a servile coward, trembling under the frown of a brother worm of the*

Master Auld to keep him there, explaining Covey's cruelty. Auld was unmoved. Douglass could spend the night but then must return to Covey's farm.

Making his way back to the farm, Douglass feared the worst. He told himself that he would do his best to obey Covey and somehow survive the weeks ahead. Arriving at the stables where he was supposed to work that day, he began his chores, when out of nowhere, like a snake, Covey slithered in, rope in hand. He lunged at Douglass, trying to get a slipknot on his leg and tie him up. He was clearly intending the thrashing to end all thrashings.

Risking an even more intense beating, Douglass pushed Covey away and, without hitting him, kept him from getting the rope around his leg. At that moment something clicked in Douglass's head. Every defiant thought that had been suffocated by his months of brutal labor came back to him. He was not afraid. Covey could kill him, but it was better to go down fighting for his life.

Suddenly a cousin came to Covey's aid, and, finding himself surrounded, Douglass did the unthinkable: he swung hard at the man and knocked him to the ground. Hitting a white man would most likely lead to his hanging. A "fighting madness" came over Douglass. He returned Covey's blows. The struggle went on for two hours until, bloodied, exhausted, and gasping for breath, Covey gave up and slowly staggered back to his house.

Douglass could only assume that Covey would now come after him with a gun or find some other way to kill him. It never happened. Slowly it dawned on Douglass: to kill him, or punish him in some powerful way, posed too great a risk. Word would get out that Covey had failed to break a Negro this one time, had had to resort to a gun when his terror tactics did not work. The mere hint of that would ruin his reputation far and wide, and his job depended on his perfect reputation. Better to leave the wild sixteen-year-old slave alone than risk the kind of crazy or unpredictable response Douglass had showed himself capable of. Better to let him calm down and go quietly away when his time of service there was over.

For the rest of Douglass's stay with Covey, the white man did not lay a hand on him. Douglass had noticed that slaveholders often "prefer to whip those who are most easily whipped." Now he had learned the lesson for himself: never again would he be submissive. Such weakness only encouraged the tyrants to go further. He would rather risk death, returning blow for blow with his fists or his wits.

### Interpretation

Reflecting on this moment years later in his book *My Bondage and My Freedom*, after he had escaped to the North and become a leading advocate of the abolitionist movement, Douglass wrote, "This battle with Mr. Covey . . . was the turning point in my '*life as a slave.*' . . . I was a changed being after that fight. . . . I had reached the point at which I was not *afraid to die.* This spirit made me a freeman in *fact,* while I remained a slave in *form.*" For the rest of his life, he adopted this fighting stance: by being unafraid of the consequences, Douglass gained a degree of control of his situation both physically and psychologically. Once he had rooted fear out of himself, he opened up possibilities for action—sometimes fighting back overtly, sometimes being clever and deceitful. From a slave with no control, he became a man with some options and some power, all of which he leveraged into real freedom when the time came.

To control the dynamic, you must be able to control yourself and your emotions. Getting angry and lashing out will only limit your options. And in conflict, fear is the most debilitating emotion of all. Even before anything has happened, your fear puts you on your heels, cedes the initiative to the enemy. The other side has endless possibilities for using your fear to help control you, keep you on the defensive. Those who are tyrants and domineering types can smell your anxiety, and it makes them even more tyrannical. Before anything else you must lose your fear—of death, of the consequences of a bold maneuver, of other people's opinion of you. That single moment will suddenly open up vistas of possibilities. And in the end whichever side has more possibilities for positive action has greater control.

**5.** Early in his career, the American psychiatrist Milton H. Erickson (1901–80) noticed that patients had countless ways of controlling the relationship between patient and therapist. They might withhold information from him or resist entering into a hypnotic trance (Erickson often used hypnosis in his therapy); they might question the therapist's abilities, insist he do more of the talking, or emphasize the hopelessness of their problems and the futility of therapy. These attempts at control in fact mirrored whatever their problem was in daily life: they resorted to all kinds of unconscious and passive games of domination while denying to themselves and to others they were up to such tricks. And so, over the years, Erickson developed what he called his "Utilization Technique"—literally using these patients' passive aggression, their clever manipulations, as instruments to change them.

Erickson often dealt with patients whom someone else—a partner, a parent—had forced to seek his help. Resentful of this, they would get revenge by deliberately withholding information about their lives. Erickson would begin by telling these patients that it was natural, even healthy, not to want to reveal everything to the therapist. He would insist they withhold any sensitive information. The patients would then feel trapped: by keeping secrets they were obeying the therapist, which was just the opposite of what they wanted to do. Usually by the second session, they would open up, rebelling to such an extent that they revealed everything about themselves.

One man, on his first visit to Erickson's office, began anxiously

*dust, but, my long-cowed spirit was roused to an attitude of manly independence. I had reached a point, at which I was* not afraid to die. *This spirit made me a freeman in* fact, *while I remained a slave in* form. *When a slave cannot be flogged he is more than half free. He has a domain as broad as his own manly heart to defend, and he is really* "a power on earth." *While slaves prefer their lives, with flogging, to instant death, they will always find christians enough, like unto Covey, to accommodate that preference. From this time, until that of my escape from slavery, I was never fairly whipped. Several attempts were made to whip me, but they were always unsuccessful. Bruises I did get, as I shall hereafter inform the reader; but the case I have been describing, was the end of the brutification to which slavery had subjected me.*

My Bondage and My Freedom, Frederick Douglass, 1818–1895

*It is not the same when a fighter moves because he wants to move, and another when he moves because he has to.*

JOE FRAZIER, 1944

pacing the room. By refusing to sit down and relax, he was making it impossible for Erickson to hypnotize him or work with him at all. Erickson began by asking him, "Are you willing to cooperate with me by continuing to pace the floor as you are doing now?" The patient agreed to this strange request. Then Erickson asked if he could tell the patient where to pace and how fast. The patient could see no problem with this. Minutes later Erickson began to hesitate in giving his directions; the patient waited to hear what he was to do next with his pacing. After this happened a few times, Erickson finally told him to sit in a chair, where the man promptly fell into a trance.

With those who were patently cynical about therapy, Erickson would deliberately try a method of hypnosis that would fail, and then he would apologize for using that technique. He would talk about his own inadequacies and the many times he had failed. Erickson knew that these types needed to one-up the therapist, and that once they felt they had gained the advantage, they would unconsciously open themselves up to him and fall easily into a trance.

A woman once came to Erickson complaining that her husband used his supposedly weak heart to keep her on constant alert and dominate her in every way. The doctors had found nothing wrong with him, yet he genuinely seemed weak and always believed that a heart attack was imminent. The woman felt concerned, angry, and guilty all at the same time. Erickson advised her to continue being sympathetic to his condition, but the next time he talked about a heart attack, she was to tell him politely she needed to tidy the house. She was then to place brochures she had collected from morticians all around the house. If he did it again, she was to go to the desk in the living room and begin adding up the figures in his life-insurance policies. At first the husband was furious, but soon he came to fear seeing those brochures and hearing the sound of the adding machine. He stopped talking about his heart and was forced to deal with his wife in a more direct manner.

## Interpretation

In some relationships you may have a gnawing feeling that the other person has gained control of the dynamic, yet you find it hard to pinpoint how or when this occurred. All that can be said for certain is that you feel unable to move the other person, to influence the course of the relationship. Everything you do only seems to feed the power of the controller. The reason for this is that the other person has adopted subtle, insidious forms of control that are easily disguised and yet all the more effective for being unconscious and passive. Such types exert control by being depressed, overly anxious, overburdened with work—they are the victims of constant injustice. They cannot help their situation. They demand attention, and if you fail to provide it, they make you feel guilty. They are elusive and impossible to fight because they make it appear at each turn that they are not at all looking for control. They are more willful than you, but better at disguising it. In truth, you are the one who feels helpless and confused by their guerrilla-like tactics.

To alter the dynamic, you must first recognize that there is far less helplessness in their behavior than they let on. Second, these people need to feel that everything takes place on their terms; threaten that desire and they fight back in underhanded ways. You must never inadvertently feed their rebelliousness by arguing, complaining, trying to push them in a direction. This makes them feel more under attack, more like a victim, and encourages passive revenge. Instead move within their system of control, applying Erickson's Utilization Technique. Be sympathetic to their plight, but make it seem that whatever they do, they are actually cooperating with your own desires. That will put them off balance; if they rebel now, they are playing into your hands. The dynamic will subtly shift, and you will have room to insinuate change. Similarly, if the other person wields a fundamental weakness as a weapon (the heart-attack tactic), make that threat impossible to use against you by taking it further, to the point of parody or painfulness. The only way to beat passive opponents is to outdo them in subtle control.

**Image:** The Boxer. The superior fighter does not rely on his powerful punch or quick reflexes. Instead he creates a rhythm to the fight that suits him, advancing and retreating at a pace he sets; he controls the ring, moving his opponent to the center, to the ropes, toward or away from him. Master of time and space, he creates frustration, compels mistakes, and engenders a mental collapse that precedes the physical. He wins not with his fists but by controlling the ring.

**Authority:** In order to have rest oneself it is necessary to keep the enemy occupied. This throws them back on the defensive, and once they are placed that way they cannot rise up again during the entire campaign. *—Frederick the Great (1712–1786)*

REVERSAL

This strategy has no reversal. Any effort to seem *not* to control a situation, to refuse to influence a relationship, is in fact a form of control. By ceding power to others, you have gained a kind of passive authority that you can use later for your own purposes. You are also the one determining who has control by relinquishing it to the other side. There is no escape from the control dynamic. Those who say they are doing so are playing the most insidious control game of all.

# 16

## HIT THEM WHERE IT HURTS

## THE CENTER-OF-GRAVITY STRATEGY

*Everyone has a source of power on which he or she depends. When you look at your rivals, search below the surface for that source, the center of gravity that holds the entire structure together. That center can be their wealth, their popularity, a key position, a winning strategy. Hitting them there will inflict disproportionate pain. Find what the other side most cherishes and protects—that is where you must strike.*

*Man depends on his throat for fluent breathing and the maintenance of life. When his throat is strangled, his five sense organs will lose their sensibility and no longer function normally. He will not be able to stretch his limbs, which become numb and paralyzed. The man can therefore rarely survive. Thus, when the banners of the enemy come into sight and the beating of its battle drums can be heard, we must first ascertain the positions of its back and throat. Then we can attack it from the back and strangle its throat. This is an excellent strategy to crush the enemy.*

THE WILES OF WAR: 36 MILITARY STRATEGIES FROM ANCIENT CHINA, TRANSLATED BY SUN HAICHEN, 1991

## PILLARS OF COLLAPSE

In 210 B.C. a young Roman general named Publius Scipio the Younger (later called Scipio Africanus) was sent to northeastern Spain with a simple mission: to hold the Ebro River against the powerful Carthaginian armies that were threatening to cross it and take control of the peninsula. This was Scipio's first assignment as commander, and as he looked out on the river and plotted his strategy, he felt a strange mix of emotions.

Eight years earlier the great Carthaginian commander Hannibal had crossed this river heading north. Onward he had gone into Gaul and then, catching the Romans by surprise, had crossed the Alps into Italy. Scipio, only eighteen at the time, had fought alongside his father, a general, in the first battles against Hannibal on Italian soil. He had seen the North African's battlefield skills with his own eyes: Hannibal had maneuvered his small army brilliantly, made maximum use of his superior cavalry, and through inexhaustible creativity had constantly managed to surprise the Romans and inflict on them a series of humiliating defeats, culminating in the virtual annihilation of the Roman legions at the Battle of Cannae in 216 B.C. Matching wits with Hannibal, Scipio knew, was futile. It had seemed back then that Rome itself was doomed.

Scipio also recalled two events after Cannae that had had an overwhelming effect on him. First, a Roman general named Fabius had finally conceived a strategy to keep Hannibal at bay. Keeping his legions in the hills and avoiding direct battle, Fabius had launched hit-and-run raids designed to wear down the Carthaginians, who were fighting far from their home, in what is now Tunisia. The campaign had worked as a holding action, but to Scipio it had seemed equally exhausting for the Romans to fight so long and still have the enemy at their doorstep. Also, since the plan could not lead to any real defeat of Hannibal, it was basically flawed.

Second, a year into Hannibal's invasion, the Romans had sent Scipio's father to Spain to try to knock out the Carthaginian bases there. Carthage had had colonies in Spain for many years and earned wealth from Spanish mines. It used Spain as a training ground for its soldiers and as the base for its war on Rome. For six years Scipio's father had fought the Carthaginians on the Spanish peninsula, but the campaign had ended in his defeat and death in 211 B.C.

As Scipio studied the reports coming in about the situation beyond the Ebro, a plan took root in his mind: with one bold maneuver, he could avenge his father's death the year before, demonstrate the effectiveness of a strategy he thought far better than Fabius's, and set in motion the eventual collapse not just of Hannibal but of Carthage itself. Along the coast to the south of him was the city of New Carthage (present-day Cartagena), the Carthaginians' capital in Spain. There they stored their vast wealth, their army's supplies, and the hostages they had taken from different Spanish tribes to be held as ransom in case of rebellion. At this moment the Carthaginian armies—which outnumbered the Romans two to one—were scattered about the country, trying to gain further domination over the Spanish tribes, and were all several days' march from New Carthage. Their commanders, Scipio learned, had been quarreling among themselves over power and money. Meanwhile New Carthage was garrisoned with only 1,000 men.

Disobeying his orders to take his stand at the Ebro, Scipio advanced south by ship and led a daring raid on New Carthage. This walled city was considered impregnable, but he timed his attack for the ebbing of the tide in a lagoon on the city's north side; there his men were able to scale the walls relatively easily, and New Carthage was taken. In one move, Scipio had produced a dramatic turnaround. Now the Romans commanded the central position in Spain; they had the money and supplies on which the Carthaginians in Spain depended; and they had Carthage's hostages, whom they now could use to stir revolt among the conquered tribes. Over the next few years, Scipio exploited this position and slowly brought Spain under Roman control.

In 205 B.C., Scipio returned to Rome a hero—but Hannibal was still a menace in Italy's interior. Scipio now wanted to take the war to Africa, by marching on Carthage itself. That was the only way to get Hannibal out of Italy and finally erase Carthage as a threat. But Fabius was still the commander in charge of Rome's strategy, and few saw the point of fighting Hannibal by waging war so far from him, and from Rome. Yet Scipio's prestige was high, and the Roman Senate finally gave him an army—a small, low-quality army—to use for his campaign.

Wasting no time on arguing his case, Scipio proceeded to make an alliance with Masinissa, king of the Massyles, Carthage's neighbors. Masinissa would supply him with a large and well-trained cavalry. Then, in the spring of 204 B.C., Scipio sailed for Africa and landed near Utica, not far from Carthage. Initially surprised, the Carthaginians rallied and were able to pin Scipio's troops on a peninsula outside the town. The situation looked bleak. If Scipio could somehow advance past the enemy troops that blocked his path, he would enter the heart of the enemy state and gain control of the situation, but that seemed an impossible task—he could not hope to fight his way past the tight Carthaginian cordon; trapped where he was, his supplies would eventually run out, forcing him to surrender. Scipio bargained for peace but used the negotiations as a way to spy on the Carthaginian army.

Scipio's ambassadors told him that the enemy had two camps, one for its own army and the other for its main ally, the Numidians, whose camp was rather disorganized, a swarm of reed huts. The Carthaginian camp was more orderly but made of the same combustible materials. Over the next few weeks, Scipio seemed indecisive, first breaking off ne-

*The third shogun Iemitsu was fond of sword matches. Once, when he arranged to see some of his outstanding swordsmen display their skills, he spotted among the gathering a master equestrian by the name of Suwa Bunkuro, and impulsively asked him to take part. Bunkuro responded by saying that he would be pleased to if he could fight on horseback, adding that he could defeat anyone on horseback. Iemitsu was delighted to urge the swordsmen to fight Bunkuro in the style he preferred. As it turned out, Bunkuro was right in his boasting. Brandishing a sword on a prancing horse wasn't something many swordsmen were used to, and Bunkuro easily defeated everyone who dared face him on horseback. Somewhat exasperated, Iemitsu told Munenori to give it a try. Though a bystander on this occasion, Munenori at once complied and mounted a horse. As his horse trotted up to Bunkuro's, Munenori suddenly stopped his horse and slapped the nose of Bunkuro's horse with his wooden sword. Bunkuro's horse reared, and while the famed equestrian was trying to restore his balance, Munenori struck him off his horse.*

*The Sword and The Mind*, translated by Hiroaki Sato, 1985

*Heracles did not return to Mycenae by a direct route. He first traversed Libya, whose King Antaeus, son of Poseidon and Mother Earth, was in the habit of forcing strangers to wrestle with him until they were exhausted, whereupon he killed them; for not only was he a strong and skilful athlete, but whenever he touched the earth, his strength revived. He saved the skulls of his victims to roof a temple of Poseidon. It is not known whether Heracles, who was determined to end this barbarous practice, challenged Antaeus, or was challenged by him. Antaeus, however, proved no easy victim, being a giant who lived in a cave beneath a towering cliff, where he feasted on the flesh of lions, and slept on the bare ground in order to conserve and increase his already colossal strength. Mother Earth, not yet sterile after her birth of the Giant, had conceived Antaeus in a Libyan cave, and found more reason to boast of him than even of her monstrous elder children, Typhon, Tityus, and Briareus. It would have gone ill with the Olympians if he had fought against them on the Plains of Phlegra. In preparation for the wrestling match, both combatants cast off their lion pelts, but while Heracles rubbed himself with oil in the Olympian fashion, Antaeus poured hot*

gotiations, then reopening them, confusing the Carthaginians. Then one night he made a sneak attack on the Numidian camp and set it on fire. The blaze spread quickly, and the African soldiers panicked, scattering in every direction. Awakened by the hubbub, the Carthaginians opened the gates to their own camp to come to their allies' rescue—but in the confusion the Romans were able to steal in and set fire to their camp as well. The enemy lost half their army in this battle by night, the rest managing to retreat to Numidia and Carthage.

Suddenly the Carthaginian interior lay open to Scipio's army. He proceeded to take town after town, advancing much as Hannibal had in Italy. Then he landed a contingent of troops at the port of Tunis, within sight of Carthage's walls. Now it was the Carthaginians' turn to panic, and Hannibal, their greatest general, was immediately recalled. In 202 B.C., after sixteen years of fighting on Rome's doorstep, Hannibal was finally compelled to leave Italy.

Hannibal landed his army to Carthage's south and made plans to fight Scipio. But the Roman general retreated west, to the Bagradas Valley—the most fertile farmlands of Carthage, its economic base. There he went on a rampage, destroying everything in sight. Hannibal had wanted to fight near Carthage, where he had shelter and material reinforcements. Instead he was compelled to pursue Scipio before Carthage lost its richest territory. But Scipio kept retreating, refusing battle until he had lured Hannibal to the town of Zama, where he secured a solid position and forced Hannibal to camp in a place without water. Now the two armies finally met in battle. Exhausted by their pursuit of Scipio, their cavalry neutralized by Masinissa's, the Carthaginians were defeated, and with no refuge close enough to retreat to, Hannibal was forced to surrender. Carthage quickly sued for peace, and under the harsh terms imposed by Scipio and the Senate, it was reduced to a client state of Rome. As a Mediterranean power and a threat to Rome, Carthage was finished for good.

### Interpretation

Often what separates a mediocre general from a superior one is not their strategies or maneuvers but their vision—they simply look at the same problem from a different angle. Freed from the stranglehold of convention, the superior general naturally hits on the right strategy.

The Romans were dazzled by Hannibal's strategic genius. They came to so fear him that the only strategies they could use against him were delay and avoidance. Scipio Africanus simply saw differently. At every turn he looked not at the enemy army, nor even at its leader, but at the pillar of support on which it stood—its critical vulnerability. He understood that military power was located not in the army itself but in its foundations, the things that supported it and made it possible: money, supplies, public goodwill, allies. He found those pillars and bit by bit knocked them down.

Scipio's first step was to see Spain, not Italy, as Hannibal's center of gravity. Within Spain the key was New Carthage. He did not chase the various Roman armies but took New Carthage and turned the war around. Now Hannibal, deprived of his main military base and source of supply, would have to lean more heavily on his other support base: Carthage itself, with its wealth and resources. So Scipio took the war to Africa. Trapped near Utica, he examined what gave the enemy its power in this situation, and saw that it was not the armies themselves but the position they had taken: get them to move out of that position without wasting men in a frontal battle and Carthage's soft underbelly would be exposed. By burning the camps, Scipio moved the armies. Then, instead of marching on the city of Carthage—a glittering prize that would have drawn most generals like a magnet—he hit what would hurt the Carthaginian state most: the fertile farming zone that was the source of its wealth. Finally, instead of chasing Hannibal, he made Hannibal come after him, to an area in the middle of the country where he would be deprived of reinforcements and support. Now that Scipio had unbalanced the Carthaginians so completely, their defeat at Zama was definitive.

Power is deceptive. If we imagine the enemy as a boxer, we tend to focus on his punch. But still more than he depends on his punch, he depends on his legs; once they go weak, he loses balance, he cannot escape the other fighter, he is subject to grueling exchanges, and his punches gradually diminish in force until he is knocked out. When you look at your rivals, do not be distracted by their punch. To engage in any exchange of punches, in life or in war, is the height of stupidity and waste. Power depends on balance and support; so look at what is holding your enemy up, and remember that what holds him up can also make him fall. A person, like an army, usually gets his or her power from three or four simultaneous sources: money, popularity, skillful maneuvering, some particular advantage he has fostered. Knock out one and he will have to depend more on the others; knock out those and he is lost. Weaken a boxer's legs and he will reel and stagger, and when he does, be merciless. No power can stand without its legs.

*When the vanes are removed from an arrow, even though the shaft and tip remain it is difficult for the arrow to penetrate deeply.*

*—Ming dynasty strategist Chieh Hsüan (early seventeenth century A.D.)*

## KEYS TO WARFARE

It is natural in war to focus on the physical aspect of conflict—bodies, equipment, matériel. Even an enlightened strategist will tend to look first

*sand over his limbs lest contact with the earth through the soles of his feet alone should prove insufficient. Heracles planned to preserve his strength and wear Antaeus down, but after tossing him full length on the ground, he was amazed to see the giant's muscles swell and a healthy flush suffuse his limbs as Mother Earth revived him. The combatants grappled again, and presently Antaeus flung himself down of his own accord, not waiting to be thrown; upon which, Heracles, realizing what he was at, lifted him high into the air, then cracked his ribs and, despite the hollow groans of Mother Earth, held him aloft until he died.*

THE GREEK MYTHS, VOL. 2, ROBERT GRAVES, 1955

*The people of Israel cried out to the Lord their God, for their courage failed, because all their enemies had surrounded them and there was no way of escape from them. The whole Assyrian army, their infantry, chariots, and cavalry, surrounded them for thirty-four days, until all the vessels of water belonging to every inhabitant of Bethulia were empty. . . . Their children lost heart, and the women and young men fainted from thirst and fell down in the streets of the city and in the passages through the gates; there was no strength left in them any longer. . . . When Judith had ceased crying out to the God of Israel, and had ended all these words, she rose from where she lay prostrate and called her maid and went down into the house where she lived on sabbaths and on her feast days; and she removed the sackcloth which she had been wearing, and took off her widow's garments, and bathed her body with water, and anointed herself with precious ointment, and combed her hair and put on a tiara, and arrayed herself in her gayest apparel, which she used to wear while her husband Manasseh was living. And she put sandals on her feet, and put on her anklets and bracelets and rings, and her earrings and all her ornaments, and made herself very beautiful, to entice the eyes of all men who might see her. . . . . . . The women went straight on through the valley; and an Assyrian patrol met her and took her into custody, and asked her, "To what people do you belong, and where are you going?" She replied, "I am a daughter of the Hebrews, but I am fleeing from them, for they are about to be handed over to you to be devoured. I am on my way to the presence of Holofernes the commander of your*

at the enemy's army, firepower, mobility, reserves. War is a visceral, emotional affair, an arena of physical danger, and it takes great effort to rise above this level and ask different questions: What makes the enemy army move? What gives it impetus and endurance? Who guides its actions? What is the underlying source of its strength?

Most people have the problem of seeing war as a separate activity unrelated to other realms of human life. But in fact war is a form of power—Carl von Clausewitz called it "politics by other means"—and all forms of power share the same essential structures.

The most visible thing about power is its outward manifestation, what its witnesses see and feel. An army has its size, its weaponry, its shows of discipline, its aggressive maneuvers; individuals have many ways of showing their position and influence. It is the nature of power to present a forceful front, to seem menacing and intimidating, strong and decisive. But this outward display is often exaggerated or even downright deceptive, since power does not dare show its weaknesses. And beneath the display is the support on which power rests—its "center of gravity." The phrase is von Clausewitz's, who elaborated it as "the hub of all power and movement, on which everything depends." This is the part that governs the whole, a kind of nerve center.

To attack this center of gravity, to neutralize or destroy it, is the ultimate strategy in war, for without it the whole structure will collapse. The enemy may have great generals and strong armies, like Hannibal and his invincible army in Italy, but without a center of gravity those armies cannot move and have no force or coherence. Hitting the center will have devastating psychological effects, throwing the enemy off balance and inducing a creeping panic. If conventional generals look at the physical aspect of the enemy army, focusing on its weaknesses and trying to exploit them, superior strategists look behind and beyond, to the support system. The enemy's center of gravity is where an injury will hurt him most, his point of greatest vulnerability. Hitting him there is the best way to end a conflict definitively and economically.

The key is analyzing the enemy force to determine its centers of gravity. In looking for those centers, it is crucial not to be misled by the intimidating or dazzling exterior, mistaking the outward appearance for what sets it in motion. You will probably have to take several steps, one by one, to uncover this ultimate power source, peeling away layer after layer. Remember Scipio, who saw first that Hannibal depended on Spain, then that Spain depended on Carthage, then that Carthage depended on its material prosperity, which itself had particular sources. Strike at Carthage's prosperity, as Scipio eventually did, and the whole thing would fall apart.

To find a group's center of gravity, you must understand its structure and the culture within which it operates. If your enemies are individuals, you must fathom their psychology, what makes them tick, the structure of their thinking and priorities.

In crafting a strategy to defeat the United States in the Vietnam War, General Vo Nguyen Giap determined that the real center of gravity in the American democracy was the political support of its citizens. Given that support—the kind of support the military had had during World War II—the country could prosecute a war with the utmost effectiveness. Without that support, though, the effort was doomed. Through the Tet Offensive of 1968, Giap was able to undermine the American public's support for the war. He had gained an understanding of American culture that allowed him to aim at the right target.

The more centralized the enemy, the more devastating becomes a blow at its leader or governing body. Hernán Cortés was able to conquer Mexico with a handful of soldiers by capturing Moctezuma, the Aztec emperor. Moctezuma was the center around which everything revolved; without him Aztec culture quickly collapsed. When Napoleon invaded Russia in 1812, he assumed that by taking Moscow, the capital, he could force the Russians to surrender. But the true center of gravity in this authoritarian nation was the czar, who was determined to continue the war. The loss of Moscow only steeled his resolve.

A more decentralized enemy will have several separate centers of gravity. The key here is to disorganize them by cutting off communication between them. That was what General Douglas MacArthur did in his remarkable campaign in the Pacific during World War II: he skipped some islands but took key ones, keeping the Japanese extended over a vast area and making it impossible for them to communicate with each other. It is almost always strategically wise to disrupt your enemy's lines of communication; if the parts cannot communicate with the whole, chaos ensues.

Your enemy's center of gravity can be something abstract, like a quality, concept, or aptitude on which he depends: his reputation, his capacity to deceive, his unpredictability. But such strengths become critical vulnerabilities if you can make them unattractive or unusable. In fighting the Scythians in what is now modern-day Iran, a tribe that no one could figure out how to defeat, Alexander the Great saw the center of gravity as their complete mobility on horseback and their fluid, almost chaotic style of fighting. He simply plotted to neutralize the source of this power by luring them on to enclosed ground in which they could not use their cavalry and pell-mell tactics. He defeated them with ease.

To find an enemy's center of gravity, you have to erase your own tendency to think in conventional terms or to assume that the other side's center is the same as your own. When Salvador Dalí came to the United States in 1940, intent on conquering the country as an artist and making his fortune, he made a clever calculation. In the European art world, an artist had to win over the critics and make a name as "serious." In America, though, that kind of fame would doom an artist to a ghetto, a limited circle. The real center of gravity was the American media. By wooing the newspapers, he would gain access

*army, to give him a true report; and I will show him a way by which he can go and capture all the hill country without losing one of his men, captured or slain." When the man heard her words, and observed her face—she was in their eyes marvelously beautiful—they said to her ". . . Go at once to his tent. . . . And when you stand before him, do not be afraid in your heart, but tell him just what you have said, and he will treat you well." . . . Her words pleased Holofernes and all his servants, and they marveled at her wisdom and said, "There is no such a woman from one end of the earth to the other, either for beauty of face or wisdom of speech!" . . . On the fourth day Holofernes held a banquet for his slaves only, and did not invite any of his officers. And he said to Bagoas, the eunuch who had charge of all his personal affairs, "Go now and persuade the Hebrew woman who is in your care to join us and eat and drink with us." . . . So Bagoas went out from the presence of Holofernes, and approached her . . . And Judith said, "Who am I, to refuse my lord? Surely whatever pleases him I will do at once, and it will be a joy to me until the day of my death!" So she got up and arrayed herself in all her woman's finery. Then Judith came in and lay down, and olofernes' heart was ravished with her and he was moved with great desire to possess her; for he had been waiting for an opportunity to deceive her, ever since the day he first saw her. So Holofernes said to her, "Drink now, and be merry with us!" Judith said, "I will drink now, my lord, because my life means more to me today than in all the days since I was born." . . . And Holofernes was greatly pleased with her, and drank a great quantity of wine, much more than he had ever drunk in any one day since he was born. When evening came, his slaves quickly withdrew, and Bagoas closed the tent from outside and shut out the attendants from his master's presence; and they went to bed. . . . So Judith was left alone in the tent, with Holofernes stretched out on his bed, for he was overcome with wine. . . . [Judith] went to the post at the end of the bed, above Holofernes' head, and took down his sword that hung there. She came close to his bed and took ahold of the hair on his head. . . . And she struck his neck twice with all her might and severed his head from his body. Then she tumbled his body off the bed and pulled the canopy from the posts; after a moment she went out, and gave Holofernes' head to her maid, who placed it in her food bag.*

to the American public, and the American public would make him a star.

Again, in the civil war between Communists and Nationalists for control of China in the late 1920s and early '30s, most of the Communists focused on taking cities, as the Bolsheviks had done in Russia. But Mao Tse-tung, an outsider within the dogmatic Chinese Communist Party, was able to look at China in a clear light and see China's center of gravity as its vast peasant population. Win them to his side, he believed, and the revolution could not fail. That single insight proved the key to the Communists' success. Such is the power of identifying the center of gravity.

We often hide our sources of power from view; what most people consider a center of gravity is often a front. But sometimes an enemy will reveal his center of gravity by what he protects the most fervently. In bringing the Civil War into Georgia, General William Tecumseh Sherman discovered that the South was particularly anxious to protect Atlanta and the areas around it. That was the South's industrial center of gravity. Like Sherman, attack what the enemy most treasures, or threaten it to make the enemy divert forces to defend itself.

In any group, power and influence will naturally devolve to a handful of people behind the scenes. That kind of power works best when it is not exposed to the light of day. Once you discover this coterie holding the strings, win it over. As president during the Depression, Franklin Roosevelt faced problems from so many sides that it was difficult for him to know where his energy should go. In the end he decided that the key to enacting his reforms was winning over Congress. Then, within Congress, there were particular leaders who held the real power. He concentrated on wooing and seducing these leaders with his great charm. It was one of the secrets to his success.

What ultimately guides a group is the command-and-control center, the operational brain that takes in information, then makes the crucial decisions. Disrupting the functioning of that brain will cause dislocation throughout the enemy army. Before almost every battle, Alexander the Great would examine the enemy's organization, pinpointing as best he could the location of the command structure, then either attacking it or isolating it, making it impossible for the brain to communicate with the body.

Even in a sport as physical as boxing, Muhammad Ali, in crafting a strategy to defeat his archnemesis Joe Frazier, took aim at Frazier's mind, the ultimate center of gravity for any individual. Before every fight, Ali would get under Frazier's skin, riling him up by calling him an Uncle Tom, a tool of the white man's media. He would keep going during the fight itself, taunting Frazier mercilessly in the ring. Frazier became obsessed with Ali, could not think about him without bursting with anger. Controlling Frazier's mind was the key to controlling his body.

In any interaction with people, you must train yourself to focus on their strength, the source of their power, whatever it is that gives them their most crucial support. That knowledge will afford you many strategic options, many angles from which to attack, subtly or not so subtly undermining their strength rather than hitting it head-on. You can create no greater sense of panic in your enemies than that of being unable to use their strengths.

*. . . So Bagoas . . . went into the bedchamber and found him thrown down on the platform dead, with his head cut off and missing. And he cried out with a loud voice and wept and groaned. . . . Then he went to the tent where Judith had stayed, and when he did not find her he rushed out to the people and shouted, "The slaves have tricked us! One Hebrew woman has brought disgrace upon the house of Nebuchadnezzar! For look, here is Holofernes lying on the ground, and his head is not on him!" When the leaders of the Assyrian army heard this, they rent their tunics and were greatly dismayed, and their loud cries and shouts arose in the midst of the camp. . . . When the men in the tents heard it, they were amazed at what had happened. Fear and trembling came over them, so that they did not wait for one another, but with one impulse all rushed out and fled by every path across the plain and through the country. . . . Then the men of Israel, every one that was a soldier, rushed out upon them. . . . And when the Israelites heard it, with one accord they fell upon the enemy, and cut them down as far as Choba.*

JUDITH 7:19–15:7

**Image:** The Wall. Your opponents stand behind a wall, which protects them from strangers and intruders. Do not hit your head against the wall or lay siege to it; find the pillars and supports that make it stand and give it strength. Dig under the wall, sapping its foundations until it collapses on its own.

**Authority:** The first principle is that the ultimate substance of enemy strength must be traced back to the fewest possible sources, and ideally to one alone. The attack on these sources must be compressed into the fewest possible actions. . . . By constantly seeking out the center of his power, by daring all to win all, will one really defeat the enemy.

*—Carl von Clausewitz,* On War *(1780–1831)*

## REVERSAL

Every living creature has a center of gravity. Even the most decentralized group has to communicate and depends on a network that is vulnerable to attack. There is no reversal to this principle.

# 17

# DEFEAT THEM IN DETAIL

## THE DIVIDE-AND-CONQUER STRATEGY

*When you look at your enemies, do not be intimidated by their appearance. Instead look at the parts that make up the whole. By separating the parts, sowing dissension and division from within, you can weaken and bring down even the most formidable foe. In setting up your attack, work on their minds to create internal conflict. Look for the joints and links, the things that connect the people in a group or connect one group to another. Division is weakness, and the joints are the weakest part of any structure. When you are facing troubles or enemies, turn a large problem into small, eminently defeatable parts.*

*There were, however, many occasions when the French were faced not by one but by two or a whole series of enemy armies within supporting distance of one another. Faced with such a difficult situation, Napoleon often adopted a second system of maneuver—the "strategy of the central position." Very often under these circumstances the French found themselves operating at a numerical disadvantage against the combined strength of their opponents, but could procure superior numbers against any one part of their adversaries' forces. It was this second factor that the system was designed to exploit to the full. "The art of generalship consists in, when actually inferior in numbers to the enemy (overall), being superior to him on the battlefield." In brief, Napoleon set himself the task of isolating one part of the enemy armament, concentrating a stronger force to ensure its defeat and if possible its destruction, and then turning with his full strength to attack the second enemy army; that is to say, instead of a single decisive blow, he planned a series of smaller blows against scattered adversaries and set out to destroy them in detail. How could this be done? Once again, the sequence of the*

## THE CENTRAL POSITION

One day in early August of 490 B.C., the citizens of Athens received word that a massive Persian fleet had just landed some twenty-four miles to the north, along the coastal plains of Marathon. A mood of doom quickly spread. Every Athenian knew Persia's intentions—to capture their city; destroy its young democracy and restore a former tyrant, Hippias, to the throne; and sell many of its citizens into slavery. Some eight years earlier, Athens had sent ships to support the Greek cities of Asia Minor in a rebellion against King Darius, ruler of the Persian Empire. The Athenians had sailed home after a few battles—they soon saw that this business was hopeless—but they had participated in burning down the city of Sardis, an unforgivable outrage, and Darius wanted revenge.

The Athenians' predicament seemed desperate. The Persian army was enormous, some 80,000 men strong, transported by hundreds of ships; it had excellent cavalry and the best archers in the world. The Athenians, meanwhile, had only infantry, some 10,000 strong. They had sent a runner to Sparta urgently requesting reinforcements, but the Spartans were celebrating their moon festival and it was taboo to fight during such a time. They would send troops as soon as they could, within a week—but that would probably be too late. Meanwhile a group of Persian sympathizers within Athens—mostly from wealthy families—despised the democracy, looked forward to Hippias's return, and were doing their best to sow dissension and betray the city from within. Not only would the Athenians have to fight the Persians alone, but they were divided into factions among themselves.

The leaders of democratic Athens gathered to discuss the alternatives, all of which seemed bad. The majority argued for concentrating the Athenian forces outside the city in a defensive cordon. There they could wait to fight the Persians on terrain they knew well. The Persian army, however, was large enough to surround the city by both land and sea, choking it off with a blockade. So one leader, Miltiades, made a very different proposal: to march the entire Athenian army immediately toward Marathon, to a place where the road to Athens passed through a narrow pass along the coast. That would leave Athens itself unprotected; in trying to block the Persian advance on land, it would open itself to an attack by sea. But Miltiades argued that occupying the pass was the only way to avoid being surrounded. He had fought the Persians in Asia Minor and was the Athenians' most experienced soldier. The leaders voted for his plan.

And so a few days later, the 10,000 Athenian infantrymen began the march north, slaves carrying their heavy body armor, mules and donkeys transporting their food. When they reached the pass looking down on the plains of Marathon, their hearts sank: as far as the eye could see, the long strip of land was filled with tents, horses, and soldiers from all over the Persian Empire. Ships cluttered the coast.

For several days neither side moved. The Athenians had no choice but to hold their position; without cavalry and hopelessly outnumbered, how could they do battle at Marathon? If enough time went by, perhaps the Spartans would arrive as reinforcements. But what were the Persians waiting for?

Before dawn on August 12, some Greek scouts ostensibly working for the Persians slipped across to the Athenian side and reported startling news: under cover of darkness, the Persians had just sailed for the Bay of Phaleron outside Athens, taking most of their cavalry with them and leaving a holding force of some 15,000 soldiers in the plains of Marathon. They would take Athens from the sea, then march north, squeezing the Athenian army at Marathon between two larger forces.

Of the Athenian army's eleven commanders, Miltiades alone seemed calm, even relieved: this was their opportunity. As the sun was getting ready to rise, he argued for an immediate attack on the Persians at Marathon. Some of the other commanders resisted this idea: the enemy still had more men, some cavalry, and plenty of archers. Better to wait for the Spartans, who would surely arrive soon. But Miltiades countered that the Persians had divided their forces. He had fought them before and knew that the Greek infantryman was superior in discipline and spirit. The Persians at Marathon now only slightly outnumbered the Greeks; they could fight them and win.

Meanwhile, even with a good wind, it would take the Persian ships ten to twelve hours to round the coast and arrive at the Bay of Phaleron. Then they would need more time to disembark the troops and horses. If the Athenians defeated the Persians at Marathon quickly, they would have just enough time to run back to Athens and defend the city the same day. If instead they opted to wait, the Spartans might never arrive; the Persians would surround them, and, more ominously, the Persian sympathizers within Athens would probably betray the city from within and open its walls to the barbarians. It was now or never. By a vote of six to five, the commanders decided to attack at dawn.

At six in the morning, the Athenians began their charge. A hail of arrows from the Persian archers rained down on them, but they closed in on the enemy so quickly that the battle now had to be fought hand-to-hand—and, as Miltiades had foreseen, in close combat the Athenians were superior. They pushed the Persians back into the marshes at the north end of the plain, where thousands drowned. The waters reddened with blood. By nine in the morning, the Athenians had control of the plains, having lost fewer than two hundred men.

Although emotionally spent by this battle, the Athenians now had only around seven hours to make the twenty-four miles back to Athens in time to stop the Persians. There was simply no time to rest; they ran, as fast their feet could take them, loaded down in their heavy armor, impelled by the thought of the imminent dangers facing their families and fellow citizens. By four in the afternoon, the fastest among them

*Napoleonic attack reveals the formula. First of all the Emperor would accumulate as much information about the forces facing him from captured newspapers, deserters and most especially from the indications brought in by his probing cavalry patrols. From the data thus provided, he would carefully plot the known dispositions of his foes on the map, and then select the place where their respective army boundaries converged. This was the "hinge" or "joint" of the enemy's strategic dispositions, and as such was vulnerable to attack. This point would be selected by Napoleon for his initial blitzkrieg attack, carried out as often as not in full strength. Shielded by the cavalry screen, the French army would perform a crash concentration and fall like a thunderbolt on the handful of troops defending this central point. Invariably this initial onslaught would be successful. Immediately Napoleon had massed his army at this newly captured point, he was master of the "central position"—that is to say, he had successfully interposed his concentrated army between the forces of his enemies who, ideally, would have staggered back under the impact of the surprise blow in such a way as to increase the distance between their respective armies. This would inevitably mean that the foe would have to operate on "exterior lines" (i.e., have greater distances to march from one flank to the other) while the better-positioned French would have a shorter distance to travel to reach either enemy.*

THE CAMPAIGNS OF NAPOLEON, DAVID G. CHANDLER, 1966

had straggled to a point overlooking the Bay of Phaleron. The rest soon followed. Within a matter of minutes after their arrival, the Persian fleet sailed into the bay to see a most unwelcome sight: thousands of Athenian soldiers, caked in dust and blood, standing shoulder to shoulder to fight the landing.

The Persians rode at anchor for a few hours, then headed out to sea, returning home. Athens was saved.

## Interpretation

The victory at Marathon and race to Athens were perhaps the most decisive moments in Athenian history. Had the soldiers not come in time, the Persians would have taken the city, then certainly all of Greece, and eventually they would have expanded throughout the Mediterranean, for no other power in existence at the time could have stopped them. History would have been altered irrevocably.

Miltiades' plan worked by the narrowest of margins, but it was based on sound and timeless principles. When a powerful foe attacks you in strength, threatening your ability to advance and take the initiative, you must work to make the enemy divide its forces and then defeat these smaller forces one by one—"in detail," as the military say.

The key to Miltiades' strategy was his intuition to take the battle to Marathon. By placing himself at the pass that led to Athens, he occupied the central position in the war instead of the southern periphery. With the entire army holding the pass, the Persians would have a bloody time forcing their way through, so they decided to divide their forces before the Spartan reinforcements arrived. Once divided, and with their cavalry diluted, they lost their advantage and the central position from which they could dominate the war.

For the Athenians it was imperative to fight the smallest force first, the one they faced at Marathon. That done, and having taken the central position, they had the shorter route to Athens, while the invaders had to round the coast. Arriving first at Phaleron, the Athenians allowed no safe place to disembark. The Persians could have returned to Marathon, but the arrival of the bloodied Athenian soldiers from the north must have told them they had already lost the battle there, and their spirits were broken. Retreat was the only option.

There will be times in life when you face a powerful enemy—a destructive opponent seeking your undoing, a slew of seemingly insurmountable problems hitting you at once. It is natural to feel intimidated in these situations, which may paralyze you into inaction or make you wait in the vain hope that time will bring a solution. But it is a law of war that by allowing the larger force to come to you, at full strength and unified, you increase the odds against you; a large and powerful army on the move will gain an irresistible momentum if left unchecked. You will find yourself quickly overwhelmed. The wisest course is to take a risk, meet the enemy before it comes to you, and try to blunt its momentum by forcing or enticing it to divide. And the best way to make an enemy divide is to occupy the center.

Think of battle or conflict as existing on a kind of chessboard. The chessboard's center can be physical—an actual place like Marathon—or more subtle and psychological: the levers of power within a group, the support of a critical ally, a troublemaker at the eye of the storm. Take the center of the chessboard and the enemy will naturally break into parts, trying to hit you from more than one side. These smaller parts are now manageable, can be defeated in detail or forced to divide yet again. And once something large is divided, it is prone to further division, to being splintered into nothingness.

*As your army faces the enemy and the enemy appears powerful, try to attack the enemy in one particular spot. If you are successful in crumbling that one particular spot, leave that spot and attack the next, and so on and so forth, as if you were going down a winding road.*
—*Miyamoto Musashi (1584–1645)*

## ATTACKING THE JOINTS

As a young man, Samuel Adams (1722–1803) of colonial-era Boston developed a dream: the American colonies, he believed, should one day win complete independence from England and establish a government based on the writings of the English philosopher John Locke. According to Locke, a government should reflect the will of its citizens; a government that did not do so had lost its right to exist. Adams had inherited a brewery from his father, but he did not care about business, and while the brewery veered toward bankruptcy, he spent his time writing articles on Locke and the need for independence. He was an excellent writer, good enough to get his articles published, but few took his ideas seriously: he seemed to rant, to be somewhat out of touch with the world. He had that obsessive glint in the eye that makes people think you're a crackpot. The problem was that the ties between England and America were strong; the colonists did have their grievances, but there was hardly a clamor for independence. Adams began to have bouts of depression; his self-appointed mission seemed hopeless.

The British desperately needed money from the colonies, and in 1765 they passed a law called the Stamp Act: to make any document legal, American businesses would be required to purchase and affix to it a stamp of the British crown. The colonists were growing ticklish about the taxes they paid to England; they saw the Stamp Act as a new kind of tax in disguise, and a few disgruntled voices were raised in urban taverns. Even so, for most the issue seemed minor—but Adams saw the Stamp Act as the opportunity he had been waiting for his whole life. It gave him something tangible to attack, and he flooded newspapers throughout the colonies with editorials, all fulminating against the act.

*A novice chess player soon learns that it is a good idea to control the center of the board. This recognition will recur, in novel disguises, in situations far from the chess-board. It may help to seek the equivalent of the center of the board in any situation, or to see that the role of the center has migrated to the flanks, or to realize that there is no board and no singular topology. . . .*

CLAUSEWITZ ON STRATEGY, TIHA VON GHYZY, BOLKO VON OETINGER, CHRISTOPHER BASSFORD, EDS., 2001

Without consulting the colonies, he wrote, England was imposing a new kind of tax, and this, in a memorable phrase, was taxation without representation, the first step toward tyranny.

These editorials were so well written and so audacious in their criticisms that many began to take a closer look at the Stamp Act, and they did not like what they saw. Adams had never previously gone beyond writing articles, but now that he had lit this fire of discontent, he saw the urgency in stoking it further with action. For many years he had fraternized with working-class people considered riffraff by polite society—dockworkers and the like; now he banded these men into an organization called the Sons of Liberty. The group marched through the streets of Boston shouting a slogan Adams had coined: "Liberty, property, and no stamps!" They burned effigies of political figures who had promoted the Stamp Act. They distributed pamphlets containing Adams's arguments against the act. They also worked to intimidate the future distributors of the stamps, even going so far as to destroy one of their offices. The more dramatic the action, the more publicity Adams would earn, publicity into which he could insert arguments against the act.

Having gained momentum, the relentless Adams would not stop. He organized a statewide work stoppage for the day the act was to become law: shops would close, the courts would be empty. Since no business would be conducted in Massachusetts, no stamps would be purchased. The boycott was massively successful.

Adams's articles, demonstrations, and boycott made a splash in England, and there were members of Parliament who sympathized with the colonists and spoke out against the Stamp Act. Finally King George III had had enough, and in April 1766 the act was repealed. Americans rejoiced at their first show of power. The British were smarting from their defeat, however, and the following year they sneaked in another series of indirect taxes known as the Townshend System.

Clearly they had underestimated their enemy: Adams went to war. As he had with the Stamp Act, he wrote countless articles on the nature of the taxes the English had tried to disguise, once again stirring up anger. He also organized further demonstrations by the Sons of Liberty, now more menacing and violent than ever—in fact, the English were forced to send troops to Boston to keep the peace. This had been Adams's goal all along; he had ratcheted up the tension. Belligerent encounters between the Sons of Liberty and the English troops put the soldiers on edge, and finally a nervous group of them fired into a crowd, killing several Bostonians. Adams called this the Boston Massacre and spread fiery word of it throughout the colonies.

With the people of Boston now bubbling with anger, Adams organized another boycott: no citizen of Massachusetts, not even a prostitute, would sell anything to British soldiers. No one would rent them lodgings. They were shunned in the streets and taverns; even eye contact was avoided. All of this had a demoralizing effect on the British soldiers.

Feeling isolated and antagonized, many of them began to desert or find ways to be sent home.

News of the problems in Massachusetts spread north and south; colonists everywhere began to talk about Britain's actions in Boston, its use of force, its hidden taxes, its patronizing attitude. Then, in 1773, Parliament passed the Tea Act, on the surface a rather harmless attempt to solve the economic problems of the East India Company by giving it a virtual monopoly on the sale of tea in the colonies. The law also levied a nominal tax, but, even so, it would have made tea cheaper in the colonies, because the middlemen—the colonial importers—were to be cut out. The Tea Act, however, was deceptive in its effect, and confusing, and Adams saw in it a chance to apply the coup de grâce: it would ruin many colonial tea importers, and it did include a hidden tax, yet another form of taxation without representation. In exchange for cheaper tea, the English were making a mockery of democracy. In language more fiery than ever, Adams began to turn out articles opening up the old wounds from the Stamp Act and the Boston Massacre.

*Every kingdom divided against itself is laid waste, and a divided household falls. And if Satan also is divided against himself, how will his kingdom stand?*

LUKE 11:14

When East India Company ships began to arrive in Boston at the end of that year, Adams helped to organize a nationwide boycott of their tea. No dockworker would unload the cargo, no warehouse would store it. Then, one night in mid-December, after Adams had addressed a town meeting about the Tea Act, a group of members of the Sons of Liberty—disguised as Mohawk Indians, body paint and all—erupted in war whoops, charged to the wharves, boarded the tea ships, and destroyed their cargo, cutting open the cases of tea and pouring them into the harbor, all of this done with great revelry.

This provocative act, which later became known as the Boston Tea Party, was the turning point. The British could not tolerate it and quickly closed down Boston harbor and imposed military law on Massachusetts. Now all doubt vanished: pushed into a corner by Adams, the British were acting just as tyrannically as he had prophesied they would. The heavy military presence in Massachusetts was predictably unpopular, and it was only a matter of months before violence erupted: in April 1775, English soldiers fired on Massachusetts militiamen in Lexington. This "shot heard 'round the world" became the spark for the war that Adams had so diligently worked to kindle out of nothing.

### Interpretation

Before 1765, Adams labored under the belief that well-reasoned arguments would be enough to convince the colonists of the rightness of his cause. But as the years of failure piled up, he confronted the reality that the colonists retained a deep emotional attachment to England, as children do to a parent. Liberty meant less to them than did England's provision of protection and a sense of belonging in a threatening environment. When Adams realized this, he reformulated his goals: instead of preaching independence and the ideas of John Locke, he set to work to sever the colonists' ties with England. He made the children distrust the parent, whom they came to see not as a protector but as a domineering overlord exploiting them for its profit. The bond with England loosened, Adams's arguments for independence began to resonate. Now the colonists began to look for their sense of identity not to Mother England but to themselves.

With the Stamp Act campaign, then, Adams discovered strategy, the bridge between his ideas and reality. His writings now aimed at stirring up anger. The demonstrations he organized—pure theater—were also designed to create and build anger among the middle and lower classes, key components of the future revolution. Adams's innovative use of boycotts was calibrated to infuriate the British and bait them into rash action. Their violent response contrasted brilliantly with the relatively peaceful methods of the colonists, making them seem as tyrannical as he had said they were. Adams also worked to stir dissension among the English themselves, weakening the bond on all sides. The Stamp Act and Tea Act were actually rather trivial, but Adams strategically manipulated them to manufacture outrage, making them into wedges driven between the two sides.

Understand: rational arguments go in one ear and out the other. No one is changed; you are preaching to the converted. In the war to win people's attention and influence them, you must first separate them from whatever ties them to the past and makes them resist change. You must realize that these ties are generally not rational but emotional. By appealing to people's emotions, you can make your targets see the past in a new light, as something tyrannical, boring, ugly, immoral. Now you have room to infiltrate new ideas, shift people's vision, make them respond to a new sense of their self-interest, and sow the seeds for a new cause, a new bond. To make people join you, separate them from their past. When you size up your targets, look for what connects them to the past, the source of their resistance to the new.

A joint is the weakest part of any structure. Break it and you divide people internally, making them vulnerable to suggestion and change. Divide their minds in order to conquer them.

*Make the enemy believe that support is lacking; . . . cut off, flank, turn, in a thousand ways make his men believe themselves isolated. Isolate in like manner his squadrons, battalions, brigades and divisions; and victory is yours.*

*—Colonel Ardant du Picq (1821–1870)*

## KEYS TO WARFARE

Thousands of years ago, our primitive ancestors were prone to feelings of great weakness and vulnerability. To survive in the hostile environment of our early world, animals had speed, teeth and claws, fur against winter cold, and other advantages of power and protection. Humans had none of this and must have felt terrifyingly exposed and alone. The only way to compensate for such weakness was to form groups.

The group or tribe offered a defense against predators and greater effectiveness in the hunt. In the group there were enough people to watch your back. The larger the group, the more it allowed its members to refine that great human invention, the division of labor, and the more different individuals in the group were freed from the immediate needs of survival, the more time and energy they could devote to higher tasks. These different roles were mutually supportive and reinforcing, and the result was a net increase in human strength.

Over the centuries groups became ever larger and more complex. By learning to live in towns and settlements, people found that they could escape the feeling of imminent danger and need. Living with others also offered more subtle psychological protections. In time humans began to forget the fear that had made them form tribes in the first place. But in one group—the army—that primal terror remained as strong as ever.

The standard mode of ancient warfare was hand-to-hand combat, a frightening drama in which individuals were at all times exposed to death from behind and to each side. Military leaders learned early on to form their soldiers into tight, cohesive ranks. Trusting his fellows on either side of him not to retreat and leave him exposed, a soldier could fight the man in front of him with more spirit and confidence. The Romans extended this strategy by placing the youngest, most impetuous fighters in the front ranks, the most experienced and best fighters in the rear, and everyone else in the center. This meant that the weakest soldiers—the ones most prone to panic—were surrounded by those who were braver and steadier, giving them a powerful sense of security. No army went into battle with more cohesion and trust than the Roman legions.

In studying ancient warfare, the great nineteenth-century military writer Colonel Ardant du Picq noticed a peculiar phenomenon: in some of the most celebrated battles (Hannibal's victory over the Romans at Cannae and Julius Caesar's over Pompey at Pharsalus, for example), the losses on each side were fantastically disproportionate—a few hundred for the victors, thousands upon thousands among the vanquished. According to du Picq, what had happened in these cases was that through maneuver the ultimately victorious army had managed to surprise the enemy and splinter its lines into parts. Seeing their ranks breaking up, losing their sense of solidarity and support, and feeling isolated, soldiers panicked, dropped their weapons, and fled—and a soldier who turned his back on the enemy was an easy soldier to kill. Thousands were slaughtered this way. These great victories, then, were essentially psychological. Hannibal was vastly outnumbered at Cannae, but by making the Romans feel vulnerable and isolated, he made them overreact and retreat in confusion: easy pickings.

THE THREE OXEN AND THE LION

*There were three oxen who always grazed together. A lion had his designs upon them and wanted to eat them, but he could never get at one of them because they were always together. So he set them against each other with slanderous talk and managed to get them separated, whereupon they were isolated and he was able to eat them one after the other.*

*FABLES*, AESOP, SIXTH CENTURY B.C.

*Roosevelt . . . disliked being completely committed to any one person. He enjoyed being at the center of attention and action, and the system made him the focus through which the main lines of action radiated. . . . . . . The main reason for Roosevelt's methods, however, involved a tenacious effort to keep control of the executive branch in the face of the centrifugal forces of the American political system. By establishing in an agency one power center that counteracted another, he made each official more dependent on White House support; the President in effect became the necessary ally and partner of each. He lessened bureaucratic tendencies toward self-aggrandizement; he curbed any attempt to gang up on him. He was, in effect, adapting the old method of divide and conquer to his own purposes. . . . . . . His technique was curiously like that of Joseph Stalin, who used the overlapping delegation of function, a close student of his methods has said, to prevent "any single chain of command from making major decisions without confronting other arms of the state's bureaucracy and thus bringing the issues into the open at a high level." Roosevelt, like Stalin, was a political administrator in the*

The phenomenon is timeless: the soldier who feels he is losing the support of those around him is borne back into an intolerable primitive terror. He fears he will face death alone. Many great military leaders have turned this terror into strategy. Genghis Khan was a master at it: using the mobility of his Mongol cavalry to cut off his enemies' communications, he would isolate parts of their armies to make them feel alone and unprotected. He worked consciously to instill terror. The divide-and-isolate strategy was also used to great effect by Napoleon and the guerrilla forces of Mao Tse-tung, among many others.

Our nature has not changed. Lurking deep in even the most civilized among us is the same basic fear of being alone, unsupported, and exposed to danger. People today are more dispersed and society is less cohesive than ever before, but that only increases our need to belong to a group, to have a strong network of allies—to feel supported and protected on all sides. Take away this feeling and we are returned to that primitive sensation of terror at our own vulnerability. The divide-and-conquer strategy has never been more effective than it is today: cut people off from their group—make them feel alienated, alone, and unprotected—and you weaken them enormously. That moment of weakness gives you great power to maneuver them into a corner, whether to seduce or to induce panic and retreat.

Throughout the 1960s, one of Mao Tse-tung's most loyal and trusted followers was his minister of defense, Lin Biao. No one praised the Chinese ruler more fulsomely than Lin. And yet by 1970 Mao had come to suspect that the flattery was a ruse to disguise his intentions: Lin was plotting to be his successor. And what made Lin particularly dangerous was that, as minister of defense, he had accumulated allies in the military.

Mao went to work with great subtlety. In public he went out of his way to support Lin, as if he, too, saw the minister as his successor. That soothed the natural wariness of the plotter. At the same time, however, Mao also attacked and demoted some of Lin's most important supporters in the military. Lin was a bit of a radical, veering left on most issues; Mao urged him to propose some of his more extreme ideas for restructuring the military, secretly knowing that these ideas would prove unpopular. Lin's support among the higher branches of the military slowly began to thin.

Lin finally realized what Mao was up to, but it was too late. He had lost his power base. Frustrated and scared, he resorted to plotting a coup d'état, a desperate act that played straight into Mao's hands. In 1971, Lin died under suspicious circumstances in a plane crash.

As Mao understood, in political environments people depend on their connections even more than on their talents. In such a world, a person whose career seems to be waning is one whom few will want to know. And people who feel isolated will often overreact and do something desperate—which of course just makes them more isolated. So Mao created the impression that Lin was losing his connections. Had he attacked Lin directly, he would have gotten bogged down in an ugly fight. Dividing the minister from his power base, and in the process making him appear to be on the decline, was much more effective.

*sense that his first concern was power—albeit for very different ends.*

*ROOSEVELT: THE LION AND THE FOX,*
JAMES MACGREGOR BURNS,
1956

Before you launch an outright attack on your enemies, it is always wise first to weaken them by creating as much division in their ranks as possible. One good place to drive a wedge is between the leadership and the people, whether soldiers or citizenry; leaders function poorly when they lose their support among the people. So work to make them look authoritarian or out of touch. Or steal their foundation, as the Republican president Richard Nixon did in 1972 by wooing the blue-collar types who had traditionally voted Democrat: he split the Democrats' base. (The Republicans have been doing the same thing ever since.) Remember: once your enemy begins to splinter in any way, the rupture will tend to gain momentum. Division usually leads to more division.

In 338 B.C., Rome defeated its greatest enemy at the time, the Latin League—a confederation of Italian cities that had formed to block Rome's expansion. With this victory, however, the Romans faced a new problem: how to govern the region. If they crushed the league's members, they would leave a power vacuum, and down the road another enemy would emerge that might prove a still-greater threat. If they simply swallowed up the cities of the league, they would dilute the power and prestige of Rome, giving themselves too large an area to protect and police.

The solution the Romans came up with, which they would later call *divide et impera* (divide and rule), was to become the strategy by which they forged their empire. Essentially they broke up the league but did not treat all of its parts equally. Instead they created a system whereby some of its cities were incorporated into Roman territory and their residents given full privileges as Roman citizens; others were deprived of most of their territory but granted near-total independence; and others still were broken up and heavily colonized with Roman citizens. No single city was left powerful enough to challenge Rome, which retained the central position. (As the saying goes, all roads led to Rome.)

The key to the system was that if an independent city proved itself loyal enough to Rome or fought well enough for Rome, it won the chance of being incorporated into the empire. The individual cities now saw it as more in their interest to gain Rome's favor than to ally themselves elsewhere. Rome held out the prospect of great power, wealth, and protection, while isolation from Rome was dangerous. And so the once proud members of the Latin League now competed against one another for Rome's attention.

Divide and rule is a powerful strategy for governing any group. It is based on a key principle: within any organization people naturally form smaller groups based on mutual self-interest—the primitive desire to find strength in numbers. These subgroups form power bases that, left unchecked, will threaten the organization as a whole. The formation of

*"Do not think that I have come to bring peace on earth; I have not come to bring peace, but a sword. For I have come to set a man against his father, and a daughter against her mother, and a daughter-in-law against her mother-in-law; and a man's foes will be those of his own household. He who loves father or mother more than me is not worthy of me; and he who loves son or daughter more than me is not worthy of me; and he who does not take his cross and follow me is not worthy of me."*

MATTHEW 10:34

parties and factions can be a leader's greatest threat, for in time these factions will naturally work to secure their own interests before those of the greater group. The solution is to divide to rule. To do so you must first establish yourself as the center of power; individuals must know they need to compete for your approval. There has to be more to be gained by pleasing the leader than by trying to form a power base within the group.

When Elizabeth I became queen, England was a nation divided. The remnants of feudalism entailed many competing power centers, and the court itself was full of factions. Elizabeth's solution was to weaken the nobility by deliberately pitting one family against another. At the same time, she occupied the center, making herself a symbol of England itself, the hub around which everything revolved. Within the court, too, she made sure that no individual—except of course herself—gained ascendancy. When she saw that first Robert Dudley and then the Earl of Essex believed themselves her favorites, she quickly cut them loose.

The temptation to maintain a favorite is understandable but dangerous. Better to rotate your stars, occasionally making each one fall. Bring in people with different viewpoints and encourage them to fight it out. You can justify this as a healthy form of democracy, but the effect is that while those below you fight to be heard, you rule.

The film director Alfred Hitchcock faced enemies on all sides—writers, set designers, actors, producers, marketers—any of them quite capable of putting their egos before the quality of the film. Writers wanted to show off their literary skills, actors wanted to look like stars, producers and marketers wanted the movie to be commercial—the whole crew had competing interests. Hitchcock's solution, like Queen Elizabeth's, was to take the central position, in a variant of divide and rule. His carefully crafted role as a public celebrity was part of this: his movies' publicity campaigns always involved him as spokesperson, and he made bit-part appearances in most of his films, becoming an instantly recognizable, endearingly humorous figure. He put himself in the middle of every aspect of production, from writing the script before the shoot began to editing the film when the shoot was finished. At the same time, he kept all the filmmaking departments, even that of the producer, a little out of the loop; information about every detail of the film was kept in his head, his drawings, and his notes. No one could bypass him; every decision went through him. Before the film was shot, for example, Hitchcock would set out in detail the look of the leading lady's costumes. If the costume designer wanted to change anything, she would have to go through him or be caught out in rank insubordination. In essence, he was like Rome: all roads led to Hitchcock.

Within your group, factions may emerge quite subtly by virtue of the fact that people who are experts in their area may not tell you everything they're doing. Remember: they see only the small picture; you are in charge of the whole production. If you are to lead, you must occupy the center. Everything must flow through you. If information is to be withheld, you are the one to do it. That is divide and rule: if the different parts of the operation lack access to all the information, they will have to come to you to get it. It is not that you micromanage but that you keep overall control of everything vital and isolate any potential rival power base.

Throughout the 1950s and '60s, Major General Edward Lansdale was considered America's principal expert in counterinsurgency. Working with President Ramon Magsaysay of the Philippines, he had crafted a plan that had defeated the country's Huk guerrilla movement in the early 1950s. Counterinsurgency requires a deft hand, more political than military, and for Lansdale the key to success was to stamp out government corruption and bring the people close to the government through various popular programs. That would deny the insurgents their cause, and they would die of isolation. Lansdale thought it folly to imagine that leftist rebels could be defeated by force; in fact, force just played into their hands, giving them a cause they could use to rally support. For insurgents, isolation from the people is death.

Think of the people in your group who are working primarily for their own interests as insurgents. They are Cassius types who thrive on discontent in the organization, fanning it into dissension and factionalism. You can always work to divide such factions once you know about them, but the better solution is to keep your soldiers satisfied and contented, giving the insurgents nothing to feed on. Bitter and isolated, they will die off on their own.

The divide-and-rule strategy is invaluable in trying to influence people verbally. Start by seeming to take your opponents' side on some issue, occupying their flank. Once there, however, create doubt about some part of their argument, tweaking and diverting it a bit. This will lower their resistance and maybe create a little inner conflict about a cherished idea or belief. That conflict will weaken them, making them vulnerable to further suggestion and guidance.

Japan's great seventeenth-century swordsman Miyamoto Musashi on several occasions faced bands of warriors determined to kill him. The sight of such a group would intimidate most people, or at least make them hesitate—a fatal flaw in a samurai. Another tendency would be to lash out violently, trying to kill as many of the attackers as possible all at once, but at the risk of losing control of the situation. Musashi, however, was above all else a strategist, and he solved these dilemmas in the most rational way possible. He would place himself so that the men would have to come at him in a line or at an angle. Then he would focus on killing the first man and move swiftly down the line. Instead of being overwhelmed or trying too hard, he would break the band into parts. Then he just had to kill opponent number one, while leaving himself in position to deal with opponent number two and preventing his mind from being clouded and confused by the other attackers awaiting him. The effect was that he could retain his focus while keeping his opponents off balance, for as he proceeded down the line, they would become the ones who were intimidated and flustered.

Whether you are beset by many small problems or by one giant problem, make Musashi the model for your mental process. If you let the complexity of the situation confuse you and either hesitate or lash out without thought, you will lose mental control, which will only add momentum to the negative force coming at you. Always divide up the issue at hand, first placing yourself in a central position, then proceeding down the line, killing off your problems one by one. It is often wise to begin with the smallest problem while keeping the most dangerous one at bay. Solving that one will help you create momentum, both physical and psychological, that will help you overwhelm all the rest.

The most important thing is to move quickly against your enemies, as the Athenians did at Marathon. Waiting for troubles to come to you will only multiply them and give them a deadly momentum.

**Image:** The Knot. It is large, hopelessly entangled, and seemingly impossible to unravel. The knot consists of thousands of smaller knots, all twisted and intertwined. Let time go by and the knot will get only worse. Instead of trying to pick it apart from this side or that, take up your sword and cut it in half with one blow. Once divided, it will come undone on its own.

**Authority:** In antiquity those who were referred to as excelling in the employment of the army were able to keep the enemy's forward and rear forces from connecting; the many and few from relying on each other; the noble and lowly from coming to each other's rescue; the upper and lower ranks from trusting each other; the troops to be separated, unable to reassemble, or when assembled, not to be well-ordered. —*Sun-tzu (fourth century B.C.)*

## REVERSAL

Dividing your forces as a way of creating mobility can be a powerful strategy, as Napoleon demonstrated with his flexible system of corps, which let him hit his enemy unpredictably from many different angles. But to make his system work, Napoleon needed precise coordination of its parts and overall control over their movements—and his goal was ultimately to bring the parts together to strike a major blow. In guerrilla warfare a commander will disperse his forces to make them harder to hit, but this, too, demands coordination: a guerrilla army cannot succeed if the parts are unable to communicate with each other. In general, any division of your forces must be temporary, strategic, and controlled.

In attacking a group in order to sow division, be careful that your blow is not too strong, for it can have the opposite effect, causing people to unite in times of great danger. That was Hitler's miscalculation during the London Blitz, his bombing campaign designed to push England out of World War II. Intended to demoralize the British public, the Blitz only made them more determined: they were willing to suffer short-term danger in order to beat him in the long run. This bonding effect was partly the result of Hitler's brutality, partly the phenomenon of a culture willing to suffer for the greater good.

Finally, in a divided world, power will come from keeping your own group united and cohesive, and your own mind clear and focused on your goals. The best way to maintain unity may seem to be the creation of enthusiasm and high morale, but while enthusiasm is important, in time it will naturally wane, and if you have come to depend on it, you will fail. Far greater defenses against the forces of division are knowledge and strategic thinking. No army or group can be divided if it is aware of the enemy's intentions and makes an intelligent response. As Samuel Adams discovered, strategy is your only dependable sword and shield.

THE PLOUGHMAN'S QUARRELSOME SONS

*A ploughman's sons were always quarrelling. He scolded them to no avail—his words did nothing to change their ways. So he decided to teach them a practical lesson. He asked them to bring him a load of firewood. As soon as they had done this he gave a bundle to each and told them to break it all up for him. But, in spite of all their efforts, they were unable to do so. The ploughman therefore undid the bundles and handed each of his sons a stick at a time. These they broke without any trouble.*
*"So!" said the father, "you too, my children, if you stay bound together, can be invincible to your enemies. But if you are divided you will be easy to defeat."*

*FABLES*, AESOP, SIXTH CENTURY B.C.

# 18

# EXPOSE AND ATTACK YOUR OPPONENT'S SOFT FLANK

## THE TURNING STRATEGY

*When you attack people directly, you stiffen their resistance and make your task that much harder. There is a better way: distract your opponents' attention to the front, then attack them from the side, where they least expect it. By hitting them where they are soft, tender, and unprotected, you create a shock, a moment of weakness for you to exploit. Bait people into going out on a limb, exposing their weakness, then rake them with fire from the side. The only way to get stubborn opponents to move is to approach them indirectly.*

*The Emperor [Napoleon Bonaparte], while he was quite prepared "to break eggs to make omelettes," as von Clausewitz puts it, was always eager to gain total victory for a minimum expenditure of manpower and effort. Consequently he disliked having to force a full-scale, fully arrayed frontal battle—that is to say, marching directly against the enemy to fight him on ground of his (the adversary's) choosing, for such battles were inevitably expensive and rarely conclusive (Borodino in 1812 is a case in point). Instead, whenever possible, after pinning the foe frontally by a feint attack, he marched his main army by the quickest possible "safe" route, hidden by the cavalry screen and natural obstacles, to place himself on the rear or flank of his opponent. Once this move had been successfully achieved, he occupied a natural barrier or "strategical curtain" (usually a river line or mountain range), ordered the blocking of all crossings, and thus isolated his intended victim from his rear depots and reduced his chances of reinforcement. Thereafter, Napoleon advanced relentlessly toward the foe's army, offering him only two alternatives—to fight for survival on ground*

## TURNING THE FLANK

In 1793, Louis XIV and his wife, Marie Antoinette, the king and queen of France, were beheaded by order of the new government put in place after the French Revolution. Marie Antoinette was the daughter of Maria Theresa, the empress of Austria, and as a result of her death the Austrians became determined enemies of France. Early in 1796 they prepared to invade the country from northern Italy, which at the time was an Austrian possession.

In April of that year, the twenty-six-year-old Napoleon Bonaparte was given command of the French army in Italy and charged with a simple mission: to prevent these Austrian armies from entering France. Under Napoleon, for the first time since the revolution not only were the French able to hold a defensive position, but they successfully went on the offensive, pushing the Austrians steadily east. Shocking as it was to lose to the revolutionary army, it was downright humiliating to be defeated by an unknown general on his first campaign. For six months the Austrians sent armies to defeat Napoleon, but he forced each one to retreat into the fortress of Mantua, until finally this stronghold was crammed with Austrian soldiers.

Leaving a force at Mantua to pin down the Austrians, Napoleon established his base to the north, in the pivotal city of Verona. If the Austrians were to win the war, they would somehow have to push him out of Verona and free up the starving soldiers trapped in Mantua. And they were running out of time.

In October 1796, Baron Joseph d'Alvintzi was given command of some 50,000 Austrian soldiers and the urgent mission of expelling the French from Verona. An experienced commander and clever strategist, d'Alvintzi studied Napoleon's Italian campaign carefully and came to respect his enemy. To defeat this brilliant young general, the Austrians would have to be more flexible, and d'Alvintzi thought he had the solution: he would divide his army into two columns, one under himself, the other under the Russian general Paul Davidovich. The columns would separately march south, converging at Verona. At the same time, d'Alvintzi would launch a campaign of deception to make Napoleon think that Davidovich's army was small (it was in fact 18,000 men strong), merely a holding force to protect the Austrian lines of communication. If Napoleon underestimated Davidovich, the Russian general would face less opposition and his way to Verona would be smooth. D'Alvintzi's plan was to trap Napoleon between the jaws of these two armies.

The Austrians entered northern Italy in early November. To d'Alvintzi's delight, Napoleon seemed to have fallen for their trick; he sent a relatively light force against Davidovich, who promptly gave the French in Italy their first real defeat and began his advance toward Verona. Meanwhile d'Alvintzi himself advanced all the way to a point not far from Verona and was poised to fall on the city from the east. As he pored over his maps, d'Alvintzi took pleasure in his plan. If Napoleon sent more men to stop Davidovich, he would weaken Verona against d'Alvintzi. If he tried to block d'Alvintzi's entrance from the east, he would weaken Verona against Davidovich. If he sought reinforcements from his troops at Mantua, he would free up the 20,000 Austrian soldiers trapped there and they would gobble him up from the south. D'Alvintzi also knew that Napoleon's men were exhausted and hungry. Having fought for six months without rest, they were at a breaking point. Not even a young genius like Napoleon could escape this trap.

*not of his own choosing, or to surrender. The advantages afforded by such a strategy are obvious. The enemy army would be both taken by surprise and almost certainly demoralized by the sudden apparition of the enemy army in its rear, cutting its communications.*

THE CAMPAIGNS OF NAPOLEON, DAVID G. CHANDLER, 1966

A few days later, d'Alvintzi advanced to the village of Caldiero, at Verona's doorstep. There he inflicted another defeat on the French troops sent to stop him. After a string of victories, Napoleon had now lost two battles in a row; the pendulum had swung against him.

As d'Alvintzi prepared for his final pounce on Verona, he received confusing news: against all prediction Napoleon had in fact divided his army in Verona, but instead of sending parts of it against either d'Alvintzi or Davidovich, he had marched a sizable force somewhere to the southeast. The next day this army appeared outside the town of Arcola. If the French crossed the river to Arcola and advanced a few miles north, they would directly cross d'Alvintzi's line of communications and of retreat, and they would be able to seize his supply depots at Villa Nova. Having this large French army to his rear was more than alarming; d'Alvintzi was forced to forget about Verona for the moment and hastily marched east.

He had retreated in the nick of time and was able to halt the French before they could cross the river and attack Villa Nova. For several days the two armies settled into a fiercely contested battle for the bridge at Arcola. Napoleon himself led several charges and was nearly killed. A portion of the troops blocking Mantua were dispatched north to reinforce the French at Arcola, but d'Alvintzi's army hunkered down, and the battle turned into a stalemate.

On the third day of fighting, d'Alvintzi's soldiers—their lines thinned by relentless French attacks—were preparing for another battle for the bridge when they suddenly heard trumpets blaring from their southern flank. A French force had somehow crossed the river below the bridge and was marching toward the Austrian flank at Arcola. The sound of trumpets was quickly replaced by shouts and the whizzing of bullets. The sudden appearance of the French on their flank was too much for the wearied Austrians; not waiting to see the size of the French force, they panicked and fled the scene. The French poured across the river. D'Alvintzi gathered up his men as best he could and managed to lead them east to safety. But the battle for Verona was lost, and with it the doom of Mantua was sealed.

Somehow Napoleon had managed to snatch victory from defeat. The battle of Arcola helped forge the legend of his invincibility.

*Now came the critical problem of judging the correct moment for the enveloping force to reveal its disconcerting position on the enemy flank. For maximum effect, it was important that this should not occur before the enemy had committed all or most of his reserves to the frontal battle, and this need for accurate timing of the flank attack called for the greatest judgment on the part of Napoleon and his key subordinates. The former had to judge the moment when all the enemy troops were indeed committed to the frontal battle (and with the billowing clouds of black-powder smoke obliterating the scene this was no easy matter); the latter had the task of keeping their eager troops "on the leash" so as to avoid any premature attack disclosing their presence. Then, when the exact moment came, Napoleon would give the signal. . . . Then the* attaque débordante *would spring to life. A roar of cannon away on his hitherto secure flank would cause the enemy to look apprehensively over his shoulder, and before long the spyglasses of his anxious staff would be able to detect a line of dust and smoke crawling ever nearer from the flank or rear. This threat to his communications and line of retreat could not be ignored. The enemy general might now theoretically adopt one of two courses (but in practice only one). He could either order an immediate general retreat to slip out of the trap before it shut behind his army (although this was generally out of the question, as Napoleon would of course launch a general frontal attack against all sectors of the enemy line to coincide with the unmasking of his flanking force and thus pin the foe still tighter to the ground he was holding); or he would be compelled to find troops from somewhere to form a new line at right angles to his main position to face the new onslaught and protect his flank. As all reserves were (ideally) already committed to battle, this could be easily and quickly effected only by deliberately weakening those frontal*

### Interpretation

Napoleon was no magician, and his defeat of the Austrians in Italy was deceptively simple. Facing two armies converging on him, he calculated that d'Alvintzi's was the more imminent danger. The fight for Caldiero encouraged the Austrians to think that Verona would be defended through direct, frontal confrontation. But Napoleon instead divided his army and sent the larger portion of it to threaten the Austrian supply depot and lines of communication and retreat. Had d'Alvintzi ignored the threat and advanced on Verona, he would have moved farther away from his critical base of operations and put himself in great jeopardy; had he stayed put, Napoleon would have squeezed him between two armies. In fact, Napoleon knew d'Alvintzi would have to retreat—the threat was too real—and once he had done so, he would have relinquished the initiative. At Arcola, sensing that the enemy was tiring, Napoleon sent a small contingent to cross the river to the south and march on the Austrian flank, with instructions to make as much noise as possible—trumpets, shouts, gunfire. The presence of this attacking force, small though it was, would induce panic and collapse. The ruse worked.

This maneuver—the *manoeuvre sur les derrières*, Napoleon called it—would become a favorite strategy of his. Its success was based on two truths: First, generals like to place their armies in a strong frontal position, whether to make an attack or to meet one. Napoleon would often play on this tendency to face forward in battle by seeming to engage the enemy frontally; in the fog of battle, it was hard to tell that really only half of his army was deployed here, and meanwhile he would sneak the other half to the side or rear. Second, an army sensing attack from the flank is alarmed and vulnerable and must turn to face the threat. This moment of turning contains great weakness and confusion. Even an army in the stronger position, like d'Alvintzi's at Verona, will almost always lose cohesion and balance as it turns.

Learn from the great master himself: attacking from the front is rarely wise. The soldiers facing you will be tightly packed in, a concentration of force that will amplify their power to resist you. Go for their flank, their vulnerable side. This principle is applicable to conflicts or encounters of any scale.

Individuals often show their flank, signal their vulnerability, by its opposite, the front they show most visibly to the world. This front can be an aggressive personality, a way of dealing with people by pushing them around. Or it can be some obvious defense mechanism, a focus on keeping out intruders to maintain stability in their lives. It can be their most cherished beliefs and ideas; it can be the way they make themselves liked. The more you get people to expose this front, to show more of themselves and the directions they tend to move in, the more their unprotected flanks will come into focus—unconscious desires, gaping insecurities, precarious alliances, uncontrollable compulsions. Once you move on their flanks, your targets will turn to face you and lose their equilibrium. All enemies are vulnerable from their sides. There is no defense against a well-designed flanking maneuver.

*Opposition to the truth is inevitable, especially if it takes the form of a new idea, but the degree of resistance can be diminished—by giving thought not only to the aim but to the method of approach. Avoid a frontal attack on a long-established position; instead, seek to turn it by flank movement, so that a more penetrable side is exposed to the thrust of truth.*
—*B. H. Liddell Hart (1895–1970)*

## OCCUPYING THE FLANK

As a young man, Julius Caesar (100–44 B.C.) was once captured by pirates. They asked for a ransom of twenty talents; laughing, he replied that a man of his nobility was worth fifty talents, and he volunteered to pay that sum. His attendants were sent for the money, and Caesar was left alone with these bloodthirsty pirates. For the weeks he remained among them, he participated in their games and revelry, even playing a little rough with them, joking that he would have them crucified someday.

Amused by this spirited yet affectionate young man, the pirates practically adopted him as their own. But once the ransom was paid and Caesar was freed, he proceeded to the nearest port, manned some ships at his own expense, then went after the pirates and surprised them in their lair. At first they welcomed him back—but Caesar had them arrested, took back the money he had given them, and, as promised, had them crucified. In the years to come, many would learn—whether to their delight or to their horror—that this was how Caesar did battle.

Caesar, however, did not always exact retribution. In 62 B.C., during a religious ceremony in Caesar's home, a young man named Publius Clodius was caught among the female celebrants, dressed as a woman and cavorting with Caesar's wife, Pompeia. This was considered an outrage, and Caesar immediately divorced Pompeia, saying, "My wife must be above suspicion." Yet when Clodius was arrested and tried for sacrilege, Caesar used his money and influence to get the youth acquitted. He was more than repaid a few years later, when he was preparing to leave Rome for wars in Gaul and needed someone to protect his interests while he was away. He used his clout to get Clodius named to the political office of tribune, and in that position Clodius doggedly supported Caesar's interests, stirring up so much trouble in the Senate with his obnoxious maneuvers that no one had the time or inclination to intrigue against the absent general.

The three most powerful men in Rome at the time were Caesar, Crassus, and Pompey. Fearing Pompey, a popular and famously successful general, Crassus tried to form a secret alliance with Caesar, but Caesar balked; instead, a few years later, he approached the wary Pompey

*sectors closest to the new threat. This thinning out of the enemy front is what Napoleon termed "the Event"—and was of course exactly what he intended to have happen. The curtain on the first act would now fall; the enemy was reacting as required; the destruction of the cohesion of his line, the final ruination of his equilibrium, could now be undertaken with practically a guarantee of ultimate success.*

THE CAMPAIGNS OF NAPOLEON,
DAVID G. CHANDLER,
1966

*During this survey one impression became increasingly strong—that, throughout the ages, effective results in war have rarely been attained unless the approach has had such indirectness as to ensure the opponent's unreadiness to meet it. The indirectness has usually been physical, and always psychological. In strategy, the longest way round is often the shortest way home.*
*More and more clearly has the lesson emerged that a direct approach to one's mental object, or physical objective, along the "line of natural expectation" for the opponent, tends to produce negative results. The reason has been expressed vividly in Napoleon's dictum that "the moral is to the*

(who was suspicious of and hostile toward Caesar as a possible future rival) and suggested they form their own alliance. In return he promised to support some of Pompey's political proposals, which had been stalled in the Senate. Surprised, Pompey agreed, and Crassus, not wanting to be left out, agreed to join the group to form the First Triumvirate, which was to rule Rome for the next several years.

In 53 B.C., Crassus was killed in battle in Syria, and a power struggle quickly emerged between Pompey and Caesar. Civil war seemed inevitable, and Pompey had more support in the Senate. In 50 B.C., the Senate ordered that both Caesar (who was fighting in Gaul at the time) and Pompey should send one of their legions to Syria to support the Roman army fighting there. But since Pompey had already lent Caesar a legion for the war in Gaul, he proposed to send that one to Syria—so that Caesar would have lost two legions instead of one, weakening him for the impending war.

Caesar did not complain. He sent off the two legions, one of which, however—as he had expected—did not go to Syria but was conveniently quartered near Rome, at Pompey's disposal. Before the two legions left, Caesar paid each soldier handsomely. He also instructed their officers to spread the rumor in Rome that his troops still in Gaul were exhausted and that, should he dare to send them against Pompey, they would switch sides as soon as they had crossed the Alps. Coming to believe these false reports, and expecting massive defections, Pompey did not trouble to recruit more soldiers for the imminent war, which he would later regret.

In January of 49 B.C., Caesar crossed the Rubicon, the river between Gaul and Italy, a dramatic, unexpected move that initiated the Civil War. Caught by surprise, Pompey fled with his legions to Greece, where he began to prepare a major operation. As Caesar marched south, many of Pompey's supporters, left behind in Rome, were terrified. Caesar had established a reputation in Gaul for brutal treatment of the enemy, leveling whole towns and killing their inhabitants. Yet when Caesar took the key town of Corfinium, capturing important senators and army officers who had fought there alongside troops loyal to Pompey, he did not punish these men; in fact, he returned to them the monies his soldiers had looted in taking the town. This remarkable act of clemency became the model for his treatment of Pompey's supporters. Instead of Caesar's men switching allegiance to Pompey, it was Pompey's who now became the most ardent followers of Caesar. As a result, Caesar's march on Rome was quick and bloodless.

Next, although Pompey had established his base in Greece, Caesar decided to first attack his flank: the large army he had quartered in Spain. Over several months of campaigning, he completely outmaneuvered this force, led by Pompey's generals Afranius and Petreius, and finally cornered them. They were surrounded, the situation was hopeless, and Afranius and many of the soldiers, knowing of Caesar's gentle treatment of his enemies, sent word that they were ready to surrender; but

*physical as three to one." It may be expressed scientifically by saying that, while the strength of an opposing force or country lies outwardly in its numbers and resources, these are fundamentally dependent upon stability of control, morale, and supply.*

*To move along the line of natural expectation consolidates the opponent's balance and thus increases his resisting power. In war, as in wrestling, the attempt to throw the opponent without loosening his foothold and upsetting his balance results in self-exhaustion, increasing in disproportionate ratio to the effective strain put upon him. Success by such a method only becomes possible through an immense margin of superior strength in some form—and, even so, tends to lose decisiveness. In most campaigns the dislocation of the enemy's psychological and physical balance has been the vital prelude to a successful attempt at his overthrow.*

STRATEGY, B. H. LIDDELL HART, 1954

Petreius, horrified at this betrayal, ordered that any soldier who supported Caesar be slaughtered. Then, determined to go down fighting, he led his remaining men out of the camp for battle—but Caesar refused to engage. The soldiers were unable to fight.

Finally, desperately low in supplies, Pompey's men surrendered. This time they could expect the worst, for Caesar knew about the massacre in the camp—yet once again he pardoned Petreius and Afranius and simply disbanded their army, giving the soldiers supplies and money for their return to Rome. Hearing of this, the Spanish cities still loyal to Pompey quickly changed sides. In a matter of three months, Roman Spain had been conquered through a combination of maneuver and diplomacy, and with barely a drop of blood spilled.

In the following months, Pompey's political support in Rome evaporated. All he had left was his army. His defeat by Caesar at the Battle of Pharsalus, in northern Greece, a year later merely put the seal on his inevitable destruction.

### Interpretation

Caesar discovered early on in his political life that there are many ways to conquer. Most people advance more or less directly, attempting to overpower their opponents. But unless they kill the foes they beat this way, they are merely creating long-term enemies who harbor deep resentment and will eventually make trouble. Enough such enemies and life becomes dangerous.

Caesar found another way to do battle, taking the fight out of his enemies through strategic and cunning generosity. Disarmed like this, enemy becomes ally, negative becomes positive. Later on, if necessary, when the former foe's guard is down, you can exact retribution, as Caesar did with the pirates. Behave more gently, though, and your enemy may become your best follower. So it was with Publius Clodius, who, after disgracing Caesar's home, became the devoted agent of the general's dirty work.

When the Civil War broke out, Caesar understood that it was a political phenomenon as much as a military one—in fact, what mattered most was the support of the Senate and the Romans. His acts of mercy were part of a calculated campaign to disarm his enemies and isolate Pompey. In essence, what Caesar was doing here was occupying his enemies' flank. Instead of attacking them frontally and engaging them directly in battle, he would take their side, support their causes, give them gifts, charm them with words and favors. With Caesar apparently on their side, both politically and psychologically they had no front to fight against, nothing to oppose. In contact with Caesar, all hostility toward him melted away. This way of waging war allowed him to defeat the militarily superior Pompey.

Life is full of hostility—some of it overt, some clever and underhanded. Conflict is inevitable; you will never have total peace. Instead of

THE TENTH LABOUR: THE CATTLE OF GERYON

*Heracles' Tenth Labour was to fetch the famous cattle of Geryon from Erytheia, an island near the Ocean stream, without either demand or payment. Geryon, a son of Chrysaor and Callirrhoë, a daughter of the Titan Oceanus, was the King of Tartessus in Spain, and reputedly the strongest man alive. He had been born with three heads, six hands, and three bodies joined together at the waist. Geryon's shambling red cattle, beasts of marvellous beauty, were guarded by the herdsman Eurytion, son of Ares, and by the two-headed watchdog Orthrus—formerly Atlas' property—born of Typhoon and Echidne. . . .*
*On his arrival, [Hercules] ascended Mount Abas. The dog Orthrus rushed at him, barking, but Heracles' club struck him lifeless; and Eurytion, Geryon's herdsman, hurrying to Orthrus' aid, died in the same manner. Heracles then proceeded to drive away the cattle. Menoetes, who was pasturing the cattle of Hades near by—but Heracles had left these untouched—took the news to Geryon. Challenged to battle, Heracles ran to Geryon's flank and shot him sideways through all three bodies with a single arrow. . . . As*

*Hera hastened to Geryon's assistance, Heracles wounded her with an arrow in the right breast, and she fled. Thus he won the cattle, without either demand or payment.*

THE GREEK MYTHS, VOL. 2, ROBERT GRAVES, 1955

imagining you can avoid these clashes of will, accept them and know that the way you deal with them will decide your success in life. What good is it to win little battles, to succeed in pushing people around here and there, if in the long run you create silent enemies who will sabotage you later? At all cost you must gain control of the impulse to fight your opponents directly. Instead occupy their flank. Disarm them and make them your ally; you can decide later whether to keep them on your side or to exact revenge. Taking the fight out of people through strategic acts of kindness, generosity, and charm will clear your path, helping you to save energy for the fights you cannot avoid. Find their flank—the support people crave, the kindness they will respond to, the favor that will disarm them. In the political world we live in, the flank is the path to power.

*Your gentleness shall force*
*More than your force move us to gentleness.*

AS YOU LIKE IT, WILLIAM SHAKESPEARE, 1564–1616

*Let us see if by moderation we can win all hearts and secure a lasting victory, since by cruelty others have been unable to escape from hatred and maintain their victory for any length of time. . . . This is a new way of conquering, to strengthen one's position by kindness and generosity.*

*—Julius Caesar (100–44 B.C.)*

*When, in the course of studying a long series of military campaigns, I first came to perceive the superiority of the indirect over the direct approach, I was looking merely for light upon strategy. With deepened reflection, however, I began to realize that the indirect approach had a much wider application—that it was a law of life in all spheres: a truth of philosophy. Its fulfillment was seen to be the key to practical achievement in dealing with any problem where the human factor predominates, and a conflict of wills tends to spring from an underlying concern for all interests. In all such cases, the direct assault of new ideas provokes a stubborn resistance, thus intensifying the difficulty of producing*

## KEYS TO WARFARE

The conflict and struggle we go through today are astounding—far greater than those faced by our ancestors. In war the passages of armies are marked with arrows on maps. If we had to map the battles of our own daily lives, we would draw thousands of those arrows, a constant traffic of moves and maneuvers—not to speak of the arrows actually hitting us, the people trying to persuade us of one thing or another, to move us in a particular direction, to bend us to their will, their product, their cause.

Because so many people are constantly shifting for power, our social world becomes blanketed in barely disguised aggression. In this situation it requires time and patience to be indirect; in the daily rush to move and influence people, the subtle approach is too difficult and time-consuming, so people tend to take the direct route to what they want. To convince us of the correctness of their ideas, they use argument and rhetoric, growing ever louder and more emotional. They push and pull with words, actions, and orders. Even those more passive players who use the tools of manipulation and guilt are quite direct, not in the least subtle, in the paths they choose; witness a few of their maneuvers and they are rather easy to figure out.

The result of all of this is twofold: we have all become more defensive, resistant to change. To maintain some peace and stability in our lives, we build our castle walls ever higher and thicker. Even so, the increasingly direct brutality of daily life is impossible to avoid. All those arrows hitting us infect us with their energy; we cannot help but try to give back what we get. Reacting to direct maneuvers, we find ourselves dragged into head-to-head arguments and battles. It takes effort to step away from this vicious arena and consider another approach.

You must ask yourself this question: what is the point of being direct and frontal if it only increases people's resistance, and makes them more certain of their own ideas? Directness and honesty may give you a feeling of relief, but they also stir up antagonism. As tactics they are ineffective. In war itself—blood war, not the interpersonal wars of everyday life—frontal battles have become rare. Military officers have come to realize that direct attack increases resistance, while indirection lowers it.

The people who win true power in the difficult modern world are those who have learned indirection. They know the value of approaching at an angle, disguising their intentions, lowering the enemy's resistance, hitting the soft, exposed flank instead of butting horns. Rather than try to push or pull people, they coax them to turn in the direction they desire. This takes effort but pays dividends down the road in reduced conflict and greater results.

The key to any flanking maneuver is to proceed in steps. Your initial move cannot reveal your intentions or true line of attack. Make Napoleon's *manoeuvre sur les derrières* your model: First hit them directly, as Napoleon did the Austrians at Caldiero, to hold their attention to the front. Let them come at you mano a mano. An attack from the side now will be unexpected and hard to combat.

At a palace reception in Paris in 1856, all eyes were on a new arrival on the scene: an eighteen-year-old Italian aristocrat called the Countess de Castiglione. She was stunningly beautiful and more: she carried herself like a Greek statue come to life. Emperor Napoleon II, a notorious womanizer, could not help but take notice and be fascinated, but for the moment that was all—he tended to prefer more hot-blooded women. Yet as he saw her again over the months that followed, he became intrigued despite himself.

In events at court, Napoleon and the countess would exchange glances and occasional remarks. She always left before he could engage her in conversation. She wore stunning dresses, and long after the evening was over, her image would return to his mind.

What drove the emperor crazy was that he apparently didn't excite her—she seemed only modestly interested in him. He began to court her assiduously, and after weeks of assault, she finally succumbed. Yet even now that she was his mistress, he still sensed her coldness, still had to pursue her, was never sure of her feelings. At parties, too, she would draw men's attention like a magnet, making him furiously jealous. The affair went on, but before too long the emperor naturally tired of the countess and moved on to another woman. Even so, while it lasted, he could think of no one else.

In Paris at the time was Victor-Emmanuel, the king of Piedmont, the

*a change of outlook. Conversion is achieved more easily and rapidly by unsuspected infiltration of a different idea or by an argument that turns the flank of instinctive opposition. The indirect approach is as fundamental to the realm of politics as to the realm of sex. In commerce, the suggestion that there is a bargain to be secured is far more important than any direct appeal to buy. And in any sphere, it is proverbial that the surest way of gaining a superior's acceptance of a new idea is to weaken resistance before attempting to overcome it; and the effect is best attained by drawing the other party out of his defences.*

*STRATEGY*,
B. H. LIDDELL HART,
1954

*Six in the fifth place means:*
*The tusk of a gelded boar.*
*Good fortune.*
*Here the restraining of the impetuous forward drive is achieved in an indirect way. A boar's tusk is in itself dangerous, but if the boar's nature is altered, the tusk is no longer a menace. Thus also where men are concerned, wild force should not be combated directly.*

*THE I CHING*,
CHINA, CIRCA EIGHTH CENTURY B.C.

*After this meeting a story about Mao's methods went the rounds of Shanghai's remaining executive suites. Mao called in Liu [Shaoqi] and Zhou [Enlai]. He had a question for them: "How would you make a cat eat pepper?" Liu spoke up first. "That's easy," said the number-two man. "You get somebody to hold the cat, stuff the pepper in its mouth, and push it down with a chopstick." Mao raised his hands in horror at such a made-in-Moscow solution. "Never use force. . . . Everything must be voluntary." Zhou had been listening. Mao inquired what the premier would do with the cat. "I would starve the cat," replied the man who had often walked the tightrope of opportunity. "Then I would wrap the pepper with a slice of meat. If the cat is sufficiently hungry it will swallow it whole." Mao did not agree with Zhou any more than with Liu. "One must not use deceit either—never fool the people." What, then would the Chairman himself do? "Easy," he said—concurring with Liu at least on that. "You rub the pepper thoroughly into the cat's backside. When it burns, the cat will lick it off—and be happy that it is permitted to do so."*

*MAO: A BIOGRAPHY*, ROSS TERRILL, 1999

countess's home. Italy was divided into small states like this one at the time, but with France's support it would soon become a unified nation, and Victor-Emmanuel harbored the secret desire to become its first king. In her conversations with Napoleon, the countess would occasionally talk of the king of Piedmont, praising his character and describing his love of France and his strength as a leader. The emperor could only agree: Victor-Emmanuel would make the perfect king of Italy. Soon Napoleon was broaching this idea with his advisers, then actively promoting Victor-Emmanuel for the throne as if it were his own idea—and eventually he made this happen. Little did he know: his affair with the countess had been set up by Victor-Emmanuel and his clever adviser, the Count di Cavour. They had planted her in Paris to seduce Napoleon and slowly insinuate the idea of Victor-Emmanuel's promotion.

The countess's seduction of the emperor had been planned like an elaborate military campaign, right down to the dresses she would wear, the words she would say, the glances she would throw. Her discreet way of roping him in was a classic flanking attack, a seductive *manoeuvre sur les derrières.* The countess's cold beauty and fascinating manner drew the emperor on until he had advanced so far that he was convinced it was he who was the aggressor. Holding his attention to the front, the countess worked to the side, subtly conjuring the idea of crowning Victor-Emmanuel. Had she pursued the emperor directly or suggested the crowning of the king in so many words, not only would she have failed, but she would have pushed the emperor in the opposite direction. Drawn forward frontally by his weakness for a beautiful woman, he was vulnerable to gentle persuasion on his flank.

Maneuvers like this one should be the model for your attempts at persuasion. Never reveal your intentions or goals; instead use charm, pleasant conversation, humor, flattery—whatever works—to hold people's attention to the front. Their focus elsewhere, their flank is exposed, and now when you drop hints or suggest subtle changes in direction, the gates are open and the walls are down. They are disarmed and maneuverable.

Think of people's ego and vanity as a kind of front. When they are attacking you and you don't know why, it is often because you have inadvertently threatened their ego, their sense of importance in the world. Whenever possible, you must work to make people feel secure about themselves. Again, use whatever works: subtle flattery, a gift, an unexpected promotion, an offer of alliance, a presentation of you and they as equals, a mirroring of their ideas and values. All these things will make them feel anchored in their frontal position relative to the world, lowering their defenses and making them like you. Secure and comfortable, they are now set up for a flanking maneuver. This is particularly devastating with a target whose ego is delicate.

A common way of using the flanking maneuver in war is to get your enemies to expose themselves on a weak salient. This means maneuvering them onto ground or luring them to advance in such a way that their front is narrow and their flanks are long—a delicious target for a side attack.

In 1519, Hernán Cortés landed with a small army in eastern Mexico, planning to realize his dream of conquering the Aztec Empire. But first he had to conquer his own men, particularly a small yet vocal group of supporters of Diego de Velázquez, the governor of Cuba, who had sent Cortés on no more than a scouting mission and who coveted the conquest of Mexico himself. Velázquez's supporters caused trouble for Cortés at every step, constantly conspiring against him. One bone of contention was gold, which the Spanish were to collect for delivery to the king of Spain. Cortés had been letting his soldiers barter for gold but then had been using that gold to buy food. This practice, Velázquez's men argued, must end.

Appearing to concede, Cortés suggested the Velázquez men appoint a treasurer. They quickly named one of their own, and with their help this man began to collect everyone's gold. This policy, naturally, proved extremely unpopular with the soldiers, who were braving enormous dangers for little benefit. They complained bitterly—but Cortés just pointed to the men who had insisted on this policy in the name of the governor of Cuba. He personally, of course, had never been in favor of it. Soon the Velázquez men were universally hated, and Cortés, at the urgent request of the other soldiers, gladly rescinded the policy. From then on, the conspirators could get nowhere with the men. They were exposed and despised.

Cortés used this strategy often to deal with dissenters and troublemakers. At first he would seem to go along with their ideas, would even encourage them to take things further. In essence, he would get his enemies to expose themselves on a weak salient, where their selfish or unpopular ideas could be revealed. Now he had a target to hit.

When people present their ideas and arguments, they often censor themselves, trying to appear more conciliatory and flexible than is actually the case. If you attack them directly from the front, you end up not getting very far, because there isn't much there to aim at. Instead try to make them go further with their ideas, giving you a bigger target. Do this by standing back, seeming to go along, and baiting them into moving rashly ahead. (You can also make them emotional, pushing their buttons, getting them to say more than they had wanted to.) They will expose themselves on a weak salient, advancing an indefensible argument or position that will make them look ridiculous. The key is never to strike too early. Give your opponents time to hang themselves.

In a political world, people are dependent on their social position. They need support from as many sources as possible. That support, the base of most people's power, presents a rich flank to expose and attack. Franklin D. Roosevelt knew that a politician's vulnerable flank was the electorate, the people who might or might not vote for him in his next

*Inner truth. Pigs and fishes.*
*Good fortune.*
*It furthers one to cross the great water.*
*Perseverance furthers.*
*Pigs and fishes are the least intelligent of all animals and therefore the most difficult to influence. The force of inner truth must grow great indeed before its influence can extend to such creatures. In dealing with persons as intractable and as difficult to influence as a pig or a fish, the whole secret of success depends on finding the right way of approach. One must first rid oneself of all prejudice and, so to speak, let the psyche of the other person act on one without restraint. Then one will establish contact with him, understand and gain power over him. When a door has been thus opened, the force of one's personality will influence him. If in this way one finds no obstacles insurmountable, one can undertake even the most dangerous things, such as crossing the great water, and succeed.*

*The I Ching*, China, circa eighth century B.C.

race. Roosevelt could get a politician to sign off on a bill or support a nomination, whatever his real thoughts about the issues, by threatening a maneuver that would injure the other man's popularity with his constituents. A flanking attack on someone's social status and reputation will make him or her turn to face this menace, giving you ample room to maneuver the opponent in other directions.

The more subtle and indirect your maneuvers in life, the better. In 1801, Napoleon suddenly offered Russia the chance to become the protector of the island of Malta, then under French control. That would give the Russians an important base in the Mediterranean. The offer seemed generous, but Napoleon knew that the English would soon take control of the island, for they coveted it and had the forces in place to take it, and the French navy was too weak to hold it. The English and the Russians were allies, but their alliance would be endangered by a squabble over Malta. That discord was Napoleon's goal all along.

The ultimate evolution of strategy is toward more and more indirection. An opponent who cannot see where you are heading is at a severe disadvantage. The more angles you use—like a cue ball in billiards caroming off several sides of the table—the harder it will be for your opponents to defend themselves. Whenever possible, calculate your moves to produce this caroming effect. It is the perfect disguise for your aggression.

*The* Book of Changes (I Ching) *is often considered the Oriental apotheosis of adaptation, of flexibility. In this book the recurring theme is one of observing life and blending with its flow in order to survive and develop. In effect, the theme of this work is that everything in existence can be a source of conflict, of danger, and, ultimately, of violence if opposed from the wrong angle or in the wrong manner—that is, if confronted directly at the point of its maximum strength, since this approach renders the encounter potentially devastating. By the same token, any and every occurrence can be dealt with by approaching it from the right angle and in the proper manner—that is, at its source, before it can develop full power, or from the sides (the vulnerable "flanks of a tiger").*

SECRETS OF THE SAMURAI, OSCAR RATTI AND ADELE WESTBROOK, 1973

**Image:** The Lobster. The creature seems intimidating and impenetrable, with its sharp claws quick to grab, its hard protective shell, its powerful tail propelling it out of danger. Handle it directly and you will pay the price. But turn it over with a stick to reveal its tender underside and the creature is rendered helpless.

**Authority:** It is by turning the enemy, by attacking his flank, that battles are won. *—Napoleon Bonaparte (1769–1821)*

## REVERSAL

In politics, occupying the flank by taking a similar position to the other side, co-opting its ideas for your own purposes, is a powerful ploy, one that President Clinton used to great effect in his triangulations with the Republicans. This gives the opponent nothing to strike at, no room to maneuver. But staying too long on the opponent's flank can bring a price: the public—the real soft flank for any politician—loses its sense of what the triangulator stands for, what sets him and his party apart from the other side. Over time this can prove dangerous; polarity (see chapter 1)—creating the appearance of sharp differences—is more effective in the long run. Beware of occupying the opponent's flank at the expense of exposing your own.

# 19

## ENVELOP THE ENEMY

## THE ANNIHILATION STRATEGY

*People will use any kind of gap in your defenses to attack you or revenge themselves on you. So offer no gaps. The secret is to envelop your opponents—create relentless pressure on them from all sides, dominate their attention, and close off their access to the outside world. Make your attacks unpredictable to create a vaporous feeling of vulnerability. Finally, as you sense their weakening resolve, crush their willpower by tightening the noose. The best encirclements are psychological—you have surrounded their minds.*

*Legend has it that Shaka altered the nature of fighting in the region for ever, by inventing a heavy, broad-bladed spear designed to withstand the stresses of close-quarter combat. Perhaps he did: certainly both Zulu sources and the accounts of white travellers and officials in the nineteenth century credit him with this achievement. . . . His military innovations made an impact on Zulu folklore, if nothing else, for Shaka certainly developed fighting techniques to an unprecedented degree, and there is a wealth of stories concerning his prowess as a warrior: he may, indeed, have been one of the great military geniuses of his age. In place of the loose skirmishing tactics with light throwing spears, Shaka trained his warriors to advance rapidly in tight formations and engage hand-to-hand, battering the enemy with larger war-shields, then skewering their foes with the new spear as they were thrown off balance. If the results are anything to judge by, Shaka's capacity for conquest must have been dramatic. By 1824 the Zulus had eclipsed all their rivals, and had extended their influence over an area many times larger than their original homeland.*

THE ANATOMY OF THE ZULU ARMY, IAN KNIGHT, 1995

## THE HORNS OF THE BEAST

In December 1878 the British declared war on the Zulus, the warrior tribe of present-day South Africa. The rather flimsy pretext was border troubles between Zululand and the British state of Natal; the real aim was to destroy the Zulu army, the last remaining native force threatening British interests in the area, and to absorb Zulu territories into a British-run confederation of states. The British commander, Lieutenant General Lord Chelmsford, drafted a plan to invade Zululand with three columns, the central one aimed at the capital of Ulundi, the heart of the kingdom.

Many Englishmen in Natal were thrilled at the prospect of war and at the potential benefits of taking over Zululand, but no one was as excited as forty-eight-year-old Colonel Anthony William Durnford. For years Durnford had bounced from one lonely British Empire outpost to another, finally ending up in Natal. In all his years of military service, Durnford had not once seen action. He yearned to prove his valor and worth as a soldier, but he was approaching the age when such youthful dreams could no longer be fulfilled. Now, suddenly, the impending war was sending the opportunity his way.

Eager to impress, Durnford volunteered to organize an elite force of native soldiers from Natal to fight alongside the British. His offer was accepted, but as the British invaded Zululand in early January 1879, he found himself cut out of the main action. Lord Chelmsford did not trust him, thinking his hunger for glory made him impetuous; also, for someone with no battle experience, he was old. So Durnford and his company were stationed at Rorke's Drift, in western Zululand, to help monitor the border areas with Natal. Dutifully but bitterly, Durnford followed his orders.

In the first days after the invasion, the British failed to locate the main Zulu army, only trickles of men here and there. They were growing frustrated. On January 21, Chelmsford took half of the central column, which was encamped at the foot of a mountain called Isandlwana, and led it east in search of the Zulus. Once he had found the enemy, he would bring the rest of his army forward—but the elusive Zulus might attack the camp while he was away, and the men at Rorke's Drift were the closest reserves. Needing to reinforce Isandlwana, he sent word to Durnford to bring his company there. As colonel, Durnford would now be the highest-ranking officer at the camp, but Chelmsford could not worry about Durnford's leadership qualities—the impending battle was the only thing on his mind.

Early on the morning of January 22, Durnford received the news he had been waiting for all his life. Barely able to contain his excitement, he lead his four hundred men east to Isandlwana, arriving at the camp at around 10:00 A.M. Surveying the land, he understood why Chelmsford had put his main camp here: to the east and south were miles of rolling grassland—Zulus approaching from that direction would be seen well in advance. To the north was Isandlwana, and beyond it the plains of

Nqutu. This side was a little less secure, but scouts had been placed at key points in the plains and at the mountain passes; attack from that direction would almost certainly be detected in time.

Shortly after his arrival, Durnford received a report that a seemingly large Zulu force had been spotted on the plains of Nqutu heading east, perhaps to attack Chelmsford's half of the central column from the rear. Chelmsford had left explicit orders to keep the 1,800 men at Isandlwana together. In case of attack, they had enough firepower to defeat the entire Zulu army—as long as they stayed concentrated and kept their lines in order. But to Durnford it was more important to find the main Zulu force. The British soldiers were beginning to grow edgy, not knowing where this vaporous enemy was. The Zulus had no cavalry, and many of them fought with spears; once their hiding place was uncovered, the rest would be easy—the superior weaponry and discipline of the British soldiers would prevail. Durnford thought Chelmsford was too cautious. As senior officer at the camp, he decided to disobey orders and lead his 400 men northeast, parallel to the plains of Nqutu, to find out what the Zulus were up to.

As Durnford marched out of the camp, a scout on the plains of Nqutu saw a few Zulus herding cattle some four miles away. He gave chase on his horse, but the Zulus disappeared into thin air. Riding to the point where they had vanished, he stopped his horse just in time: below him lay a wide, deep ravine, completely hidden from the surface of the plains, and crowded into the ravine, as far as he could see in both directions, were Zulu warriors in full war regalia, an eerie intensity in their eyes. They seemed to have been meditating on the imminent battle. For a second the horseman was too stunned to move, but as hundreds of spears were suddenly aimed at him, he turned and galloped away. The Zulus quickly rose and began clambering out of the ravine.

Soon the other scouts on the plains saw the same terrifying sight: a wide line of Zulus filling the horizon, some 20,000 men strong. Even from a distance, it was clear that they were moving in formation, each end of their line coming forward in a shape resembling horns. The scouts quickly brought word to the camp that the Zulus were coming. By the time Durnford received the news, he could look up to the ridge above him and see a line of Zulus streaming down the slope. He quickly formed his own men into lines to fight them off while retreating to the camp. The Zulus maneuvered with incredible precision. What Durnford could not see was that the men in the left tip of the horn were moving through the tall grass toward the rear of the camp, to link up with the other end of the horn and complete the encirclement.

The Zulus facing Durnford and his men seemed to grow out of the earth, emerging from behind boulders or from out of the grass in ever-greater numbers. A knot of five or six of them would suddenly charge, throwing spears or firing rifles, then disappear back into the grass. Whenever the British stopped to reload, the Zulus would advance ever closer,

*The careful use by the Zulus of cover during their advance was observed time and again by the British. Another anonymous survivor of Isandlwana noted that as the Zulus crested the Nyoni ridge and came within sight of camp, they "appeared almost to grow out of the earth. From rock and bush on the heights above started scores of men; some with rifles, others with shields and assegais." Lieutenant Edward Hutton of the 60th left a rather more complete description of the Zulu army deploying for the attack at Gingindlovu: "The dark masses of men, in open order and under admirable discipline, followed each other in quick succession, running at a steady pace through the long grass. Having moved steadily round so as exactly to face our front, the larger portion of the Zulus broke into three lines, in knots and groups of from five to ten men, and advanced towards us. . . . [They] continued to advance, still at a run, until they were about 800 yards from us, when they began to open fire. In spite of the excitement of the moment we could not but admire the perfect manner in which these Zulus skirmished. A knot of five or six would rise and dart through the long grass, dodging from side to side with heads down, rifles and shields kept low and out of sight. They would then suddenly sink into the long grass, and nothing but puffs of curling smoke would show their whereabouts. Then they advance again. . . ." The speed of this final advance was terrifying. When the British gave the order to cease firing and fall back at Isandlwana, the Zulus were pinned down some two or three hundred yards from the British position. Lieutenant Curling of the Artillery noted that in the time it took for his experienced men to limber his guns, the Zulus had rushed in so quickly that one gunner had actually been stabbed as he mounted the axletree seat. A Zulu veteran of the battle, uMhoti of the uKhandempemvu, thought the final charge so swift that "like a flame the whole Zulu force sprang to its feet and darted upon them."*

*THE ANATOMY OF THE ZULU ARMY,* IAN KNIGHT, 1995

occasionally one reaching Durnford's lines and disemboweling a British soldier with the powerful Zulu spear, which made an unbearable sucking sound as it went in and out.

Durnford managed to get his men back into camp. The British were surrounded, but they closed ranks and fired away, killing scores of Zulus and keeping them at bay. It was like target practice: as Durnford had predicted, their superior weaponry was making the difference. He looked around; the fight had turned into a stalemate, and his soldiers were responding with relative confidence. Almost imperceptibly, though, Durnford noticed a slight slackening in their fire. Soldiers were running out of ammunition, and in the time it took them to open a new crate and reload, the Zulus would tighten the circle and a wave of fear would ripple through the men as here and there a soldier in the front lines would be impaled. The Zulus fought with an intensity the British had never seen; rushing forward as if bullets could not harm them, they seemed to be in a trance.

Suddenly, sensing the turning point in the battle, the Zulus began to rattle their spears against their shields and emit their war cry: *"Usuthu!"* It was a terrifying din. At the northern end of the camp, a group of British soldiers gave way—just a few, panicking at the sight and sound of the Zulus, now only a few yards distant, but the Zulus poured through the gap. As if on cue, those in the circle between the two horns rained spears on the British, killing many and making havoc of their lines. From out of nowhere, a reserve force rushed forward, fanning around the circle and doubling its squeezing power. Durnford tried to maintain order, but it was too late: in a matter of seconds, panic. Now it was every man for himself.

Durnford ran to the one gap in the encirclement and tried to keep it open so that his remaining men could retreat to Rorke's Drift. Minutes later he was impaled by a Zulu spear. Soon the battle at Isandlwana was over. A few hundred managed to escape through the gap that Durnford had died in securing; the rest, over fourteen hundred men, were killed.

After such a devastating defeat, the British forces quickly retreated out of Zululand. For the time being, the war was indeed over, but not as the British had expected.

### Interpretation

A few months after the defeat at Isandlwana, the British mounted a larger invasion and finally defeated the Zulus. But the lesson of Isandlwana remains instructive, particularly considering the incredible discrepancy in technology.

The Zulu way of fighting had been perfected earlier in the nineteenth century by King Shaka Zulu, who by the 1820s had transformed what had been a relatively minor tribe into the region's greatest fighting force. Shaka invented the heavy, broad-bladed Zulu spear, the assegai, that was so devastating in battle. He imposed a rigorous discipline, training the Zulus to advance and encircle their enemies with machine-like precision. The circle was extremely important in Zulu culture—as a symbol of their national unity, a motif in their artwork, and their dominant pattern in warfare. The Zulus could not fight for extended periods, since their culture required lengthy cleansing rituals after the shedding of blood in battle. During these rituals they were completely vulnerable to attack—no Zulu could fight again, or even rejoin the tribe, until he had been cleansed. The immense Zulu army was also costly to maintain in the field. Once mobilized, then, the army not only had to defeat its enemies in battle, it had to annihilate every last one of them, eliminating the possibility of a counterattack during the vulnerable cleansing period and allowing a speedy demobilization. Encirclement was the Zulu method of obtaining this complete kind of victory.

Before any battle, the Zulus would scout the terrain for places to hide. As one looks out over the grasslands and plains of South Africa, they seem to offer wide visibility, but they often conceal ravines and gullies undetectable from any distance. Even up close, grasses and boulders provide excellent coverage. The Zulus would move quickly to their hiding places, their feet tough as leather from years of running over the grasslands. They would send out scouting parties as distractions to hide the movements of the main force.

Once they emerged from their hiding place and headed into battle, the Zulus would form what they called the "horns, chest, and loins." The chest was the central part of the line, which would hold and pin the enemy force. Meanwhile the horns to either side would encircle it, moving in to the sides and rear. Often the tip of one horn would stay hidden behind tall grass or boulders; when it emerged to complete the encirclement it would at the same time give the enemy a nasty psychological shock. The loins were a reserve force kept back to be thrown in for the coup de grâce. These men often actually stood with their backs to the battle, so as not to grow overly excited and rush in before the right moment.

Years after Isandlwana a commission laid the blame for the disaster on Durnford, but in reality it was not his fault. It was true that the British had let themselves be surrounded, but they managed to form lines in decent order and fought back bravely and well. What destroyed them was what destroyed every opponent of the Zulus: the terror created by the precision of their movements, the feeling of being encircled in an ever-tightening space, the occasional sight of a fellow soldier succumbing to the horrible Zulu spear, the war cries, the spears that rained down at the moment of greatest weakness, the nightmarish sight of a reserve force suddenly joining the circle. For all the superiority of their weaponry, the British collapsed under this calculated psychological pressure.

We humans are extremely clever creatures: in disaster or setback, we often find a way to adapt, to turn the situation around. We look for any gap and often find it; we thrive on hope, craftiness, and will. The

*As soon as it grew light, Hannibal sent forward the Balearics and the other light infantry. He then crossed the river in person and as each division was brought across he assigned it its place in the line. The Gaulish and Spanish horse he posted near the bank on the left wing in front of the Roman cavalry; the right wing was assigned to the Numidian troopers. The centre consisted of a strong force of infantry, the Gauls and Spaniards in the middle, the Africans at either end of them. . . . These nations, more than any other, inspired terror by the vastness of their stature and their frightful appearance: the Gauls were naked above the waist, the Spaniards had taken up their position wearing white tunics embroidered with purple, of dazzling brilliancy. The total number of infantry in the field [at Cannae] was 40,000, and there were 10,000 cavalry. Hasdrubal was in command of the left wing, Marhabal of the right; Hannibal himself with his brother Mago commanded the centre. It was a great convenience to both armies that the sun shone obliquely on them, whether it was that they purposely so placed themselves, or whether it happened by accident, since the Romans faced the north, the Carthaginians the* history of war is littered with stories of dramatic adjustments and reversals, except in one place: the envelopment. Whether physical or psychological, this is the only true exception to the possibility of turning things around.

When properly executed, this strategy gives your opponents no gaps to exploit, no hope. They are surrounded, and the circle is tightening. In the abstract space of social and political warfare, encirclement can be any maneuver that gives your opponents the feeling of being attacked from all sides, being pushed into a corner and denied hope of making a counterattack. Feeling surrounded, their willpower will weaken. Like the Zulus, keep a force in reserve, the loins to work with your horns—you hit them with these forces when you sense their weakness growing. Let the hopelessness of their situation encircle their minds.

*You must make your opponent acknowledge defeat from the bottom of his heart.*
*—Miyamoto Musashi (1584–1645)*

## KEYS TO WARFARE

Thousands of years ago, we humans lived a nomadic life, wandering across deserts and plains, hunting and gathering. Then we shifted into living in settlements and cultivating our food. The change brought us comfort and control, but in a part of our spirit we remain nomads: we cannot help but associate the room to roam and wander with a feeling of freedom. To a cat, tight, enclosed spaces may mean comfort, but to us they conjure suffocation. Over the centuries this reflex has become more psychological: the feeling that we have options in a situation, a future with prospects, translates into something like the feeling of open space. Our minds thrive on the sense that there is possibility and strategic room to maneuver.

Conversely, the sense of psychological enclosure is deeply disturbing to us, often making us overreact. When someone or something encircles us—narrowing our options, besieging us from all sides—we lose control of our emotions and make the kinds of mistakes that render the situation more hopeless. In history's great military sieges, the greater danger almost always comes from the panic and confusion within. Unable to see what is happening beyond the siege, losing contact with the outside world, the defenders also lose their grip on reality. An animal that cannot observe the world around it is doomed. When all you can see are Zulus closing in, you succumb to panic and confusion.

The battles of daily life occur not on a map but in a kind of abstract space defined by people's ability to maneuver, act against you, limit your power, and cut into your time to respond. Give your opponents any room in this abstract or psychological space and they will exploit it, no matter how powerful you are or how brilliant your strategies—so make them feel surrounded. Shrink their possibilities of action and close off

*south. The wind, called by the inhabitants the Vulturnus, was against the Romans, and blew great clouds of dust into their faces, making it impossible for them to see in front of them. When the battle [at Cannae] was raised, the auxiliaries ran forward, and the battle began with the light infantry. Then the Gauls and Spaniards on the left engaged the Roman cavalry on the right; the battle was not at all like a cavalry fight, for there was no room for maneuvering, the river on the one side and the infantry on the other hemming them in, compelled them to fight face to face. Each side tried to force their way straight forward, till at last the horses were standing in a closely pressed mass, and the riders seized their opponents and tried to drag them from their horses. It had become mainly a struggle of infantry, fierce but short, and the Roman cavalry was repulsed and fled. Just as this battle of the cavalry was finished, the infantry became engaged, and as long as the Gauls and Spaniards kept their ranks unbroken, both sides were equally matched in strength and courage. At length after long and repeated efforts the Romans closed up their ranks, echeloned their front, and by the sheer weight of their deep column bore down the division* their escape routes. Just as the inhabitants of a city under siege may slowly lose their minds, your opponents will be maddened by their lack of room to maneuver against you.

There are many ways to envelop your opponents, but perhaps the simplest is to put whatever strength or advantage you naturally have to maximum use in a strategy of enclosure.

In his struggle to gain control of the chaotic American oil industry in the 1870s, John D. Rockefeller—founder and president of Standard Oil—worked first to gain a monopoly on the railroads, which were then oil's main transport. Next he moved to gain control over the pipelines that connected the refineries to the railroads. Independent oil producers responded by banding together to fund a pipeline of their own that would run from Pennsylvania to the coast, bypassing the need for railroads and Rockefeller's network of pipelines. Rockefeller tried buying up the land that lay in the path of the project, being built by a company called Tidewater, but his opponents worked around him, building a zigzag pipeline all the way to the sea.

Rockefeller was faced with a classic paradigm in war: a motivated enemy was utilizing every gap in his defenses to avoid his control, adjusting and learning how to fight him along the way. His solution was an enveloping maneuver. First, Rockefeller built his own pipeline to the sea, a larger one than Tidewater's. Then he began a campaign to buy up stock in the Tidewater company, gaining a minority interest in it and working from within to damage its credit and stir dissension. He initiated a price war, undermining interest in the Tidewater pipeline. And he purchased refineries before they could become Tidewater clients. By 1882 his envelopment was complete: Tidewater was forced to work out a deal that gave Standard Oil even more control over the shipping of oil than it had had before this war.

Rockefeller's method was to create relentless pressure from as many directions as possible. The result was confusion on the part of the independent oil producers—they could not tell how far his control extended, but it seemed enormous. They still had options at the point when they surrendered, but they had been worn down and made to believe the fight was hopeless. The Tidewater envelopment was made possible by the immense resources at Rockefeller's disposal, but he used these resources not just practically but psychologically, generating an impression of himself as a relentless foe who would leave no gaps for the enemy to sneak through. He won not only by how much he spent but by his use of his resources to create psychological pressure.

To envelop your enemies, you must use whatever you have in abundance. If you have a large army, use it to create the appearance that your forces are everywhere, an encircling pressure. That is how Toussaint l'Ouverture ended slavery in what today is called Haiti, at the end of the eighteenth century, and liberated the island from France: he used his greater numbers to create the feeling among the whites on the island that

*of the enemy which was stationed in front of Hannibal's line, and was too thin and weak to resist the pressure. Without a moment's pause they followed up their broken and hastily retreating foe till they took to headlong flight. Cutting their way through to the mass of fugitives, who offered no resistance, they penetrated as far as the Africans who were stationed on both wings, somewhat further back than the Gauls and Spaniards who had formed the advanced centre. As the latter fell back, the whole front became level, and as they continued to give ground, it became concave and crescent-shaped, the Africans at either end forming the horns. As the Romans rushed on incautiously between them, they were enfiladed by the two wings, which extended and closed round them in the rear. On this, the Romans, who had fought one battle to no purpose, left the Gauls and Spaniards, whose rear they had been slaughtering, and commenced a fresh struggle with the Africans. The contest was a very one-sided one, for not only were they hemmed in on all sides, but wearied with the previous fighting they were meeting fresh and vigorous opponents.*

*The History of Rome*,
Livy,
59 B.C.–A.D. 17

they were hopelessly engulfed by a hostile force. No minority can withstand such a feeling for long.

Remember: the power of envelopment is ultimately psychological. Making the other side *feel* vulnerable to attack on many sides is as good as enveloping them physically.

In the Ismaili Shiite sect during the eleventh and twelfth centuries A.D., a group later known as the Assassins developed the strategy of killing key Islamic leaders who had tried to persecute the sect. Their method was to infiltrate an Assassin into the target's inner circle, perhaps even joining his bodyguard. Patient and efficient, the Assassins were able over the years to instill the fear that they could strike at any time and at any person. No caliph or vizier felt secure. The technique was a masterpiece of economy, for in the end the Assassins actually killed quite a few people, yet the threat they posed gave the Ismailis great political power.

A few well-timed blows to make your enemies feel vulnerable in multiple ways and from multiple directions will do the same thing for you. Often, in fact, less is more here: too many blows will give you a shape, a personality—something for the other side to respond to and develop a strategy to combat. Instead seem vaporous. Make your maneuvers impossible to anticipate. Your psychological encirclement will be all the more sinister and complete.

The best encirclements are those that prey on the enemy's preexisting, inherent vulnerabilities. Be attentive, then, to signs of arrogance, rashness, or other psychological weakness. Once Winston Churchill saw the paranoid streak in Adolf Hitler, he worked to create the impression that the Axis might be attacked from anywhere—the Balkans, Italy, western France. Churchill's resources were meager; he could only hint at these possibilities through deception. But that was enough: a man like Hitler could not bear the thought of being vulnerable from any direction. By 1942 his forces were stretched across vast parts of Europe, and Churchill's ploys made him stretch them even thinner. At one point a mere feint at the Balkans made him hold back forces from the invasion of Russia, which in the end cost him dearly. Feed the fears of the paranoid and they will start to imagine attacks you hadn't even thought of; their overheated brains will do much of the encirclement for you.

When the Carthaginian general Hannibal was planning what turned out to be perhaps the most devastating envelopment in history—his victory at the Battle of Cannae in 216 B.C.—he heard from his spies that one of the opposing Roman generals, Varro, was a hothead, arrogant and contemptuous. Hannibal was outnumbered two to one, but he made two strategic decisions that turned this around. First, he lured the Romans onto tight terrain, where their greater numbers would find it hard to maneuver. Second, he weakened the center of his lines, placing his best troops and cavalry at the lines' outer ends. Led by the rash Varro, the Romans charged into the center, which gave way. The Romans pushed farther and farther. Then, just as the Zulus would encircle the British

*That night Ren Fu stationed the [Song army] troops by the Haoshui River, while Zhu Guan and Wu Ying made camp on a tributary of the river. They were about five* li *apart. Scouts reported that the Xia forces were inferior in number and seemed rather fearful. At this Ren Fu lost his vigilance and grew contemptuous of the men of Xia. He did not stop his officers and men from pursuing the Xia army and capturing its abandoned provisions. Geng Fu reminded him that the men of Xia had always been deceptive and advised him to bring the troops under discipline and advance slowly in a regular formation. Scouts should also be dispatched to probe further into the surrounding areas in order to find out what tricks the enemy was up to. However, Ren Fu ignored this advice. He made arrangements with Zhu Guan to proceed by separate routes to pursue the enemy and join forces at the mouth of Haoshui River the next day.*

*The Xia horsemen feigned defeat, emerging now and then four or five* li *in front of the Song army. Ren Fu and Zhu Guan marched swiftly in a hot pursuit, eventually arriving to the north of the city of Longgan. There the Xia soldiers suddenly vanished from sight.*

within two horns, the outer ends of the Carthaginian line pushed inward, enclosing the Romans in a tight and fatal embrace.

The impetuous, violent, and arrogant are particularly easy to lure into the traps of envelopment strategies: play weak or dumb and they will charge ahead without stopping to think where they're going. But any emotional weakness on the opponent's part, or any great desire or unrealized wish, can be made an ingredient of encirclement.

That is how the Iranians enveloped the administration of President Ronald Reagan in 1985–86, in what became known as the Iran-Contra Affair. America was leading an international embargo on the sale of weapons to Iran. In fighting this boycott, the Iranians saw two American weaknesses: first, Congress had cut off U.S. funding for the war of the Contras against the Sandinista government in Nicaragua—a cause dear to the Reagan government—and second, the administration was deeply disturbed about the growing number of Americans held hostage in the Middle East. Playing on these desires, the Iranians were able to lure the Americans into a Cannae-like trap: they would work for the release of hostages and secretly fund the Contras, in exchange for weapons.

It seemed too good to resist, but as the Americans entered further into this web of duplicity (backroom deals, secret meetings), they could sense their room to maneuver slowly narrowing: the Iranians were able to ask for more in exchange for less. In the end they got plenty of weapons, while the Americans got only a handful of hostages and not enough money to make a difference in Nicaragua. Worse, the Iranians openly told other diplomats about these "secret" dealings, closing their encirclement by ensuring that it would be revealed to the American public. For the government officials who had been involved in the affair, there was no possible escape route from the mess they had been drawn into. Feeling intense pressure from all sides as news of the deal became public, their attempts to cover it up or explain it away only made the situation worse.

In luring your enemies into such a trap, always try to make them feel as if they are in control of the situation. They will advance as far as you want them to. Many of the Americans involved in Iran-Contra believed they were the ones conning the naïve Iranians.

Finally, do not simply work to envelop your opponents' forces or immediate emotions, but rather envelop their whole strategy—indeed, their whole conceptual framework. This ultimate form of envelopment involves first studying the rigid, predictable parts of your opponents' strategy, then crafting a novel strategy of your own that goes outside their experience. Taking on the armies of Islam, Russia, Poland, Hungary, and the Teutonic Order, the Mongols did not merely defeat them, they annihilated them—by inventing a new brand of mobile warfare to use against an enemy mired in centuries-old methods of fighting. This kind of strategic mismatch can lead to victory not just in any given battle but in large-scale campaigns—the ultimate goal in any form of war.

*Ren Fu realized at last that he had been deceived and decided to pull the troops out of the mountainous region.*
*The next day Ren Fu led his men to move westward along the Haoshui River. They finally got out of the Liupan Mountains and proceeded towards the city of Yangmulong. At this juncture Ren Fu got reports of enemy activity in the vicinity. He had to call the troops to a halt about five* li *from the city and array them in a defensive formation. Just then, several large wooden boxes were discovered lying by the road. The boxes were tightly sealed and rustling sound came from within. Curiously, Ren Fu ordered the boxes to be opened. All of a sudden, dozens of pigeons fluttered out of the boxes and flew high into the sky, with loud tinkling sounds coming from the small bells attached to their claws. All the Song soldiers looked up in astonishment, when large hosts of Xia soldiers appeared in every direction to form a complete encirclement. On hearing the pigeon bells, Yuanhao knew that the Song army had entered his ambush ring. Thereupon he sent an assistant general with fifty thousand men to surround and assault the band of troops led by Zhu Guan and led the other half of his troops in*

*person to attack Ren Fu, whom he considered a tougher opponent than Zhu Guan. . . . The Song soldiers failed to penetrate the encirclement and were compelled to continue the tangled fight. Many were killed and some even threw themselves down the precipice in despair. Ren Fu himself was hit by over a dozen arrows. One of his guards urged him to surrender, which seemed the only way to save his life and the remnants of his men. But Ren Fu sighed and said, "I am a general of the Song and shall pay for this defeat with my life." With this he brandished his mace and fought fiercely until he was mortally injured on the face by a spear. Then he took his own life by strangling himself. All of Ren Fu's subordinate officers died in combat, and his army was completely wiped out.*

*THE WILES OF WAR: 36 MILITARY STRATEGIES FROM ANCIENT CHINA*, TRANSLATED BY SUN HAICHEN, 1991

**Image:**
The Noose. Once it is in place, there is no escape, no hope. At the mere thought of being caught in it, the enemy will grow desperate and struggle, its frantic efforts to escape only hastening its destruction.

**Authority:** Place a monkey in a cage, and it is the same as a pig, not because it isn't clever and quick, but because it has no place to freely exercise its capabilities. —Huainanzi *(second century B.C.)*

## REVERSAL

The danger of envelopment is that unless it is completely successful, it may leave you in a vulnerable position. You have announced your plans. The enemy knows that you are trying to annihilate it, and unless you can quickly deliver your knockout punch, it will work furiously not only to defend itself but to destroy you—for now your destruction is its only safeguard. Some armies that have failed in their envelopments have found themselves later encircled by their enemies. Use this strategy only when you have a reasonable chance of bringing it to the conclusion you desire.

# 20

# MANEUVER THEM INTO WEAKNESS

## THE RIPENING-FOR-THE-SICKLE STRATEGY

*No matter how strong you are, fighting endless battles with people is exhausting, costly, and unimaginative. Wise strategists generally prefer the art of maneuver: before the battle even begins, they find ways to put their opponents in positions of such weakness that victory is easy and quick. Bait enemies into taking positions that may seem alluring but are actually traps and blind alleys. If their position is strong, get them to abandon it by leading them on a wild-goose chase. Create dilemmas: devise maneuvers that give them a choice of ways to respond—all of them bad. Channel chaos and disorder in their direction. Confused, frustrated, and angry opponents are like ripe fruit on the bough: the slightest breeze will make them fall.*

*Warfare is like hunting. Wild animals are taken by scouting, by nets, by lying in wait, by stalking, by circling around, and by other such stratagems rather than by sheer force. In waging war we should proceed in the same way, whether the enemy be many or few. To try simply to overpower the enemy in the open, hand to hand and face to face, even though you might appear to win, is an enterprise which is very risky and can result in serious harm. Apart from extreme emergency, it is ridiculous to try to gain a victory which is so costly and brings only empty glory. . . .*

BYZANTINE EMPEROR MAURIKIOS, A.D. 539–602

## MANEUVER WARFARE

Throughout history two distinct styles of warfare can be identified. The most ancient is the war of attrition: the enemy surrenders because you have killed so many of its men. A general fighting a war of attrition will calculate ways to overwhelm the other side with larger numbers, or with the battle formation that will do the most damage, or with superior military technology. In any event, victory depends on wearing down the other side in battle. Even with today's extraordinary technology, attrition warfare is remarkably unsophisticated, playing into humanity's most violent instincts.

Over many centuries, and most notably in ancient China, a second method of waging war developed. The emphasis here was not destroying the other side in battle but weakening and unbalancing it before the battle began. The leader would maneuver to confuse and infuriate and to put the enemy in a bad position—having to fight uphill, or with the sun or wind in its face, or in a cramped space. In this kind of war, an army with mobility could be more effective than one with muscle.

The maneuver-warfare philosophy was codified by Sun-tzu in his *Art of War*, written in China's Warring States period, in the fifth to third century B.C.—over two hundred years of escalating cycles of warfare in which a state's very survival depended on its army and strategists. To Sun-tzu and his contemporaries, it was obvious that the costs of war went far beyond its body counts: it entailed a loss of resources and political goodwill and a lowering of morale among soldiers and citizens. These costs would mount over time until eventually even the greatest warrior nation would succumb to exhaustion. But through adroit maneuvering a state could spare itself such high costs and still emerge victorious. An enemy who had been maneuvered into a weak position would succumb more easily to psychological pressure; even before the battle had begun, it had imperceptibly started to collapse and would surrender with less of a fight.

Several strategists outside Asia—most notably Napoleon Bonaparte—have made brilliant use of maneuver warfare. But in general, attrition warfare is deeply engrained in the Western way of thinking—from the ancient Greeks to modern America. In an attrition culture, thoughts naturally gravitate toward how to overpower problems, obstacles, those who resist us. In the media, emphasis is placed on big battles, whether in politics or in the arts—static situations in which there are winners and losers. People are drawn to the emotional and dramatic quality in any confrontation, not the many steps that lead to such confrontation. The stories that are told in such cultures are all geared toward such battlelike moments, a moral message preached through the ending (as opposed to the more telling details). On top of it all, this way of fighting is deemed more manly, honorable, honest.

More than anything, maneuver war is a different way of thinking. What matters here is process—the steps toward battle and how to manipulate them to make the confrontation less costly and violent. In the maneuver universe, nothing is static. Battles are in fact dramatic illusions, short moments in the larger flow of events, which is fluid, dynamic, and susceptible to alteration through careful strategy. This way of thinking finds no honor or morality in wasting time, energy, and lives in battles. Instead wars of attrition are seen as lazy, reflecting the primitive human tendency to fight back reactively, without thinking.

In a society full of attrition fighters, you will gain an instant advantage by converting to maneuver. Your thought process will become more fluid, more on the side of life, and you will be able to thrive off the rigid, battle-obsessed tendencies of the people around you. By always thinking first about the overall situation and about how to maneuver people into positions of weakness rather than fight them, you will make your battles less bloody—which, since life is long and conflict is endless, is wise if you want a fruitful and enduring career. And a war of maneuver is just as decisive as a war of attrition. Think of weakening your enemies as ripening them like grain, ready to be cut down at the right moment.

The following are the four main principles of maneuver warfare:

**Craft a plan with branches.** Maneuver warfare depends on planning, and the plan has to be right. Too rigid and you leave yourself no room to adjust to the inevitable chaos and friction of war; too loose and unforeseen events will confuse and overwhelm you. The perfect plan stems from a detailed analysis of the situation, which allows you to decide on the best direction to follow or the perfect position to occupy and suggests several effective options (branches) to take, depending on what the enemy throws at you. A plan with branches lets you outmaneuver your enemy because your responses to changing circumstances are faster and more rational.

**Give yourself room to maneuver.** You cannot be mobile, you cannot maneuver freely, if you put yourself in cramped spaces or tie yourself down to positions that do not allow you to move. Consider the ability to move and keeping open more options than your enemy has as more important than holding territories or possessions. You want open space, not dead positions. This means not burdening yourself with commitments that will limit your options. It means not taking stances that leave you nowhere to go. The need for space is psychological as well as physical: you must have an unfettered mind to create anything worthwhile.

**Give your enemy dilemmas, not problems.** Most of your opponents are likely to be clever and resourceful; if your maneuvers simply present them with a problem, they will inevitably solve it. But a dilemma is different: whatever they do, however they respond—retreat, advance, stay still—they are still in trouble. Make every option bad: if you maneuver quickly to a point, for instance, you can force your enemies either to fight before they are ready or to retreat. Try constantly to put them in positions that seem alluring but are traps.

**Create maximum disorder.** Your enemy depends on being able to read you, to get some sense of your intentions. The goal of your maneuvers should be to make that impossible, to send the enemy on a wild-goose chase for meaningless information, to create ambiguity as to which way you are going to jump. The more you break down people's ability to reason about you, the more disorder you inject into their system. The disorder you create is controlled and purposeful, at least for you. The disorder the enemy suffers is debilitating and destructive.

*So to win a hundred victories in a hundred battles is not the highest excellence; the highest excellence is to subdue the enemy's army without fighting at all.*
—*Sun-tzu (fourth century* B.C.*)*

## HISTORICAL EXAMPLES

**1.** On November 10, 1799, Napoleon Bonaparte completed the coup d'état that brought him to power as first consul, giving him near-complete control of the French state. For over ten years, France had been convulsed with revolution and war. Now that Napoleon was leader, his most pressing need was peace, to give the country time to recoup and himself time to consolidate his power—but peace would not come easily.

France had a bitter enemy in Austria, which had put two large armies in the field, ready to move against Napoleon: one to the east of the Rhine and the other in northern Italy under General Michael Melas. The Austrians were clearly planning a major campaign. Waiting was too dangerous; Napoleon had to seize the initiative. He had to defeat at least one of these armies if he were to force Austria to negotiate peace on his terms. The one trump card he had was that several months earlier a French army had gained control of Switzerland. There were also French troops in northern Italy, which Napoleon had taken from the Austrians several years earlier.

To plan for the first real campaign under his direction, Napoleon holed himself up in his office for several days. His secretary, Louis de Bourienne, would recall seeing him lying on giant maps of Germany, Switzerland, and Italy laid out wall to wall on the floor. The desks were piled high with reconnaissance reports. On hundreds of note cards organized into boxes, Napoleon had calculated the Austrians' reactions to the feints he was planning. Muttering to himself on the floor, he mulled over every permutation of attack and counterattack.

By the end of March 1800, Napoleon had emerged from his office with a plan for a campaign in northern Italy that went far beyond any-

*"Addicts of attrition," as Simpkin calls them, generally cannot think beyond the battle, and they consider that the only way—or at least the preferred way—to defeat an enemy is to destroy the physical components of his army, especially the combat portions (armored fighting vehicles, troops, guns, etc.). If the attrition addict appreciates war's intangibles at all (such as morale, initiative, and shock), he sees them only as combat multipliers with which to fight the attrition battle better. If the attrition warrior learns about maneuver, he sees it primarily as a way to get to the fight. In other words, he moves in order to fight. Maneuver theory, on the other hand, attempts to defeat the enemy through means other than simple destruction of his mass. Indeed, the highest and purest application of maneuver theory is to* preempt *the enemy, that is, to disarm or neutralize him before the fight. If such is not possible, the maneuver warrior seeks to* dislocate *the enemy forces, i.e., removing the enemy from the decisive point, or vice versa, thus rendering them useless and irrelevant to the fight. If the enemy cannot be preempted or dislocated, then the maneuver-warfare practitioner will attempt to* disrupt *the*

thing his lieutenants had ever seen before. In the middle of April, a French army under General Jean Moreau would cross the Rhine and push the eastern Austrian army back into Bavaria. Then Napoleon would lead a 50,000-man force, already in place in Switzerland, into northern Italy through several different passes in the Alps. Moreau would then release one of his divisions to move south and follow Napoleon into Italy. Moreau's initial move into Bavaria, and the subsequent scattered dispatch of divisions into Italy, would confuse the Austrians as to Napoleon's intentions. And if the Austrian army at the Rhine was pushed east, it would be too distant to support the Austrian army in northern Italy.

*enemy, i.e., destroy or neutralize his center of gravity, preferably by attacking with friendly strengths through enemy weaknesses.*

*THE ART OF MANEUVER,* ROBERT R. LEONHARD, 1991

Once across the Alps, Napoleon would concentrate his forces and link up with the divisions under General André Massena already stationed in northern Italy. He would then move much of his army to the town of Stradella, cutting off communications between Melas in northern Italy and command headquarters in Austria. With Melas's troops now isolated and the mobile French army within reach of them, Napoleon would have many excellent options for dislocating and destroying them. At one point, as he described this plan to Bourienne, Napoleon lay down on the giant map on his floor and stuck a pin next to the town of Marengo, in the center of the Italian theater of war. "I will fight him here," he said.

A few weeks later, as Napoleon began to position his armies, he received some troubling news: Melas had beaten him to the punch by attacking Massena's army in Northern Italy. Massena was forced back to Genoa, where the Austrians quickly surrounded him. The danger here was great: if Massena surrendered, the Austrians could sweep into southern France. Also, Napoleon had been counting on Massena's army to help him beat Melas. Yet he took the news with surprising calm and simply made some adjustments: he transferred more men to Switzerland and sent word to Massena that he must do whatever he could to hold out for at least eight weeks, keeping Melas busy while Napoleon moved into Italy.

Within a week there was more irritating news. After Moreau had begun the campaign to push the Austrians back from the Rhine, he refused to transfer the division that Napoleon had counted on for Italy, claiming he could not spare it. Instead he sent a smaller, less experienced division. The French army in Switzerland had already begun the dangerous crossings through the Alps. Napoleon had no choice but to take what Moreau gave him.

By May 24, Napoleon had brought his army safely into Italy. Absorbed with the siege at Genoa, Melas ignored reports of French movements to the north. Next Napoleon advanced to Milan, close to Stradella, where he cut Austrian communications as planned. Now, like a cat stealing up on its prey, he could wait for Melas to notice the trap he was in and try to fight his way out of it near Milan.

On June 8, however, once again more bad news reached Napoleon:

*Aptitude for maneuver is the supreme skill in a general; it is the most useful and rarest of gifts by which genius is estimated.*

NAPOLEON BONAPARTE, 1769–1821

two weeks before he had hoped, Massena had surrendered. Napoleon now had fewer men to work with, and Melas had won a strong base in Genoa. Since its inception the campaign had been plagued with mistakes and unforeseen events—the Austrians attacking early, Massena retreating into a trap at Genoa, Moreau disobeying orders, and now Massena's surrender. Yet while Napoleon's lieutenants feared the worst, Napoleon himself not only stayed cool, he seemed oddly excited by these sudden twists of fortune. Somehow he could discern opportunities in them that were invisible to everyone else—and with the loss of Genoa, he sensed the greatest opportunity of all. He quickly altered his plan; instead of waiting at Milan for Melas to come to him, he suddenly cast his divisions in a wide net to the west.

Watching his prey closely, Napoleon sensed that Melas was mesmerized by the movements of the French divisions—a fatal hesitation. Napoleon moved one division west to Marengo, close to the Austrians at Genoa, almost baiting them to attack. Suddenly, on the morning of June 14, they took the bait, and in surprising force. This time it was Napoleon who had erred; he had not expected the Austrian attack for several days, and his divisions were scattered too widely to support him. The Austrians at Marengo outnumbered him two to one. He dispatched urgent messages in all directions for reinforcements, then settled into battle, hoping to make his small forces hold ground until they came.

The hours went by with no sign of aid. Napoleon's lines grew weaker, and at three in the afternoon the Austrians finally broke through, forcing the French to retreat. This was the ultimate downturn in the campaign, and it was yet again Napoleon's moment to shine. He seemed encouraged by the way the retreat was going, the French scattering and the Austrians pursuing them, without discipline or cohesion. Riding among the men who had retreated the farthest, he rallied them and prepared them to counterattack, promising them that reinforcements would arrive within minutes—and he was right. Now French divisions were coming in from all directions. The Austrians, meanwhile, had let their ranks fall into disorder, and, stunned to find themselves facing new forces in this condition, they halted and then gave ground to a quickly organized French counterattack. By 9:00 P.M. the French had routed them.

Just as Napoleon had predicted with his pin on the map, he met and defeated the enemy at Marengo. A few months later, a treaty was signed that gave France the peace it so desperately needed, a peace that was to last nearly four years.

### Interpretation

Napoleon's victory at Marengo might seem to have depended on a fair amount of luck and intuition. But that is not at all the case. Napoleon believed that a superior strategist could create his own luck—through calculation, careful planning, and staying open to change in a dynamic situation. Instead of letting bad fortune face him down, Napoleon incorporated it into his plans. When he learned that Massena had been forced back to Genoa, he saw that the fight for the city would lock Melas into a static position, giving Napoleon time to move his men into place. When Moreau sent him a smaller division, Napoleon sent it through the Alps by a narrower, more obscure route, throwing more sand in the eyes of the Austrians trying to figure out how many men he had available. When Massena unexpectedly surrendered, Napoleon realized that it would be easier now to bait Melas into attacking his divisions, particularly if he moved them closer. At Marengo itself he knew all along that his first reinforcements would arrive sometime after three in the afternoon. The more disorderly the Austrian pursuit of the French, the more devastating the counterattack would be.

Napoleon's power to adjust and maneuver on the run was based in his novel way of planning. First, he spent days studying maps and using them to make a detailed analysis. This was what told him, for example, that putting his army at Stradella would pose a dilemma for the Austrians and give him many choices of ways to destroy them. Then he calculated contingencies: if the enemy did *x,* how would he respond? If part *y* of his plan misfired, how would he recover? The plan was so fluid, and gave him so many options, that he could adapt it infinitely to whatever situation developed. He had anticipated so many possible problems that he could come up with a rapid answer to any of them. His plan was a mix of detail and fluidity, and even when he made a mistake, as he did in the early part of the encounter at Marengo, his quick adjustments kept the Austrians from taking advantage of it—before they'd figured out what to do, he was already somewhere else. His devastating freedom of maneuver cannot be separated from his methodical planning.

Understand: in life as in war, nothing ever happens just as you expect it to. People's responses are odd or surprising, your staff commits outrageous acts of stupidity, on and on. If you meet the dynamic situations of life with plans that are rigid, if you think of only holding static positions, if you rely on technology to control any friction that comes your way, you are doomed: events will change faster than you can adjust to them, and chaos will enter your system.

In an increasingly complex world, Napoleon's way of planning and maneuvering is the only rational solution. You absorb as much information and as many details as possible; you analyze situations in depth, trying to imagine the enemy's responses and the accidents that might happen. You do not get lost in this maze of analysis but rather use it to formulate a free-flowing plan with branches, one that puts you in positions with the possibility of maneuver. You keep things loose and adjustable. Any chaos that comes your way is channeled toward the enemy. In practicing this policy, you will come to understand Napoleon's dictum that luck is something you create.

*Now the army's disposition of force* (hsing) *is like water. Water's configuration* (hsing) *avoids heights and races downward. . . . Water configures* (hsing) *its flow in accord with the terrain; the army controls its victory in accord with the enemy. Thus the army does not maintain any constant strategic configuration of power* (shih), *water has no constant shape* (hsing). *One who is able to change and transform in accord with the enemy and wrest victory is termed spiritual.*

*THE ART OF WAR,*
SUN-TZU,
FOURTH CENTURY B.C.

THE REED AND THE OLIVE

*The reed and the olive tree were arguing over their steadfastness, strength and ease. The olive taunted the reed for his powerlessness and pliancy in the face of all the winds. The reed kept quiet and didn't say a word. Then, not long after this, the wind blew violently. The reed, shaken and bent, escaped easily from it, but the olive tree, resisting the wind, was snapped by its force.* The story shows that people who yield to circumstances and to superior power have the advantage over their stronger rivals.

*FABLES,* AESOP, SIXTH CENTURY B.C.

**2.** As the Republicans prepared their convention to pick a presidential candidate in 1936, they had reason to hope. The sitting president, the Democrat Franklin D. Roosevelt, was certainly popular, but America was still in the Depression, unemployment was high, the budget deficit was growing, and many of Roosevelt's New Deal programs were mired in inefficiency. Most promising of all, many Americans had become disenchanted with Roosevelt as a person—in fact, they had even come to hate him, thinking him dictatorial, untrustworthy, a socialist at heart, perhaps even un-American.

Roosevelt was vulnerable, and the Republicans were desperate to win the election. They decided to tone down their rhetoric and appeal to traditional American values. Claiming to support the spirit of the New Deal but not the man behind it, they pledged to deliver the needed reforms more efficiently and fairly than Roosevelt had. Stressing party unity, they nominated Alf M. Landon, the governor of Kansas, as their presidential candidate. Landon was the perfect moderate. His speeches tended to be a little dull, but he seemed so solid, so middle class, a comfortable choice, and this was no time to be promoting a radical. He had supported much of the New Deal, but that was fine—the New Deal was popular. The Republicans nominated Landon because they thought he had the best chance to defeat Roosevelt, and that was all that mattered to them.

During the nominating ceremony, the Republicans staged a western pageant with cowboys, cowgirls, and covered wagons. In his acceptance speech, Landon did not talk about specific plans or policy but about himself and his American values. Where Roosevelt was associated with unpleasant dramas, he would bring stability. It was a feel-good convention.

The Republicans waited for Roosevelt to make his move. As expected, he played the part of the man above the fray, keeping his public appearances to a minimum and projecting a presidential image. He talked in vague generalities and struck an optimistic note. After the Democratic convention, he departed for a long vacation, leaving the field open to the Republicans, who were only too happy to fill the void: they sent Landon out on the campaign trail, where he made stump speeches about how he was the one to enact reforms in a measured, rational way. The contrast between Landon and Roosevelt was one of temperament and character, and it seemed to resonate: in the polls, Landon pulled into the lead.

Sensing that the election would be close and feeling that this was their great chance, the Republicans escalated their attacks, accusing Roosevelt of class warfare and painting a bleak picture of his next term. The anti-Roosevelt newspapers published a slew of editorials attacking him in personal terms. The chorus of criticism grew, and the Republicans watched gleefully as many in Roosevelt's camp seemed to panic. One poll had Landon building a substantial lead.

Not until late September, a mere six weeks before the election, did

Roosevelt finally start his campaign—and then, to everyone's shock, he dropped the nonpartisan, presidential air that he had worn so naturally. Positioning himself clearly to Landon's left, he drew a sharp contrast between the two candidates. He quoted with great sarcasm Landon's speeches supporting the New Deal but claiming to be able to do it better: why vote for a man with basically the same ideas and approach but with no experience in making them work? As the days went by, Roosevelt's voice grew louder and clearer, his gestures more animated, his oratory even biblical in tone: he was David facing the Goliath of the big-business interests that wanted to return the country to the era of monopolies and robber barons.

The Republicans watched in horror as Roosevelt's crowds swelled. All those whom the New Deal had helped in any way showed up in the tens of thousands, and their response to Roosevelt was almost religious in its fervor. In one particularly rousing speech, Roosevelt catalogued the moneyed interests arrayed against him: "Never before in our history," he concluded, "have these forces been so united against one candidate as they stand today. They are unanimous in their hate for me—and I welcome their hatred. . . . I should like to have it said of my second administration that in it these forces met their master."

Landon, sensing the great change in the tide of the election, came out with sharper attacks and tried to distance himself from the New Deal, which he had earlier claimed to support—but all of this only seemed to dig him a deeper hole. He had changed too late, and clearly in reaction to his waning fortunes.

On Election Day, Roosevelt won by what at the time was the greatest popular margin in U.S. electoral history; he won all but two states, and the Republicans were reduced to sixteen seats in the Senate. More amazing than the size of his unprecedented victory was the speed with which he had turned the tide.

### Interpretation

As Roosevelt followed the Republican convention, he clearly saw the line they would take in the months to come—a centrist line, emphasizing values and character over policy. Now he could lay the perfect trap by abandoning the field. Over the weeks to come, Landon would pound his moderate position into the public's mind, committing himself to it further and further. Meanwhile the more right-wing Republicans would attack the president in bitter, personal terms. Roosevelt knew that a time would come when Landon's poll numbers would peak. The public would have had its fill of his bland message and the right's vitriolic attacks.

Sensing that moment in late September, he returned to the stage and positioned himself clearly to Landon's left. The choice was strategic, not ideological; it let him draw a sharp distinction between Landon and himself. In a time of crisis like the Depression, it was best to look resolute

*Of course this beautiful simplicity of strategic movement, with its infinite flexibility, is extremely deceptive. The task of correlating and coordinating the daily movements of a dozen or more major formations, all moving along separate routes, of ensuring that every component is within one or, at most, two days' marching distance of its immediate neighbors, and yet at the same time preserving the appearance of an arbitrary and ill-coordinated "scatter" of large units in order to deceive the foe concerning the true gravity of his situation—this is the work of a mathematical mind of no common caliber. It is in fact the hallmark of genius—that "infinite capacity for taking pains." . . . The ultimate aim of all this carefully considered activity was to produce the greatest possible number of men on the battlefield, which on occasion had been chosen months in advance of the actual event. Bourienne gives his celebrated . . . eyewitness account of the First Consul, in the early days of the Italian Campaign of 1800, lying full-length on the floor, pushing colored pins into his maps, and saying, "I shall fight him here—on the plain of the Scrivia," with that uncanny prescience which was in reality the product of mental calculations of*

and strong, to stand for something firm, to oppose a clear enemy. The attacks from the right gave him that clear enemy, while Landon's milquetoast posturing made him look strong by contrast. Either way he won.

Now Landon was presented with a dilemma. If he kept going with his centrist appeal, he would bore the public and seem weak. If he moved to the right—the choice he actually took—he would be inconsistent and look desperate. This was pure maneuver warfare: Begin by taking a position of strength—in Roosevelt's case his initial, presidential, bipartisan pose—that leaves you with open options and room to maneuver. Then let your enemies show their direction. Once they commit to a position, let them hold it—in fact, let them trumpet it. Now that they are fixed in place, maneuver to the side that will crowd them, leaving them only bad options. By waiting to make this maneuver until the last six weeks of the presidential race, Roosevelt both denied the Republicans any time to adjust and kept his own strident appeal from wearing thin.

Everything is political in the world today, and politics is all about positioning. In any political battle, the best way to stake out a position is to draw a sharp contrast with the other side. If you have to resort to speeches to make this contrast, you are on shaky ground: people distrust words. Insisting that you are strong or well qualified rings as self-promotion. Instead make the opposing side talk and take the first move. Once they have committed to a position and fixed it in other people's minds, they are ripe for the sickle. Now you can create a contrast by quoting their words back at them, showing how different you are—in tone, in attitude, in action. Make the contrast deep. If they commit to some radical position, do not respond by being moderate (moderation is generally weak); attack them for promoting instability, for being power-hungry revolutionaries. If they respond by toning down their appeal, nail them for being inconsistent. If they stay the course, their message will wear thin. If they become more strident in self-defense, you make your point about their instability.

Use this strategy in the battles of daily life, letting people commit themselves to a position you can turn into a dead end. Never *say* you are strong, *show* you are, by making a contrast between yourself and your inconsistent or moderate opponents.

**3.** The Turks entered World War I on the side of Germany. Their main enemies in the Middle Eastern theater were the British, who were based in Egypt, but by 1917 they had arrived at a comfortable stalemate: the Turks controlled a strategic eight-hundred-mile stretch of railway that ran from Syria in the north to the Hejaz (the southwestern part of Arabia) in the south. Due west of the central part of this railway line was the town of Aqaba, on the Red Sea, a key Turkish position from which they could quickly move armies north and south to protect the railway.

The Turks had already beaten back the British at Gallipoli (see chapter 5), a huge boost to their morale. Their commanders in the Middle

East felt secure. The English had tried to stir up a revolt against the Turks among the Arabs of the Hejaz, hoping the revolt would spread north; the Arabs had managed a few raids here and there but had fought more among themselves than against the Turks. The British clearly coveted Aqaba and plotted to take it from the sea with their powerful navy, but behind Aqaba was a mountain wall marked by deep gorges. The Turks had converted the mountain into a fortress. The British knew that even if their navy took Aqaba, they would be unable to advance inland, rendering the city's capture useless. Both the British and the Turks saw the situation the same way, and the stalemate endured.

In June 1917 the Turkish commanders of the forts guarding Aqaba received reports of strange enemy movements in the Syrian deserts to the northeast. It seemed that a twenty-nine-year-old British liaison officer to the Arabs named T. E. Lawrence had trekked across hundreds of miles of desolate terrain to recruit an army among the Howeitat, a Syrian tribe renowned for fighting on camels. The Turks dispatched scouts to find out more. They already knew a little about Lawrence: unusually for British officers of the time, he spoke Arabic, mixed well with the local people, and even dressed in their style. He had also befriended Sherif Feisal, a leader of the Arab revolt. Could he be raising an army to attack Aqaba? To the extent that this was possible, he was worth watching carefully. Then word came that Lawrence had imprudently told an Arab chief, secretly in Turkish pay, that he was heading for Damascus to spread the Arab revolt. This was the Turks' great fear, for a revolt in the more populated areas of the north would be unmanageable.

The army Lawrence had recruited could not have numbered more than 500, but the Howeitat were great fighters on camel, fierce and mobile. The Turks alerted their colleagues in Damascus and dispatched troops to hunt Lawrence down, a difficult task given the mobility of the Arabs and the vastness of the desert.

In the next few weeks, the Englishman's movements were baffling, to say the least: his troops moved not north toward Damascus but south toward the railway town of Ma'an, site of a storage depot used to supply Aqaba, forty miles away. No sooner had Lawrence appeared in the area of Ma'an, however, than he disappeared, reemerging over a hundred miles north to lead a series of raids on the railway line between Amman and Damascus. Now the Turks were doubly alarmed and sent 400 cavalry from Amman to find him.

For a few days, there was no sign of Lawrence. In the meantime an uprising several miles to the south of Ma'an surprised the Turks. An Arab tribe called the Dhumaniyeh had seized control of the town of Abu el Lissal, directly along the route from Ma'an to Aqaba. A Turkish battalion dispatched to take the town back found the blockhouse guarding it destroyed and the Arabs gone. Then, suddenly, something unexpected and quite disturbing occurred: out of nowhere Lawrence's Howeitat army emerged on the hill above Abu el Lissal.

*computer-like complexity. After considering every possible course of action open to the Austrian Melas, Bonaparte eliminated them one by one, made allowance for the effect of chance on events, and came up with the answer—subsequently borne out by the events of June 14 on the field of Marengo, which lies, surely enough, on the plain bounded by the rivers Bormida and Scrivia.*

*The Campaigns of Napoleon*, David G. Chandler, 1966

*The warrior and the statesman, like the skillful gambler, do not make their luck but prepare for it, attract it, and seem almost to determine their luck. Not only are they, unlike the fool and the coward, adept at making use of opportunities when these occur; they know furthermore how to take advantage, by means of precautions and wise measures, of such and such an opportunity, or of several at once. If one thing happens, they win; if another, they are still the winners; the same circumstance often makes them win in a variety of ways. These prudent men may be praised for their good fortune as well as their good management, and rewarded for their luck as well as for their merits.*

CHARACTERS, JEAN DE LA BRUYÈRE, 1645–1696

Distracted by the local uprising, the Turks had lost track of Lawrence. Now, linking up with the Dhumaniyeh, he had trapped a Turkish army at Abu el Lissal. The Arabs rode along the hill with enormous speed and dexterity, goading the Turks into wasting ammunition by firing on them. Meanwhile the midday heat took its toll on the Turkish riflemen, and, having waited until the Turks were sufficiently tired, the Arabs, Lawrence among them, charged down the hill. The Turks closed their ranks, but the swift-moving camel cavalry took them from the flank and rear. It was a massacre: 300 Turkish soldiers were killed and the rest taken prisoner.

Now the Turkish commanders at Aqaba finally saw Lawrence's game: he had cut them off from the railway line on which they depended for supplies. Also, seeing the Howeitat's success, other Arab tribes around Aqaba joined up with Lawrence, creating a powerful army that began to wend its way through the narrow gorges toward Aqaba. The Turks had never imagined an army coming from this direction; their fortifications faced the other way, toward the sea and the British. The Arabs had a reputation for ruthlessness with enemies who resisted, and the commanders of the forts in back of Aqaba began to surrender. The Turks sent out their 300-man garrison from Aqaba to put a stop to this advance, but they were quickly surrounded by the swelling number of Arabs.

On July 6 the Turks finally surrendered, and their commanders watched in shock as Lawrence's ragtag army rushed to the sea to take what had been thought to be an impregnable position. With this one blow, Lawrence had completely altered the balance of power in the Middle East.

## Interpretation

The fight between Britain and Turkey during World War I superbly demonstrates the difference between a war of attrition and a war of maneuver. Before Lawrence's brilliant move, the British, fighting by the rules of attrition warfare, had been directing the Arabs to capture key points along the railway line. This strategy had played into Turkish hands: the Turks had too few men to patrol the entire line, but once they saw the Arabs attacking at any one place, they could quickly move the men they had and use their superior firepower to either defend it or take it back. Lawrence—a man with no military background, but blessed with common sense—saw the stupidity in this right away. Around the railway line were thousands of square miles of desert unoccupied by the Turks. The Arabs had been masters at a mobile form of warfare on camelback since the days of the prophet Mohammed; vast space at their disposal gave them infinite possibilities for maneuvers that would create threats everywhere, forcing the Turks to bunker themselves in their forts. Frozen in place, the Turks would wither from lack of supplies and would be unable to defend the surrounding region. The key to the overall war was to spread the revolt north, toward Damascus, allowing the Arabs to threaten the entire railway line. But to spread the revolt north, they needed a base in the center. That base was Aqaba.

The British were as hidebound as the Turks and simply could not picture a campaign of a group of Arabs led by a liaison officer. Lawrence would have to do it on his own. Tracing a series of great loops in the vast spaces of the desert, he left the Turks bewildered as to his purpose. Knowing that the Turks feared an attack on Damascus, he deliberately spread the lie that he was aiming for it, making the Turks send troops on a wild-goose chase to the north. Then, exploiting their inability to imagine an Arab attack on Aqaba from the landward side (a failing they shared with his British countrymen), he caught them off guard. Lawrence's subsequent capture of Aqaba was a masterpiece of economy: only two men died, on his side. (Compare this to the unsuccessful British attempt to take Gaza from the Turks that same year in head-on battle, in which over three thousand British soldiers were killed.) The capture of Aqaba was the turning point in Britain's eventual defeat of the Turks in the Middle East.

The greatest power you can have in any conflict is the ability to confuse your opponent about your intentions. Confused opponents do not know how or where to defend themselves; hit them with a surprise attack and they are pushed off balance and fall. To accomplish this you must maneuver with just one purpose: to keep them guessing. You get them to chase you in circles; you say the opposite of what you mean to do; you threaten one area while shooting for another. You create maximum disorder. But to pull this off, you need room to maneuver. If you crowd yourself with alliances that force your hand, if you take positions that box you into corners, if you commit yourself to defending one fixed position, you lose the power of maneuver. You become predictable. You are like the British and the Turks, moving in straight lines in defined areas, ignoring the vast desert around you. People who fight this way deserve the bloody battles they face.

**4.** Early in 1937, Harry Cohn, longtime chief of Columbia Pictures, faced a crisis. His most successful director, Frank Capra, had just left the studio, and profits were down. Cohn needed a hit and a replacement for Capra. And he believed he had found the right formula with a comedy called *The Awful Truth* and a thirty-nine-year-old director named Leo McCarey. McCarey had directed *Duck Soup*, with the Marx Brothers, and *Ruggles of Red Gap*, with Charles Laughton, two different but successful comedies. Cohn offered McCarey *The Awful Truth.*

McCarey said he did not like the script, but he would do the picture anyway for a hundred thousand dollars—a huge sum in 1937 dollars. Cohn, who ran Columbia like Mussolini (in fact, he kept a picture of Il Duce in his office), exploded at the price. McCarey got up to go, but as he was leaving, he noticed the producer's office piano. McCarey was a frustrated songwriter. He sat down and began to play a show tune. Cohn

*Expanding on the issue of directive control, Lind introduces the reader to a decision-making model known as the Boyd cycle. Named for Col. John Boyd, the term refers to the understanding that war consists of the repeated cycle of observation, orientation, decision, and action. Colonel Boyd constructed his model as a result of his observations of fighter combat in the Korean War. He had been investigating why American fighter pilots had been consistently able to best enemy pilots in dogfights. His analysis of opposing aircraft led to some startling discoveries. Enemy fighters typically outperformed their American counterparts in speed, climb, and turning ability. But the Americans had the advantage in two subtly critical aspects. First, the hydraulic controls allowed for faster transition from one maneuver to another. Second, the cockpit allowed for a wide field of view for the pilot. The result was that the American pilots could more rapidly observe and orient to the tactical situation moment by moment. Then, having decided what to do next, they could quickly change maneuvers. In battle, this ability to rapidly pass through the observation-orientation-decision-action loop (the Boyd*

had a weakness for such music, and he was entranced: "Anybody who likes music like that has got to be a talented man," he said. "I'll pay that exorbitant fee. Report to work tomorrow."

In the days to come, Cohn was going to regret his decision.

Three stars were cast for *The Awful Truth*—Cary Grant, Irene Dunne, and Ralph Bellamy. All had problems with their roles as written in the script, none of them wanted to do the picture, and, as time went by, their unhappiness only grew. Revisions to the script began to come in: McCarey had apparently junked the original and was starting over, but his creative process was peculiar—he would sit in a parked car on Hollywood Boulevard with the screenwriter Viña Delmar and verbally improvise scenes with her. Later, when shooting began, he would walk on the beach and scribble the next day's setups on torn pieces of brown paper. His style of directing was equally upsetting to the actors. One day, for instance, he asked Dunne whether she played the piano and Bellamy whether he could sing. Both answered, "Not very well," but McCarey's next step was to have Dunne play "Home on the Range" as best she could while Bellamy sang off key. The actors did not enjoy this rather humiliating exercise, but McCarey was delighted and filmed the entire song. None of this was in the script, but all of it ended up in the film.

Sometimes the actors would wait on set while McCarey would mess around on the piano, then suddenly come up with an idea for what to shoot that day. One morning Cohn visited the set and witnessed this odd process. "I hired you to make a great comedy so I could show up Frank Capra. The only one who's going to laugh at this picture is Capra!" he exclaimed. Cohn was disgusted and basically wrote the whole thing off. His irritation grew daily, but he was contractually bound to pay Dunne forty thousand dollars for the film, whether it was shot or not. He could not fire McCarey at this point without creating greater problems, nor could he have him go back to the original script, since McCarey had already begun shooting and only he seemed to know where the film was going.

Yet as the days went on, the actors began to see some method in McCarey's madness. He would shoot them in long takes in which much of their work was only loosely guided; the scenes had spontaneity and liveliness. Casual as he seemed, he knew what he wanted and would reshoot the simplest shot if the look on the actors' faces was not loving enough. His shoot days were short and to the point.

One day, after many days' absence, Cohn showed up on set to find McCarey serving drinks to the cast. Cohn was about to explode when the director told him they were drinking to celebrate—they had just finished shooting. Cohn was shocked and delighted; McCarey had finished ahead of schedule and two hundred thousand dollars under budget. Then, to his surprise as well, the picture came together in the editing room like a strange puzzle. It was good, very good. Test audiences roared with laughter. Premiering in 1937, *The Awful Truth* was a complete success and won McCarey the best director Oscar. Cohn had found his new Frank Capra.

Unfortunately, McCarey had seen his boss's dictatorial tendencies all too clearly, and though Cohn made lucrative offers, McCarey never worked for Columbia again.

Interpretation

Leo McCarey, one of the great directors of Hollywood's golden era, was essentially a frustrated composer and songwriter. He had gone to work directing slapstick comedies—McCarey was the man who paired Laurel with Hardy—only because he was unable to make a living in music. *The Awful Truth* is considered one of the greatest screwball comedies ever made, and both its style and the way McCarey worked on it stemmed from his musical instincts: he composed the film in his head in just the same loose yet logical way that he would tinker with a tune on the piano. To create a film this way required two things: room to maneuver and the ability to channel chaos and confusion into the creative process.

McCarey kept his distance from Cohn, the actors, the screenwriters —in fact, everyone—as best he could. He would not let himself be boxed in by anyone's idea of how to shoot a film. Given room to maneuver, he could improvise, experiment, move fluidly in different directions in any scene, yet keep everything perfectly controlled—he always seemed to know what he wanted and what worked. And because filmmaking this way made every day a fresh challenge, the actors had to respond with their own energy, rather than simply regurgitating words from a script. McCarey allowed room for chance and the random events of life to enter his creative scheme without being overwhelmed by chaos. The scene he was inspired to create when he learned of Dunne's and Bellamy's lack of musical skill, for example, seems unrehearsed and lifelike because it really was. Had it been scripted, it would have been far less funny.

Directing a film—or any project, artistic or professional or scientific —is like fighting a war. There is a certain strategic logic to the way you attack a problem, shape your work, deal with friction and the discrepancy between what you want and what you get. Directors or artists often start out with great ideas but in the planning create such a straitjacket for themselves, such a rigid script to follow and form to fit in, that the process loses all joy; there's nothing left to explore in the creation itself, and the end result seems lifeless and disappointing. On the other side, artists may start with a loose idea that seems promising, but they are too lazy or undisciplined to give it shape and form. They create so much space and confusion that in the end nothing coheres.

The solution is to plan, to have a clear idea what you want, then put

*cycle) gave the American pilots a slight time advantage. If one views a dogfight as a series of Boyd cycles, one sees that the Americans would repeatedly gain a time advantage each cycle, until the enemy's actions become totally inappropriate to the changing situations. Hence, the American pilots were able to "out-Boyd cycle" the enemy, thus outmaneuvering him and finally shooting him down. Colonel Boyd and others then began to question whether this pattern might be applicable to other forms of warfare as well.*

THE ART OF MANEUVER, ROBERT R. LEONHARD, 1991

*Mobility, defined as the ability to project power over distance, is another characteristic of good chess. It is the goal of a good chess player to ensure that each of his pieces can exert pressure upon a maximum number of squares, rather than being bottled up in a corner, surrounded by other pieces. Hence, the chess master looks forward to pawn exchanges (infantry battles, if you will), not because he is trying to wear down the enemy, but because he knows that he can project the power of his rooks (mechanized forces) down the resulting open files. In this manner, the chess master fights in order to move. This idea is central to maneuver-warfare theory.*

*THE ART OF MANEUVER*, ROBERT R. LEONHARD, 1991

yourself in open space and give yourself options to work with. You direct the situation but leave room for unexpected opportunities and random events. Both generals and artists can be judged by the way they handle chaos and confusion, embracing it yet guiding it for their own purposes.

**5.** One day in the Japan of the 1540s, in a ferryboat crowded with farmers, merchants, and craftsmen, a young samurai regaled all who would listen with tales of his great victories as a swordsman, wielding his three-foot-long sword as he spoke to demonstrate his prowess. The other passengers were a little afraid of this athletic young man, so they feigned interest in his stories to avoid trouble. But one older man sat to the side, ignoring the young boaster. The older man was obviously a samurai himself—he carried two swords—but no one knew that this was in fact Tsukahara Bokuden, perhaps the greatest swordsman of his time. He was in his fifties by then and liked to travel alone and incognito.

Bokuden sat with his eyes closed, seemingly deep in meditation. His stillness and silence began to annoy the young samurai, who finally called out, "Don't you like this kind of talk? You don't even know how to wield a sword, old man, do you?" "I most certainly do," answered Bokuden. "My way, however, is not to wield my sword in such inconsequential circumstances as these." "A way of using a sword that doesn't use a sword," said the young samurai. "Don't talk gibberish. What is your school of fighting called?" "It is called Mutekatsu-ryu [style that wins without swords or fighting]," replied Bokuden. "What? Mutekatsu-ryu? Don't be ridiculous. How can you defeat an opponent without fighting?"

By now the young samurai was angry and irritated, and he demanded that Bokuden demonstrate his style, challenging him to a fight then and there. Bokuden refused to duel in the crowded boat but said he would show the samurai Mutekatsu-ryu at the nearest shore, and he asked the ferryman to guide the boat to a tiny nearby island. The young man began to swing his sword to loosen up. Bokuden continued to sit with his eyes closed. As they approached the island, the impatient challenger shouted, "Come! You are as good as dead. I will show you how sharp my sword is!" He then leaped onto the shore.

Bokuden took his time, further infuriating the young samurai, who began to hurl insults. Bokuden finally handed the ferryman his swords, saying, "My style is Mutekatsu-ryu. I have no need for a sword"—and with those words he took the ferryman's long oar and pushed it hard against the shore, sending the boat quickly out into the water and away from the island. The samurai screamed, demanding the boat's return. Bokuden shouted back to him, "This is what is called victory without fighting. I dare you to jump into the water and swim here!"

Now the passengers on the boat could look back at the young samurai receding into the distance, stranded on the island, jumping up and down, flailing his arms as his cries became fainter and fainter. They began to laugh: Bokuden had clearly demonstrated Mutekatsu-ryu.

Interpretation

The minute Bokuden heard the arrogant young samurai's voice, he knew there would be trouble. A duel on a crowded boat would be a disaster, and a totally unnecessary one; he had to get the young man off the boat without a fight, and to make the defeat humiliating. He would do this through maneuver. First, he remained still and quiet, drawing the man's attention away from the innocent passengers and drawing him toward Bokuden like a magnet. Then he confused the man with a rather irrational name for a school of fighting, overheating the samurai's rather simple mind with a perplexing concept. The flustered samurai tried to cover up with bluster. He was now so angry and mentally off balance that he leaped to the shore alone, failing to consider the rather obvious meaning of Mutekatsu-ryu even once he got there. Bokuden was a samurai who always depended on setting up his opponents first and winning the victory easily, by maneuver rather than brute force. This was the ultimate demonstration of his art.

The goal of maneuver is to give you easy victories, which you do by luring opponents into leaving their fortified positions of strength for unfamiliar terrain where they must fight off balance. Since your opponents' strength is inseparable from their ability to think straight, your maneuvers must be designed to make them emotional and befuddled. If you are too direct in this maneuvering, you run the risk of revealing your game; you must be subtle, drawing opponents toward you with enigmatic behavior, slowly getting under their skin with provocative comments and actions, then suddenly stepping back. When you feel that their emotions are engaged, that their frustration and anger are mounting, you can speed up the tempo of your maneuvers. Properly set up, your opponents will leap onto the island and strand themselves, giving you the easy victory.

**Image:**
The Sickle.
The simplest of
instruments. To
cut the tall grass or
unripened fields of
wheat with it is ex-
hausting labor. But
let the stalks turn
golden brown, hard
and dry, and in that
brief time even the
dullest sickle will
mow the wheat
with ease.

NO. 71. THE VICTORY IN THE MIDST OF A HUNDRED ENEMIES

*To priest Yozan, the 28th teacher at Enkakuji, came for an interview a samurai named Ryozan, who practised Zen. The teacher said: "You are going into the bath-tub, stark naked without a stitch on. Now a hundred enemies in armour, with bows and swords, appear all around you. How will you meet them? Will you crawl before them and beg for mercy? Will you show your warrior birth by dying in combat against them? Or does a man of the Way get some special holy grace?" Ryozan said, "Let me win without surrendering and without fighting."*

*Test*
*Caught in the midst of the hundred enemies, how will you manage to win without surrendering and without fighting?*

*SAMURAI ZEN: THE WARRIOR KOANS,* TREVOR LEGGETT, 1985

**Authority:** Battles are won by slaughter and maneuver. The greater the general, the more he contributes in maneuver, the less he demands in slaughter. . . . Nearly all the battles which are regarded as masterpieces of the military art . . . have been battles of maneuver in which very often the enemy has found himself defeated by some novel expedient or device, some queer, swift, unexpected thrust or stratagem. In such battles the losses of the victors have been small. —*Winston Churchill (1874–1965)*

## REVERSAL

There is neither point nor honor in seeking direct battle for its own sake. That kind of fighting, however, may have value as part of a maneuver or strategy. A sudden envelopment or powerful frontal blow when the enemy is least expecting it can be crushing.

The only danger in maneuver is that you give yourself so many options that you yourself get confused. Keep it simple—limit yourself to the options you can control.

# 21

## NEGOTIATE WHILE ADVANCING

## THE DIPLOMATIC-WAR STRATEGY

*People will always try to take from you in negotiation what they could not get from you in battle or direct confrontation. They will even use appeals to fairness and morality as a cover to advance their position. Do not be taken in: negotiation is about maneuvering for power or placement, and you must always put yourself in the kind of strong position that makes it impossible for the other side to nibble away at you during your talks. Before and during negotiations, you must keep advancing, creating relentless pressure and compelling the other side to settle on your terms. The more you take, the more you can give back in meaningless concessions. Create a reputation for being tough and uncompromising, so that people are back on their heels before they even meet you.*

## WAR BY OTHER MEANS

After Athens was finally defeated by Sparta in the Peloponnesian War of 404 B.C., the great city-state fell into steady decline. In the decades that followed, many citizens, including the great orator Demosthenes, began to dream of a revival of the once dominant Athens.

In 359 B.C. the king of Macedonia, Perdiccas, was killed in battle, and a power struggle emerged for his succession. The Athenians saw Macedonia as a barbaric land to the north, its only importance its proximity to Athenian outposts that helped secure their supplies of corn from Asia and of gold from local mines. One such outpost was the city of Amphipolis, a former Athenian colony, which, however, had lately fallen into Macedonian hands. A plan emerged among the politicians of Athens to support one of the claimants to the Macedonian throne (a man named Argaeus) with ships and soldiers. If he won, he would be indebted to Athens and would return to them the valuable city of Amphipolis.

Unfortunately, the Athenians backed the wrong horse: Perdiccas's twenty-four-year-old brother, Philip, easily defeated Argaeus in battle and became king. To the Athenians' surprise, however, Philip did not push his advantage but stepped back, renouncing all claim to Amphipolis and making the city independent. He also released without ransom all the Athenian soldiers he had captured in battle. He even discussed forming an alliance with Athens, his recent enemy, and in secret negotiations proposed to reconquer Amphipolis in a few years and deliver it to Athens in exchange for another city still under Athenian control, an offer too good to refuse.

The Athenian delegates at the talks reported that Philip was an amiable sort and that beneath his rude exterior he was clearly an admirer of Athenian culture—indeed, he invited Athens's most renowned philosophers and artists to reside in his capital. Overnight, it seemed, the Athenians had gained an important ally to the north. Philip set about fighting barbaric tribes on other borders, and peace ruled between the two powers.

A few years later, as Athens was racked by an internal power struggle of its own, Philip marched on and captured Amphipolis. Following their agreement, the Athenians dispatched envoys to negotiate, only to find, to their surprise, that Philip no longer offered them the city but merely made vague promises for the future. Distracted by their problems at home, the envoys had no choice but to accept this. Now, with Amphipolis securely under his control, Philip had unlimited access to the gold mines and rich forests in the area. It seemed that he had been playing them all along.

Now Demosthenes came forward to rail against the duplicitous Philip and warn of the danger he posed to all of Greece. Urging the citizens of Athens to raise an army to meet the threat, the orator recalled their victories in the past over other tyrants. Nothing happened then, but a few years later, when Philip maneuvered to take the pass at

Thermopylae—the narrow gateway that controlled movement from central to southern Greece—Athens indeed sent an army to defend it. Philip retreated, and the Athenians congratulated themselves on their victory.

In the years to come, the Athenians watched warily as Philip extended his domain to the north, the east, and well into central Greece. Then, in 346 B.C., he suddenly proposed to negotiate a treaty with Athens. He had proved he could not be trusted, of course, and many of the city's politicians had sworn never to deal with him again, but the alternative was to risk war with Macedonia at a time when Athens was ill prepared for it. And Philip seemed absolutely sincere in his desire for a solid alliance, which, at the very least, would buy Athens a period of peace. So, despite their reservations, the Athenians sent ambassadors to Macedonia to sign a treaty called the Peace of Philocrates. By this agreement Athens relinquished its rights to Amphipolis and in exchange received promises of security for its remaining outposts in the north.

The ambassadors left satisfied, but on the way home they received news that Philip had marched on and taken Thermopylae. Challenged to explain himself, Philip responded that he had acted to secure his interests in central Greece from a temporary threat by a rival power, and he quickly abandoned the pass. But the Athenians had had enough—they had been humiliated. Time and time again, Philip had used negotiations and treaties to cover nefarious advances. He was not honorable. He might have abandoned Thermopylae, but it did not matter: he was always taking control over larger territories, then making himself look conciliatory by giving some of his acquisitions back—but only some, and he often retook the conceded lands later anyway. The net effect was inevitably to enlarge his domain. Mixing war with deceptive diplomacy, he had slowly made Macedonia the dominant power in Greece.

Demosthenes and his followers were now on the ascendant. The Peace of Philocrates was obviously a disgrace, and everyone involved in it was thrown out of office. The Athenians began to make trouble in the country to the east of Amphipolis, trying to secure more outposts there, even provoking quarrels with Macedonia. In 338 B.C. they engaged in an alliance with Thebes to prepare for a great war against Philip. The two allies met the Macedonians in battle at Chaeronea, in central Greece—but Philip won the battle decisively, his son Alexander playing a key role.

Now the Athenians were in panic: barbarians from the north were about to descend on their city and burn it to the ground. And yet again they were proved wrong. In a most generous peace offer, Philip promised not to invade Athenian lands. In exchange he would take over the disputed outposts in the east, and Athens would become an ally of Macedonia. As proof of his word, Philip released his Athenian prisoners from the recent war without asking for payment of any ransom. He also had his son Alexander lead a delegation to Athens bearing the ashes of all the Athenian soldiers who had died at Chaeronea. Overwhelmed with gratitude, the Athenians granted citizenship to both Alexander and his father and erected a statue of Philip in their agora.

Later that year Philip convened a congress of all the Greek city-states (except for Sparta, which refused to attend) to discuss an alliance to form what would be called the Hellenic League. For the first time, the Greek city-states were united in a single confederation. Soon after the terms of the alliance were agreed upon, Philip proposed a united war against the hated Persians. The proposal was happily accepted, with Athens leading the way. Somehow everyone had forgotten how dishonorable Philip had been; they only remembered the king who had recently been so generous.

In 336 B.C., before the war against Persia got under way, Philip was assassinated. It would be his son Alexander who would lead the league into war and the creation of an empire. And through it all, Athens would remain Macedonia's most loyal ally, its critical anchor of stability within the Hellenic League.

*Lord Aberdeen, the British ambassador to Austria, proved even easier to deal with. Only twenty-nine years old, barely able to speak French, he was not a match for a diplomat of Metternich's subtlety. His stiffness and self-confidence only played into Metternich's hands. "Metternich is extremely attentive to Lord Aberdeen," reported Cathcart. The results were not long delayed. Metternich had once described the diplomat's task as the art of seeming a dupe, without being one, and he practised it to the fullest on the high-minded Aberdeen. "Do not think Metternich such a formidable personage . . . ," Aberdeen wrote to Castlereagh. "Living with him at all times . . . , is it possible I should not know him? If indeed he were the most subtle of mankind, he might certainly impose on one little used to deceive, but this is not his character. He is, I repeat it to you, not a very clever man. He is vain . . . but he is to be trusted. . . ." For his mixture of condescension and gullibility, Aberdeen earned himself Metternich's sarcastic epithet as the "dear simpleton of diplomacy."*

A WORLD RESTORED, HENRY KISSINGER, 1957

### Interpretation

On one level, war is a relatively simple affair: you maneuver your army to defeat your enemy by killing enough of its soldiers, taking enough of its land, or making yourself secure enough to proclaim victory. You may have to retreat here and there, but your intention is eventually to advance as far as possible. Negotiation, on the other hand, is almost always awkward. On the one hand, you need both to secure your existing interests and to get as much on top of them as you can; on the other hand, you need to bargain in good faith, make concessions, and gain the opposing side's trust. To mix these needs is an art, and an almost impossible one, for you can never be sure that the other side is acting in good faith. In this awkward realm between war and peace, it is easy to misread the opponent, leading to a settlement that is not in your long-term interest.

Philip's solution was to see negotiation not as separate from war but rather as an extension of it. Negotiation, like war, involved maneuver, strategy, and deception, and it required you to keep advancing, just as you would on the battlefield. It was this understanding of negotiation that led Philip to offer to leave Amphipolis independent while promising to take it for Athens later on, a promise he never meant to keep. This opening maneuver bought him friendship and time, and kept the pesky Athenians out of his hair while he dealt with his enemies elsewhere. The Peace of Philocrates similarly covered his moves in central Greece and kept the Athenians off balance. Having decided at some point that his ultimate goal was to unite all of Greece and lead it on a crusade against Persia, Philip determined that Athens—with its noble history—would have to function as a symbolic center of the Hellenic League. His generous peace terms were calculated to purchase the city's loyalty.

Philip never worried about breaking his word. Why should he

sheepishly honor his agreements when he knew the Athenians would find some excuse later on to extend their outposts to the north at his expense? Trust is not a matter of ethics, it is another maneuver. Philip saw trust and friendship as qualities for sale. He would buy them from Athens later on, when he was powerful and had things to offer it in exchange.

Like Philip, you must see any negotiating situation in which your vital interests are at stake as a realm of pure maneuver, warfare by other means. Earning people's trust and confidence is not a moral issue but a strategic one: sometimes it is necessary, sometimes it isn't. People will break their word if it serves their interests, and they will find any moral or legal excuse to justify their moves, sometimes to themselves as well as to others.

Just as you must always put yourself in the strongest position before battle, so it is with negotiation. If you are weak, use negotiations to buy yourself time, to delay battle until you are ready; be conciliatory not to be nice but to maneuver. If you are strong, take as much as you can before and during negotiations—then later you can give back some of what you took, conceding the things you least value to make yourself look generous. Do not worry about your reputation or about creating distrust. It is amazing how quickly people will forget your broken promises when you are strong and in a position to offer them something in their self-interest.

> *Therefore, a prudent ruler ought not to keep faith when by so doing it would be against his interest. . . . If men were all good, this precept would not be a good one; but as they are bad, and would not observe their faith with you, so you are not bound to keep faith with them. Nor have legitimate grounds ever failed a prince who wished to show colorable excuse for the nonfulfillment of his promise.*
> —*Niccolò Machiavelli,* The Prince *(1469–1527)*

## JADE FOR TILE

Early in 1821 the Russian foreign minister, Capo d'Istria, heard news he had long been awaiting: a group of Greek patriots had begun a rebellion against the Turks (Greece was then part of the Ottoman Empire), aiming to throw them out and establish a liberal government. D'Istria, a Greek nobleman by birth, had long dreamed of involving Russia in Greek affairs. Russia was a growing military power; by supporting the revolution—assuming the rebels won—it would gain influence over an independent Greece and Mediterranean ports for its navy. The Russians also saw themselves as the protectors of the Greek Orthodox Church, and Czar Alexander I was a deeply religious man; leading a crusade against the Islamic Turks would satisfy his moral consciousness as well as Russian political interests. It was all too good to be true.

Only one obstacle stood in d'Istria's way: Prince Klemens von

Metternich, the Austrian foreign minister. A few years earlier, Metternich had brought Russia into an alliance with Austria and Prussia called the Holy Alliance. Its goal was to protect these nations' governments from the threat of revolution and to maintain peace in Europe after the turmoil of the Napoleonic Wars. Metternich had befriended Alexander I. Sensing that the Russians might intervene in Greece, he had sent the czar hundreds of reports claiming that the revolution was part of a Europe-wide conspiracy to get rid of the continent's monarchies. If Alexander came to Greece's aid, he would be the revolutionaries' dupe and would be violating the purpose of the Holy Alliance.

D'Istria was no fool: he knew that what Metternich really wanted was to prevent Russia from expanding its influence in the Mediterranean, which would upset England and destabilize Europe, Metternich's greatest fear. To d'Istria it was simple: he and Metternich were at war over who would have ultimate influence over the czar. And d'Istria had the advantage: he saw the czar often and could counteract Metternich's persuasive powers through constant personal contact.

The Turks inevitably moved to suppress the Greek rebellion, and as their atrocities against the Greeks mounted, it seemed almost certain that the czar would intervene. But in February 1822, as the revolution was reaching a boiling point, the czar made what in d'Istria's eyes was a fatal mistake: he agreed to send an envoy to Vienna to discuss the crisis with Metternich. The prince loved to lure negotiators to Vienna, where he would charm them to death. D'Istria felt the situation slipping out of his hands. Now he had just one option: to choose the envoy who would go to Vienna and brief him in detail.

D'Istria's choice was a man called Taticheff, who had been Russia's ambassador to Spain. Taticheff was a shrewd, experienced negotiator. Called in for a meeting shortly before he was to leave, he listened carefully as d'Istria laid out the dangers: Metternich would try to charm and seduce Taticheff; to prevent the czar from intervening, he would offer to negotiate a settlement between the Russians and Turks; and, of course, he would call for a European conference to discuss the issue. This last was Metternich's favorite ploy: he was always able to dominate these conferences and somehow get what he wanted. Taticheff was not to fall under his spell. He was to give Metternich a note from d'Istria arguing that Russia had a right to come to the aid of fellow Christians suffering at the hands of the Turks. And on no account was he to agree to Russia's participation in a conference.

On the eve of his departure for Vienna, Taticheff was unexpectedly called in for a meeting with the czar himself. Alexander was nervous and conflicted. Unaware of d'Istria's instructions, he told Taticheff to tell Metternich that he wanted both to act in accordance with the alliance and to meet his moral obligation in Greece. Taticheff decided he would have to delay giving this message as long as he could—it would make his work far too confusing.

In his first meeting with Metternich in Vienna, Taticheff took measure of the Austrian minister. He saw him as rather vain, apparently more interested in fancy-dress balls and young girls than in Greece. Metternich seemed detached and somewhat ill-informed; the little he said about the situation in Greece betrayed confusion. Taticheff read d'Istria's note to him, and, as if without thinking, Metternich asked if these were the czar's instructions as well. Put on the spot, Taticheff could not lie. His hope now was that the czar's rather contradictory instructions would further confuse the prince, letting Taticheff stay one step ahead.

In the days to come, Taticheff had a splendid time in the delightful city of Vienna. Then he had another meeting with Metternich, who asked him if they could begin negotiations based on the instructions of the czar. Before Taticheff could think, Metternich next asked what Russia's demands might be in this situation. That seemed fair, and Taticheff replied that the Russians wanted to make Greece a protectorate state, to get the alliance's approval for Russian intervention in Greece, on and on. Metternich turned down every proposal, saying his government would never agree to such things, so Taticheff asked him to suggest alternate ideas. Instead Metternich launched into an abstract discussion of revolution, of the importance of the Holy Alliance, and other irrelevancies. Taticheff left confused and rather annoyed. He had wanted to stake out a position, but these discussions were informal and shapeless; feeling lost, he had been unable to steer them in the direction he wanted.

A few days later, Metternich called Taticheff in again. He looked uncomfortable, even pained: the Turks, he said, had just sent him a note claiming that the Russians were behind the trouble in Greece and asking him to convey to the czar their determination to fight to the death to hold on to what was theirs. In solemn tones suggesting that he was angry at the Turks' lack of diplomacy, Metternich said he thought it beneath his country's dignity to pass this disgraceful message to the czar. He added that the Austrians considered Russia their staunchest ally and would support Russia's conditions for resolving the crisis. Finally, if the Turks refused to concede, Austria would break off relations with them.

Taticheff was quite moved by this sudden emotional display of solidarity. Perhaps the Russians had misread the prince—perhaps he was really on their side. Fearing that d'Istria would misunderstand, Taticheff reported this meeting to the czar alone. A few days later, Alexander responded that from now on, Taticheff was to report only to him; d'Istria was to be excluded from the negotiations.

The pace of the meetings with Metternich picked up. Somehow the two men discussed only diplomatic solutions to the crisis; Russia's right to intervene in Greece militarily was no longer mentioned. Finally, Metternich invited the czar to attend a conference on the question in Verona, Italy, a few months later. Here Russia would lead the debate on how best to settle the matter; it would be at the center of attention, with the czar rightly celebrated as Europe's savior in the crusade against revolution. The czar happily agreed to attend.

Back in St. Petersburg, d'Istria fumed and ranted to anyone who would listen, but shortly after Taticheff got home, the Russian foreign minister was kicked out of office for good. And at the later conference in Verona, just as he had predicted, the Greek crisis was resolved in precisely the way that best served Austria's interests. The czar was the star of the show, but apparently he did not care or notice that he had signed a document essentially precluding Russia from intervening unilaterally in the Balkans, thereby conceding a right insisted upon by every Russian leader since Peter the Great. Metternich had won the war with d'Istria more completely than the former minister had ever imagined possible.

## Interpretation

Metternich's goal was always a settlement that would best serve Austria's long-term interests. Those interests, he decided, involved not just preventing Russian intervention in Greece but maneuvering the czar into permanently relinquishing the right to send troops into the Balkans, an enduring source of instability in Europe. So Metternich looked at the relative forces on both sides. What leverage did he have over the Russians? Very little; in fact, he had the weaker hand. But Metternich possessed a trump card: his years-long study of the czar's rather strange personality. Alexander was a highly emotional man who would act only in a state of exaltation; he had to turn everything into a crusade. So, right at the beginning of the crisis, Metternich planted the seed that the real crusade here was one not of Christians against Turks but of monarchies against revolution.

Metternich also understood that his main enemy was d'Istria and that he would have to drive a wedge between d'Istria and the czar. So he lured an envoy to Vienna. In one-on-one negotiations, Metternich was a chess player on the grand-master level. With Taticheff as with so many others, he first lowered his opponent's suspicions by playing the foppish, even dim-witted aristocrat. Next he drew out the negotiations, miring them in abstract, legalistic discussions. That made him seem even more stupid, further misleading Taticheff but also confusing and irritating him. A confused and riled negotiator is prone to make mistakes—such as reveal too much about what he is after, always a fatal error. A confused negotiator is also more easily seduced by emotional demonstrations. In this case Metternich used the note from the Turks to stage a little drama in which he appeared to reveal a sudden change in his sympathies. That put Taticheff—and through him the czar—completely under his spell.

From then on, it was child's play to reframe the discussion to suit Metternich's purpose. The offer to stage a conference at which the czar would shine was dazzling and alluring, and it also seemed to offer Russia the chance of greater influence in European affairs (one of Alexander's deepest desires). In fact the result was the opposite: Alexander ended up signing a document that cut Russia out of the Balkans—Metternich's goal all along. Knowing how easily people are seduced by appearances, the Austrian minister gave the czar the appearance of power (being the center of attention at the conference), while he himself retained its substance (having the signed document). It is what the Chinese call giving someone a gaudy piece of painted tile in exchange for jade.

As Metternich so often demonstrated, success in negotiation depends on the level of preparation. If you enter with vague notions as to what you want, you will find yourself shifting from position to position depending on what the other side brings to the table. You may drift to a position that seems appropriate but does not serve your interests in the end. Unless you carefully analyze what leverage you have, your maneuvers are likely to be counterproductive.

Before anything else you must anchor yourself by determining with utmost clarity your long-term goals and the leverage you have for reaching them. That clarity will keep you patient and calm. It will also let you toss people meaningless concessions that seem generous but actually come cheap, for they do not hurt your real goals. Before the negotiations begin, study your opponents. Uncovering their weaknesses and unfulfilled desires will give you a different kind of leverage: the ability to confuse them, make them emotional, seduce them with pieces of tile. If possible, play a bit of the fool: the less people understand you and where you are headed, the more room you have to maneuver them into corners.

> *Everyone wants something without having any idea how to obtain it, and the really intriguing aspect of the situation is that nobody quite knows how to achieve what he desires. But because I know what I want and what the others* are capable of *I am completely prepared.*
>
> —*Prince Klemens von Metternich (1773–1859)*

## KEYS TO WARFARE

Conflict and confrontation are generally unpleasant affairs that churn up unpleasant emotions. Out of a desire to avoid such unpleasantness, people will often try to be nice and conciliatory to those around them, in the belief that that will elicit the same response in return. But so often experience proves this logic to be wrong: over time, the people you treat nicely will take you for granted. They will see you as weak and exploitable. Being generous does not elicit gratitude but creates either a spoiled child or someone who resents behavior perceived as charity.

Those who believe against the evidence that niceness breeds niceness in return are doomed to failure in any kind of negotiation, let alone in the game of life. People respond in a nice and conciliatory way only when it is in their interest and when they have to do so. Your goal is to create that imperative by making it painful for them to fight. If you ease up the pressure out of a desire to be conciliatory and gain their trust, you

*In gratitude for his acquittal, Orestes dedicated an altar to Warlike Athene; but the Erinnyes threatened, if the judgement were not reversed, to let fall a drop of their own hearts' blood which would bring barrenness upon the soil, blight the crops, and destroy all the offspring of Athens. Athene nevertheless soothed their anger by flattery: acknowledging them to be far wiser than herself, she suggested that they should take up residence in a grotto at Athens, where they would gather such throngs of worshippers as they could never hope to find elsewhere. Hearth-altars proper to Underworld deities should be theirs, as well as sober sacrifices, torchlight libations, first-fruits offered after the consummation of marriage or the birth of children, and even seats in the Erechtheum. If they accepted this invitation she would decree that no house where worship was withheld from them might prosper; but they, in return, must undertake to invoke fair winds for her ships, fertility for her land, and fruitful marriages for her people—also rooting out the impious, so that she might see fit to grant Athens victory in war. The Erinnyes, after a short deliberation, graciously agreed to these proposals.*

THE GREEK MYTHS, VOL. 2, ROBERT GRAVES, 1955

only give them an opening to procrastinate, deceive, and take advantage of your niceness. That is human nature. Over the centuries those who have fought wars have learned this lesson the hard way.

When nations have violated this principle, the results are often tragic. In June 1951, for example, the U.S. military halted its extremely effective offensive against the Chinese People's Liberation Army in Korea because the Chinese and the North Koreans had signaled they were ready to negotiate. Instead they drew out the talks as long as they could while they recovered their forces and strengthened their defenses. When the negotiation failed and the war was resumed, the American forces found that their battlefield advantage was lost. This pattern was repeated in the Vietnam War and to some extent in the Gulf War of 1991 as well. The Americans acted partly out of a desire to reduce casualties, partly to be seen as trying to bring these wars to an end as soon as possible, to appear conciliatory. What they did not realize was that the enemy's incentive to negotiate in good faith was lost in the process. In this case, trying to be conciliatory and save lives led to much longer wars, more bloodshed, real tragedy. Had the United States continued to advance in Korea in 1951, it could have compelled the Koreans and Chinese to negotiate on its own terms; had it continued its bombing campaigns in Vietnam, it could have forced the North Vietnamese to negotiate instead of procrastinate; had it continued its march all the way to Baghdad in 1991, it could have forced Saddam Hussein out of office as a condition of peace, preventing a future war and saving countless lives.

The lesson is simple: by continuing to advance, by keeping up unrelenting pressure, you force your enemies to respond and ultimately to negotiate. If you advance a little further every day, attempts to delay negotiation only make their position weaker. You are demonstrating your resolve and determination, not through symbolic gestures but by administering real pain. You do not continue to advance in order to grab land or possessions but to put yourself in the strongest possible position and win the war. Once you have forced them to settle, you have room to make concessions and give back some of what you've taken. In the process you might even seem nice and conciliatory.

Sometimes in life you will find yourself holding the weak hand, the hand without any real leverage. At those times it is even more important to keep advancing. By demonstrating strength and resolve and maintaining the pressure, you cover up your weaknesses and gain footholds that will let you manufacture leverage for yourself.

In June 1940, shortly after the German blitzkrieg had destroyed France's defenses and the French government had surrendered, General Charles de Gaulle fled to England. He hoped to establish himself there as the leader of Free France, the legitimate government in exile, as opposed to the German-dominated Vichy government that now ruled much of the country. The odds were stacked heavily against de Gaulle: he had never been a high-profile figure within France. Many better-known

French soldiers and politicians could claim the role he wanted; he had no leverage to make the allies recognize him as the leader of Free France, and without their recognition he would be powerless.

From the beginning de Gaulle ignored the odds and presented himself to one and all as the only man who could save France after its disgraceful surrender. He broadcast stirring speeches to France over the radio. He toured England and the United States, making a show of his sense of purpose, casting himself as a kind of latter-day Joan of Arc. He made important contacts within the French Resistance. Winston Churchill admired de Gaulle but often found him unbearably arrogant, and Franklin Roosevelt despised him; time and again the two leaders tried to persuade him to accept shared control of Free France. But his response was always the same: he would not compromise. He would not accept anything less than sole leadership. In negotiating sessions he was downright rude, to the point where he would sometimes walk out, making it clear that for him it was all or nothing.

Churchill and Roosevelt cursed de Gaulle's name, ruing the day they had let him take any position at all. They even talked about demoting him and forcing him out of the picture. But they always backed down, and in the end they gave him what he wanted. To do otherwise would mean a public scandal in delicate times and would disrupt their relations with the French underground. They would be demoting a man whom much of the public had come to revere.

Understand: if you are weak and ask for little, little is what you will get. But if you act strong, making firm, even outrageous demands, you will create the opposite impression: people will think that your confidence must be based on something real. You will earn respect, which in turn will translate into leverage. Once you are able to establish yourself in a stronger position, you can take this further by refusing to compromise, making it clear that you are willing to walk away from the table—an effective form of coercion. The other side may call your bluff, but you make sure there's a price to pay for this—bad publicity, for instance. And if in the end you do compromise a little, it will still be a lot less than the compromises they would have forced on you if they could.

The great British diplomat and writer Harold Nicholson believed there were two kinds of negotiators: warriors and shopkeepers. Warriors use negotiations as a way to gain time and a stronger position. Shopkeepers operate on the principle that it is more important to establish trust, to moderate each side's demands and come to a mutually satisfying settlement. Whether in diplomacy or in business, the problem arises when shopkeepers assume they are dealing with another shopkeeper only to find they are facing a warrior.

It would be helpful to know beforehand which kind of negotiator you face. The difficulty is that skillful warriors will make themselves masters of disguise: at first they will seem sincere and friendly, then will reveal their warrior nature when it is too late. In resolving a conflict with an enemy you do not know well, it is always best to protect yourself by playing the warrior yourself: negotiate while advancing. There will always be time to back off and fix things if you go too far. But if you fall prey to a warrior, you will be unable to recoup anything. In a world in which there are more and more warriors, you must be willing to wield the sword as well, even if you are a shopkeeper at heart.

**Image:** The Big Stick. You may speak softly and nicely, but the other side sees that you hold something fearsome in your hand. He does not have to feel the actual pain of it striking his head; he knows the stick is there, that it is not going away, that you have used it before, and that it hurts. Better to end the argument and negotiate a settlement, at whatever price, than risk a painful thwack.

**Authority:** Let us not consider ourselves victorious until the day *after* battle, nor defeated until four days later. . . . Let us always carry the sword in one hand and the olive branch in the other, always ready to negotiate but negotiating only while advancing. —*Prince Klemens von Metternich (1773–1859)*

## REVERSAL

In negotiation as in war, you must not let yourself get carried away: there is a danger in advancing too far, taking too much, to the point where you create an embittered enemy who will work for revenge. So it was after World War I with the Allies, who imposed such harsh conditions on Germany in negotiating the peace that they arguably laid the foundations for World War II. A century earlier, on the other hand, when Metternich negotiated, it was always his goal to prevent the other side from feeling wronged. Your purpose in any settlement you negotiate is never to satisfy greed or to punish the other side but to secure your own interests. In the long run, a punitive settlement will only win you insecurity.

# 22

# KNOW HOW TO END THINGS

## THE EXIT STRATEGY

*You are judged in this world by how well you bring things to an end. A messy or incomplete conclusion can reverberate for years to come, ruining your reputation in the process. The art of ending things well is knowing when to stop, never going so far that you exhaust yourself or create bitter enemies that embroil you in conflict in the future. It also entails ending on the right note, with energy and flair. It is not a question of simply winning the war but the way you win it, the way your victory sets you up for the next round. The height of strategic wisdom is to avoid all conflicts and entanglements from which there are no realistic exits.*

*If one overshoots the goal, one cannot hit it. If a bird will not come to its nest but flies higher and higher, it eventually falls into the hunter's net. He who in times of extraordinary salience of small things does not know how to call a halt, but restlessly seeks to press on and on, draws upon himself misfortune at the hands of gods and men, because he deviates from the order of nature.*

*The I Ching*, China, circa eighth century B.C.

## NO EXIT

For the most senior members of the Soviet Politburo—General Secretary Leonid Brezhnev, KGB head Yuri Andropov, and Defense Minister Dmitri Ustinov—the late 1960s and early '70s seemed a golden era. These men had survived the nightmare of the Stalin years and the bumbling reign of Khrushchev. Now, finally, there was some stability in the Soviet empire. Its satellite states in Eastern Europe were relatively docile, particularly after an uprising in Czechoslovakia in 1968 had been squashed. Its archnemesis, the United States, had received a black eye from the Vietnam War. And, most promising of all, the Russians had slowly been able to expand their influence in the Third World. The future looked bright.

A key country in the Russians' plans for expansion was Afghanistan, on their southern border. Afghanistan was rich in natural gas and other minerals and had ports on the Indian Ocean; to make it a Soviet satellite would be a dream come true. The Russians had been insinuating themselves into the country since the 1950s, helping to train its army, building the Salang Highway from Kabul north to the Soviet Union, and trying to modernize this backward nation. All was going according to plan until the early to mid 1970s, when Islamic fundamentalism began to become a political force across Afghanistan. The Russians saw two dangers: first, that the fundamentalists would come to power and, seeing communism as godless and loathsome, would cut off ties with the Soviets; and second, that fundamentalist unrest would spill over from Afghanistan into the southern Soviet Union, which had a large Islamic population.

In 1978, to prevent such a nightmare scenario, Brezhnev secretly supported a coup that brought the Afghan Communist Party to power. But the Afghan Communists were hopelessly factionalized, and only after a long power struggle did a leader emerge: Hafizullah Amin, whom the Soviets distrusted. On top of that, the Communists were not popular in Afghanistan, and Amin resorted to the most brutal means to maintain his party's power. This only fed the fundamentalist cause. All around the country, insurgents—the mujahideen—began to rebel, and thousands of Afghan soldiers defected from the army to them.

By December 1979 the Communist government in Afghanistan was on the verge of collapse. In Russia the senior members of the Politburo met to discuss the crisis. To lose Afghanistan would be a devastating blow and a source of instability after so much progress had been made. They blamed Amin for their problems; he had to go. Ustinov proposed a plan: Repeating what the Soviets had done in quelling rebellions in Eastern Europe, he advocated a lightning strike by a relatively small Soviet force that would secure Kabul and the Salang Highway. Amin would then be ousted, and a Communist named Babrak Karmal would take his place. The Soviet army would assume a low profile, and the Afghan army would be beefed up to take over from it. During the course of some ten years, Afghanistan would be modernized and would slowly become a stable member of the Soviet Bloc. Blessed with peace and prosperity, the Afghan people would see the great benefits of socialism and embrace it.

A few days after the meeting, Ustinov presented his plan to the army's chief of staff, Nikolai Orgakov. Told that the invading army would not exceed 75,000 men, Orgakov was shocked: that force, he said, was far too small to secure the large, mountainous expanses of Afghanistan, a very different world from Eastern Europe. Ustinov countered that a giant invading force would generate bad publicity for the Soviets in the Third World and would give the insurgents a rich target. Orgakov responded that the fractious Afghans had a tradition of suddenly uniting to throw out an invader—and that they were fierce fighters. Calling the plan reckless, he said it would be better to attempt a political solution to the problem. His warnings were ignored.

The plan was approved by the Politburo and on December 24 was put in motion. Some Red Army forces flew into Kabul while others marched down the Salang Highway. Amin was quietly taken away and killed while Karmal was shuffled into power. Condemnation poured in from all over the world, but the Soviets figured that would eventually die down—it usually did.

In February 1980, Andropov met with Karmal and instructed him on the importance of winning the support of the Afghan masses. Presenting a plan for that purpose, he also promised aid in money and expertise. He told Karmal that once the borders were secured, the Afghan army built up, and the people reasonably satisfied with the government, Karmal should politely *ask* the Soviets to leave.

The invasion itself went more easily than the Soviets had expected, and for this military phase their leaders could confidently declare "mission accomplished." But within weeks of Andropov's visit, they had to adjust this assessment: the mujahideen were not intimidated by the Soviet army, as the Eastern Europeans had been. In fact, since the invasion their power only seemed to grow, their ranks swelling with both Afghan recruits and outsiders. Ustinov funneled more soldiers into Afghanistan and ordered a series of offensives in parts of the country that were sheltering the mujahideen. The Soviets' first major operation was that spring, when they moved into the Kunar Valley with heavy weaponry, leveling entire villages and forcing the inhabitants to flee to refugee camps in Pakistan. Having cleared the area of rebels, they withdrew.

A few weeks later, reports came in that the mujahideen had quietly returned to the Kunar Valley. All the Soviets had done was leave the Afghans more embittered and enraged, making it easier for the mujahideen to recruit. But what could the Soviets do? To let the rebels alone was to give the mujahideen the time and space to grow more dangerous, yet the army was too small to occupy whole regions. Its answer was to repeat its police operations again and again, but with more violence, hoping to intimidate the Afghans—but, as Orgakov had predicted, this only emboldened them.

*Solitudinem faciunt pacem appellant (They create desolation and call it peace).*

TACITUS,
CIRCA A.D. 55–CIRCA 120

*ALL'S WELL THAT ENDS WELL:*
*still the fine's the crown;*
*Whate'er the course, the end is the renown.*

*ALL'S WELL THAT ENDS WELL,*
WILLIAM SHAKESPEARE,
1564–1616

*Ten thousand Muslims then marched through the mountain valleys upon Mecca. Muhammad divided his force into four columns. . . . Muhammad gave strict orders that no violence was to be used. His own tent was pitched on high ground immediately overlooking the town. Eight years before, he had fled from Mecca under cover of darkness, and lain hidden three days in a cave on Mount Thor, which from his tent he could now see rising beyond the city. Now ten thousand warriors were ready to obey his least command and his native town lay helpless at his feet. After a brief rest, he remounted his camel and entered the town, reverently touched the black stone and performed the seven ritual circuits of the kaaba. . . . Muhammad the Conqueror was not vindictive. A general amnesty was proclaimed, from which less than a dozen persons were excluded, only four being actually executed. Ikrima, the son of Abu Jahal, escaped to the Yemen, but his wife appealed to the Apostle, who agreed to forgive him. . . . The Muslim occupation of Mecca was thus virtually bloodless. The fiery Khalid ibn al Waleed killed a few people at the southern gate and was sharply reprimanded by Muhammad for doing so. Although the Apostle had himself*

Meanwhile Karmal initiated programs to teach literacy, to give more power to women, to develop and modernize the country—all to peel off support from the rebels. But the Afghans preferred their traditional way of life by a vast majority, and the Communist Party's attempts to expand its influence had the opposite effect.

Most ominous of all, Afghanistan quickly became a magnet for other countries eager to exploit the situation there against the Soviets. The United States in particular saw an opportunity to revenge itself on Russia for supplying the North Vietnamese during the Vietnam War. The CIA funneled vast sums of money and matériel to the mujahideen. In neighboring Pakistan, President Zia ul-Haq viewed the invasion as a gift from heaven: having come to power a few years earlier in a military coup, and having recently earned worldwide condemnation by executing his prime minister, Zia saw a way to gain favor with both the United States and the Arab nations by allowing Pakistan to serve as a base for the mujahideen. The Egyptian president Anwar Sadat, who had recently signed a controversial peace treaty with Israel, likewise saw a golden opportunity to shore up his Islamic support by sending aid to fellow Muslims.

With Soviet armies stretched thin in Eastern Europe and around the world, Ustinov refused to send in more men; instead he armed his soldiers with the latest weaponry and worked to enlarge and strengthen the Afghan army. But none of this translated into progress. The mujahideen improved their ambushes of Soviet transports and used the latest Stinger missiles acquired from the Americans to great effect. Years passed, and morale in the Soviet army dropped precipitously: the soldiers felt the hatred of the local population and were stuck guarding static positions, never knowing when the next ambush would come. Abuse of drugs and alcohol became widespread.

As the costs of the war rose, the Russian public began to turn against it. But the Soviet leaders could not afford to pull out: besides creating a dangerous power vacuum in Afghanistan, that would deliver a sharp blow to their global reputation as a superpower. And so they stayed, each year supposedly the last. The senior members of the Politburo slowly died off—Brezhnev in 1982, Andropov and Ustinov in 1984—without seeing the slightest progress.

In 1985, Mikhail Gorbachev became general secretary of the Soviet Union. Having opposed the war from the beginning, Gorbachev started phased withdrawals of troops from Afghanistan. The last soldiers left early in 1989. In all, over 14,000 Soviet soldiers died in the conflict, but the hidden costs—to the delicate Russian economy, to the people's slender faith in their government—were far greater. Only a few years later, the entire system would come tumbling down.

### Interpretation

The great German general Erwin Rommel once made a distinction between a gamble and a risk. Both cases involve an action with only a chance of success, a chance that is heightened by acting with boldness. The difference is that with a risk, if you lose, you can recover: your reputation will suffer no long-term damage, your resources will not be depleted, and you can return to your original position with acceptable losses. With a gamble, on the other hand, defeat can lead to a slew of problems that are likely to spiral out of control. With a gamble there tend to be too many variables to complicate the picture down the road if things go wrong. The problem goes further: if you encounter difficulties in a gamble, it becomes harder to pull out—you realize that the stakes are too high; you cannot afford to lose. So you try harder to rescue the situation, often making it worse and sinking deeper into a hole that you cannot get out of. People are drawn into gambles by their emotions: they see only the glittering prospects if they win and ignore the ominous consequences if they lose. Taking risks is essential; gambling is foolhardy. It can be years before you recover from a gamble, if you recover at all.

The invasion of Afghanistan was a classic gamble. The Soviets were drawn in by the irresistible lure of possessing a client state in the region. Dazzled by that prospect, they ignored the reality: the mujahideen and outside powers had too much at stake to ever allow the Soviets to leave behind a secure Afghanistan. There were too many variables beyond their control: the actions of the United States and Pakistan, the mountainous border areas impossible to seal off, and more. An occupying army in Afghanistan involved a double bind: the larger the military presence, the more it would be hated, and the more it was hated, the larger it would have to be to protect itself, and so on indefinitely.

Yet the Soviets took their gamble and made their mess. Now, too late, they realized that the stakes had been raised: to pull out—to lose—would be a devastating blow to their prestige. It would mean the expansion of American interests and a cancerous insurgency on their border. Since they should never have invaded in the first place, they had no rational exit strategy. The best they could do would be to cut their losses and run—but that is nearly impossible with a gamble, for gambling is governed by emotions, and once the emotions are engaged, it is difficult to retreat.

The worst way to end anything—a war, a conflict, a relationship—is slowly and painfully. The costs of such an ending run deep: loss of self-confidence, unconscious avoidance of conflict the next time around, the bitterness and animosity left breeding—it is all an absurd waste of time. Before entering any action, you must calculate in precise terms your exit strategy. How exactly will the engagement end, and where it will leave you? If the answers to those questions seem vague and full of speculation, if success seems all too alluring and failure somewhat dangerous, you are more than likely taking a gamble. Your emotions are leading you into a situation that could end up a quagmire.

Before that happens, catch yourself. And if you do find you have made this mistake, you have only two rational solutions: either end the

*been persecuted in the city and although many of his bitterest opponents were still living there, he won all hearts by his clemency on his day of triumph. Such generosity, or statesmanship, was particularly remarkable among Arabs, a race to whom revenge has always been dear. His success had been won by policy and diplomacy rather than by military action. In an age of violence and bloodshed, he had realized that ideas are more powerful than force.*

*The Great Arab Conquests*,
John Bagot Glubb,
1963

*Aut non tentaris, aut perfice (Either don't attempt it, or carry it through to the end).*

OVID, 43 B.C.–A.D. 17

conflict as quickly as you can, with a strong, violent blow aimed to win, accepting the costs and knowing they are better than a slow and painful death, or cut your losses and quit without delay. Never let pride or concern for your reputation pull you farther into the morass; both will suffer far greater blows by your persistence. Short-term defeat is better than long-term disaster. Wisdom is knowing when to end.

*To go too far is as bad as to fall short.*
*—Confucius (551?–479 B.C.)*

*Indeed, deepening study of past experience leads to the conclusion that nations might often have come nearer to their object by taking advantage of a lull in the struggle to discuss a settlement than by pursuing the war with the aim of "victory." History reveals, also, that in many cases a beneficial peace could have been obtained if the statesmen of the warring nations had shown more understanding of the elements of psychology in their peace "feelers." Their attitude has commonly been too akin to that seen in the typical domestic quarrel; each party is afraid to appear yielding, with the result that when one of them shows any inclination towards conciliation this is usually expressed in language that is too stiff, while the other is apt to be slow to respond—partly from pride or obstinacy and partly from a tendency to interpret such a gesture as a sign of weakening when it may be a sign of returning common sense. Thus*

## ENDING AS BEGINNING

As a young man, Lyndon B. Johnson had just one dream: to climb the ladder of politics and become president. When Johnson was in his mid-twenties, the goal was starting to seem unreachable. A job as the secretary of a Texas congressman had allowed him to meet and make an impression on President Franklin D. Roosevelt, who had named him the Texas director of the National Youth Administration, a post promising excellent political connections. But Texas voters were extremely loyal, often returning congressmen to their seats for decades, or until they died. Johnson urgently wanted a seat in Congress. If he did not get one soon enough, he would be too old to climb the ladder, and he burned with ambition.

On February 22, 1937, out of the blue, the chance of a lifetime opened up: the Texas congressman James Buchanan suddenly died. The seat he left empty, that of Texas's Tenth District, was a rare opportunity, and the state's eligible political heavyweights immediately threw their hats in the ring. The many contenders included Sam Stone, a popular county judge; Shelton Polk, an ambitious young Austin attorney; and C. N. Avery, Buchanan's former campaign manager, the favorite to win. Avery had the support of Tom Miller, mayor of Austin, the Tenth District's only large city. With Miller's backing he could count on almost enough votes to win the election.

Johnson was faced with a terrible predicament. If he entered the race, the odds would be absurdly against him: he was young—only twenty-eight—and in the district he was unknown and poorly connected. A bad loss would damage his reputation and set him far back on the road to his long-term goal. If he chose not to run, on the other hand, he might wait ten years for another chance. With all this in mind, he threw caution to the winds and entered the race.

Johnson's first step was to call to his side the dozens of young men and women whom he had helped or hired over the years. His campaign strategy was simple: he would separate himself from the other contenders by presenting himself as Roosevelt's staunchest supporter. A vote for Johnson was a vote for the president, the popular architect of the New Deal. And since Johnson could not compete in Austin, he decided to aim his army of volunteers at the countryside, the sparsely populated Hill Country. This was the district's poorest area, a place where candidates rarely ventured. Johnson wanted to meet every last farmer and sharecropper, shake every possible hand, win the votes of people who had never voted before. It was the strategy of a desperate man who recognized that this was his best and only chance for victory.

One of Johnson's most loyal followers was Carroll Keach, who would serve as his chauffeur. Together the two men drove every square mile of the Hill Country, tracing every dirt path and cow trail. Spotting some out-of-the-way farmhouse, Johnson would get out of the car, walk to the door, introduce himself to the startled inhabitants, listen patiently to their problems, then leave with a hearty handshake and a gentle plea for their vote. Convening meetings in dusty towns consisting mainly of a church and a gas station, he would deliver his speech, then mingle with the audience and spend at least a few minutes with everyone present. He had an incredible memory for faces and names: if he happened to meet the same person twice, he could recall everything he or she had said the first time around, and he often impressed strangers by knowing someone who knew them. He listened intensely and was always careful to leave people with the feeling that they would see him again, and that if he won they would finally have someone looking out for their interests in Washington. In bars, grocery stores, and gas stations all through the Hill Country, he would talk with the locals as if he had nothing else to do. On leaving he would make sure to buy something—candy, groceries, gasoline—a gesture they greatly appreciated. He had the gift of creating a connection.

As the race ran on, Johnson went days without sleep, his voice turning hoarse, his eyes drooping. As Keach drove the length of the district, he would listen in amazement as the exhausted candidate in the car muttered to himself about the people he had just met, the impression he had made, what he could have done better. Johnson never wanted to seem desperate or patronizing. It was that last handshake and look in the eye that mattered.

The polls were deceptive: they continued to show Johnson behind, but he knew he had won votes that no poll would register. And in any case he was slowly catching up—by the last week he had crept into third place. Now, suddenly, the other candidates took notice. The election turned nasty: Johnson was attacked for his youth, for his blind support of Roosevelt, for anything that could be dug up. Trying to win a few votes in Austin, Johnson came up against the political machine of Mayor Miller, who disliked him and did everything possible to sabotage his campaign. Undeterred, Johnson personally visited the mayor several times in that last week to broker some kind of truce. But Miller saw through his charm. His personal appeal might have won over the district's poorest voters, but the other candidates saw a different side of him: he was ruthless and capable of slinging mud. As he rose in the polls, he made more and more enemies.

*the fateful moment passes, and conflict continues—to the common damage. Rarely does a continuation serve any good purpose where the two parties are bound to go on living under the same roof. This applies even more to modern war than to domestic conflict, since the industrialization of nations has made their fortunes inseparable.*

*STRATEGY*, B. H. LIDDELL HART, 1954

*If you concentrate exclusively on victory, with no thought for the after-effect, you may be too exhausted to profit by the peace, while it is almost certain that the peace will be a bad one, containing the germs of another war. This is a lesson supported by abundant experience.*

*STRATEGY*, B. H. LIDDELL HART, 1954

*It is even possible that the attacker, reinforced by the psychological forces peculiar to attack, will in spite of his exhaustion find it less difficult to go on than to stop—like a horse pulling a load uphill. We believe that this demonstrates without inconsistency how an attacker can overshoot the point at which, if he stopped and assumed the defensive, there would still be a chance of success—that is, of equilibrium. It is therefore important to calculate this point correctly when planning the campaign. An attacker may otherwise take on more than he can manage and, as it were, get into debt; a defender must be able to recognize this error if the enemy commits it, and exploit it to the full. In reviewing the whole array of factors a general must weigh before making his decision, we must remember that he can gauge the direction and value of the most important ones only by considering numerous other possibilities—some immediate, some remote. He must* guess, *so to speak: guess whether the first shock of battle will steel the enemy's resolve and stiffen his resistance, or whether, like a Bologna flask, it will shatter as soon as its surface is scratched; guess the extent of debilitation and paralysis that the drying-up of the particular sources of*

On Election Day, Johnson pulled off one of the greatest upsets in American political history, outdistancing his nearest rival by three thousand votes. Exhausted by the grueling pace he had set, he was hospitalized, but the day after his victory he was back at work—he had something extremely important to do. From his hospital bed, Johnson dictated letters to his rivals in the race. He congratulated them for running a great campaign; he also described his own victory as a fluke, a vote for Roosevelt more than for himself. Learning that Miller was visiting Washington, Johnson telegraphed his connections in the city to chaperone the mayor and treat him like royalty. As soon as Johnson left the hospital, he paid visits to his rivals and acted with almost embarrassing humility. He even befriended Polk's brother, driving him around town to run errands.

A mere eighteen months later, Johnson had to stand for reelection, and these onetime opponents and bitter enemies suddenly turned into the most fervent Johnson believers, donating money, even campaigning on his behalf. And Mayor Miller, the one man who had hated Johnson the most, now became his strongest supporter and remained so for years to come.

### Interpretation

For most of us, the conclusion of anything—a project, a campaign, an attempt at persuasion—represents a kind of wall: our work is done, and it is time to tally our gains and losses and move on. Lyndon Johnson looked at the world much differently: an ending was not like a wall but more like a door, leading to the next phase or battle. What mattered to him was not gaining a victory but where it left him, how it opened onto the next round. What good would it do to win the election of 1937 if he were thrown out of office eighteen months later? That would be a devastating setback to his dream of the presidency. If, after the election, he had basked in his moment of triumph, he would have sown the seeds of failure in the next election. He had made too many enemies—if they didn't run against him in 1938, they would stir up trouble while he was away in Washington. So Johnson immediately worked to win these men over, whether with charm, with meaningful gestures, or with clever appeals to their self-interest. He kept his eye on the future, and on the kind of success that would keep him moving forward.

Johnson used the same approach in his efforts to win over voters. Instead of trying to persuade people to support him with speeches and fancy words (he was not a good orator anyway), he focused on the feeling he left people with. He knew that persuasion is ultimately a process of the emotions: words can sound nice, but if a politician leaves people suspecting him of being insincere, of merely plugging for votes, they will close off to him and forget him. So Johnson worked to establish an emotional connection with voters, and he would close his conversations with them with a hearty handshake and with a look in his eye, a tremor in his voice, that sealed the bond between them. He left them feeling that they would see him again, and he stirred emotions that would erase any suspicion he might be insincere. The end of the conversation was in fact a kind of beginning, for it stayed in their minds and translated into votes.

Understand: in any venture, your tendency to think in terms of winning or losing, success or failure, is dangerous. Your mind comes to a stop, instead of looking ahead. Emotions dominate the moment: a smug elation in winning, dejection and bitterness in losing. What you need is a more fluid and strategic outlook on life. Nothing ever really ends; how you finish something will influence and even determine what you do next. Some victories are negative—they lead nowhere—and some defeats are positive, working as a wake-up call or lesson. This fluid kind of thinking will force you to put more strategic emphasis on the quality and mood of the ending. It will make you look at your opponents and decide whether you might do better to be generous to them at the end, taking a step back and transforming them into allies, playing on the emotions of the moment. Keeping your eyes on the aftermath of any encounter, you will think more of the feeling you leave people with—a feeling that might translate into a desire to see more of you. By understanding that any victory or defeat is temporary, and that what matters is what you do with them, you will find it easier to keep yourself balanced during the thousands of battles that life entails. The only real ending is death. Everything else is a transition.

*As Yasuda Ukyo said about offering up the last wine cup, only the end of things is important. One's whole life should be like this. When guests are leaving, the mood of being reluctant to say farewell is essential.*

—*Yamamoto Tsunetomo,* Hagakure: The Book of the Samurai *(1659–1720)*

## KEYS TO WARFARE

There are three kinds of people in the world. First, there are the dreamers and talkers, who begin their projects with a burst of enthusiasm. But this burst of energy quickly peters out as they encounter the real world and the hard work needed to bring any project to an end. They are emotional creatures who live mainly in the moment; they easily lose interest as something new grabs their attention. Their lives are littered with half-finished projects, including some that barely make it beyond a daydream.

Then there are those who bring whatever they do to a conclusion, either because they have to or because they can manage the effort. But they cross the finish line with distinctly less enthusiasm and energy than they had starting out. This mars the end of the campaign. Because they are impatient to finish, the ending seems hurried and patched together. And it leaves other people feeling slightly unsatisfied; it is not memorable, does not last, has no resonance.

Both of these types begin each project without a firm idea of how to

*supply and the severing of certain lines of communication will cause in the enemy; guess whether the burning pain of the injury he has been dealt will make the enemy collapse with exhaustion or, like a wounded bull, arouse his rage; guess whether the other powers will be frightened or indignant, and whether and which political alliances will be dissolved or formed. When we realize that he must hit upon all this and much more by means of his discreet judgement, as a marksman hits a target, we must admit that such an accomplishment of the human mind is no small achievement. Thousands of wrong turns running in all directions tempt his perception; and if the range, confusion and complexity of the issues are not enough to overwhelm him, the dangers and responsibilities may.*
*This is why the great majority of generals will prefer to stop well short of their objective rather than risk approaching it too closely, and why those with high courage and an enterprising spirit will often overshoot it and so fail to attain their purpose. Only the man who can achieve great results with limited means has really hit the mark.*

*ON WAR,*
CARL VON CLAUSEWITZ,
1780–1831

*The great prizefighter Jack Dempsey was once asked, "When you are about to hit a man, do you aim for his chin or his nose?" "Neither," Dempsey replied. "I aim for the back of his head."*

QUOTED IN *THE MIND OF WAR*, GRANT T. HAMMOND, 2001

end it. And as the project progresses, inevitably differing from what they had imagined it would be, they become unsure how to get out of it and either give up or simply rush to the end.

The third group comprises those who understand a primary law of power and strategy: the end of something—a project, a campaign, a conversation—has inordinate importance for people. It resonates in the mind. A war can begin with great fanfare and can bring many victories, but if it ends badly, that is all anyone remembers. Knowing the importance and the emotional resonance of the ending of anything, people of the third type understand that the issue is not simply finishing what they have started but finishing it well—with energy, a clear head, and an eye on the afterglow, the way the event will linger in people's minds. These types invariably begin with a clear plan. When setbacks come, as setbacks will, they are able to stay patient and think rationally. They plan not just to the end but past it, to the aftermath. These are the ones who create things that last—a meaningful peace, a memorable work of art, a long and fruitful career.

The reason it is hard to end things well is simple: endings inspire overpowering emotions. At the end of a bitter conflict, we have a deep desire for peace, an impatience for the truce. If the conflict is bringing us victory, we often succumb to delusions of grandeur or are swept by greed and grab for more than we need. If the conflict has been nasty, anger moves us to finish with a violent, punitive strike. If we lose, we are left with a burning desire for revenge. Emotions like these can ruin all of our prior good work. There is in fact nothing harder in the realm of strategy than keeping our head on straight all the way to the end and past the end—yet nothing is more necessary.

Napoleon Bonaparte was perhaps the greatest general that ever lived. His strategies were marvels of combined flexibility and detail, and he planned all the way to the end. But after defeating the Austrians at Austerlitz and then the Prussians at Jena-Auerstadt—his two greatest victories—he imposed on these nations harsh terms intended to make them weakened satellites of France. Accordingly, in the years after the treaties, both countries harbored a powerful desire for revenge. They secretly built up their armies and waited for the day when Napoleon would be vulnerable. That moment came after his disastrous retreat from Russia in 1812, when they pounced on him with horrible fervor.

Napoleon had allowed petty emotion—the desire to humiliate, revenge himself, and force obedience—to infect his strategy. Had he stayed focused on his long-term interests, he would have known that it was better to weaken Prussia and Austria psychologically rather than physically—to seduce them with apparently generous terms, transforming them into devoted allies instead of resentful satellites. Many in Prussia had initially seen Napoleon as a great liberator. Had he only kept Prussia as a happy ally, he would have survived the debacle in Russia and there would have been no Waterloo.

Learn the lesson well: brilliant plans and piled-up conquests are not enough. You can become the victim of your own success, letting victory seduce you into going too far, creating hard-bitten enemies, winning the battle but losing the political game after it. What you need is a strategic third eye: the ability to stay focused on the future while operating in the present and ending your actions in a way that will serve your interests for the next round of war. This third eye will help you counteract the emotions that can insidiously infect your clever strategies, particularly anger and the desire for revenge.

*Victory seems to have been achieved. There remains merely a remnant of the evil resolutely to be eradicated as the time demands. Everything looks easy. Just there, however, lies the danger. If we are not on guard, evil will succeed in escaping by means of concealment, and when it has eluded us new misfortunes will develop from the remaining seeds, for evil does not die easily.*

*THE I CHING*, CHINA, CIRCA EIGHTH CENTURY B.C.

The critical question in war is knowing when to stop, when to make your exit and come to terms. Stop too soon and you lose whatever you might have gained by advancing; you allow too little time for the conflict to show you where it is heading. Stop too late and you sacrifice your gains by exhausting yourself, grabbing more than you can handle, creating an angry and vengeful enemy. The great philosopher of war Carl von Clausewitz analyzed this problem, discussing what he called "the culminating point of victory"—the optimum moment to end the war. To recognize the culminating point of victory, you must know your own resources, how much you can handle, the morale of your soldiers, any signs of a slackening effort. Fail to recognize that moment, keep fighting past it, and you bring on yourself all kinds of unwanted consequences: exhaustion, escalating cycles of violence, and worse.

At the turn of the twentieth century, the Japanese watched as Russia made advances into China and Korea. In 1904, hoping to stem Russian expansion, they launched a surprise attack on the Russian-held town of Port Arthur, on the coast of Manchuria. Since they were clearly the smaller country and had fewer military resources, they hoped that a quick offensive would work in their favor. The strategy—the brainchild of Baron Gentaro Kodama, vice chief of Japan's general staff—was effective: by stealing the initiative, the Japanese were able to bottle up the Russian fleet at Port Arthur while they landed armies in Korea. That allowed them to defeat the Russians in key battles on land and at sea. Momentum was clearly on their side.

In April 1905, however, Kodama began to see great danger in his own success. Japan's manpower and resources were limited; Russia's were vast. Kodama convinced the Japanese leaders to consolidate the gains they had made and sue for peace. The Treaty of Portsmouth, signed later that year, granted Russia more-than-generous terms, but Japan solidified its position: the Russians moved out of Manchuria and Korea and left Port Arthur to Japan. Had the Japanese been carried along by their momentum, they would surely have passed the culminating point of victory and had all their gains wiped out by the inevitable counterattack.

On the other side of the scale, the Americans ended the Gulf War of 1991 too soon, allowing much of the Iraqi army to escape its encirclement. That left Saddam Hussein still strong enough to brutally put

*CENTCOM's lightning war [Desert Storm] was over. It had been billed as a 100-hour blitz, but three years later it was still an unfinished war. Recalled Gordon Brown, the foreign service officer who served as Schwarzkopf's chief foreign policy advisor at CENTCOM, "We never did have a plan to terminate the war."*

*THE GENERAL'S WAR: THE INSIDE STORY OF THE CONFLICT IN THE GULF,* MICHAEL R. GORDON AND GENERAL BERNARD E. TRAINOR, 1995

down the Shiite and Kurdish uprisings that erupted after his defeat in Kuwait and to hang on to power. The allied forces were held back from completing the victory by their desire not to appear to be beating up on an Arab nation and by the fear of a power vacuum in Iraq. Their failure to finish led to far greater violence in the long run.

Imagine that everything you do has a moment of perfection and fruition. Your goal is to end your project there, at such a peak. Succumb to tiredness, boredom, or impatience for the end and you fall short of that peak. Greed and delusions of grandeur will make you go too far. To conclude at this moment of perfection, you must have the clearest possible sense of your goals, of what you really want. You must also command an in-depth knowledge of your resources—how far can you practicably go? This kind of awareness will give you an intuitive feel for the culminating point.

Endings in purely social relationships demand a sense of the culminating point as much as those in war. A conversation or story that goes on too long always ends badly. Overstaying your welcome, boring people with your presence, is the deepest failing: you should leave them wanting more of you, not less. You can accomplish this by bringing the conversation or encounter to an end a moment before the other side expects it. Leave too soon and you may seem timid or rude, but do your departure right, at the peak of enjoyment and liveliness (the culminating point), and you create a devastatingly positive afterglow. People will still be thinking of you long after you are gone. In general, it is always best to end with energy and flair, on a high note.

Victory and defeat are what you make of them; it is how you deal with them that matters. Since defeat is inevitable in life, you must master the art of losing well and strategically. First, think of your own mental outlook, how you absorb defeat psychologically. See it as a temporary setback, something to wake you up and teach you a lesson, and even as you lose, you end on a high note and with an edge: you are mentally prepared to go on the offensive in the next round. So often, those who have success become soft and imprudent; you must welcome defeat as a way to make yourself stronger.

Second, you must see any defeat as a way to demonstrate something positive about yourself and your character to other people. This means standing tall, not showing signs of bitterness or becoming defensive. Early in his term as president, John F. Kennedy embroiled the country in the Bay of Pigs fiasco, a failed invasion of Cuba. While he accepted full responsibility for the debacle, he did not overdo his apologies; instead he went to work on correcting the mistake, making sure it would not happen again. He kept his composure, showing remorse but also strength. In doing so he won public and political support that helped him immensely in his future fights.

Third, if you see that defeat is inevitable, it is often best to go down swinging. That way you end on a high note even as you lose. This helps to rally the troops, giving them hope for the future. At the Battle of the Alamo in 1836, every last American fighting the Mexican army died—but they died heroically, refusing to surrender. The battle became a rallying cry—"Remember the Alamo!"—and an inspired American force under Sam Houston finally defeated the Mexicans for good. You do not have to experience physical martyrdom, but a display of heroism and energy makes defeat into a moral victory that will soon enough translate into a concrete one. Planting the seeds of future victory in present defeat is strategic brilliance of the highest order.

Finally, since any ending is a kind of beginning of the next phase, it is often wise strategy to end on an ambivalent note. If you are reconciling with an enemy after a fight, subtly hint that you still have a residue of doubt—that the other side must still prove itself to you. When a campaign or project comes to an end, leave people feeling that they cannot foresee what you will do next—keep them in suspense, toying with their attention. By ending on a note of mystery and ambiguity—a mixed signal, an insinuating comment, a touch of doubt—you gain the upper hand for the next round in a most subtle and insidious fashion.

*Knowing how to end. Masters of the first rank are recognized by the fact that in matters great and small they know how to find an end perfectly, be it at the end of a melody or a thought; of a tragedy's fifth act or an act of state. The best of the second rank always get restless toward the end, and do not fall into the sea with such proud and calm balance as do, for example, the mountains at Portofino—where the bay of Genoa finishes its melody.*

*The Gay Science*, Friedrich Nietzsche, 1882

**Image:**
The Sun. When it finishes its course and sets below the horizon, it leaves behind a brilliant and memorable afterglow. Its return is always desired.

**Authority:** To conquer is nothing. One must profit from one's success. —*Napoleon Bonaparte (1769–1821)*

REVERSAL

There can be no value in ending anything badly. There is no reversal.

PART

# V

# UNCONVENTIONAL (DIRTY) WARFARE

A general fighting a war must constantly search for an advantage over the opponent. The greatest advantage comes from the element of surprise, from hitting enemies with strategies that are novel, outside their experience, completely unconventional. It is in the nature of war, however, that over time any strategy with any possible application will be tried and tested, so that the search for the new and unconventional has an innate tendency to become more and more extreme. At the same time, moral and ethical codes that governed warfare for centuries have gradually loosened. These two effects dovetail into what we today call "dirty war," where anything goes, down to the killing of thousands of unwarned civilians. Dirty war is political, deceptive, and supremely manipulative. Often the last recourse of the weak and desperate, it uses any means available to level the playing field.

The dynamic of the dirty has filtered into society and the culture at large. Whether in politics, business, or society, the way to defeat your opponents is to surprise them, to come at them from an unexpected angle. And the increasing pressures of these daily wars make dirty strategies inevitable. People go underground: they seem nice and decent but use slippery, devious methods behind the scenes.

The unconventional has its own logic that you must understand. First, nothing stays new for long. Those who depend on novelty must constantly come up with some fresh idea that goes against the orthodoxies of the time. Second, people who use unconventional methods are very hard to fight. The classic, direct route—the use of force and strength—does not work. You must use indirect methods to combat indirection, fight fire with fire, even at the cost of going dirty yourself. To try to stay clean out of a sense of morality is to risk defeat.

The chapters in this section will initiate you into the various forms of the unorthodox. Some of these are strictly unconventional: deceiving your opponents and working against their expectations. Others are more political and slippery: making morality a strategic weapon, applying the arts of guerrilla warfare to daily life, mastering the insidious forms of passive aggression. And some are unapologetically dirty: destroying the enemy from within, inflicting terror and panic. These chapters are designed to give you a greater understanding of the diabolical psychology involved in each strategy, helping to arm you with the proper defense.

# 23

# WEAVE A SEAMLESS BLEND OF FACT AND FICTION

## MISPERCEPTION STRATEGIES

*Since no creature can survive without the ability to see or sense what is going on around it, you must make it hard for your enemies to know what is going on around* them, *including what you are doing. Disturb their focus and you weaken their strategic powers. People's perceptions are filtered through their emotions; they tend to interpret the world according to what they want to see. Feed their expectations, manufacture a reality to match their desires, and they will fool themselves. The best deceptions are based on ambiguity, mixing fact and fiction so that the one cannot be disentangled from the other. Control people's perceptions of reality and you control them.*

*In war-time, truth is so precious that she should always be attended by a bodyguard of lies.*

WINSTON CHURCHILL, 1874–1965

## THE FALSE MIRROR

On November 3, 1943, Adolf Hitler had a document distributed to his top generals: Directive 51, which discussed his conviction that the Allies would invade France the following year and explained how to beat them. For years Hitler had depended on a kind of intuition in making his most important strategic decisions, and time and again his instincts had been right; the Allies had tried before to make him believe that an invasion of France was imminent, but each time Hitler had seen through the deception. This time he was not only sure that the invasion was coming, he felt he knew exactly where it would come: the Pas de Calais, the region of France along the English Channel that was the country's closest point to Britain.

*Dudley Clarke was always clear—and a little later it will be shewn to have been a pity that others were not equally so—that you can never, by deception, persuade an enemy of anything not according with his own expectations, which usually are not far removed from his hopes. It is only by using your knowledge of them that you are able to hypnotize him, not just into thinking, but doing what you want.*

*MASTER OF DECEPTION,* DAVID MURE, 1980

The Pas de Calais had a number of major ports, and the Allies would need a port to land their troops. The region was also where Hitler planned to place his V-1 and V-2 rockets, soon to be operational; with these jet-propelled unmanned missiles so close to London, he could bomb Britain into submission. The English knew he was putting missiles there, and that provided them yet another reason to invade France at the Pas de Calais, before Hitler could begin his bombing campaign.

In Directive 51, Hitler warned his commanders to expect the Allies to wage a major deception campaign to cloak the time and place of the invasion. The Germans had to see through these deceptions and repel the landing, and despite recent setbacks in the German war effort, Hitler felt supremely confident they could. Several years earlier he had commissioned the construction of the Atlantic Wall, a line of forts up and down the coast from France to Norway, and he had over 10 million soldiers at his disposal, a million of them in France alone. The German armaments industry was churning out ever more and better weapons. Hitler also controlled most of Europe, giving him enormous resources and endless options for moving his troops here and there.

*Themistocles therefore had two urgent and simultaneous problems to solve. He must take* effective *action, not only to block any projected withdrawal by the Peloponnesian contingents, but also to ensure that they fought where and when he planned that they should; and he must somehow tempt Xerxes into making the one move which might lead to a Greek victory—that is, ordering his fleet to attack* in the Salamis channel. . . .

Finally, to invade France the Allies would need a massive armada, which, once assembled, would be impossible to conceal. Hitler had infiltrated agents into all levels of the British military, who supplied him with excellent intelligence—they would forward to him the time and location of the invasion. The Allies would not surprise him. And once he had defeated them on the shores of France, England would have to sue for peace; Roosevelt would certainly lose the upcoming U.S. presidential election. Hitler could then concentrate his entire army against the Soviet Union and finally defeat it. In truth, the invasion of France was the opportunity he craved to turn the war around.

Hitler's commander in Western Europe was Field Marshal Gerd von Runstedt, Germany's most respected general. To further solidify the defensive position in France, Hitler made General Erwin Rommel the commander of the forces along the French coast. Rommel proceeded to make improvements in the Atlantic Wall, turning it into a "devil's garden" of minefields and fire zones. Rommel and Runstedt also asked for more troops to ensure that the Germans could repel the Allies at the water's edge. But the Führer denied their request.

Hitler had lately come to mistrust his top staff. In the past few years, he had survived several assassination attempts that had clearly originated among his officers. His generals were increasingly arguing with his strategies, and in his mind they had botched several battles in the Russian campaign; he saw many of them as incompetents or traitors. He began to spend less time with his officers and more time holed up in his Bavarian mountain retreat at Berchtesgaden, with his mistress, Eva Braun, and his beloved dog, Biondi. There he pored over maps and intelligence reports, determined to make the important decisions himself and to manage the entire war effort more directly.

This caused a change in his way of thinking: instead of making quick, intuitive choices, he was trying to foresee every possibility and was taking longer to make up his mind. Now he thought Rommel and Runstedt—in their request for more troops to be transferred to France—were being overly cautious and even panicky. He alone would have to foil the Allied invasion; it was up to him to see through his generals' weaknesses and the enemy's deceptions. The only downside to this was that his workload had increased tenfold, and he was more tired than ever. At night he took sleeping pills, by day whatever he could get his hands on to keep him alert.

Early in 1944 key information arrived in Hitler's hands: a German agent in Turkey stole classified documents confirming that the Allies would invade France that year. The documents also indicated plans for an imminent invasion of the Balkans. Hitler was particularly sensitive to any threats to the Balkans, a valuable source of resources for Germany; a loss there would be devastating. The threat of such an attack made it impossible to transfer troops from there to France. Hitler's agents in England also discovered plans to invade Norway, and here Hitler actually reinforced his troops to ward off the threat.

By April, as Hitler pored over intelligence reports, he began to feel increasingly excited: he discerned a pattern in the enemy activity. As he had thought, everything pointed toward an invasion of the Pas de Calais. One sign particularly stood out: indications of an enormous army forming in southeastern England under the command of General George Patton. This army, called FUSAG (First United States Army Group), was clearly positioned for a crossing to the Pas de Calais. Of all the Allies' generals, Hitler feared Patton the most. He had proven his military skill in North Africa and Sicily. He would be the perfect commander for the invasion.

Hitler demanded more information on Patton's army. High-flying reconnaissance planes photographed enormous military camps, docking equipment, thousands of tanks moving through the countryside, a pipeline being built to the coast. When a captured German general who had been imprisoned in England was finally repatriated, he caught

*The device Themistocles finally adopted—what Plutarch calls "his celebrated trick with Sicinnus"—is one of the most enigmatic episodes in all Greek history. Evidence for it goes back as far as Aeschylus's* Persians, *performed only eight years after Salamis. . . . What seems to have happened was this. At some point during the long argument over final strategy, Themistocles, anticipating defeat, slipped away from the conference and sent for his children's tutor, "the most faithful of his slaves," an Asiatic Greek named Sicinnus. This man was given a carefully prepared message, or letter, to deliver to Xerxes, and sent off across the straits in a small boat, probably just before dawn on 19 September. . . . The substance of the message was as follows. Themistocles sent it under his own name, as commander of the Athenian contingent: he had, he told Xerxes, changed sides, and was now ardently desirous of a Persian victory. (No real reason is given for this* volte-face, *though disgust at the attitude of the Peloponnesian contingents would provide a strong enough motive to carry conviction.) The Greek allies were at each other's throats, and would offer no serious opposition—"on the contrary, you will see the pro-Persians amongst them fighting the rest." Furthermore, they were planning a general withdrawal from Salamis under cover of darkness, to be carried out the following night. . . . If Xerxes struck at once, on the divide-and-rule principle, he could forestall such a move. "Attack them and destroy their naval power, while they are still disorganized and before they have joined forces with their land army"* [*Plut.* Them. *12.4*]. *The conquest of the Peloponnese would then become a comparatively simple matter. On the other hand, if Xerxes allowed the various Greek contingents to slip through his fingers and disperse homewards, the war might drag on indefinitely, since he would have to deal with each separate city-state in turn. Sicinnus's arguments impressed the Persian admirals, and they duly passed them on to the Great King himself. Xerxes, we are told, believed the report because it "was in itself plausible"—and also because it was just what he* wanted *to hear: there was trouble brewing in Ionia and the empire, and the sooner this Greek expedition was wound up, the better. Themistocles, always a shrewd judge of human nature, knew very well that after so many days of delay and frustration, the Great King would grasp at anything which seemed to offer a quick solution to his problem.*

THE GRECO-PERSIAN WARS, PETER GREEN, 1996

glimpses of massive activity in the FUSAG area on his trip from his internment camp to London. Agents in Switzerland reported that every map of the Pas de Calais area had been mysteriously bought up. The pieces of a giant puzzle were coming together.

Now only one question remained: when would it happen? As April turned to May, Hitler was deluged with all kinds of conflicting reports, rumors, and sightings. The information was confusing, taxing his strained mind, but two nuggets of intelligence seemed to clarify the picture. First, a German agent in England reported that the Allies would attack Normandy, southeast of the Pas de Calais, between June 5 and 7. But the Germans had strong indications that this man was operating as a double agent, and his report was clearly part of an Allied disinformation campaign. The attack would probably be coming at the end of June or beginning of July, when the weather was generally more predictable. Then, later in May, a series of more reliable German spies spotted Britain's top general, Sir Bernard Montgomery, in Gibraltar and then in Algiers. Montgomery would certainly command a large part of any invading force. The invasion could not be imminent if he was so far away.

On the night of June 5, Hitler pored over the maps. Maybe he was wrong—maybe the plan was for Normandy all along. He had to consider both options; he would not be fooled in what might be the most decisive battle of his life. The British were tricky; he had to keep his forces mobile in case it was Normandy after all. He would not commit himself until he knew for sure. Reading the weather reports for the Channel—stormy that evening—he took his usual sleeping pill and went to bed.

Early the next morning, Hitler woke to startling news: a massive invasion was under way—in southern Normandy. A large armada had left England in the middle of the night, and hundreds of parachutists had been dropped near the Normandy coast. As the day progressed, the reports became more exact: the Allies had landed on the beaches southeast of Cherbourg.

A critical moment had come. If some of the forces stationed in the Pas de Calais were hurried to the beaches of Normandy, the Allies could be pinned and thrown back into the sea. This was the recommendation of Rommel and Runstedt, who anxiously awaited Hitler's approval. But through the night and into the following day, Hitler hesitated. Then, just as he was on the verge of sending reinforcements to Normandy, he received word of increased Allied activity in the FUSAG area. Was Normandy in fact a giant diversion? If he moved his reserves there, would Patton immediately cross the Channel to the Pas de Calais? No, Hitler would wait to see if the attack was real. And so the days went by, with Rommel and Runstedt fuming at his indecision.

After several weeks Hitler finally accepted that Normandy was the real destination. But by then he was too late. The Allies had established a beachhead. In August they broke out of Normandy, sending the Germans into full retreat. To Hitler the disaster was yet another indication of the incompetence of those around him. He had no idea how deeply and decisively he had been fooled.

### Interpretation

In trying to deceive Hitler about the Normandy invasions, the Allies were faced with a problem: not only was the Führer suspicious and wary by nature, he knew of previous attempts to mislead him and knew that the Allies would have to try to deceive him again. How could the Allies possibly disguise the actual goal of a vast armada from a man who had reason to believe they would try to mislead him and was scrutinizing their every move?

Fortunately, British intelligence had been able to provide the planners of the D-Day landings, including Prime Minister Winston Churchill, with information that would prove invaluable to them. First, they knew that Hitler was growing paranoid; he was isolated and overworked, his imagination overheated. He was prone to emotional outbursts, and he was suspicious of everyone and everything. Second, they knew of his belief that the Allies would try to invade the Balkans before France and that the landing site in France would be in the Pas de Calais. He almost seemed to want these invasions to happen, as proof of his superior reasoning powers and foresight.

Fooling Hitler into keeping his forces dispersed across Europe and France would give the Allies a slim margin of time in which to establish a beachhead. The key was to present him a picture, composed of many different kinds of evidence, that would tell him the Allies were doing just what he had thought they would. But this picture could not be made up of all kinds of flashing signs pointing to the Balkans and the Pas de Calais—that would reek of deception. Instead they had to create something that had the weight and feel of reality. It had to be subtle, a mix of banal truths with little falsehoods stitched in. If Hitler saw that in its outlines it supported his expectations, his overactive mind would fill in the rest. This is how the Allies wove such a picture.

By late 1943 the British had secretly identified all of the German agents active in England. The next step was to turn them into unwitting double agents by feeding them false information—about Allied plans for an attack on the Balkans and Norway, say, and the massing of a fictional army—commanded by Patton, the American general Hitler so feared—opposite the Pas de Calais. (This army, FUSAG, existed only in piles of phony paperwork and wireless transmissions that mimicked a normal army.) German agents were allowed to steal FUSAG documents and intercept transmissions—carefully misleading messages but at the same time banal and bureaucratic ones, too banal to be seen as fake. Working with film designers, the Allies built an elaborate set of rubber, plastic, and wood that from German reconnaissance planes would look like an enormous camp of tents, airplanes, and tanks. The German general who saw FUSAG with his own eyes was misled about the direction he was

*At the end of the war, Allied Intelligence Officers discovered in captured files of the German Secret Service the text of two hundred and fifty messages received from agents and other sources before D-Day. Nearly all mentioned July and the Calais sector. One message alone gave the exact date and place of the invasion. It had come from a French colonel in Algiers. The Allies had discovered this officer was working for the Abwehr and he was arrested and subsequently turned round. He too was used to mislead Berlin—used and abused. The Germans were so often deceived by him that they ended by treating all his information as valueless. But they kept in contact, for it is always useful to know what the enemy wants you to believe. Allied Intelligence, with great boldness and truly remarkable perversity, had the colonel announce that the Invasion would take place on the coast of Normandy on the 5th, 6th or 7th June. For the Germans, his message was absolute proof that the invasion was to be on any day* except the 5th, 6th or 7th June, *and on any part of the coast except Normandy.*

THE SECRETS OF D-DAY, GILLES PERRAULT, 1965

*Now Ravana said to himself, "These are all petty weapons. I should really get down to proper business." And he invoked the one called "Maya"—a weapon which created illusions and confused the enemy. With proper incantations and worship, he sent off this weapon and it created an illusion of reviving all the armies and its leaders—Kumbakarna and Indrajit and the others—and bringing them back to the battlefield. Presently Rama found all those who, he thought, were no more, coming on with battle cries and surrounding him. Every man in the enemy's army was again up in arms. They seemed to fall on Rama with victorious cries. This was very confusing and Rama asked Matali, whom he had by now revived, "What is happening now? How are all these coming back? They were dead." Matali explained, "In your original identity you are the creator of illusions in this universe. Please know that Ravana has created phantoms to confuse you. If you make up your mind, you can dispel them immediately." Matali's explanation was a great help. Rama at once invoked a weapon called "Gnana"—which means "wisdom" or "perception." This was a very rare weapon, and he sent it forth. And all the terrifying armies who seemed to have come on in such a great mass suddenly evaporated into thin air.*

*THE RAMAYANA*, VALMIKI, INDIA, CIRCA FOURTH CENTURY B.C.

taking toward London: he had actually passed the real army to the west of FUSAG's supposed site, massing for the invasion of Normandy.

As the date of the invasion drew near, the Allies left clues combining fact and fiction still more intricately. The real time and place of the invasion were planted with an agent whom the Germans completely mistrusted, giving Hitler the feeling that he had seen through a deception when in fact he was staring at the truth. Now, if real information on the timing of the invasion somehow leaked out, Hitler would not know what to believe. The Allies knew that reports on the buying up of Pas de Calais maps in Switzerland would reach Hitler, and this would have its own realistic logic. As for the Montgomery sightings in Gibraltar, little did the German agents know they were seeing a look-alike, a man trained to act like the general. In the end the picture the Allies painted was so real to Hitler that well into July he believed in it, long after D-Day had actually happened. Through such subtle deceptions they had compelled him to keep his forces dispersed—perhaps the decisive factor in the success of the invasion.

In a competitive world, deception is a vital weapon that can give you a constant advantage. You can use it to distract your opponents, send them on goose chases, waste valuable time and resources in defending attacks that never come. But more than likely your concept of deception is wrong. It does not entail elaborate illusions or all sorts of showy distractions. People are too sophisticated to fall for such things. Deception should mirror reality. It can be elaborate, as the British deception around D-Day was, but the effect should be of reality only subtly, slightly altered, not completely transformed.

To mirror reality you must understand its nature. Above all, reality is subjective: we filter events through our emotions and preconceptions, seeing what we want to see. Your false mirror must conform to people's desires and expectations, lulling them to sleep. (If the Allies had wanted to attack the Pas de Calais, as Hitler suspected, and tried to convince Hitler the attack was coming to Normandy, that would have been a great deal harder than playing on his preexisting belief.) Your false mirror must incorporate things that are visibly true. It must seem somewhat banal, like life itself. It can have contradictory elements, as the D-Day deception did; reality is often contradictory. In the end, like an Escher painting, you must blend truth and illusion to the point where they become indistinguishable, and your false mirror is taken for reality.

*What we wish, we readily believe, and what we ourselves think, we imagine others think also.*
*—Julius Caesar (100–44 B.C.)*

## KEYS TO WARFARE

In the early history of warfare, military leaders were faced with the following predicament: The success of any war effort depended on the ability to know as much about the other side—its intentions, its strengths and weaknesses—as possible. But the enemy would never willingly disclose this information. In addition, the enemy often came from an alien culture, with its peculiar ways of thinking and behaving. A general could not really know what was going on in the mind of the opposing general. From the outside the enemy represented something of an impenetrable mystery. And yet, lacking some understanding of the other side, a general would be operating in the dark.

The only solution was to scrutinize the enemy for outward signs of what was going on within. A strategist might count the cooking fires in the enemy camp, for example, and the changes in that number over time; that would show the army's size and whether it was increasing as reserves arrived or decreasing as it was split, or perhaps as soldiers deserted. To see where the army was heading, or whether it was readying for battle, he would look for signs of movement or changes in its formation. He would try to get agents and spies to report on its activities from within. A leader who picked up enough of these signs and deciphered them correctly could piece together a reasonably clear picture.

The leader also knew that just as he was watching the other side, the other side was doing the same with him. In pondering this back-and-forth game of reading appearances, certain enlightened strategists in cultures around the world had a similar epiphany: Why not deliberately distort the signs the enemy was looking at? Why not mislead by playing with appearances? If the enemy is counting our cooking fires, just as we are counting theirs, why not light more fires, or fewer, to create a false impression of our strength? If they are following our army's every move, why not move it in deceptive patterns or send part of it in a direction as a decoy? If the enemy has sent spies and agents into our ranks, why not feed them false information? An enemy that thinks it knows our size and intentions, and is unaware that it has been misled, will act on its false knowledge and commit all kinds of mistakes. It will move its men to fight an enemy that is not there. It will fight with shadows.

Thinking in this way, these ancient strategists created the art of organized deception, an art that would eventually filter beyond warfare into politics and society at large. In essence, military deception is about subtly manipulating and distorting signs of our identity and purpose to control the enemy's vision of reality and get them to act on their misperceptions. It is the art of managing appearances, and it can create a decisive advantage for whichever side uses it better.

In war, where the stakes are so high, there is no moral taint in using deception. It is simply an added weapon to create an advantage, much as some animals use camouflage and other tricks to help them survive. To refuse this weapon is a form of unilateral disarmament, giving the other

*The real impact of such a strategy is the dissipation of resources, the creation of both self-fulfilling and suicidal prophecies, and the destruction of truth and trust. It maximizes confusion and disorder and destroys the organization's resilience, adaptability, core values, and ability to respond. The key to such a strategy, says [Colonel John] Boyd, is less deception (the creation of a false order) and more ambiguity (confusion about reality itself). You want to combine fact and fiction to create ambiguity for an adversary, for the combination creates more problems, requires longer to sort out, and calls more into question than merely inserting false information. As an example, he recalled the story of a group of Germans after the Normandy invasion who had stolen some American uniforms and jeeps. They went around the French countryside changing all the road signs to confuse the allies as they advanced through the area. Soon, the Americans figured out that the directions had been reversed and simply did the opposite of whatever the signs indicated. How much more effective it would have been if the Germans had changed only a portion of the signs, a third to a half, and created even more problems for the Americans. Creating ambiguity about the signs' accuracy and prolonging the time it* side a clearer view of the field—an advantage that can translate into victory. And there is no morality or goodness in losing a war.

We face a similar dynamic in our daily battles in life. We are social creatures, and our happiness, even our survival, depends on our ability to understand what other people are intending and thinking. But because we cannot get inside their heads, we are forced to read the signs in their outward behavior. We ponder their past actions as indications of what they might do in the future. We examine their words, their looks, the tone in their voice, certain actions that seem laden with significance. Everything a person does in the social realm is a sign of some sort. At the same time, we are aware that a thousand pairs of eyes are in turn watching us, reading us, and trying to sense our intentions.

It is a never-ending battle over appearance and perception. If other people can read what we are up to, predict what we are going to do, while we have no clue about them, they have a constant advantage over us that they cannot help but exploit. That is why, in the social realm, we learn from an early age to use deception—we tell others what they want to hear, concealing our real thoughts, hedging with the truth, misleading to make a better impression. Many of these deceptions are entirely unconscious.

Since appearances are critical and deception is inevitable, what you want is to elevate your game—to make your deceptions more conscious and skillful. You need the power to cloak your maneuvers, to keep people off balance by controlling the perceptions they have of you and the signs you give out. In this sense there is a lot you can learn from the military arts of deception, which are based on timeless laws of psychology and are infinitely applicable to the battles of daily life.

To master this art, you must embrace its necessity and find creative pleasure in manipulating appearances—as if you were directing a film. The following are the six main forms of military deception, each with its own advantage.

**The false front.** This is the oldest form of military deception. It originally involved making the enemy believe that one was weaker than in fact was the case. A leader would feign a retreat, say, baiting a trap for the enemy to rush into, luring it into an ambush. This was a favorite tactic of Sun-tzu's. The *appearance* of weakness often brings out people's aggressive side, making them drop strategy and prudence for an emotional and violent attack. When Napoleon found himself outnumbered and in a vulnerable strategic position before the Battle of Austerlitz, he deliberately showed signs of being panicked, indecisive, and scared. The enemy armies abandoned their strong position to attack him and rushed into a trap. It was his greatest victory.

Controlling the front you present to the world is the most critical deceptive skill. People respond most directly to what they see, to what is most visible to their eyes. If you seem clever—if you seem deceptive—their guard will be up and it will be impossible to mislead them. Instead

*would take to discover the problem would have been far more effective than changing all the signs in a consistent fashion.*

THE MIND OF WAR, GRANT T. HAMMOND, 2001

*And the Lord said to Joshua, "Do not fear or be dismayed; take all the fighting men with you, and arise, go up to Ai; see, I have given into your hand the king of Ai, and his people, his city, and his land; and you shall do to Ai and its king as you did to Jericho and its king; only its spoil and its cattle you shall take as booty for yourselves; lay an ambush against the city, behind it." . . . So Joshua arose, and all the fighting men, to go up to Ai; and Joshua chose thirty thousand mighty men of valor, and sent them forth by night. And he commanded them, "Behold, you shall lie in ambush against the city, behind it; do not go very far from the city, but hold yourselves all in readiness; and I, and all the people who are with me, will approach the city. And when they come out against us, as before, we shall flee before them; and they will come out after us, till we have drawn them away from the city; for they will say, 'They are fleeing from us, as before.' So we will flee from them; then you shall rise up from the ambush, and* you need to present a front that does the opposite—disarms suspicions. The best front here is weakness, which will make the other side feel superior to you, so that they either ignore you (and being ignored is very valuable at times) or are baited into an aggressive action at the wrong moment. Once it is too late, once they are committed, they can find out the hard way that you are not so weak after all.

In the battles of daily life, making people think they are better than you are—smarter, stronger, more competent—is often wise. It gives you breathing space to lay your plans, to manipulate. In a variation on this strategy, the front of virtue, honesty, and uprightness is often the perfect cover in a political world. These qualities may not seem weak but serve the same function: they disarm people's suspicions. In that situation, though, it is important not to get caught doing something underhanded. Appearing as a hypocrite will set you far back in the deception game.

In general, as strategists advocated in the days of ancient China, you should present a face to the world that promises the opposite of what you are actually planning. If you are getting ready to attack, seem unprepared for a fight or too comfortable and relaxed to be plotting war. Appear calm and friendly. Doing this will help you gain control over your appearance and sharpen your ability to keep your opponents in the dark.

**The decoy attack.** This is another ruse dating back to ancient times, and it remains perhaps the military's most common deceptive ploy. It began as a solution to a problem: if the enemy knew you were going to attack point A, they would put all their defenses there and make your job too difficult. But to deceive them on that score was not easy: even if before battle you were able to disguise your intentions and fool them out of concentrating their forces at point A, the minute they actually saw your army headed there, they would rush to its defense. The only answer was to march your army toward point B or, better, to send part of your army in that direction while holding troops in reserve for your real objective. The enemy would now have to move some or all of its army to defend point B. Do the same with points C and D and the enemy would have to disperse all over the map.

The key to this tactic is that instead of relying on words or rumors or planted information, the army really moves. It makes a concrete action. The enemy forces cannot afford to guess whether a deception is in the works: if they guess wrong, the consequences are disastrous. They have to move to cover point B, no matter what. It is in any case almost impossible to doubt the reality of actual troop movements, with all the time and energy those involve. So the decoy attack keeps the enemy dispersed and ignorant of your intentions—the ultimate dream of any general.

The decoy attack is also a critical strategy in daily life, where you must retain the power to hide your intentions. To keep people from defending the points you want to attack, you must follow the military model and make real gestures toward a goal that does not interest you.

*seize the city; for the Lord your God will give it into your hand. And when you have taken the city, you shall set the city on fire, doing as the Lord has bidden; see, I have commanded you." . . . . . . And when the king of Ai saw this, he and all his people, the men of the city, made haste and went out early to the descent toward the Arabah to meet Israel in battle; but he did not know that there was an ambush against him behind the city. And Joshua and all Israel made a pretense of being beaten before them, and fled in the direction of the wilderness. So all the people who were in the city were called together to pursue them, and as they pursued Joshua they were drawn away from the city. There was not a man left in Ai or Bethel, who did not go out after Israel; they left the city open, and pursued Israel. . . . And the ambush rose quickly out of their place, and . . . they ran and entered the city and took it; and they made haste to set the city on fire. So when the men of Ai looked back, behold, the smoke of the city went up to heaven; and they had no power to flee this way or that, for the people that fled to the wilderness turned back upon the pursuers. And when Joshua and all Israel saw that the ambush had taken the city, and that the smoke of the city went up, then they turned back and smote the men of Ai.*

JOSHUA 8: 1–9, 14–23

*The principle is also employed in less tortuous circumstances, but with the same purpose of getting an individual to act naturally in a role because, in fact, he does not know that he is playing a false one. For example, take the design of the "Man Who Never Was" operation during World War II—wherein a high-level courier carrying secret papers containing misdirections regarding the Mediterranean invasion was to be washed up on the coast of Spain. After the "Major" was dropped in Spanish waters, the British attaché in Spain was "confidentially" told that papers of great importance had been lost, and that he should discreetly determine whether the courier's briefcase had been recovered. The attaché was thus able to act out his part in the fake-out in a very convincing manner by virtue of the fact that for him it wasn't an act.*

THE SECRETS OF DDAY, GILLES PERRAULT, 1965

You must seem to be investing time and energy to attack that point, as opposed to simply trying to signal the intention with words. Actions carry such weight and seem so real that people will naturally assume that is your real goal. Their attention is distracted from your actual objective; their defenses are dispersed and weakened.

**Camouflage.** The ability to blend into the environment is one of the most terrifying forms of military deception. In modern times Asian armies have proven particularly adept in this art: at the battles of Guadalcanal and Iwo Jima during World War II, American soldiers were astounded at the ability of their Japanese foes to blend into the various terrains of the Pacific theater. By sewing grass, leaves, twigs, and foliage to their uniforms and helmets, the Japanese would merge with the forest—but the forest would incrementally advance, undetected until it was too late. Nor could the Americans pinpoint the Japanese guns, for their barrels were concealed in natural rock crevices or were hidden under removable camouflage covers. The North Vietnamese were equally brilliant at camouflage, reinforcing their skills by the use of tunnels and underground chambers that allowed armed men to pop up seemingly anywhere. Worse, in a different kind of camouflage, they could blend into the civilian population. Preventing your enemies from seeing you until it is too late is a devastating way to control their perceptions.

The camouflage strategy can be applied to daily life in two ways. First, it is always good to be able to blend into the social landscape, to avoid calling attention to yourself unless you choose to do so. When you talk and act like everyone else, mimicking their belief systems, when you blend into the crowd, you make it impossible for people to read anything particular in your behavior. (Appearances are all that count here—dress and talk like a businessman and you must *be* a businessman.) That gives you great room to move and plot without being noticed. Like a grasshopper on a leaf, you cannot be picked from your context—an excellent defense in times of weakness. Second, if you are preparing an attack of some sort and begin by blending into the environment, showing no sign of activity, your attack will seem to come out of nowhere, doubling its power.

**The hypnotic pattern:** According to Machiavelli, human beings naturally tend to think in terms of patterns. They like to see events conforming to their expectations by fitting into a pattern or scheme, for schemes, whatever their actual content, comfort us by suggesting that the chaos of life is predictable. This mental habit offers excellent ground for deception, using a strategy that Machiavelli calls "acclimatization"—deliberately creating some pattern to make your enemies believe that your next action will follow true to form. Having lulled them into complacency, you now have room to work against their expectations, break the pattern, and take them by surprise.

In the Six-Day War of 1967, the Israelis submitted their Arab enemies to a devastating and lightning-fast defeat. In doing so they confirmed all their preexisting military beliefs: the Arabs were undisciplined, their weaponry was outdated, and their strategies were stale. Six years later the Egyptian president Anwar Sadat exploited these prejudices in signaling that his army was in disarray and still humbled by its defeat in 1967, and that he was squabbling with his Soviet patrons. When Egypt and Syria attacked Israel on Yom Kippur in 1973, the Israelis were caught almost totally by surprise. Sadat had tricked them into letting down their guard.

This tactic can be extended indefinitely. Once people feel you have deceived them, they will expect you to mislead them again, but they usually think you'll try something different next time. No one, they will tell themselves, is so stupid as to repeat the exact same trick on the same person. That, of course, is just when to repeat it, following the principle of always working against your enemy's expectations. Remember the example of Edgar Allan Poe's short story "The Purloined Letter": hide something in the most obvious place, because that is where no one will look.

Betrayer's masterpiece.—*To express to a fellow conspirator the grievous suspicion that one is going to be betrayed by him, and to do so at precisely the moment one is oneself engaged in betrayal, is a masterpiece of malice, because it keeps the other occupied with himself and compels him for a time to behave very openly and unsuspiciously, thus giving the actual betrayer full freedom of action.*

HUMAN, ALL TOO HUMAN, FRIEDRICH NIETZSCHE, 1878

**Planted information.** People are much more likely to believe something they see with their own eyes than something they are told. They are more likely to believe something they discover than something pushed at them. If you plant the false information you desire them to have—with third parties, in neutral territory—when they pick up the clues, they have the impression *they* are the ones discovering the truth. The more you can make them dig for their information, the more deeply they will delude themselves.

During World War I, in addition to the infamous standoff on the Western Front, the Germans and the British fought a lesser-known battle for control of East Africa, where both sides had colonies. The man in charge of English intelligence in the area was Colonel Richard Meinhertzhagen, and his main rival on the German side was an educated Arab. Meinhertzhagen's job included feeding the Germans misinformation, and he tried hard to deceive this Arab, but nothing seemed to work—the two men were equals at the game. Finally Meinhertzhagen sent his opponent a letter. He thanked the Arab for his services as a double agent and for the valuable information he had supplied to the British. He enclosed a large sum of money and entrusted the letter's delivery to his most incompetent agent. Sure enough, the Germans captured this agent en route and found the letter. The agent, under torture, assured them that his mission was genuine—because he believed it was; Meinhertzhagen had kept him out of the loop. The agent was not acting, so he was more than believable. The Germans quietly had the Arab shot.

No matter how good a liar you are, when you deceive, it is hard to be completely natural. Your tendency is to try so hard to seem natural and sincere that it stands out and can be read. That is why it is so effective

*Agamemnon had sent Odysseus on a foraging expedition to Thrace, and when he came back empty-handed, Palamedes son of Nauplius upbraided him for his sloth and cowardice. "It was not my fault," cried Odysseus, "that no corn could be found. If Agamemnon had sent you in my stead, you would have had no greater success." Thus challenged, Palamedes set sail at once and presently reappeared with a ship-load of grain. . . .*
*After days of tortuous thought, Odysseus at last hit upon a plan by which he might be revenged on Palamedes; for his honour was wounded. He sent word to Agamemnon: "The*

*gods have warned me in a dream that treachery is afoot: the camp must be moved for a day and a night." When Agamemnon gave immediate orders to have this done, Odysseus secretly buried a sackfull of gold at the place where Palamedes's tent had been pitched. He then forced a Phrygian prisoner to write a letter, as if from Priam to Palamedes, which read: "The gold that I have sent is the price you asked for betraying the Greek camp." Having then ordered the prisoner to hand Palamedes this letter, Odysseus had him killed just outside the camp, before he could deliver it. Next day, when the army returned to the old site, someone found the prisoner's corpse and took the letter to Agamemnon. Palamedes was court-martialled and, when he hotly denied having received gold from Priam or anyone else, Odysseus suggested that his tent should be searched. The gold was discovered, and the whole army stoned Palamedes to death as a traitor.*

THE GREEK MYTHS, VOL. 2, ROBERT GRAVES, 1955

to spread your deceptions through people whom you keep ignorant of the truth—people who believe the lie themselves. When working with double agents of this kind, it is always wise to initially feed them some true information—this will establish the credibility of the intelligence they pass along. After that they will be the perfect conduits for your lies.

**Shadows within shadows.** Deceptive maneuvers are like shadows deliberately cast: the enemy responds to them as if they were solid and real, which in and of itself is a mistake. In a sophisticated, competitive world, however, both sides know the game, and the alert enemy will not necessarily grasp at the shadow you have thrown. So you have to take the art of deception to a level higher, casting shadows *within* shadows, making it impossible for your enemies to distinguish between fact and fiction. You make everything so ambiguous and uncertain, spread so much fog, that even if you are suspected of deceit, it does not matter—the truth cannot be unraveled from the lies, and all their suspicion gives them is torment. Meanwhile, as they strain to figure out what you are up to, they waste valuable time and resources.

During the World War II desert battles in North Africa, the English lieutenant Dudley Clarke ran a campaign to deceive the Germans. One of his tactics was to use props—dummy tanks and artillery—to make it impossible for the Germans to figure out the size and location of the English army. From high-flying reconnaissance aircraft, these dummy weapons would photograph like the real thing. A prop that worked particularly well was the fake airplane made of wood; Clarke dotted bogus landing fields filled up with rows of these around the landscape. At one point a worried officer told him that intelligence had been intercepted revealing that the Germans had figured out a way to distinguish the fake planes from the real ones: they simply looked for the wooden struts holding up the wings of the dummy planes (enlarged photos could reveal this). They would now have to stop using the dummies, said the officer. But Clarke, one of the great geniuses of modern deception, had a better idea: he decided to put struts under the wings of real aircraft as well as phony ones. With the original deception, the Germans were confused but could eventually uncover the truth. Now, however, Clarke took the game to a higher level: the enemy could not distinguish the real from the fake in general, which was even more disconcerting.

If you are trying to mislead your enemies, it is often better to concoct something ambiguous and hard to read, as opposed to an outright deception—that deception can be uncovered and enemies can turn their discovery to their advantage, especially if you think they are still fooled and act under that belief. You are the one doubly deceived. By creating something that is simply ambiguous, though, by making everything blurry, there is no deception to uncover. They are simply lost in a mist of uncertainty, where truth and falsehood, good and bad, all merge into one, and it is impossible to get one's bearings straight.

**Image:** Fog. It makes the shape and color of objects impossible to know. Learn to create enough of it and you free yourself of the enemy's intrusive gaze; you have room to maneuver. You know where you are headed, while the enemy goes astray, deeper and deeper into the fog.

**Authority:** One who is good at combating the enemy fools it with inscrutable moves, confuses it with false intelligence, makes it relax by concealing one's strength, . . . deafens its ears by jumbling one's orders and signals, blinds its eyes by converting one's banners and insignias, . . . confounds its battle plan by providing distorted facts. —*Tou Bi Fu Tan,* A Scholar's Dilettante Remarks on War *(16th century A.D.)*

*Appearance and intention inevitably ensnare people when artfully used, even if people sense that there is an ulterior intention behind the overt appearance. When you set up ploys and opponents fall for them, then you win by letting them act on your ruse. As for those who do not fall for a ploy, when you see they won't fall for the open trap, you have another set. Then even if opponents haven't fallen for your original ploy, in effect they actually have.*

*FAMILY BOOK ON THE ART OF WAR,* YAGYU MUNENORI, 1571–1646

## REVERSAL

To be caught in a deception is dangerous. If you don't know that your cover is blown, now, suddenly, your enemies have more information than you do and you become their tool. If the discovery of your deceit is public, on the other hand, your reputation takes a blow, or worse: the punishments for spying are severe. You must use deception with utmost caution, then, employing the least amount of people as possible, to avoid the inevitable leaks. You should always leave yourself an escape route, a cover story to protect you should you be exposed. Be careful not to fall in love with the power that deception brings; the use of it must always be subordinate to your overall strategy and kept under control. If you become known as a deceiver, try being straightforward and honest for a change. That will confuse people—because they won't know how to read you, your honesty will become a higher form of deception.

# 24

# TAKE THE LINE OF LEAST EXPECTATION

## THE ORDINARY-EXTRAORDINARY STRATEGY

*People expect your behavior to conform to known patterns and conventions. Your task as a strategist is to upset their expectations. Surprise them and chaos and unpredictability—which they try desperately to keep at bay—enter their world, and in the ensuing mental disturbance, their defenses are down and they are vulnerable. First, do something ordinary and conventional to fix their image of you, then hit them with the* extraordinary. *The terror is greater for being so sudden. Never rely on an unorthodox strategy that worked before—it is conventional the second time around. Sometimes the ordinary is extraordinary because it is unexpected.*

## UNCONVENTIONAL WARFARE

Thousands of years ago, military leaders—aware of the incredibly high stakes involved in war—would search high and low for anything that could bring their army an advantage on the battlefield. Some generals who were particularly clever would devise novel troop formations or an innovative use of infantry or cavalry: the newness of the tactic would prevent the enemy from anticipating it. Being unexpected, it would create confusion in the enemy. An army that gained the advantage of surprise in this way could often leverage it into victory on the battlefield and perhaps a string of victories.

The enemy, however, would work hard to come up with a defense against the new strategy, whatever it was, and would often find one quite fast. So what once brought brilliant success and was the epitome of innovation soon no longer worked and in fact became conventional. Furthermore, in the process of working out a defense against a novel strategy, the enemy itself would often be forced to innovate; now it was their turn to introduce something surprising and horribly effective. And so the cycle would go on. War has always been ruthless; nothing stays unconventional for long. It is either innovate or die.

In the eighteenth century, nothing was more startling than the tactics of the Prussian king Frederick the Great. To top Frederick's success, French military theorists devised radical new ideas that were finally tested on the battlefield by Napoleon. In 1806, Napoleon crushed the Prussians—who were still using the once unconventional tactics of Frederick the Great, now grown stale—at the Battle of Jena-Auerstadt. The Prussians were humiliated by their defeat; now it was up to them to innovate. They studied in depth Napoleon's success, adapted his best strategies, and took them further, creating the seeds for the formation of the German General Staff. This new Prussian army played a large role in the defeat of Napoleon at Waterloo and went on to dominate the military scene for decades.

In modern times the constant challenge to top the enemy with something new and unconventional has taken a turn into dirty warfare. Loosening the codes of honor and morality that in the past limited what a general could do (at least to some extent), modern armies have slowly embraced the idea that anything goes. Guerrilla and terrorist tactics have been known since ancient times; now they have become not only more common but more strategic and refined. Propaganda, disinformation, psychological warfare, deception, and political means of waging war have all become active ingredients in any unconventional strategy. A counterstrategy usually develops to deal with the latest in dirty warfare, but it often involves falling to the enemy's level, fighting fire with fire. The dirty enemy adapts by sinking to a dirtier level still, creating a downward spiral.

This dynamic is particularly intense in warfare but it permeates every aspect of human activity. If you are in politics and business and your opponents or competitors come up with a novel strategy, you must adapt it for your own purposes or, better, top it. Their once new tactic becomes conventional and ultimately useless. Our world is so fiercely competitive that one side will almost always end up resorting to something dirty, something outside earlier codes of accepted behavior. Ignore this spiral out of a sense of morality or pride and you put yourself at a severe disadvantage; you are called to respond—in all likelihood to fight a little dirty yourself.

The spiral dominates not just politics or business but culture as well, with its desperate search for the shocking and novel to gain attention and win momentary acclaim. Anything goes. The speed of the process has grown exponentially with time; what was unconventional in the arts a few years ago now seems unbearably trite and the height of conformity.

What we consider unconventional has changed over the years, but the laws that make unconventionality effective, being based on elemental psychology, are timeless. And these immutable laws are revealed in the history of warfare. Almost twenty-five hundred years ago, the great Chinese strategist Sun-tzu expressed their essence in his discussion of ordinary and extraordinary means; his analysis is as relevant to modern politics and culture as it is to warfare, whether clean or dirty. And once you understand the essence of unconventional warfare, you will be able to use it in your daily life.

Unconventional warfare has four main principles, as gleaned from the great practitioners of the art.

*Everything which the enemy least expects will succeed the best. If he relies for security on a chain of mountains that he believes impracticable, and you pass these mountains by roads unknown to him, he is confused to start with, and if you press him he will not have time to recover from his consternation. In the same way, if he places himself behind a river to defend the crossing and you find some ford above or below on which to cross unknown to him, this surprise will derange and confuse him. . . .*

FREDERICK THE GREAT, 1712–86

**Work outside the enemy's experience.** Principles of war are based on precedent: a kind of canon of strategies and counterstrategies develops over the centuries, and since war is so dangerously chaotic, strategists come to rely on these principles for lack of anything else. They filter what's happening now through what happened in the past. The armies that have shaken the world, though, have always found a way to operate outside the canon, and thus outside the enemy's experience. This ability imposes chaos and disorder on the enemy, which cannot orient itself to novelty and collapses in the process.

Your task as a strategist is to know your enemies well, then use your knowledge to contrive a strategy that goes outside their experience. What they might have read or heard about matters less than their personal experience, which dominates their emotional lives and determines their responses. When the Germans invaded France in 1940, the French had secondhand knowledge of their blitzkrieg style of warfare from their invasion of Poland the year before but had never experienced it personally and were overwhelmed. Once a strategy is used and is no longer outside your enemy's experience, though, it will not have the same effect if repeated.

**Unfold the extraordinary out of the ordinary.** To Sun-tzu and the ancient Chinese, doing something extraordinary had little effect without a

*Make a false move, not to pass it for a genuine one but to transform it into a genuine one after the enemy has been convinced of its falsity.*

*The Wiles of War: 36 Military Strategies from Ancient China*, translated by Sun Haichen, 1991

setup of something ordinary. You had to mix the two—to fix your opponents' expectations with some banal, ordinary maneuver, a comfortable pattern that they would then expect you to follow. With the enemy sufficiently mesmerized, you would then hit it with the extraordinary, a show of stunning force from an entirely new angle. Framed by the predictable, the blow would have double the impact.

The unconventional maneuver that confused enemies, though, would have become conventional the second or third time around. So the wily general might then go back to the ordinary strategy that he had used earlier to fix their attention and use it for his main attack, for that would be the last thing the enemy would expect. And so the ordinary and the extraordinary are effective only if they play off each other in a constant spiraling manner. This applies to culture as much as to war: to gain attention with some cultural product, you have to create something new, but something with no reference to ordinary life is not in fact unconventional, but merely strange. What is truly shocking and extraordinary unfolds out of the ordinary. The intertwining of the ordinary and extraordinary is the very definition of surrealism.

**Act crazy like a fox.** Despite appearances, a lot of disorder and irrationality lurks beneath the surface of society and individuals. That is why we so desperately strain to maintain order and why people acting irrationally can be terrifying: they are demonstrating that they have lost the walls we build to keep out the irrational. We cannot predict what they will do next, and we tend to give them a wide berth—it is not worth mixing it up with such sources of chaos. On the other hand, these people can also inspire a kind of awe and respect, for secretly we all desire access to the irrational seas churning inside us. In ancient times the insane were seen as divinely possessed; a residue of that attitude survives. The greatest generals have all had a touch of divine, strategic madness.

The secret is to keep this streak under control. Upon occasion you allow yourself to operate in a way that is deliberately irrational, but less is more—do this too much and you may be locked up. You will in any case frighten people more by showing an occasional flash of insanity, just enough to keep everyone off balance and wondering what will come next. As an alternative, act somewhat randomly, as if what you did were determined by a roll of the dice. Randomness is thoroughly disturbing to humans. Think of this behavior as a kind of therapy—a chance to indulge occasionally in the irrational, as a relief from the oppressive need to always seem normal.

**Keep the wheels in constant motion.** The unconventional is generally the province of the young, who are not comfortable with conventions and take great pleasure in flouting them. The danger is that as we age, we need more comfort and predictability and lose our taste for the unorthodox. This is how Napoleon declined as a strategist: he came to rely more on the size of his army and on its superiority in weapons than on novel strategies and fluid maneuvers. He lost his taste for the spirit of strategy and succumbed to the growing weight of his accumulating years. You must fight the psychological aging process even more than the physical one, for a mind full of stratagems, tricks, and fluid maneuvers will keep you young. Make a point of breaking the habits you have developed, of acting in a way that is contrary to how you have operated in the past; practice a kind of unconventional warfare on your own mind. Keep the wheels turning and churning the soil so that nothing settles and clumps into the conventional.

*No one is so brave that he is not disturbed by something unexpected.*
*—Julius Caesar (100–44 B.C.)*

## HISTORICAL EXAMPLES

**1.** In 219 B.C., Rome decided it had had enough of the Carthaginians, who had been stirring up trouble in Spain, where both city-states had valuable colonies. The Romans declared war on Carthage and prepared to send an army to Spain, where the enemy forces were led by the twenty-eight-year-old general Hannibal. Before the Romans could reach Hannibal, though, they received the startling news that he was coming to them—he had already marched east, crossing the most treacherous part of the Alps into northern Italy. Because Rome had never imagined that an enemy would attack from that direction, there were no garrisons in the area, and Hannibal's march south toward Rome was unimpeded.

His army was relatively small; only some 26,000 soldiers had survived the crossing of the Alps. The Romans and their allies could field an army of close to 750,000 men; their legions were the most disciplined and feared fighters in the world, and they had already defeated Carthage in the First Punic War, twenty-odd years earlier. But an alien army marching into Italy was a novel surprise, and it stirred the rawest emotions. They had to teach these barbarians a lesson for their brazen invasion.

Legions were quickly dispatched to the north to destroy Hannibal. After a few skirmishes, an army under the Roman consul Sempronius Longus prepared to meet the Carthaginians in direct battle near the river Trebia. Sempronius burned with both hatred and ambition: he wanted to crush Hannibal and also to be seen as the savior of Rome. But Hannibal was acting strangely. His light cavalry would cross the river as if to attack the Romans, then retreat back: Were the Carthaginians afraid? Were they ready to make only minor raids and sorties? Finally Sempronius had had enough and went in pursuit. To make sure he had sufficient forces to defeat the enemy, he brought his entire army across the freezing-cold river (it was wintertime), all of which took hours and was exhausting. Finally, however, the two armies met just to the west of the river.

*It is assumed that Alexander encamped at Haranpur; opposite him on the eastern back of the Hydaspes was Porus, who was seen to have with him a large number of elephants. . . . . . . Because all fords were held by pickets and elephants, Alexander realized that his horses could neither be swum nor rafted across the river, because they would not face the trumpeting of the elephants and would become frantic when in the water or on their rafts. He resorted to a series of feints. While small parties were dispatched to reconnoitre all possible crossing places, he divided his army into columns, which he marched up and down the river as if he sought a place of crossing. Then, when shortly before the summer solstice the rains set in and the river became swollen, he had corn conveyed from all quarters to his camp so that Porus might believe that he had resolved to remain where he was until the dry weather. In the meantime he reconnoitred the river with his ships and ordered tent skins to be stuffed with hay and converted into rafts. Yet, as Arrian writes, "all the time he was waiting in ambush to see whether by rapidity of movement he could not steal a passage anywhere without being observed." At length, and we may be certain after a close personal reconnaissance, Alexander resolved to make the attempt at the headland*

At first, as Sempronius had expected, his tough, disciplined legions fared well against the Carthaginians. But on one side the Roman lines were made up of Gallic tribesmen fighting for the Romans, and here, suddenly, the Carthaginians unleashed a group of elephants ridden by archers. The tribesmen had never seen such beasts; they panicked and fell into a chaotic retreat. At the same time, as if out of nowhere, some 2,000 Carthaginians, hidden in dense vegetation near the river, fell on the Romans' rear. The Romans fought bravely to get out of the trap that Hannibal had laid for them, but thousands of them drowned in the frigid waters of the Trebia.

The battle was a disaster, and back in Rome emotions turned from outrage to anxiety. Legions were quickly dispatched to block the most accessible passes in the Apennines, the mountains that run across central Italy, but once again Hannibal defied expectations: he crossed the Apennines at their most unlikely, most inhospitable point, one that no army had ever passed through before because of the treacherous marshes on the other side. But after four days of trudging through soft mud, Hannibal brought the Carthaginians to safe ground. Then, in yet another clever ambush, he defeated a Roman army at Lake Trasimene, in present-day Umbria. Now his path to Rome was clear. In a state of near panic, the Roman republic resorted to the ancient tradition of appointing a dictator to lead them through the crisis. Their new leader, Fabius Maximus, quickly built up the city's walls and enlarged the Roman army, then watched perplexed as Hannibal bypassed Rome and headed south into Apulia, the most fertile part of Italy, and began to devastate the countryside.

Determined first and foremost to protect Rome, Fabius came up with a novel strategy: he would post his legions in mountainous areas where Hannibal's cavalry would be harmless, and he would harass the Carthaginians in a guerrilla-style campaign, denying them supplies and isolating them in their position so far from home. Avoiding direct battle with their formidable leader at all costs, he would defeat them by exhausting them. But many Romans saw Fabius's strategy as disgraceful and unmanly. Worse, as Hannibal continued to raid the countryside, he hit none of Fabius's many properties, making it seem as if the two were in cahoots. Fabius became more and more unpopular.

Having razed Apulia, Hannibal entered a fertile plain in Campania, to Rome's south—terrain that Fabius knew well. Finally deciding he had to act or be thrown out of power, the dictator devised a trap: he stationed Roman armies at all the exit points from the plain, each army close enough to support the other. But Hannibal had entered Campania through the eastern mountain pass of Allifae, and Fabius had noticed that he never left by the same route he entered. Although Fabius kept a sufficiently large Roman garrison at Allifae just in case, he reinforced the other passes in greater numbers. The beast, he thought, was caged. Eventually Hannibal's supplies would run out, and he would be forced to try to break through. Fabius would wait.

In the weeks to come, Hannibal sent his cavalry north, perhaps trying to break out in that direction. He also plundered the richest farms in the area. Fabius saw through his tricks: he was trying to bait the Romans into a battle of his choice. But Fabius was determined to fight on his own terms, and only when the enemy tried to retreat from the trap. Anyway, he knew Hannibal would try to break through to the east, the only direction that afforded him a clean break, into country the Romans did not control.

One night the Roman soldiers guarding the pass at Allifae saw sights and heard sounds that made them think they were losing their minds: an enormous army, signaled by thousands of torches, seemed to be heading up the pass, covering its slopes, accompanied by loud bellowing sounds as if possessed by some evil demon. The army seemed irresistible—far larger than the maximum estimate of Hannibal's strength. Afraid that it would climb above them and surround them, the Romans fled from their garrison, abandoning the pass, too scared even to look behind them. And a few hours later, Hannibal's army came through, escaping from Fabius's cordon.

No Roman leader could figure out what Hannibal had conjured up on the slopes that night—and by the following year Fabius was out of power. The consul Terentius Varro burned to avenge the disgrace of Allifae. The Carthaginians were encamped near Cannae, in southeastern Italy not far from modern Bari. Varro marched to face them there, and as the two armies arrayed themselves in ranks to meet in battle, he could only have felt supremely confident: the terrain was clear, the enemy was in full view, there could be no hidden armies or last-minute tricks—and the Romans outnumbered the Carthaginians by two to one.

The battle began. At first the Romans seemed to have the edge: the center of the Carthaginian line proved surprisingly weak and easily gave ground. The Romans attacked this center with force, hoping to break through and indeed pushing forward—when, to their shock and horror, they looked behind them and saw the two outer ends of the Carthaginian lines moving around to encircle them. They were trapped in a lethal embrace; it was a slaughter. Cannae would go down in history as Rome's most devastating and humiliating defeat.

The war with Hannibal would drag on for years. Carthage never sent him the reinforcements that might have turned the tide, and the much larger and more powerful Roman army was able to recover from its many defeats at his hands. But Hannibal had earned a terrifying reputation. Despite their superior numbers, the Romans became so frightened of Hannibal that they avoided battle with him like the plague.

### Interpretation

Hannibal must be considered the ancient master of the military art of the unorthodox. In attacking the Romans on their own soil, he never intended to take Rome itself; that would have been impossible. Its walls

*and island described by Arrian, and in preparation he decided on a manoeuvre almost identical with that adopted by General Wolfe in his 1759 Quebec campaign. Under cover of night he sent out his cavalry to various points along the western bank of the river with orders to make a clamour, and from time to time to raise the battle-cry; for several nights Porus marched his elephants up and down the eastern bank to block an attempted crossing until he got tired of it, kept his elephants in camp, and posted scouts along the eastern bank. Then "when Alexander had brought it about that the mind of Porus no longer entertained any fear of his nocturnal attempts, he devised the following stratagem": Upstream and along the western bank he posted a chain of sentries, each post in sight and hearing of the next one, with orders to raise a din and keep their picket fires burning, while visible preparations were made at the camp to effect a crossing. . . . . . . When Porus had been lulled into a sense of false security and all preparations were completed at the camp and the crossing place, Alexander set out secretly and kept at some distance from the western bank of the river so that his march would not be observed. . . .*

*The Generalship of Alexander the Great*, J.F.C. Fuller, 1960

*To cross the sea without heaven's knowledge, one had to move openly over the sea but act as if one did not intend to cross it. Each military maneuver has two aspects: the superficial move and the underlying purpose. By concealing both, one can take the enemy completely by surprise. . . . [If] it is highly unlikely that the enemy can be kept ignorant of one's actions, one can sometimes play tricks right under its nose.*

THE WILES OF WAR: 36 MILITARY STRATEGIES FROM ANCIENT CHINA, TRANSLATED BY SUN HAICHEN, 1991

were high, its people fierce and united in their hatred of him, and his forces were small. Rather, Hannibal's goal was to wreak havoc on the Italian peninsula and to undermine Rome's alliances with neighboring city-states. Weakened at home, Rome would have to leave Carthage alone and put a stop to its imperial expansion.

To sow this kind of chaos with the tiny army he had been able to bring over the Alps, Hannibal had to make his every action unexpected. A psychologist before his time, he understood that an enemy that is caught by surprise loses its discipline and sense of security. (When chaos strikes those who are particularly rigid and orderly to begin with, such as the people and armies of Rome, it has double the destructive power.) And surprise can never be mechanical, repetitive, or routine; that would be a contradiction in terms. Surprise takes constant adaptation, creativity, and a mischievous pleasure in playing the trickster.

So Hannibal always took the route that Rome least expected him to take—the road through the Alps, for example, considered impassable to an army and therefore unguarded. Eventually, inevitably, the Romans caught on and began to expect him to take the least obvious route; at that point it was the obvious that was unexpected, as at Allifae. In battle, Hannibal would fix the enemy's attention on a frontal assault—the ordinary, usual way armies fought at the time—then unleash the extraordinary in the form of elephants or a reserve force hidden to the enemy's rear. In his raids in the Roman countryside, he deliberately protected Fabius's property, creating the impression that the two men were in collusion and ultimately forcing the embarrassed leader to take action—an unorthodox use of politics and extramilitary means in war. At Allifae, Hannibal had bundles of kindling tied to the horns of oxen, then lit them and sent the terrified, bellowing animals up the slopes to the pass at night—creating an indecipherable image to the Roman sentries, literally in the dark, and a terrifying one.

At Cannae, where the Romans were by this time expecting the unorthodox, Hannibal disguised his stratagem in broad daylight, lining up his army like any other army of the period. The Roman force was already impelled by the violence of the moment and the desire for revenge; he let them make quick progress through his deliberately weak center, where they became crowded together. Then the swift-moving outer wings of his line closed in and choked them. On and on he went, each one of Hannibal's ingeniously unorthodox maneuvers flowering out of the other in a constant alternation between the uncanny and the banal, the hidden and the obvious.

Adapting Hannibal's method to your own daily battles will bring you untold power. Using your knowledge of your enemies' psychology and way of thinking, you must calculate your opening moves to be what they least expect. The line of least expectation is the line of least resistance; people cannot defend themselves against what they cannot foresee. With less resistance in your path, the progress you make will inflate their impression of your power; Hannibal's small army seemed to the Romans much larger than it really was. Once they come to expect some extraordinary maneuver on your part, hit them with the ordinary. Establish a reputation for the unconventional and you set your opponents on their heels: knowing to anticipate the unexpected is not the same thing as knowing what the unexpected will be. Before long your opponents will give way to your reputation alone.

*Chaos—where brilliant dreams are born.*

*THE I CHING*, CHINA, CIRCA EIGHTH CENTURY B.C.

**2.** In 1962, Sonny Liston became the heavyweight boxing champion of the world by defeating Floyd Patterson. Shortly afterward he turned up to watch a young hotshot on the scene, Cassius Clay, take on and beat rather decisively the veteran Archie Moore. After the fight, Liston paid a visit to Clay's dressing room. He put his arm around the boy's shoulder—at twenty, Clay was ten years younger than Liston—and told him, "Take care, kid. I'm gonna need you. But I'm gonna have to beat you like I'm your daddy." Liston was the biggest, baddest fighter in the world, and to those who understood the sport, he seemed invincible. But Liston recognized Clay as a boxer just crazy enough to want to fight him down the road. It was best to instill a touch of fear in him now.

The fear did not take: as Liston had guessed he would, Clay soon began to clamor for a fight with the champion and to brag to one and all that he would beat him in eight rounds. On television and radio shows, he taunted the older boxer: maybe it was Liston who was afraid to take on Cassius Clay. Liston tried to ignore the upstart; "If they ever make the fight," he said, "I'll be locked up for murder." He considered Clay too pretty, even effeminate, to be a heavyweight champion.

Time passed, and Clay's antics provoked a desire for the fight in the public: most people wanted to see Liston beat the daylights out of Clay and shut him up. Late in 1963 the two men met to sign on for a championship fight in Miami Beach the following February. Afterward Clay told reporters, "I'm not afraid of Liston. He's an old man. I'll give him talking lessons and boxing lessons. What he needs most is falling-down lessons." As the fight grew closer, Clay's rhetoric became still more insulting and shrill.

Of the sportswriters polled on the upcoming fight, most of them predicted that Clay would not be able to walk on his own after it was over. Some worried that he would be permanently injured. "I guess it's quite hard to tell Clay not to fight this monster now," said the boxer Rocky Marciano, "but I'm sure he'll be more receptive after he's been there with Liston." What worried the experts most of all was Cassius Clay's unusual fighting style. He was not the typical heavyweight bruiser: he would dance in place with his hands down at his side; he rarely put his full body into his punches, instead hitting just from the arms; his head was constantly moving, as if he wanted to keep his pretty face unscathed; he was reluctant to go inside, to brawl and pummel the body—the usual way to wear down a heavyweight. Instead Clay preferred to dance and shuffle, as if his fights were ballet, not boxing. He was too small to be a heavyweight, lacked the requisite killer instinct—the press critique ran on.

At the weigh-in on the morning of the bout, everyone was waiting for Clay's usual prefight antics. He exceeded their expectations. When Liston got off the scales, Clay began to shout at him: "Hey, sucker, you're a chump. You've been tricked, chump. . . . You are too ugly. . . . I'm going to whup you so bad." Clay jumped and screamed, his whole body shaking, his eyes popping, his voice quivering. He seemed possessed. Was he afraid or downright crazy? For Liston this was quite simply the last straw. He wanted to kill Clay and shut the challenger up for good.

As they stood in the ring before the opening bell, Liston tried to stare down Clay as he had stared down others, giving him the evil eye. But unlike other boxers Clay stared back. Bobbing up and down in place, he repeated, "Now I've got you, chump." The fight began, and Liston charged forward at his prey, throwing a long left jab that missed by a mile. He kept coming, a look of intense anger on his face—but Clay shuffled back from each punch, even taunting Liston at one point by lowering his hands. He seemed able to anticipate Liston's every move. And he returned Liston's stare: even after the round ended and both men were in their corners, his eyes never left his opponent's.

The second round was more of the same, except that Liston, instead of looking murderous, began to look frustrated. The pace was far faster than in any of his earlier fights, and Clay's head kept bobbing and orbiting in disturbing patterns. Liston would move in to strike his chin, only to miss or find Clay hitting Liston's chin instead, with a lightning-quick jab that made him wobble on his feet. At the end of the third round, a flurry of punches came out of nowhere and opened a deep gash under Liston's left eye.

Now Clay was the aggressor and Liston was fighting to survive. In the sixth round, he began taking punches from all angles, opening more wounds and making Liston look weak and sad. When the bell for the seventh round rang, the mighty Liston just sat on his stool and stared—he refused to get up. The fight was over. The boxing world was stunned: Was it a fluke? Or—since Liston had seemed to fight as if under some spell, his punches missing, his movements tired and listless—had he just had an off night? The world would have to wait some fifteen months to find out, until the two boxers' rematch in Lewiston, Maine, in May 1965.

Consumed with a hunger for revenge, Liston trained like a demon for this second fight. In the opening round, he went on the attack, but he seemed wary. He followed Clay—or rather Muhammad Ali, as he was now known—around the ring, trying to reach him with jabs. One of these jabs finally grazed Ali's face as he stepped back, but, in a move so fast that few in the audience even saw it, Ali countered with a hard right that sent Liston to the canvas. He lay there for a while, then staggered to his feet, but too late—he had been down for more than ten seconds, and the referee called the fight. Many in the crowd yelled fix, claiming that no punch had landed. Liston knew otherwise. It may not have been the most powerful blow, but it caught him completely by surprise, before he could tense his muscles and prepare himself. Coming from nowhere, it floored him.

Liston would continue to fight for another five years, but he was never the same man again.

### Interpretation

Even as a child, Muhammad Ali got perverse pleasure out of being different. He liked the attention it got him, but most of all he just liked being himself: odd and independent. When he began to train as a boxer, at the age of twelve, he was already refusing to fight in the usual way, flouting the rules. A boxer usually keeps his gloves up toward his head and upper body, ready to parry a blow. Ali liked to keep his hands low, apparently inviting attack—but he had discovered early on that he was quicker than other boxers, and the best way to make his speed work for him was to lure the opponent's chin just close enough for Ali to snap a jab at him that would cause a lot more pain for being so close and so quick. As Ali developed, he also made it harder for the other boxer to reach him by working on his legs, even more than on the power of his punch. Instead of retreating the way most fighters did, one foot at a time, Ali kept on his toes, shuffling back and dancing, in perpetual motion to his own peculiar rhythm. More than any other boxer, he was a moving target. Unable to land a punch, the other boxer would grow frustrated, and the more frustrated he was, the more he would reach for Ali, opening up his guard and exposing himself to the jab from nowhere that might knock him out. Ali's style ran counter to conventional boxing wisdom in almost every way, yet its unorthodoxy was exactly what made it so difficult to combat.

Ali's unconventional tactics in the first Liston fight began well before the bout. His irritating antics and public taunts—a form of dirty warfare—were designed to infuriate the champion, cloud his mind, fill him with a murderous hatred that would make him come close enough for Ali to knock him out. Ali's behavior at the weigh-in, which seemed genuinely insane, was later revealed as pure performance. Its effect was to make Liston unconsciously defensive, unsure of what this man would do in the ring. In the opening round, as in so many of his subsequent fights, Ali lulled Liston by fighting defensively, an ordinary tactic when facing a boxer like Liston. That drew Liston in closer and closer—and now the extraordinary move, the speedy punch out of nowhere, had double the force. Unable to reach Ali with his punches, disconcerted by the dancing, the lowering of the hands, the irritating taunting, Liston made mistake after mistake. And Ali feasted on his opponents' mistakes.

Understand: as children and young adults, we are taught to conform

*One who studies ancient tactics and employs the army in accord with their methods is no different from someone who glues up the tuning stops and yet tries to play a zither. I have never heard of anyone being successful. The acumen of strategists lies in penetrating the subtle amid unfolding change and discerning the concordant and contrary. Now whenever mobilizing you must first employ spies to investigate whether the enemy's commanding general is talented or not. If instead of implementing tactics, he merely relies on courage to employ the army, you can resort to ancient methods to conquer him. However, if the commanding general excels in employing ancient tactics, you should use tactics that contradict the ancient methods to defeat him.*

HSÜ TUNG,
CHINA, 976–1018

*The chief characteristic of fashion is to impose and suddenly to accept as a new rule or norm what was, until a minute before, an exception or whim, then to abandon it again after it has become a commonplace, everybody's "thing." Fashion's task, in brief, is to maintain a continual process of standardization: putting a rarity or novelty into general and universal use, then passing on to another rarity or novelty when the first has ceased to be such. . . . Only modern art, because it expresses the avant-garde as its own extreme or supreme moment, or simply because it is the child of the romantic aesthetic of originality and novelty, can consider as the typical—and perhaps sole—form of the ugly what we might call* ci-devant *beauty, the beauty of the* ancien regime, *ex-beauty. Classical art, through the method of imitation and the practice of repetition, tends toward the ideal of renewing, in the sense of integration and perfection. But for the modern art in general, and for avant-garde in particular, the only irremediable and absolute aesthetic error is a traditional artistic creation, an art that imitates and repeats itself. From the anxious modern longing for what Remy de Gourmont chose to call, suggestively, "le beau inedit" derives that sleepless and fevered experimentation which is one of the most characteristic manifestations*

to certain codes of behavior and ways of doing things. We learn that being different comes with a social price. But there is a greater price to pay for slavishly conforming: we lose the power that comes from our individuality, from a way of doing things that is authentically our own. We fight like everyone else, which makes us predictable and conventional.

The way to be truly unorthodox is to imitate no one, to fight and operate according to your own rhythms, adapting strategies to your idiosyncrasies, not the other way around. Refusing to follow common patterns will make it hard for people to guess what you'll do next. You are truly an individual. Your unorthodox approach may infuriate and upset, but emotional people are vulnerable people over whom you can easily exert power. If your peculiarity is authentic enough, it will bring you attention and respect—the kind the crowd always has for the unconventional and extraordinary.

**3.** Late in 1862, during the American Civil War, General Ulysses S. Grant made several efforts to take the Confederate fortress at Vicksburg. The fortress was at a critical point in the Mississippi River, the lifeline of the South. If Grant's Union army took Vicksburg, it would gain control of the river, cutting the South in half. Victory here could be the turning point of the war. Yet by January 1863 the fortress's commander, General James Pemberton, felt confident he had weathered the storm. Grant had tried to take the fort from several angles to the north and had failed. It seemed that he had exhausted all possibilities and would give up the effort.

The fortress was located at the top of a two-hundred-foot escarpment on the riverbank, where any boat that tried to pass was exposed to its heavy artillery. To its west lay the river and the cliffs. To the north, where Grant was encamped, it was protected by virtually impassable swamp. Not far east lay the town of Jackson, a railroad hub where supplies and reinforcements could easily be brought in—and Jackson was firmly in Southern hands, giving the Confederacy control of the entire corridor, north and south, on the river's eastern bank. Vicksburg seemed secure from all directions, and the failure of Grant's attacks only made Pemberton more comfortable. What more could the Northern general do? Besides, he was in political hot water among President Abraham Lincoln's enemies, who saw his Vicksburg campaign as a monumental waste of money and manpower. The newspapers were portraying Grant as an incompetent drunk. The pressure was tremendous for him to give it up and retreat back to Memphis to the north.

Grant, however, was a stubborn man. As the winter dragged on, he tried every kind of maneuver, with nothing working—until, on the moonless night of April 16, Confederate scouts reported a Union flotilla of transport ships and gunboats, lights off, trying to make a run past the batteries at Vicksburg. The cannons roared, but somehow the ships got past them with minimal damage. The next few weeks saw several more runs down the river. At the same time, Union forces on the western side of the river were reported heading south. Now it was clear: Grant would use the transport ships he had sneaked past Vicksburg to cross the Mississippi some thirty miles downriver. Then he would march on the fortress from the south.

Pemberton called for reinforcements, but in truth he was not overly concerned. Even if Grant got thousands of men across the river, what could he do once there? If he moved north toward Vicksburg, the Confederacy could send armies from Jackson and points south to take him from the flank and rear. Defeat in this corridor would be a disaster, for Grant would have no line of retreat. He had committed himself to a foolhardy venture. Pemberton waited patiently for his next move.

Grant did cross the river south of Vicksburg, and in a few days his army was moving northeast, heading for the rail line from Vicksburg to Jackson. This was his most audacious move so far: if he were successful, he would cut Vicksburg off from its lifeline. But Grant's army, no different from any other, needed lines of communication and supply. These lines would have to connect to a base on the eastern side of the river, which Grant had indeed established at the town of Grand Gulf. All Pemberton had to do was send forces south from Vicksburg to destroy or even just threaten Grand Gulf, endangering Grant's supply lines. He would be forced to retreat south or risk being cut off. It was a game of chess that Pemberton could not lose.

And so, as the Northern general maneuvered his armies with speed toward the rail line between Jackson and Vicksburg, Pemberton moved on Grand Gulf. To Pemberton's utter dismay, Grant ignored him. Indeed, so far from dealing with the threat to his rear, he pushed straight on to Jackson, taking it on May 14. Instead of relying on supply lines to feed his army, he plundered the area's rich farmlands. More, he moved so swiftly and changed direction so fluidly that Pemberton could not tell which part of his army was the front, rear, or flank. Rather than struggle to defend lines of communication or supply, Grant kept none. No one had ever seen an army behave in such a manner, breaking every rule in the military playbook.

A few days later, with Jackson under his control, Grant wheeled his troops toward Vicksburg. Pemberton rushed his men back from Grand Gulf to block the Union general, but it was too late: beaten at the Battle of Champion Hill, he was forced back into the fortress, where his army was quickly besieged by the Union forces. On July 4, Pemberton surrendered Vicksburg, a blow the South would never recover from.

### Interpretation

We humans are conventional by nature. Once anyone succeeds at something with a specific strategy or method, it is quickly adopted by others and becomes hardened into principle—often to everyone's detriment

*of the avant-garde; its assiduous labor is an eternal web of Penelope, with the weave of its forms remade every day and unmade every night. Perhaps Ezra Pound intended to suggest both the necessity and the difficulty of such an undertaking when he once defined the beauty of art as "a brief gasp between one cliché and another." The connection between the avant-garde and fashion is therefore evident: fashion too is a Penelope's web; fashion too passes through the phase of novelty and strangeness, surprise and scandal, before abandoning the new forms when they become cliché, kitsch, stereotype. Hence the profound truth of Baudelaire's paradox, which gives to genius the task of creating stereotypes. And from that follows, by the principle of contradiction inherent in the obsessive cult of genius in modern culture, that the avant-garde is condemned to conquer, through the influence of fashion, that very popularity it once disdained—and this is the beginning of its end. In fact, this is the inevitable, inexorable destiny of each movement: to rise up against the newly outstripped fashion of an old avant-garde and to die when a new fashion, movement, or avant-garde appears.*

THE THEORY OF THE AVANT-GARDE, RENATO POGGIOLI, 1968

*I have forced myself to contradict myself in order to avoid conforming to my own taste.*

MARCEL DUCHAMP, 1887–1968

when it is applied indiscriminately. This habit is a particular problem in war, for war is such risky business that generals are often tempted to take the road well traveled. When so much is necessarily unsafe, what has proven safe in the past has amplified appeal. And thus for centuries the rules have been that an army must have lines of communication and supply and, in battle, must assume a formation with flanks and a front. Napoleon loosened these principles, but their hold on military thinkers remained so strong that during the American Civil War, some forty years after Napoleon's death, officers like Pemberton could not imagine an army behaving according to any other plan.

It took great courage for Grant to disobey these conventions and cut himself loose from any base, living instead off the rich lands of the Mississippi Basin. It took great courage for him to move his army without forming a front. (Even his own generals, including William Tecumseh Sherman, thought he had lost his mind.) This strategy was hidden from Pemberton's view because Grant kept up ordinary appearances by establishing a base at Grand Gulf and forming front and rear to march toward the rail line. By the time Pemberton had grasped the extraordinary nature of Grant's free-flowing attack, he had been taken by surprise and the game was over. To our eyes Grant's strategy might seem obvious, but it was completely outside Pemberton's experience.

To follow convention, to give inordinate weight to what has worked in the past, is a natural tendency. We often ignore some simple yet unconventional idea that in every sense would upset our opponents. It is a matter sometimes of cutting ourselves loose from the past and roaming freely. Going without a security blanket is dangerous and uncomfortable, but the power to startle people with the unexpected is more than worth the risk. This is particularly important when we are on the defensive or in a weakened state. Our natural tendency at such times is to be conservative, which only makes it easier for our enemies to anticipate our moves and crush us with their superior strength; we play into their hands. It is when the tide is against us that we must forget the books, the precedents, the conventional wisdom, and risk everything on the untried and unexpected.

**4.** The Ojibwa tribe of the North American plains contained a warrior society known as the Windigokan (No-flight Contraries). Only the bravest men, who had demonstrated bravery by their utter disregard for danger on the battlefield, were admitted to the Windigokan. In fact, because they had no fear of death, they were considered no longer among the living: they slept and ate separately and were not held to the usual codes of behavior. As creatures who were alive but among the dead, they spoke and acted contrarily: they called a young person an old man, and when one of them told the others to stand still, he meant charge forward. They were glum in times of prosperity, happy in the depths of winter. Although there was a clownish side to their behavior, the Windigokan could inspire great fear. No one ever knew what they would do next.

The Windigokan were believed to be inhabited by terrifying spirits called Thunderers, which appeared in the form of giant birds. That made them somehow inhuman. On the battlefield they were disruptive and unpredictable, and in raiding parties downright terrifying. In one such raid, witnessed by an outsider, they gathered first in front of the Ojibwa chief's lodge and yelled, "We are not going to war! We shall not kill the Sioux! We shall not scalp four of them and let the rest escape! We shall go in daytime!" They left camp that night, wearing costumes of rags and scraps, their bodies plastered with mud and painted with splotches of weird color, their faces covered by frightening masks with giant, beak-like noses. They made their way through the darkness, stumbling over themselves—it was hard to see through the masks—until they came upon a large Sioux war party. Although outnumbered, they did not flee but danced into the enemy's center. The grotesqueness of their dance made them seem to be possessed by demons. Some of the Sioux backed away; others drew close, curious and confused. The leader of the Windigokan shouted, "Don't shoot!" The Ojibwa warriors then pulled out guns hidden under their rags, killed four Sioux, and scalped them. Then they danced away, the enemy too terrified by this apparition to pursue them.

After such an action, the mere appearance of the Windigokan was enough for the enemy to give them a wide berth and not risk any kind of encounter.

### Interpretation

What made the Windigokan so frightening was the fact that, like the forces of nature from which they claimed to derive their powers, they could be destructive for no apparent reason. Their mounting of a raid was not governed by need or ordered by the chief; their appearance bore no relation to anything known, as if they had rolled on the ground or in trays of paint. They might wander in the dark until they chanced on an enemy. Their dancing was like nothing anyone had seen or imagined. They might suddenly start to kill and scalp, then stop at an arbitrary number. In a tribal society governed by the strictest of codes, these were spirits of random destruction and irrationality.

The use of the unconventional can startle and give you an advantage, but it does not often create a sense of terror. What will bring you ultimate power in this strategy is to follow the Windigokan and adapt a kind of randomness that goes beyond rational processes, as if you were possessed by a spirit of nature. Do this all the time and you'll be locked up, but do it right, dropping hints of the irrational and random at the opportune moment, and those around you will always have to wonder what you'll do next. You will inspire a respect and fear that will give you great power. An ordinary appearance spiced by a touch of divine madness is more shocking and alarming than an out-and-out crazy person.

*A similar vision among the Siouan tribes turns the warrior into a Heyoka, who also exhibits the clown-like behavior of the Windigokan, the use of sacking as a war shirt, and plastering the body with mud. . . . . . . Psychologically the Heyoka was of immense importance, as were similar characters among numerous other tribes. During periods of happiness and plenty he saw only gloom and despair, and could be goaded into providing hours of harmless amusement when he gorged himself on buffalo ribs while complaining there was no food in the camp, or declared he was dirty and proceeded to wash in a bath of mud. . . . . . . Yet behind this benign face of the Heyoka there lurked the ever-present fear that he was possessed by the spirit of Iktomi, and was therefore unpredictable and potentially dangerous. He, after all, was the only person who dared challenge the supernaturals even if he was in dread of a common camp dog and would run screaming in fright if one approached too close. Thus he made a mockery of the pretensions of some of the warriors, but at the same time*

Remember: your madness, like Hamlet's, must be strategic. Real madness is all too predictable.

**5.** In April 1917, New York's Society of Independent Artists prepared for its first exhibition. This was to be a grand showcase of modern art, the largest in the United States to date. The exhibition was open to any artist who had joined the society (whose dues were minimal), and the response had been overwhelming, with over twelve hundred artists contributing over two thousand pieces.

The society's board of directors included collectors like Walter Arensberg and artists like Man Ray and the twenty-nine-year-old Marcel Duchamp, a Frenchman then living in New York. It was Duchamp, as head of the Hanging Committee, who decided to make the exhibition radically democratic: he hung the works in alphabetical order, beginning with a letter drawn from a hat. The system led to cubist still lifes being hung next to traditional landscapes, amateur photographs, and the occasional lewd work by someone apparently insane. Some of the organizers loved this plan, others were disgusted and quit.

A few days before the exhibition was to open, the society received the strangest work so far: a urinal mounted on its back, with the words R. MUTT 1917 painted in large black letters on its rim. The work was called *Fountain*, and it was apparently submitted by a Mr. Mutt, along with the requisite membership fee. In viewing the piece for the first time, the painter George Bellows, a member of the society's board, claimed it was indecent and that the society could not exhibit it. Arensberg disagreed: he said he could discern an interesting work of art in its shape and presentation. "This is what the whole exhibit is about," he told Bellows. "An opportunity to allow the artist to send in anything he chooses, for the artist to decide what is art, not someone else."

Bellows was unmoved. Hours before the exhibition opened, the board met and voted by a slim margin not to show the piece. Arensberg and Duchamp immediately resigned. In newspaper articles reporting this controversy, the object was politely referred to as a "bathroom fixture." It piqued a lot of curiosity, and an air of mystery pervaded the entire affair.

At the time of the exhibition, Duchamp was one of a group of artists who published a magazine called *The Blind Man*. The magazine's second issue included a photograph of *Fountain* taken by the great photographer Alfred Stieglitz, who lit the urinal beautifully so that a shadow fell over it like a kind of veil, giving it a slightly religious appearance, along with something vaguely sexual in the arguably vaginal shape of the urinal when laid on its back. *The Blind Man* also ran an editorial, "The Richard Mutt Case," that defended the work and criticized its exclusion from the show: "Mr. Mutt's fountain is not immoral . . . no more than a bathtub is immoral. . . . Whether Mr. Mutt with his own hands made the fountain or not has no importance. He CHOSE it. He took an ordinary article of life, placed it so that its useful significance disappeared under the new title and point of view—created a new thought for that object."

It soon became clear that the "creator" of *Fountain* was none other than Duchamp. And over the years the work began to assume a life of its own, even though it mysteriously disappeared from Stieglitz's studio and was never found again. For some reason the photograph and the story of *Fountain* inspired endless ideas about art and artmaking. The work itself had strange powers to shock and compel. In 1953 the Sidney Janis Gallery, New York, was authorized by Duchamp to exhibit a replica of *Fountain* over its entrance door, a sprig of mistletoe emerging from the bowl. Soon more replicas were appearing in galleries, retrospective exhibitions of Duchamp's work, and museum collections. *Fountain* became a fetish object, something to collect. Replicas of it have sold for over $1 million.

Everyone seems to see what they want to see in the piece. Shown in museums, it often still outrages the public, some disturbed by the urinal itself, others by its presentation as art. Critics have written extended articles on the urinal, with all kinds of interpretations: in staging *Fountain*, Duchamp was urinating on the art world; he was playing with notions of gender; the piece is an elaborate verbal pun; on and on. What some of the organizers of the 1917 show believed to be merely an indecent object unworthy of being considered art has somehow turned into one of the most controversial, scandalous, and analyzed works of the twentieth century.

*emphasized the fact that the powers which guided and protected them in battle were of such strength that only a Heyoka might oppose them.*

WARRIORS: WARFARE AND THE NATIVE AMERICAN INDIAN, NORMAN BANCROFT HUNT, 1995

### Interpretation

Throughout the twentieth century, many artists wielded influence by being unconventional: the Dadaists, the surrealists, Pablo Picasso, Salvador Dalí—the list is long. But of all of them, it is Marcel Duchamp who has probably had the greatest impact on modern art, and what he called his "readymades" are perhaps the most influential of all his works. The readymades are everyday objects—sometimes exactly as they were made (a snow shovel, a bottle rack), sometimes slightly altered (the urinal laid on its back, the mustache and goatee drawn on a reproduction of *The Mona Lisa*)—"chosen" by the artist and then placed in a gallery or museum. Duchamp was giving the ideas of art priority over its images. His readymades, banal and uninteresting in themselves, inspired all kinds of associations, questions, and interpretations; a urinal may be a seedy commonplace, but to present it as art was utterly unconventional and stirred up angry, irritating, delirious ideas.

Understand: in war, politics, and culture, what is unconventional, whether it is Hannibal's elephants and oxen or Duchamp's urinal, is never material—or rather it is never *just* material. The unconventional can only arise out of the mind: something surprises, is not what we expected. We usually base our expectations on familiar conventions, clichés, habits of seeing, the ordinary. Many artists, writers, and other producers of culture seem to believe it the height of unconventionality to create images, texts, and other works that are merely weird, startling, or shocking in some way. These works may generate a momentary splash, but they have none of the power of the unconventional and extraordinary because they have no context to rub against; they do not work against our expectations. No more than strange, they quickly fade from memory.

When striving to create the extraordinary, always remember: what is crucial is the mental process, not the image or maneuver itself. What will truly shock and linger long in the mind are those works and ideas that grow out of the soil of the ordinary and banal, that are unexpected, that make us question and contest the very nature of the reality we see around us. Most definitely in art, the unconventional can only be strategic.

**Image:**
The Plow.
The ground
must be prepared.
The blades of the plow
churn the earth in constant
motion, bringing air into the
soil. The process must go on every year,
or the most pernicious weeds will take over and the
clumped soil will choke off all life. From the earth, plowed
and fertilized, the most nourishing and wondrous plants can emerge.

**Authority:** In general, in battle one engages the enemy with the orthodox and gains victory through the unorthodox. . . . The unorthodox and the orthodox mutually produce each other, just like an endless cycle. Who can exhaust them? *—Sun-tzu (fourth century B.C.)*

## REVERSAL

There is never any value in attacking opponents from a direction or in a way that they expect, allowing them to stiffen their resistance—that is, unless your strategy is suicide.

# 25

# OCCUPY THE MORAL HIGH GROUND

## THE RIGHTEOUS STRATEGY

*In a political world, the cause you are fighting for must seem more just than the enemy's. Think of this as moral terrain that you and the other side are fighting over; by questioning your enemies' motives and making them appear evil, you can narrow their base of support and room to maneuver. Aim at the soft spots in their public image, exposing any hypocrisies on their part. Never assume that the justice of your cause is self-evident; publicize and promote it. When you yourself come under moral attack from a clever enemy, do not whine or get angry; fight fire with fire. If possible, position yourself as the underdog, the victim, the martyr. Learn to inflict guilt as a moral weapon.*

## THE MORAL OFFENSIVE

In 1513 the thirty-seven-year-old Giovanni de' Medici, son of the illustrious Florentine Lorenzo de' Medici, was elected pope and assumed the name Leo X. The church that Leo now led was in many ways the dominant political and economic power in Europe, and Leo—a lover of poetry, theater, and painting, like others in his famous family—wanted to make it also a great patron of the arts. Earlier popes had begun the building of the basilica of St. Peter's in Rome, the preeminent seat of the Catholic Church, but had left the structure unfinished. Leo wanted to complete this mighty project, permanently associating it with his name, but he would need to raise a fair amount of capital to be able to pay for the best artists to work on it.

And so in 1517, Leo launched a campaign to sell indulgences. Then as now, it was Catholic practice for the faithful to confess their sins to their priest, who would enforce their contrition by assigning them a penance, a kind of worldly punishment. Today this might simply be a prayer or a counting of the rosary, but penances were once more severe, including fasts and pilgrimages—or financial payments known as indulgences. The nobility might pay an indulgence in the form of a saintly relic purchased for their church, a large expense that would translate into the promise of a reduced time spent in purgatory after death (purgatory being a kind of halfway house for those not evil enough for hell, not good enough for heaven, so forced to wait); the lower classes might pay a smaller fee to buy forgiveness for their sins. Indulgences were a major source of church income.

For this particular campaign, Leo unleashed a squadron of expert indulgence salesmen across Europe, and the money began to pour in. As his chief architect for the completion of St. Peter's he appointed the great artist Raphael, who planned to make the building a splendid work of art, Leo's lasting legacy to the world. All was going well, until, in October 1517, news reached the pope that a priest named Martin Luther (1483–1546)—some tiresome German theologian—had tacked to the doors of the castle church of Wittenberg a tract called The Ninety-five Theses. Like many important documents of the time, the tract was originally in Latin, but it had been translated into German, printed up, and passed out among the public—and within a few weeks all Germany seemed to have read it. The Ninety-five Theses was essentially an attack on the practice of selling indulgences. It was up to God, not the church, to forgive sinners, Luther reasoned, and such forgiveness could not be bought. The tract went on to say that the ultimate authority was Scripture: if the pope could cite Scripture to refute Luther's arguments, the priest would gladly recant them.

The pope did not read Luther's writings—he preferred poetry to theological discussions. And a single German priest surely posed no threat to the use of indulgences to fund worthy projects, let alone to the church itself. But Luther seemed to be challenging the church's authority in a broad sense, and Leo knew that an unchecked heresy could become the center of a sect. Within recent centuries in Europe, the church had had to put down such dissident sects by the use of force; better to silence Luther before it was too late.

Leo began relatively gently, asking the respected Catholic theologian Silvester Mazzolini, usually known as Prieras, to write an official response to Luther that he hoped would frighten the priest into submission. Prieras proclaimed that the pope was the highest authority in the church, even higher than Scripture—in fact, that the pope was infallible. He quoted various theological texts written over the centuries in support of this claim. He also attacked Luther personally, calling him a bastard and questioning his motives: perhaps the German priest was angling for a bishopric? Prieras concluded with the words, "Whoever says that the Church of Rome may not do what it is actually doing in the manner of indulgences is a heretic." The warning was clear enough.

Leo had much on his mind during these years, including turmoil in the Ottoman Empire and a plan to launch a new crusade, but Luther's response to Prieras got his attention right away. Luther wrote a text in which he mercilessly took apart Prieras's writings—the church, he argued, had failed to answer his charges and to base its arguments on Scripture. Unless its authority in granting indulgences and excommunicating heretics was rooted in the Bible, it was not spiritual in nature but worldly, political, and that kind of authority could and should be challenged. Luther published his text alongside Prieras's, allowing readers to compare the two and come to their own conclusions. His direct quotation of Prieras, his audacious and mocking tone, and his use of recently developed printing technology to spread his message far and wide—all this was quite shocking and new to church officials. They were dealing with a clever and dangerous man. It was now clear to Leo that the war between the church and Luther was a war to the death.

As the pope pondered how to get the German priest to Rome and try him as a heretic, Luther accelerated his campaign, continuing to publish at an alarming rate, his tone ever more vitriolic. In An Open Letter to the Christian Nobility of the German Nation, he claimed that Rome had used its spurious authority to bully and cow the German people for centuries, turning Germany's kingdoms into vassal states. The church, he said again, was a political power, not a spiritual one, and to prop up its worldly rule it had resorted to lies, forged documents, whatever means necessary. In On the Babylonian Captivity of the Church, he railed against the pope's lavish lifestyle, the whoring among the church hierarchy, the blasphemous art Leo funded. The pope had gone so far as to have staged an immoral and bawdy play by Machiavelli, called *Mandragola*, within the Vatican itself. Luther juxtaposed the righteous behavior advocated by the church with the way its cardinals actually lived. It was the pope and his entourage, Luther charged, who were the real heretics, not he; in fact, the pope was the Antichrist.

[Colonel John] *Boyd paid particular attention to the moral dimension and the effort to attack an adversary morally by showing the disjuncture between professed beliefs and deeds. The name of the game for a moral design for grand strategy is to use moral leverage to amplify one's spirit and strength while exposing the flaws of competing adversary systems. In the process, one should influence the uncommitted, potential adversaries and current adversaries so that they are drawn toward one's philosophy and are empathetic toward one's success.*

*The Mind of War: John Boyd and American Security*, Grant T. Hammond, 2001

*The central feature of the "exterior maneuver" is to assure for oneself the maximum freedom of action while at the same time paralysing the enemy by a multitude of deterrent checks, somewhat as the Lilliputians tied up Gulliver. As with all operations designed to deter, action will of course be primarily psychological; political, economic, diplomatic and military measures will all be combined towards the same end. The procedures employed to achieve this deterrent effect range from the most subtle to the most brutal: appeal will be made to the legal formulae of national and international law, play will be made with moral and humanitarian susceptibilities and there will be attempts to prick the enemy's conscience by making him doubtful of the justice of his cause. By these methods, opposition from some section of the enemy's internal public opinion will be roused and at the same time some sector of international public opinion will be whipped up; the result will be a real moral coalition and attempts will be made to co-opt the more unsophisticated sympathizers by arguments based upon their own preconceived ideas. This climate of opinion will be exploited at the United Nations, for instance, or at other international gatherings;*

It seemed to Leo that Luther had responded to Prieras's threat by raising the temperature. Clearly the threat had been weak; the pope had been too lenient. It was time to show real force and end this war. So Leo wrote a papal bull threatening Luther with excommunication. He also sent church officials to Germany to negotiate the priest's arrest and imprisonment. These officials, however, came back with shocking news that altered everything: in the few short years since the publication of The Ninety-five Theses, Martin Luther, an unknown German priest, had somehow become a sensation, a celebrity, a beloved figure throughout the country. Everywhere the pope's officials went, they were heckled, even threatened with stoning. Shop windows in almost every German town contained paintings of Luther with a halo over his head. "Nine-tenths of the Germans shout 'Long live Luther,' " one official reported to Leo, "and the other tenth 'Death to Rome.' " Luther had somehow aroused the German public's latent resentment and hatred of the church. And his reputation was impeccable: he was a bestselling author, yet he refused the income from his writings, clearly practicing what he preached. The more the church attacked him, the more popular Luther became. To make a martyr of him now could spark a revolution.

Nevertheless, in 1521, Leo ordered Luther to appear in the town of Worms before the Imperial Diet, a gathering of German princes, nobles, and clergy organized by the newly elected Holy Roman Emperor, Charles V. Leo hoped to get the Germans to do his dirty work, and Charles was amenable: a political creature, worried by the antiauthoritarian sentiments that Luther had sparked, he wanted the dispute over. At the Diet he demanded that the priest recant his teachings. But Luther, as usual, refused, and in dramatic fashion, uttering the memorable line "Here I stand. I cannot do otherwise. God help me." The emperor had no choice; he condemned Luther as a heretic and ordered him to return to Wittenberg to await his fate. On the way home, however, Luther was kidnapped and taken to Warburg Castle. The kidnapping had actually been planned and executed by his many supporters among the aristocracy; he was safe. Living in the castle under an assumed name, he was able to ride out the storm.

Leo died that year, and within months of his death, Luther's ideas and the reforms that he had advocated had spread throughout Germany like wildfire. By 1526 a Protestant party was officially recognized in different parts of Europe. This was the birth of the Reformation, and with it the vast worldly power of the Catholic Church, at least as Leo had inherited it, was irrevocably broken. That obscure, pedantic Wittenberg priest had somehow won the war.

## Interpretation

Luther's original intention in his Ninety-five Theses was to discuss a point of theology: the relationship, or lack of it, between God's forgiveness and papal indulgences. But when he read Prieras's response to his argument, something changed in him. The pope and his men had failed to find justification for indulgences in the Bible. There was much more they could not justify as well, such as the pope's unlimited power to excommunicate. Luther came to believe that the church needed drastic reform.

Reformation, however, would require political power. If Luther simply railed at the church's wickedness from the pulpit or among his fellow priests, he would get nowhere. The pope and his men had attacked him personally, questioning his motives; now Luther in turn would go on the offensive, fighting fire with fire.

Luther's strategy was to make the war public, transforming his moral cause into a political one. He did this by exploiting the previous century's advances in printing technology: his tracts, written in vigorous, angry language that appealed to the masses, were widely disseminated. He chose points of attack that would particularly outrage the German people: the pope's decadent lifestyle, funded through the sale of indulgences; the use of church power to meddle in German politics; on and on. Perhaps most devastating of all, Luther exposed the church's hypocrisies. Through these various tactics, he was able to spark and stoke a moral anger that spread like fire, forever tainting the public's vision not just of the pope but of the church itself.

Luther knew that Leo would respond to him not with arguments based on the Bible but with heavy-handed force, which, he also knew, would only make his cause shine all the brighter. And so with incendiary language and arguments that questioned Leo's authority, he baited the pope into rash counterattacks. Luther already led an exemplary life, but he took it further by refusing all income from his writings. This widely known move in effect made his goodness theatrical, a matter for public consumption. In a few short years, Luther gained so much support among the masses that the pope could not fight him without provoking a revolution. By using morality so consciously and publicly, he transformed it into a strategy for winning power. The Reformation was one of the greatest *political* victories in history.

Understand: you cannot win wars without public and political support, but people will balk at joining your side or cause unless it seems righteous and just. And as Luther realized, presenting your cause as just takes strategy and showmanship. First, it is wise to pick a fight with an enemy that you can portray as authoritarian, hypocritical, and power-hungry. Using all available media, you strike first with a moral offensive against the opponent's points of vulnerability. You make your language strong and appealing to the masses, and craft it, if you can, to give people the opportunity to express a hostility they already feel. You quote your enemies' own words back at them to make your attacks seem fair, almost disinterested. You create a moral taint that sticks to them like glue. Baiting them into a heavy-handed counterattack will win you even more public support. Instead of trumpeting your own goodness—which would

*primarily, however, it will be used as a threat to prevent the enemy undertaking some particular action. . . . It is a point worth noting that, just as in military operations one captures a position on the ground and thereby denies it to the enemy, on the psychological plane it is possible to take over abstract positions and equally deny them to the other side. The* [*leaders of the*] *Soviet Union for instance, . . . have turned into their own preserve the peace platform, that of the abolition of atomic weapons (while themselves continuing to develop them) and that of anti-colonialism while themselves ruling the only colonial empire still in existence. . . . It may therefore be that these ideological positions occupied by the forces of Marxism may one day be "conquered" by the West; but this presupposes that the latter in their indirect strategy have learned the value of thinking and calculating instead of merely trying to apply juridical or moral principles which their enemy can use against them at every turn.*

INTRODUCTION TO *STRATEGY*,
ANDRÉ BEAUFRE,
1963

make you seem smug and arrogant—you show it through the contrast between their unreasonable actions and your own crusading deeds. Aim at them the most withering charge of all—that they are after power, while you are motivated by something higher and selfless.

Do not worry about the manipulations you will have to resort to if you are to win this moral battle. Making a public show that your cause is more just than the enemy's will amply distract people from the means you employ.

*How should a regime pursue a counterguerrilla campaign? [Colonel John] Boyd laid out an array of tools: Undermine the guerrillas' cause and destroy their cohesion by demonstrating integrity and competence of government to represent and serve the needs of the people rather than exploit and impoverish them for the benefit of a greedy elite. (If you cannot realize such a political program, Boyd noted, you might consider changing sides now to avoid the rush later!) Take political initiative to root out and visibly punish corruption. Select new leaders with recognized competence as well as popular appeal. Ensure that they deliver justice, eliminate major grievances, and connect the government with its grass roots.*

THE MIND OF WAR: JOHN BOYD AND AMERICAN SECURITY, GRANT T. HAMMOND, 2001

*There always are concrete human groupings which fight other concrete human groupings in the name of justice, humanity, order, or peace. When being reproached for immorality and cynicism, the spectator of political phenomena can always recognize in such reproaches a political weapon used in actual combat.*

—Carl Schmitt (1888–1985)

## KEYS TO WARFARE

In almost all cultures, morality—the definition of good and evil—originated as a way to differentiate one class of people from another. In ancient Greece, for example, the word for "good" was first associated with the nobility, the higher classes who served the state and proved their bravery on the battlefield; the bad—the base, self-centered, and cowardly—were generally the lower classes. Over time a system of ethics evolved that served a similar but more sophisticated function: to keep society orderly by separating the antisocial and "evil" from the social and "good." Societies use ideas about what is and is not moral to create values that serve them well. When these values fall behind the times or otherwise cease to fit, morality slowly shifts and evolves.

There are individuals and groups, however, who use morality for a much different purpose—not to maintain social order but to extract an advantage in a competitive situation, such as war, politics or business. In their hands morality becomes a weapon they wield to attract attention to their cause while distracting attention from the nastier, less noble actions inevitable in any power struggle. They tend to play on the ambivalence we all have about conflict and power, exploiting our feelings of guilt for their purposes. For instance, they may position themselves as victims of injustice, so that opposing them seems wicked or insensitive. Or they may make such a show of moral superiority that we feel ashamed to disagree with them. They are masters at occupying the high ground and translating it into some kind of power or advantage.

Let us call these strategists "moral warriors." There are generally two types: unconscious and conscious. Unconscious moral warriors tend to be motivated by feelings of weakness. They may not be so good at the straightforward game of power, so they function by making other people feel guilty and morally inferior—an unconscious, reflexive way of leveling the playing field. Despite their apparent fragility, they are dangerous on an individual level, because they seem so sincere and can have great power over people's emotions. Conscious moral warriors are those who use the strategy knowingly. They are most dangerous on a public level, where they can take the high ground by manipulating the media. Luther was a conscious moral warrior, but, being also a genuine believer in the morality he preached, he used the strategy only to help him in his struggle with the pope; slipperier moral warriors tend to use it indiscriminately, adapting it to whatever cause they decide to take on.

*It is a world not of angels but of angles, where men speak of moral principles but act on power principles; a world where we are always moral and our enemies always immoral.*

*RULES FOR RADICALS,* SAUL D. ALINSKY, 1909–1972

The way to combat moral warriors in general is indicated by certain strategies that have evolved in modern warfare itself. The French officer and writer André Beaufre has analyzed the use of morality as a military strategy in the contexts of the French-Algerian wars of the 1950s and of the Vietnam wars fought by first France and then the United States. Both the Algerians and the North Vietnamese worked hard to frame each of their respective conflicts as a war of liberation fought by a nation struggling for its freedom against an imperialist power. Once this view was diffused in the media and accepted by many in the French and American publics, the insurgents were able to court international support, which in turn served to isolate France and the United States in the world community. Appealing directly to groups within these countries that were latently or overtly sympathetic to or at least ambivalent about their cause, they were able to sap support for the war from within. At the same time, they cleverly disguised the many nasty maneuvers to which they themselves resorted to fight their guerrilla wars. As a result, in the eyes of the world, they dominated the moral battlefield, enormously inhibiting France's and America's freedom of action. Stepping gingerly through a political and moral minefield, these powers could not fight their wars in a winnable manner.

Beaufre calls the strategic use of morality an "exterior maneuver," for it lies outside the territory being fought over and outside battlefield strategy. It takes place in its own space—its own moral terrain. For Beaufre both France and the United States made the mistake of ceding the high ground to the enemy. Because both countries had rich democratic traditions and saw their wars as justified, they assumed that others would perceive these struggles the same way. They saw no need to fight for the moral terrain—and that was a fatal mistake. Nations today must play the public game, deflecting their enemies' attempts to portray them as evil. Without appearing to whine about what the other side is doing, they must also work to expose their enemies' hypocrisies, taking the war to the moral court themselves—fighting on apparently moral terms. Cede the moral terrain to the other side and you limit your freedom of action: now anything you might have to do that is manipulative yet necessary will feed the unjust image the enemy has publicized, and you will hesitate to take such action.

This has great relevance to all forms of conflict. When your enemies

*Humanity as such cannot wage war because it has no enemy, at least not on this planet. The concept of humanity excludes the concept of the enemy, because the enemy does not cease to be a human being—and hence there is no specific differentiation in that concept. That wars are waged in the name of humanity is not a contradiction of this simple truth; quite the contrary, it has an especially intensive political meaning. When a state fights its political enemy in the name of humanity, it is not a war for the sake of humanity, but a war wherein a particular state seeks to usurp a universal concept against its military opponent. At the expense of its opponent, it tries to identify itself with humanity in the same way as one can misuse peace, justice, progress, and civilization in order to claim these as one's own and to deny the same to the enemy. The concept of humanity is an especially useful ideological instrument of imperialist expansion, and in its ethical-humanitarian form it is a specific vehicle of economic imperialism. Here one is reminded of a somewhat modified expression of Proudhon's: whoever invokes humanity wants to cheat. To confiscate the word humanity, to invoke and monopolize*

try to present themselves as more justified than you are, and therefore more moral, you must see this move for what it most often is: not a reflection of morality, of right and wrong, but a clever strategy, an exterior maneuver. You can recognize an exterior maneuver in a number of ways. First, the moral attack often comes out of left field, having nothing to do with what you imagine the conflict is about. Something you have done in a completely different arena is dredged up as a way to drain your support or inject you with guilt. Second, the attack is often ad hominem; rational argument is met with the emotional and personal. Your character, rather than the issue you are fighting over, becomes the ground of the debate. Your motives are questioned and given the darkest turn.

Once you realize you are under attack by a moral warrior using the exterior maneuver, it is vital to keep control of your emotions. If you complain or lash out angrily, you just look defensive, as if you had something to hide. The moral warrior is being strategic; the only effective response is to be strategic, too. Even if you know that your cause is just, you can never assume that the public sees it the same way. Appearances and reputation rule in today's world; letting the enemy frame these things to its liking is akin to letting it take the most favorable position on the battlefield. Once the fight for moral terrain has begun, you must fight to occupy the high ground in the same way you would in a shooting war.

Like any form of warfare, moral conflict has both offensive and defensive possibilities. When you are on the offense, you are actively working to destroy the enemy's reputation. Before and during the American Revolution, the great propagandist Samuel Adams took aim at England's reputation for being fair-minded, liberal, and civilized. He poked holes in this moral image by publicizing England's exploitation of the colonies' resources and simultaneous exclusion of their people from democratic processes. The colonists had had a high opinion of the English, but not after Adams's relentless campaign.

To succeed, Adams had to resort to exaggeration, picking out and emphasizing the cases in which the English were heavy-handed. His was not a balanced picture; he ignored the ways in which the English had treated the colonies rather well. His goal was not to be fair but to spark a war, and he knew that the colonists would not fight unless they saw the war as just and the British as evil. In working to spoil your enemy's moral reputation, do not be subtle. Make your language and distinctions of good and evil as strong as possible; speak in terms of black and white. It is hard to get people to fight for a gray area.

Revealing your opponent's hypocrisies is perhaps the most lethal offensive weapon in the moral arsenal: people naturally hate hypocrites. This will work, however, only if the hypocrisy runs deep; it has to show up in their values. Few will care about some innocuous self-contradictory comment made or vote taken long ago, but enemies who trumpet certain values as inherent to their side yet who do not always adhere to those values in reality make juicy targets. The Algerian and North Vietnamese propaganda campaigns were so destructive in part because of the discrepancy they were able to show between the values of freedom and liberty espoused by France and the United States and the actions those countries were taking to squash national independence movements. Both nations seemed hypocritical.

If a fight with your enemies is inevitable, always work to make them start it. In 1861, President Abraham Lincoln maneuvered carefully to make the South shoot first at Fort Sumter, initiating the Civil War. That put Lincoln on the moral high ground and won over many ambivalent Northerners to his side. Similarly, even if you are fighting a war of aggression, your goal to take from your enemy, find a way to present yourself not as a conqueror but as a liberator. You are fighting not for land or money but to free people suffering under an oppressive regime.

In general, in a conflict that is potentially nasty, in which you are certain the enemy will resort to almost anything, it is best that you go on the offensive with your moral campaign and not wait for their attacks. Poking holes in the other side's reputation is easier than defending your own. The more you stay on the offensive, the more you can distract the public from your own deficiencies and faults—and faults are inevitable in war. If you are physically and militarily weaker than your enemy, all the more reason to mount an exterior maneuver. Move the battle to the moral terrain, where you can hamstring and beat a stronger foe.

The best defense against moral warriors is to give them no target. Live up to your good name; practice what you preach, at least in public; ally yourself with the most just causes of the day. Make your opponents work so hard to undermine your reputation that they seem desperate, and their attacks blow up in their faces. If you have to do something nasty and not in harmony with your stated position or public image, use a cat's-paw—some agent to act for you and hide your role in the action. If that is not possible, think ahead and plan a moral self-defense. At all costs avoid actions that carry the taint of hypocrisy.

A stain on your moral reputation can spread like an infection. As you scramble to repair the damage, you often inadvertently publicize the doubts it has opened up, which simply makes things worse. So be prudent: the best defense against a moral attack is to have inoculated yourself against it beforehand, by recognizing where you may be vulnerable and taking preventive measures. When Julius Caesar crossed the Rubicon and initiated the Civil War against Pompey, he was highly vulnerable to the charge of trying to usurp the authority of the Roman Senate in order to become a dictator. He inoculated himself against these charges by acting mercifully toward his enemies in Rome, making important reforms, and going to the extreme in showing his respect for the Republic. By embracing some of the principles of his enemies, he kept their attempts at moral infection from spreading.

Wars are most often fought out of self-interest: a nation goes to war

*such a term probably has certain incalculable effects, such as denying the enemy the quality of being human and declaring him to be an outlaw of humanity; and a war can thereby be driven to the most extreme inhumanity.*

*THE CONCEPT OF THE POLITICAL,*
CARL SCHMITT,
1932

*Successful wickedness hath obtained the name virtue . . . when it is for the getting of the kingdom.*

THOMAS HOBBES, 1588–1679

to protect itself against an invading, or potentially dangerous, enemy or to seize a neighbor's land or resources. Morality is sometimes a component in the decision—in a holy war or crusade, for example—but even here self-interest usually plays a role; morality is often just a cover for the desire for more territory, more riches, more power. During World War II, the Soviet Union became a beloved ally of the United States, playing a key role in the defeat of Hitler, but after the war it became America's darkest enemy. American self-interest, not the Soviets, had changed.

Wars of self-interest usually end when the winner's interests are satisfied. Wars of morality are often longer and bloodier: if the enemy is seen as evil, as the infidel, it must be annihilated before the war can end. Wars of morality also churn up uncontrollable emotions. Luther's moral campaign against Rome generated such hatred that in the subsequent invasion of the Holy City by the troops of Charles V, in 1527, German soldiers went on a six-month rampage against the church and its officials, committing many atrocities in what came to be known as "the sack of Rome."

As in war, so in life. When you are involved in a conflict with another person or group, there is something you are fighting over, something each side wants. This could be money, power and position, on and on. Your interests are at stake, and there is no need to feel guilty about defending them. Such conflicts tend not to be too bloody; most people are at least somewhat practical and see the point in preventing a war from going on too long. But those people who fight out of a moral sense can sometimes be the most dangerous. They may be hungry for power and are using morality as a cover; they may be motivated by some dark and hidden grievance; but in any case they are after more than self-interest. Even if you beat them, or at least defend yourself against them successfully, discretion here may be the better part of valor. Avoid wars of morality if you can; they are not worth the time and dirty feelings they churn up.

**Image:** Germs. Once they get inside and attack the body, they spread quickly. Your attempts to destroy them often make them stronger and harder to root out. The best defense is prevention. Anticipate the attack and inoculate yourself against it. With such organisms you have to fight fire with fire.

**Authority:** The pivot of war is nothing but name and righteousness. Secure a good name for yourself and give the enemy a bad name; proclaim your righteousness and reveal the unrighteousness of the enemy. Then your army can set forth in a great momentum, shaking heaven and earth. —*Tou Bi Fu Tan,* A Scholar's Dilettante Remarks on War *(sixteenth century A.D.)*

## REVERSAL

A moral offensive has a built-in danger: if people can tell what you are doing, your righteous stance may disgust and alienate them. Unless you are facing a vicious enemy, it is best to use this strategy with a light touch and never seem shrill. Moral battles are for public consumption, and you must constantly gauge their effect, lowering or raising the heat accordingly.

# 26

# DENY THEM TARGETS

## THE STRATEGY OF THE VOID

*The feeling of emptiness or void—silence, isolation, nonengagement with others—is for most people intolerable. As a human weakness, that fear offers fertile ground for a powerful strategy: give your enemies no target to attack, be dangerous but elusive and invisible, then watch as they chase you into the void. This is the essence of guerrilla warfare. Instead of frontal battles, deliver irritating but damaging side attacks and pinprick bites. Frustrated at their inability to use their strength against your vaporous campaign, your opponents will grow irrational and exhausted. Make your guerrilla war part of a grand political cause—a people's war—that crests in an irresistible revolution.*

## THE LURE OF THE VOID

In 1807, Napoleon Bonaparte of France and Czar Alexander I of Russia signed a treaty of alliance. Now the period's two great military powers were linked. But this treaty was unpopular with the Russian court—among other things it allowed Napoleon nearly free rein in Poland, Russia's traditional "front yard." Russian aristocrats worked to influence the czar to repudiate it. Before too long, Alexander began to take actions that he knew would displease the French, and by August 1811, Napoleon had had enough: it was time to teach Russia a lesson. He began to lay plans for an invasion. The acquisition of this vast territory to the east would make him the ruler of the largest empire in history.

Some of Napoleon's ministers warned him of the dangers of invading such a vast country, but the emperor general felt supremely confident. The Russian army was undisciplined, and its officers were squabbling among themselves. Two forces in Lithuania were positioned to block an invasion from the west, but intelligence had revealed that they were unprepared. Napoleon would march into a central position between these forces and defeat them in detail. He would ensure victory by mobilizing an army three times larger than any he had previously led: 650,000 men would march into Russia, 450,000 as part of the main attack force, the rest to secure lines of communication and supply. With an army this size, he could dominate even the large spaces of Russia, overwhelming the feeble enemy not only with his usual brilliant maneuvers but with superior firepower.

Napoleon may have felt certain of victory, but he was not a reckless man. As always, he studied the situation from every angle. He knew, for instance, that Russian roads were notoriously bad, local food supplies were meager, the climate tended to extremes of heat and cold, and the vast distances made it much harder to encircle the enemy—there was always room to retreat. He read up on the failed invasion of Russia by the king of Sweden, Charles XII, in 1709, and anticipated that the Russians might revert to a scorched-earth policy. His army would have to be as self-sufficient as possible (the distances were too great to have extended supply lines from Europe), but, given its size, that would require incredible planning and organization.

To help provide for his army, Napoleon had vast storehouses close to the borders of Russia filled with wheat and rice. He knew it would be impossible to provide fodder for the 150,000 horses of his army, and so, thinking ahead, he decided they would have to wait until June for the invasion, when the grasses of the Russian plains would be rich and green. At the last minute, he learned Russia had very few mills to grind grain into flour, so he added to his growing list the need to bring materials to build mills along the way. With the logistical problems addressed and his usual well-devised strategy in hand, Napoleon told his ministers that he foresaw complete victory within three weeks. In the past, these predictions of Napoleon's had been uncannily accurate.

In June 1812, Napoleon's vast armada of men and supplies crossed into Russia. Napoleon always planned for the unexpected, but this time unmanageable difficulties began to pile up almost immediately: rain, the bad roads, the intense summer heat brought the army's movements to a crawl. Within days more than 10,000 horses ate rank grass and died. Supplies were failing to reach the forward troops fast enough, and they had to resort to foraging, but the uncooperative Russian peasants along the march not only refused to sell their food at any price but burned their hay rather than let the French have it. More French horses died when they were forced to feed off the thatch in the roofs of houses, only to find the houses collapsing on them. The two Russian armies in Lithuania retreated too fast to be caught, and as they went, they burned crops and destroyed all storehouses of food. Dysentery quickly spread through the French troops; some nine hundred men died each day.

In his effort to catch and destroy at least a part of his elusive enemy, Napoleon was compelled to march ever farther east. At points he came tantalizingly close to the more northern of the two Russian armies, but his exhausted men and horses could not move fast enough to meet or encircle it, and it easily escaped his traps each time. June bled into July. Now it became clear that the Russians would be able to join their two armies at Smolensk, over 200 miles east of where Napoleon had intended to fight them and a mere 280 miles from Moscow. Napoleon had to call a halt and rethink his plan.

Thousands of French soldiers had succumbed to disease and hunger without a single battle's being fought. The army was strung out along a 500-mile line, parts of which were constantly harassed by small troops of Cossacks on horseback, sowing terror with their bloodthirsty raids. Napoleon could not allow the chase to go on any longer—he would march his men to Smolensk and fight the decisive battle there. Smolensk was a holy city, with great emotional significance to the Russian people. Surely the Russians would fight to defend it rather than let it be destroyed. He knew that if he could only meet the Russians in battle, he would win.

And so the French moved on Smolensk, arriving there in mid-August, their 450,000-man attack force reduced to 150,000 and worn down by the intense heat. Finally, as Napoleon had predicted, the Russians made a stand here—but only briefly; after several days of fighting, they retreated yet again, leaving behind a burned and ruined city with nothing in it to feed on or plunder. Napoleon could not understand the Russian people, who seemed to him suicidal—they would destroy their country rather than surrender it.

Now he had to decide whether to march on Moscow itself. It might have seemed wise to wait through the winter at Smolensk, but that would give the czar time to raise a larger army that would prove too hard for Napoleon to handle with his own depleted forces. The French emperor felt certain the czar would defend Moscow, the very heart and soul of

*In addition to wasting an ever-increasing proportion of French manpower, the elusive Russian tactics also contributed to the mental as well as physical exhaustion of Napolean's forces. Tip and run raids by small bands of Cossacks were continuous and exercised a baleful influence far in excess of the military danger they represented. The French army became increasingly subject to fits of the jitters. Captain Roeder noted one typical example in his diary. The Hessian troops were mustering for parade before the Emperor's quarters at Vitebsk on August 17, when "everything was suddenly thrown into ridiculous uproar because a few Cossacks had been sighted, who were said to have carried off a forager. The entire garrison sprang to arms, and when they had ridden out it was discovered that we were really surrounded by only a few dozen Cossacks who were dodging about hither and thither. In this way they will be able to bring the whole garrison to hospital in about fourteen days without losing a single man."*

*The Campaigns of Napoleon*,
David G. Chandler,
1966

Russia. Once Moscow fell, Alexander would have to sue for peace. So Napoleon marched his haggard troops still farther east.

Now, at last, the Russians turned to face the French in battle, and on September 7 the two armies clashed near the village of Borodino, a mere seventy-five miles from Moscow. Napoleon no longer had enough forces or cavalry to attempt his usual flanking maneuver, so he was forced to attack the enemy head-on. The Russians fought bitterly, harder than any army Napoleon had ever faced. Even so, after hours of brutal fighting, the Russians retreated yet again. The road to Moscow lay open. But the Russian army was still intact, and Napoleon's forces had suffered horrific casualties.

Seven days later Napoleon's army, now reduced to 100,000 men, straggled into an undefended Moscow. A French marshal wrote to his wife that the emperor's "joy was overflowing. 'The Russians,' he thinks, 'will sue for peace, and I shall change the face of the world.' " In earlier years, when he had marched into Vienna and Berlin, he had been met as a conquering hero, with dignitaries turning over to him the keys to their cities. But Moscow was empty: no citizens, no food. A terrible fire broke out almost immediately and lasted five days; all of the city's water pumps had been removed—an elaborate sabotage to make Moscow still more inhospitable.

Napoleon sent letters to the czar, offering generous terms of peace. At first the Russians seemed willing to negotiate, but the weeks went by, and it finally became clear that they were dragging out the talks to buy time to build up their army—and to let winter grow closer.

Napoleon could not risk staying in Moscow another day; the Russians would soon be able to encircle his now meager force. On October 19 he marched the remains of his army out of the Russian capital. His goal was to get to Smolensk as fast as possible. Now those undisciplined bands of Cossacks that had harassed him on the road east had formed into larger divisions—guerrilla forces of 500 men—and every day they killed off more and more French soldiers. Marching in constant fear, Napoleon's men rarely slept. Thousands succumbed to fatigue and hunger. Napoleon was forced to lead them past the nightmarish fields of Borodino, still crowded with French corpses, many half eaten by wolves. The snow began to fall—the Russian winter set in. The French horses died from the cold, and every last soldier had to trudge through the snow on foot. Barely 40,000 made it to Smolensk.

The cold was worsening. There was no time to tarry in Smolensk. Through some deft maneuvering, Napoleon managed to get his troops across the Berezina River, allowing them a clear line of retreat to the west. Then, in early December, hearing of a failed coup d'état at home in France, he left his troops behind and headed for Paris. Of the 450,000 men in his main attack force, some 25,000 made it back. Few among the rest of the army survived as well. Napoleon had miraculously escaped to fight more wars, but he would never recoup his losses in manpower and horses. Russia was indeed his grave.

Interpretation

By the time Napoleon invaded Russia, Czar Alexander I had met him a number of times in previous years and had come to know him quite well. The emperor, Alexander saw, was an aggressive man who loved any kind of fight, even if the odds were stacked against him. He needed battles as a chance to put his genius in play. By refusing to meet him in battle, Alexander could frustrate him and lure him into a void: vast but empty lands without food or forage, empty cities with nothing to plunder, empty negotiations, empty time in which nothing happened, and finally the dead of winter. Russia's harsh climate would make a shambles of Napoleon's organizational genius. And as it played out, Alexander's strategy worked to perfection. Napoleon's inability to engage his enemy got under his skin: a few more miles east, one solid battle, and he could teach this cowardly foe a lesson. His emotions—irritation, anger, confusion—overwhelmed his ability to strategize. How could he have come to believe, for instance, that the fall of Moscow would force the czar to surrender? Alexander's army was still intact, the French had grown frighteningly weak, and winter was coming. Napoleon's mind had succumbed to the powerful pull of the void that he had entered, and that led him far astray.

Alexander's strategy wreaked havoc on the French soldiers as well, who were renowned for their superior discipline and fighting spirit. A soldier can endure almost anything except the expectation of a battle that never comes and a tension that is never relieved. Instead of battle, the French got endless raids and pinprick attacks that came out of nowhere, a continuous threat that gradually built into panic. While thousands of soldiers fell to disease, many more simply lost the will to fight.

It is human nature to not be able to endure any kind of void. We hate silence, long stretches of inactivity, loneliness. (Perhaps this is related to our fear of that final void, our own death.) We have to fill and occupy empty space. By giving people nothing to hit, being as vaporous as possible, you play upon this human weakness. Infuriated at the absence not just of a fight but of any kind of interaction at all, people will tend to chase madly after you, losing all power of strategic thought. It is the elusive side, no matter how weak or small its force, that controls the dynamic.

The bigger the enemy, the better this strategy works: struggling to reach you, the oversize opponent presents juicy targets for you to hit. To create the maximum psychological disturbance, you must make your attacks small but relentless, keeping your enemy's anger and frustration at a constant boil. Make your void complete: empty negotiations, talks leading nowhere, time passing without either victory or defeat. In a

*Such was the system Spain used against us. One hundred and fifty to two hundred guerrilla bands scattered all over Spain had sworn to kill thirty or forty Frenchmen a month each: that made six to eight thousand men a month for all guerrilla bands together. The order was never to attack soldiers traveling as a body, unless the guerrillas outnumbered them. But they fired on all stragglers, attacked small escorts, and sought to lay hands on the enemy's funds, couriers, and especially convoys. As all the inhabitants acted as spies for their fellow citizens, the guerrillas knew when the convoys would leave and how strong their escorts would be, and the bands would make sure they were twice the size. They knew the country very well, and they would attack furiously in the most favorable spot. Success often crowned the undertaking; but they always killed a lot of men, and the goal was achieved. As there are twelve months in the year, we were losing about eighty thousand men a year, without any pitched battles. The war in Spain lasted seven years, so over five hundred men were killed. . . . But that includes only those killed by the guerrillas. Add the battles of Salamanca, Talavera, and Vitoria, and several*

world of accelerated pace and activity, this strategy will have a powerfully debilitating effect on people's nerves. The less they can hit, the harder they will fall.

*Most wars are wars of contact, both forces striving to keep in touch. . . . The Arab war should be a war of detachment: to contain the enemy by the silent threat of a vast unknown desert, not disclosing themselves till the moment of attack. . . . From this theory came to be developed ultimately an unconscious habit of never engaging the enemy at all. This chimed with the numerical plea of never giving the enemy's soldier a target.*
—*T. E. Lawrence,* The Seven Pillars of Wisdom *(1926)*

## KEYS TO WARFARE

Over the centuries organized war—in all its infinite variations, from primitive to modern, Asian to Western—has always tended to follow a certain logic, which is so universal as almost to seem inherent to the process. The logic is as follows: A leader decides to take his country to war and raises an army for that purpose. That army's goal is to meet and defeat the enemy in a decisive battle that will force a surrender and favorable peace terms. The strategist guiding the campaign must deal with a specific area, the theater of war. This area is most often relatively limited; maneuvering in vast open spaces complicates the possibility of bringing the war to closure. Working within the theater of war, then, the strategist contrives to bring his army to the decisive battle in a way that will surprise the enemy or put it at a disadvantage—it is cornered, or attacked from both front and rear, or must fight uphill. To keep his forces strong enough to deliver a mortal blow, he concentrates them rather than dispersing them. Once battle begins, the army will naturally form a flank and rear that it must protect against encirclement, as well as lines of communication and supply. It may take several battles to end the war, as each side works to dominate the key positions that will give it control of the theater, but military leaders must try to end it as quickly as possible. The longer it drags on, the more the army's resources are stretched to a breaking point where the ability to fight collapses. Soldiers' morale declines with time as well.

As with any human activity, however, this positive, orderly side generates a negative, shadow side that contains its own form of power and reverse logic. The shadow side is guerrilla warfare. The rudiments of guerrilla warfare originated thousands of years ago, when smaller nations found themselves invaded by more powerful neighbors; to survive, their armies were forced to flee the invader, for any direct engagement would have destroyed them. It soon became clear that the longer they fled and eluded battle, the more they ruined the enemy's strategies and confused it by not conforming to the usual logic of engagement.

The next step was to take this further: these early guerrilla warriors learned the value of operating in small, dispersed bands as opposed to a concentrated army, keeping in constant motion, never forming a front, flank, or rear for the other side to hit. The enemy would want to keep the war confined to a particular space; better, then, to extend it over as much territory as possible, melting into the countryside, forcing the enemy to disperse itself in the chase, opening itself up to raids and pinprick attacks. The enemy would naturally want a quick end to the war, so it was desirable to drag it out as long as possible, making time an offensive weapon that consumed the enemy with friction and sagging morale.

In this way, over thousands of years and through trial and error, the art of guerrilla warfare developed and was refined into its present-day form. Conventional military training and thought revolve around concentrating for battle, maneuvering within limited areas, and straining for the quick kill. Guerrilla warfare's reversal of this natural order of war makes it impossible for a conventional army to counter—hence its power. In the shadow land of reverse warfare, where none of the normal rules apply, the conventional army flounders. Done right, guerrilla warfare is virtually unbeatable.

The word "guerrilla"—"small war" in Spanish—was coined in reference to the Peninsular War of 1808–14, which began when Napoleon invaded Spain. Melting into their country's mountains and inhospitable terrain, the Spaniards tortured the French, making it impossible for them to profit from their superior numbers and weaponry. Napoleon was bedeviled by an enemy that attacked without forming a front or rear. The Cossack fighters who undid him in Russia in 1812 had learned a lot from the Spanish and perfected the use of guerrilla warfare; their harassment caused far more damage than anything the rather incompetent Russian army could inflict.

This strategy has become a more powerful and prevalent tool in modern warfare for several reasons: First, by exploiting technological advances in weaponry and explosives, a small guerrilla band can cause disproportionate damage. Second, Napoleonic warfare greatly expanded the size of conventional armies, making them much more vulnerable to hit-and-run tactics from light and mobile forces. Finally, guerrilla war has been adopted for political purposes, to great effect. By infusing local people with the fervor of a cause, a revolutionary leader can covertly multiply his strength: his civilian supporters can sabotage the enemy's invading force, provide valuable intelligence, and turn the countryside into an armed camp.

The power of guerrilla warfare is essentially psychological. In conventional warfare everything converges on the engagement of two armies in battle. That is what all strategy is devised for and what the martial instinct requires as a kind of release from tension. By postponing this natural convergence indefinitely, the guerrilla strategist creates intense

*others that our troops lost; the sieges, . . . the fruitless attack on Cádiz; add too the invasion and evacuation of Portugal, the fevers and various illnesses that the temperature caused our soldiers to suffer, and you will see that we could add a further three hundred thousand men to that number during those seven years. . . .*
*. . . From what has been said, it will be apparent that the prime aim of this sort of war is to bring about the destruction of the enemy almost without him noticing it, and as a drop of water dripping on a stone will eventually dig a hole in the stone, patience and perseverance are needed, always following the same system. In the long run, the enemy will suffer more from this than he would from losing pitched battles.*

*On Partisans and Irregular Forces,*
J.F.A. Le Mière de Corvey,
1823

frustration. The longer this mental corrosion continues, the more debilitating it gets. Napoleon lost to the Russians because his strategic bearings were pushed off course; his mind fell before his army did.

Because it is so psychological, guerrilla strategy is infinitely applicable to social conflict. In life as in war, our thoughts and emotions naturally converge on moments of contact and engagement with others. We find people who are deliberately elusive, who evade contact, extremely disconcerting. Whether because we want to grab them and pin them down or because we are so annoyed with them that we want to hit them, they pull us toward them, so that either way the elusive one controls the dynamic. Some people take this further, attacking us in ways that are evasive and unpredictable. These opponents can gain a disturbing power over our minds, and the longer they keep it up, the more we are sucked into fighting on their terms. With advances in technology that make it easier to maintain a vaporous presence, and the use of the media as both a screen and a kind of guerrilla adjunct, the power and effectiveness of this warfare in political or social battle are greatly enhanced. In heated political times, a guerrilla-style campaign—allied with some cause—can be used to wage a people's war against large entities, corporations, entrenched powers. In this kind of public combat, everyone loves to fight on the guerrillas' side because the participants are more deeply involved in the struggle, not mere cogs in a giant machine.

Franklin Roosevelt was a kind of political guerrilla warrior. He liked to fight evasively and strategized to deny the Republicans any targets to hit. He used the media to make himself seem to be everywhere and to be waging a kind of people's war against moneyed interests. In classic guerrilla manner, he also reorganized the Democratic Party to make it less centralized, more mobile and fluid for local battles. For Roosevelt, though, the guerrilla approach was not so much a coherent strategy as a style. As many do, he unconsciously sensed the power in being evasive and fought that way to great effect—but to make this strategy really work, it is better to use it consciously and rationally. Guerrilla strategy may be the reverse side of war, but it has its own logic, backward yet rigorous. You cannot just improvise it anarchically; you must think and plan in a new way—mobile, dimensional, and abstract.

The primary consideration should always be whether a guerrilla-style campaign is appropriate for the circumstances you are facing. It is especially effective, for instance, against an opponent who is aggressive yet clever—a man like Napoleon. These types cannot stand lack of contact with an enemy. They live to maneuver, outwit, and outhit. Having nothing to strike at neutralizes their cleverness, and their aggression becomes their downfall. It is interesting to note that this strategy works in love as well as in war and that here, too, Napoleon was its victim: it was by a guerrilla-style seduction—by enticing him to chase her, giving tantalizing lures but offering him nothing solid to grasp—that Empress Josephine made him her slave.

This strategy of the void works wonders on those who are used to conventional warfare. Lack of contact is so outside their experience that it warps any strategic powers they have. Large bureaucracies are often perfect targets for a guerrilla strategy for the same reason: they are capable of responding only in the most orthodox manner. In any event, guerrilla warriors generally need an opponent that is large, slow-footed, and with bullying tendencies.

Once you have determined that a guerrilla war is appropriate, take a look at the army you will use. A large, conventional army is never suitable; fluidity and the ability to strike from many angles are what counts. The organizational model is the cell—a relatively small group of men and women, tight-knit, dedicated, self-motivated, and spread out. These cells should penetrate the enemy camp itself. This was how Mao Tse-tung organized his army in the Chinese Revolution, infiltrating the Nationalist side, causing sabotage in the cities, leaving the deceptive and terrifying impression that his men were everywhere.

When U.S. Air Force Colonel John Boyd joined the Pentagon in the late 1960s to help develop jet fighters, he faced a reactionary bureaucracy dominated by commercial interests rather than military ones. The Pentagon was in dire need of reform, but a traditional bureaucratic war—an attempt to convince key staff directly and frontally of the importance of his cause—would have been a hopeless venture: Boyd would simply have been isolated and funneled out of the system. He decided to wage a guerrilla war instead. His first and most important step was to organize cells within the Pentagon. These cells were small and hard to detect, giving the reactionaries nothing to hit at when they realized they were in a war. Boyd recruited his guerrillas from among those dissatisfied with the status quo, especially the young—young people are always more receptive to change, and they love this style of fighting.

With his cells in place, Boyd had constant intelligence as to what was going on in the Pentagon and could anticipate the timing and content of the attacks on him. He could also use these cells to spread his influence through word of mouth, infiltrating ever deeper into the bureaucracy. The main point is to avoid an organization's formal channels and tendency for bigness and concentration. Opt for mobility instead; make your army light and clandestine. You can also attach your guerrilla cells to a regular army, much as the Russian Cossacks supported the armies of Alexander. This mix of conventional and unconventional can prove highly effective.

Once you have organized your cells, you must find a way to lure the enemy into attacking you. In war this is generally accomplished by retreating, then turning to strike at the enemy with constant small raids and ambushes that cannot be ignored. This was the classic strategy pursued by T. E. Lawrence in Arabia during World War I. The nineteenth-century American financial wizard Jay Gould, a man who fought many guerrilla wars in his business life, did something similar in his daily battles. His goal was to create maximum disorder in the markets—disorder he could anticipate and exploit. One of his main adversaries was the highly aggressive mogul Commodore Cornelius Vanderbilt, whom he engaged in a war for control of the Erie Railroad in the late 1860s. Gould maintained an incredibly elusive presence; he would work back channels to gain influence in, for example, the New York State legislature, which then enacted laws undermining Vanderbilt's interests. The furious Vanderbilt would go after Gould and counterattack, but Gould would by then have moved on to some other unexpected target. To deprive Vanderbilt of the strategic initiative, Gould upset him, fed his competitive and aggressive instincts, then goaded him further by giving him no target to counterattack.

Gould also made skillful use of the media. He might plant a newspaper article that would suddenly sideswipe Vanderbilt, portraying him as an evil monopolist; Vanderbilt would have to respond, but that would only publicize the charge—and meanwhile Gould's name would be nowhere in evidence. The media in this instance are perfect as both the smoke screen concealing guerrilla tactics and the vehicle conveying them. Use the media to goad your enemies, getting them to disperse their energies in defending themselves while you watch, or find a new target to raid and ambush. Lacking a real battle to deal with, their frustration will mount and lead them to costly mistakes.

In conventional warfare the way you supply your army is a critical issue. In guerrilla warfare, on the other hand, you live off your enemies as much as possible, using their resources, their energy, and their power as a kind of supply base. Mao supplied his army mostly with captured equipment and food. Gould actually started out by infiltrating Vanderbilt's inner circle as a financial partner, then using Vanderbilt's immense resources to fund his mayhem. Using the enemy's matériel will help you to endure the longer length of any successful guerrilla campaign. In any event, you must plan to live cheaply, marshaling what you have for the long run.

In most conflicts time is a danger, bringing Murphy's Law into play: if anything can go wrong, it will. If your army is small and relatively self-sufficient, though, there is less to go wrong, and meanwhile you are working to make sure that for the enemy the passage of time is a nightmare. Morale is sinking, resources are stretched, and even great planners like Napoleon are finding themselves with problems they could never have foreseen. The effect is exponential: as unexpected problems crop up, the enemy starts making mistakes, which lead to more problems—and so it goes.

Make time an offensive weapon in your strategizing. Design your maneuvers to keep your enemies just barely going, always thinking that one more battle will do the trick. You want them to deteriorate slowly; a sudden sharp setback, a clear view of the trap you are laying for them, and they will pull out before the damage is done. Let them take key positions that give them the illusion of success. They will hold on to them tenaciously as your raids and pinprick attacks grow in number. Then, as they weaken, increase the pace of these attacks. Let them hope, let them think it is all still worth it, until the trap is set. Then break their illusion.

Just as you are extending time, contrary to convention, you are also extending space. You want to bring the fight to areas outside the theater of war, to include public and international opinion, turning the war into a political and global issue and giving the enemy too large a space to defend. Political support is invaluable to an underdog guerrilla campaign; the longer the fight is drawn out, the more the enemy seems morally unjustified and politically isolated. Always try to ally your guerrilla campaign with a cause you can defend as just and worthy.

You will win your guerrilla war in one of two ways. The first route is to increase the level of your attacks as your enemies deteriorate, then finish them off, as the Russians finished off Napoleon. The other method is by turning sheer exhaustion to your advantage: you just let the enemy give up, for the fight is no longer worth the aggravation. The latter way is the better one. It costs you less in resources, and it looks better: the enemy has fallen on his own sword. But even a guerrilla campaign cannot go on forever; at a certain point time starts to work against you as well. If the ending is taking too long, you must go on the offensive and finish the enemy off. In the Vietnam War, the North Vietnamese drew out the war to a point where it was also costing them too much. That was why they launched the Tet Offensive in 1968—to greatly accelerate the deterioration of the U.S. war effort.

The essence of guerrilla warfare is fluidity. The enemy will always try to adjust to what you are doing, attempting to find its feet in this unfamiliar terrain. You must be prepared to change and adopt whatever is contrary to expectation: this might mean occasionally fighting in a conventional manner, concentrating your army to attack here or there, then dispersing again. Your goal is maximum disorder and unfamiliarity. Remember: this war is psychological. It is more on the level of strategy than anything else that you give the enemy nothing to hold on to, nothing tangible to counter. It is the enemies' minds that are grasping at air and their minds that fall first.

**Image:** The Mosquito. Most animals present a front, back, and sides that can be attacked or threatened. Mosquitoes, though, give you nothing but an irritating whir in the ear, from all sides and angles. You cannot hit them, you cannot see them. Your flesh, meanwhile, affords them endless targets. Enough bites and you realize that the only solution is to stop fighting and move as far away as possible.

**Authority:** Anything that has form can be overcome; anything that takes shape can be countered. This is why sages conceal their forms in nothingness and let their minds soar in the void.
—Huainanzi *(second century B.C.)*

REVERSAL

A guerrilla strategy is extremely hard to counter, which is what makes it so effective. If you find yourself in a fight with guerrillas and you use conventional methods to fight them, you play into their hands; winning battles and taking territory means nothing in this kind of war. The only effective counterstrategy is to reverse the guerrillas' reversal, neutralizing their advantages. You must refuse them the freedom of time and space they need for their mayhem. You must work to isolate them—physically, politically, and morally. Above all, you must never respond in a graduated manner, by stepping up your forces bit by bit, as the United States did in the Vietnam War. You need a quick, decisive victory over such an opponent. If this seems impossible, it is better to pull out while you can than to sink into the protracted war the guerrilla fighter is trying to lure you into.

# 27

# SEEM TO WORK FOR THE INTERESTS OF OTHERS WHILE FURTHERING YOUR OWN

## THE ALLIANCE STRATEGY

*The best way to advance your cause with the minimum of effort and bloodshed is to create a constantly shifting network of alliances, getting others to compensate for your deficiencies, do your dirty work, fight your wars, spend energy pulling you forward. The art is in choosing those allies who fit the needs of the moment and fill the gaps in your power. Give them gifts, offer them friendship, help them in time of need—all to blind them to reality and put them under subtle obligation to you. At the same time, work to sow dissension in the alliances of others, weakening your enemies by isolating them. While forming convenient coalitions, keep yourself free of negative entanglements.*

THE DOG, THE COCK AND THE FOX

*A dog and a cockerel, having made friends, were strolling along a road together. As evening fell, the cockerel flew up into a tree to sleep there, and the dog went to sleep at the foot of the tree, which was hollow. According to his habit, the cockerel crowed just before daybreak. This alerted a fox nearby, who ran up to the tree and called up to the cockerel: "Do come down, sir, for I dearly wish to embrace a creature who could have such a beautiful voice as you!" The cockerel said: "I shall come down as soon as you awaken the doorkeeper who is asleep at the foot of the tree." Then, as the fox went to look for the "doorkeeper," the dog pounced briskly on him and tore him to pieces.* This fable teaches us that sensible men, when their enemies attack them, divert them to someone better able to defend them than they are themselves.

*FABLES,* AESOP, SIXTH CENTURY B.C.

## THE PERFECT ALLY

In 1467, Charles, the thirty-four-year-old Count of Charlois, received the news he had secretly longed for: his father, the Duke of Burgundy—known as Philip the Good—had died, making Charles the new duke. Father and son had clashed over the years. Philip was patient and practical and during his reign had slowly managed to expand Burgundy's already impressive holdings. Charles was more ambitious and more warlike. The empire he inherited was immense, including Flanders, Holland, Zeeland, and Luxembourg to the north of present-day France, and the important duchy of Burgundy itself in northeastern France. Now, as duke, Charles had the power and resources at his command to realize his dreams of conquest into Germany and beyond.

Two obstacles stood in his path. The first was the independent Swiss cantons to Burgundy's east. Charles would have to incorporate this territory, by force, before moving into southern Germany. The Swiss were fierce warriors who would not take kindly to any invasion. But in the end they could hardly match the size and power of the duke's army. The second obstacle was King Louis XI of France, Charles's cousin and archrival since childhood. France then was still a feudal state, composed of several duchies like Burgundy, whose dukes owed allegiance to the king. But these duchies were in fact independent powers and could form their own league if the king dared provoke them. Burgundy was the most powerful duchy of them all, and everyone knew Louis dreamed of swallowing it up and making France a united power.

Charles, however, felt confident that he could best his older cousin in both diplomacy and warfare. After all, Louis was weak, even a little soft in the head. How else to explain his strange infatuation with the Swiss cantons? Almost from the beginning of his reign, Louis had courted them assiduously, treating them almost like equals to France. There were many more powerful states he could have allied himself with to increase France's power, but he seemed obsessed with the Swiss. Perhaps he felt an affinity to their simple lifestyle; for a king, he himself had rather peasant tastes. Louis also had an aversion toward warfare, preferring to buy peace, even at a high price, than to fund an army.

It was imperative that Charles strike now, before Louis wised up and started acting more like a king. Charles formed a plan to realize his ambitions and then some: he would first move into Alsace, between France and Germany, and swallow up the weak kingdoms in the area. Then he would form an alliance with the great warrior king of England, Edward IV, whom he would persuade to land a large army at Calais. His own army would link up with the English at Reims, in central France, where Edward would be crowned the country's new king. The duke and Edward would easily dispose of Louis's weak army. The duke could then march east, across the Swiss cantons, while Edward would march south. Together they would form the dominant power in Europe.

By 1474 everything was in place. Edward had signed on to the plan.

The duke began by marching on the upper Rhine, but just as he began his maneuvers, he learned that a large Swiss army had invaded his home territory of Burgundy. This army was funded by Louis XI himself. By this action Louis and the Swiss were clearly sending a warning to the duke that they would not look kindly on any future invasion of the cantons, but Charles had enough forces in Burgundy to drive the Swiss out. He was not a man to provoke in such a way; both parties would more than pay for their rash invasion.

In the summer of 1475, the English army—the largest yet assembled for an invasion of France—landed at Calais under the personal leadership of Edward IV. Charles went to meet Edward to finalize their plans and toast their imminent conquests. He then quickly returned to his own troops, which were marching south through Lorraine in preparation for the great linkup with the English forces at Reims.

Suddenly some disturbing news reached Charles in the field: his spies at the French court reported that Louis had opened up secret negotiations with Edward. Apparently Louis had persuaded the English king that Charles was using him and could not be trusted. Knowing that England's finances were weak, Louis had offered generous terms of peace, amounting to a large annual pension paid directly to the king and his court. He had entertained the English with great feasts of food and ale. And then, to the duke's utter disgust and amazement, Edward fell for it, signed the treaty, and took his forces home.

The duke had barely had time to get over this bitter news when Louis suddenly sent him envoys to broker a long-term truce between France and Burgundy. This was typical of the king—everything he did was inconsistent and contradictory. What was he thinking? Signing the truce would mean the duke could now confidently march against the Swiss, knowing that France would not interfere. Perhaps the king was guided by his great fear of war? Charles happily approved the truce.

The Swiss were outraged: Louis had been their friend, and now, at the moment of imminent peril, he had abandoned them. But the Swiss were used to fighting on their own; they would simply have to mobilize every man available.

In the dead of the winter of 1477, the duke, impatient for victory, crossed the Jura Mountains heading east. The Swiss were waiting for him near the town of Grandson. This was the first time the duke had done battle with the Swiss, and he was caught off guard by what confronted him. It began with the alarming bellow of Swiss battle horns, which echoed in the mountains, creating a frightening din. Next, thousands of Swiss soldiers advanced down the slope toward the Burgundians. They marched with perfect precision, packed tight in phalanxes from which their enormous pikes stuck out like the spines of a giant hedgehog in motion. Their flanks and rear were protected by halberdiers swinging spiked battle-axes. It was a terrifying sight. The duke ordered attack after attack with his cavalry to break up the phalanxes, only to watch them

*Since in all her decisions, whether by chance or by choice, Rome took all steps necessary to make herself great, she did not overlook fraud. She could not at the start have been more deceitful than she was in the means she took, as we were saying just now, to acquire allies, since under this title she made them all her servants, as was the case with the Latins and other peoples round about. For she first availed herself of their arms in order to subjugate neighbouring peoples and to build up her reputation as a state, and then, having subdued them, she increased to such an extent that she could beat anyone. Nor would the Latins ever have realised that in reality they were mere slaves, if they had not seen the Samnites twice defeated and forced to accept Rome's terms.*

*The Discourses,*
Niccolò Machiavelli,
1520

*Six in the third place means:*
*He finds a comrade.*
*Now he beats the drum, now he stops.*
*Now he sobs, now he sings.*
*Here the source of a man's strength lies not in himself but in his relation to other people. No matter how close to them he may be, if his center of gravity depends on them, he is inevitably tossed to and fro between joy and sorrow. Rejoicing to high heaven, then sad unto death—this is the fate of those who depend upon an inner accord with other persons whom they love. . . .*

*The I Ching*, China, circa eighth century B.C.

being slaughtered. His artillery was hard to maneuver in the mountainous terrain. The Swiss fought with incredible fierceness, and their phalanxes were impenetrable.

A reserve Swiss force, hidden in the woods on the Burgundian right, suddenly emerged and attacked. The duke's army fell into a headlong retreat; the battle ended in a slaughter, from which the duke, however, escaped.

A few months later, it was the turn of the Swiss to go on the offensive by marching into Lorraine. In January 1478 the duke counterattacked with his now enfeebled forces; again the Burgundians were routed, and this time the duke did not escape. His body was finally identified on the field of battle, his head cloven in two by a Swiss halberd, his body pierced by pikes.

In the months after Charles's death, Louis XI swallowed up Burgundy, eliminating the last great feudal threat to a unified France. The duke had unknowingly fallen prey to Louis's elaborate plan to destroy him without wasting a single French soldier.

### Interpretation

King Louis XI would eventually become known as the Spider King, infamous for the elaborate webs he would weave to ensnare his opponents. His genius was to think far ahead and plot an indirect path to his goals—and his greatest goal was to transform France from a feudal state into a unified great power. Burgundy was his largest obstacle, and one he could not meet head-on: his army was weaker than Charles's, and he did not want to provoke a civil war. Before he became king, though, Louis had fought a short campaign against the Swiss and had seen the brutal efficiency with which their phalanxes fought and how they used their mountainous terrain to perfect advantage. He thought them unbeatable in war. Louis formed a plan to bait Charles into invading the cantons, where his military machine would be destroyed.

The strands of Louis's web were finely woven. First, he spent years courting the Swiss, forging bonds that blinded them to his ulterior purpose. This alliance also befuddled the arrogant duke, who could not imagine how Louis planned to make use of such an ally. The king also knew that by getting the Swiss to invade Burgundy in 1474, he would make the duke so enraged as to lose all patience in his desire for revenge.

When Edward landed at Calais, the king had foreseen the invasion and was ready for it. Instead of trying to fight off this mighty opponent, he worked to coax the English king away from his alliance with Burgundy by appealing to his self-interest: without risking a single battle so far from home, Edward would receive a financial payment too handsome to refuse. Again thinking far in advance, Louis knew that when he eventually swallowed up the wealthy duchy of Burgundy, he would more than recoup what he was having to pay Edward. Abandoned by the English, Charles was isolated, yet still determined to avenge the invasion of Burgundy. At this point Louis moved to sign a treaty with the duke, getting rid of the last possible obstacle in Charles's path to the Swiss cantons. This new treaty would infuriate his Swiss friends, but what did Louis care? Friendship meant little to him; the Swiss would fight to defend their land with or without him. Patient and clear about his goals, Louis used alliances as a form of bloodless warfare, crushing his opponents by getting others to do his work for him.

Almost all of us instinctively understand the importance of allies. Because we operate by feel and emotion more often than by strategy, however, we frequently make the worst kinds of alliances. A common mistake is to think that the more allies we have, the better; but quality is more important than quantity. Having numerous allies increases the chances we will become entangled in other people's wars. Going to the other extreme, we sometimes think a single powerful ally is all we need; but allies like that tend to get what they can from us and then drop us when our usefulness is exhausted, just as Louis dropped the Swiss. It is in any case a mistake to become dependent on one person. Finally, we sometimes choose those who seem the friendliest, who we think will be loyal. Our emotions lead us astray.

Understand: the perfect allies are those who give you something you cannot get on your own. They have the resources you lack. They will do your dirty work for you or fight your battles. Like the Swiss, they are not always the most obvious or the most powerful. Be creative and look for allies to whom you in turn have something to offer as well, creating a link of self-interest. To lose such allies of convenience will not destroy you or make you feel betrayed. You must think of them as temporary tools. When you no longer need such tools, there is no love lost in dumping them.

*The forces of a powerful ally can be useful and good to those who have recourse to them . . . but are perilous to those who become dependent on them.*

*—Niccolò Machiavelli,* The Prince *(1513)*

## FALSE ALLIANCES

In November 1966, Murray Bowen, a professor of clinical psychiatry at Georgetown University and one of the world's most influential family therapists, faced a brewing crisis within his own family back home in Waverly, Tennessee. Bowen was the oldest of five children. His family had operated an important business in Waverly for several generations. The third-oldest sibling, a brother nicknamed June, had been running the business for some time. Continually overworked and feeling underappreciated, June was now asking for a controlling interest in the business. Their father supported him, their mother did not. Members of the extended family were taking sides. The situation was tense.

At the same time, a death in the family of June's wife had made her

THE FOX AND THE BILLY-GOAT

*A fox, having fallen into a well, was faced with the prospect of being stuck there. But then a billy-goat came along to that same well because he was thirsty and saw the fox. He asked him if the water was good.*
*The fox decided to put a brave face on it and gave a tremendous speech about how wonderful the water was down there, so very excellent. So the billy-goat climbed down the well, thinking only of his thirst. When he had a good drink, he asked the fox what he thought was the best way to get back up again.*
*The fox said: "Well, I have a very good way to do that. Of course, it will mean our working together. If you just push your front feet up against the wall and hold your horns up in the air as high as you can, I will climb up on to them, get out, and then I can pull you up behind me."*
*The billy-goat willingly consented to this idea, and the fox briskly clambered up the legs, the shoulders, and finally the horns of his companion. He found himself at the mouth of the well, pulled himself out, and immediately scampered off. The billy-goat shouted after him, reproaching him for breaking their agreement of mutual assistance. . . .*

*FABLES,* AESOP, SIXTH CENTURY B.C.

*Heracles had performed these Ten Labours in the space of eight years and one month; but Eurystheus, discounting the Second and the Fifth, set him two more. The Eleventh Labour was to fetch fruit from the golden apple-tree, Mother Earth's wedding gift to Hera, with which she had been so delighted that she planted it in her own divine garden. This garden lay on the slopes of Mount Atlas, where the panting chariot-horses of the Sun complete their journey, and where Atlas's sheep and cattle, one thousand herds of each, wander over their undisputed pastures. When Hera found, one day, that Atlas's daughters, the Hesperides, to whom she had entrusted the tree, were pilfering the apples, she set the ever-watchful dragon Ladon to coil around the tree as its guardian. . . .*

*. . . When at last Heracles came to the Po, the river-nymphs, daughters of Zeus and Themis, showed him Nereus asleep. He seized the hoary old sea-god and, clinging to him despite his many Protean changes, forced him to prophesy how the golden apples could be won. . . .*

*. . . Nereus had advised Heracles not to pluck the apples himself, but to employ Atlas as his agent, meanwhile relieving him of his fantastic burden; therefore, on arriving at the*

so despondent that it was beginning to affect her husband's health. A ripple effect was spreading to the rest of the family, and Bowen's sister, the second-youngest and most unstable sibling, was beginning to display all kinds of nervous symptoms. Bowen feared most, though, for his father, who had a weak heart. As a family therapist, Bowen had studied a phenomenon he called an "anxiety wave," in which a peripheral event could spark enough emotional turmoil to lead to the death of the family's oldest or most vulnerable member. Somehow Bowen had to find a way to defuse this anxiety wave in his own family.

The problem for Bowen was that he was also going through a kind of personal and professional crisis at the time. One of his most influential theories held that members of a family were healthy to the extent that they could differentiate themselves from their siblings and parents, establishing their own identity, being able to make decisions on their own, while also being integrated and actively involved with the rest of the family. He saw this as a difficult psychic task for anyone. A family has a kind of group ego and an interlocking emotional network; it requires a great deal of effort and practice to establish autonomy outside this system. Yet doing so, Bowen believed, though crucial for everyone, was also professionally necessary for family therapists, who could not properly help others if they had been unable to differentiate themselves from their own families. They would carry their personal problems into their practice.

And indeed here was Professor Bowen, a man in his early fifties who had worked for years on his relationship to his family but who found himself sucked into the group dynamic, regressing emotionally, unable to think straight, every time he went home to Tennessee. That made him feel profoundly frustrated and depressed. The time had come, he decided, to attempt a radical personal experiment on his next visit home.

In late January 1967, June Bowen received a lengthy letter from his brother Murray. The two men had not written to each other for some time; in fact, June resented his brother and had avoided personal encounters with him for several years, for he felt that their mother always took Murray's side, even though it was June who was running the business. In the letter Murray passed on many gossipy stories about June that family members had told him over the years, always careful to add that Murray had better not let his "sensitive" brother hear them. Murray said he was tired of these stories and of being told how to manage his brother. It would be better, he thought, to communicate with June directly. He ended the letter by saying it would be unnecessary for the two of them to see each other on his next visit home, since he had told him everything he wanted to say. He signed the letter "Your Meddlesome Brother."

The more June thought about this letter, the angrier it made him. Murray had deliberately churned up division between June and his family. Then, a few days later, the two men's younger sister also received a letter from Murray, saying he had heard about her emotional distress and had written asking June to take care of her until he got home. He signed the letter "Your Worried Brother." This letter was as upsetting to the sister as June's letter had been to him: she was tired of people treating her as if she were sick—that only made her more anxious than she already was. After another short interval, Murray sent a third letter, this time to his mother. He mentioned the letters he had written to the others. He was trying to defuse the family crisis, he said, by attracting all attention to himself. He wrote that he had wanted to get his brother all roiled up and that he had the material to push even more buttons if necessary; but, he warned, it is never wise to share intelligence with "the enemy," so his mother should keep all this to herself. He signed the letter "Your Strategic Son." Thinking he had lost his mind, his mother burned the letter.

News of these letters passed quickly through the family, stirring up a hornet's nest of accusations, concerns, and anxieties. Everyone was in a tizzy about them, but June was the center of the storm. He showed Murray's letter to his mother, whom it greatly disturbed. June promised that on Murray's imminent visit home he would not only not avoid him but would confront him and really let him have it.

Murray arrived in Waverly in early February. On the second night of his visit, at a dinner at his sister's, June showed up with his wife; the brothers' father and mother were also present. The encounter lasted some two hours, its main participants Murray, June, and their mother. It was a bitter family confrontation. A furious June threatened a lawsuit over Murray's scurrilous stories and accused his mother of conspiring with her favorite. When Murray confirmed that he and his mother were in cahoots, that it had all been plotted years ago between him and Mom, she was outraged, denied knowledge of any plot, and said she would never tell Murray anything again. June told his own stories about his professor brother; Murray responded that they were amusing, but he knew better ones. The entire conversation centered on personal issues, and many repressed emotions came to the surface. But Murray remained strangely detached. He made sure he took no sides; no one was quite happy with what he was saying.

The next day Murray showed up at June's house—and June, for some reason, was happy to see him. Murray told more gossipy stories, including one he had heard about how well June was handling the situation considering all the stress he was under. June, feeling quite emotional, started to open up to his brother about his problems: he was really worried about their sister, he said, even thought she might be retarded. Later that day Murray visited the sister and told her what June had said about her; she was more than able to take care of herself, she replied, and had had enough of the family's intrusive concern. More visits followed with other family members. In each case, whenever someone tried to pass along some gossip or get Murray to take his or her side in the family constellation, he would either deflect the attempt with a neutral comment or pass it along to the person involved.

*Garden of the Hesperides, he asked Atlas to do him this favor. Atlas would have undertaken almost any task for the sake of an hour's respite, but he feared Ladon, whom Heracles thereupon killed with an arrow shot over the garden wall. Heracles now bent his back to receive the weight of the celestial globe, and Atlas walked away, returning presently with three apples plucked by his daughters. He found the sense of freedom delicious. "I will take these apples to Eurystheus myself without fail," he said, "if you hold up the heavens for a few moments longer." Heracles pretended to agree, but, having been warned by Nereus not to accept any such offer, begged Atlas to support the globe for only one moment more, while he put a pad on his head. Atlas, easily deceived, laid the apples on the ground and resumed his burden; whereupon Heracles picked them up and went away with an ironical farewell.*

THE GREEK MYTHS, VOL. 2, ROBERT GRAVES, 1955

*I regarded most of the people I met solely and exclusively as creatures I could use as porters in my voyages of ambition. Almost all these porters sooner or later became exhausted. Unable to endure the long marches that I forced on them at top speed and under all climatic conditions, they died on the way. I took others. To attach them to my service, I promised to get them to where I myself was going to that end-station of glory which climbers desperately want to reach. . . .*

*THE SECRET LIFE OF SALVADOR DALÍ,* SALVADOR DALÍ, 1942

The day Murray left, everyone came to say good-bye. The sister seemed more relaxed; so did the father. The family mood was noticeably altered. A week later Murray's mother sent him a letter that ended, "With all its ups and downs, your last trip home was the greatest ever." June now wrote regularly to his brother. The conflict over controlling the family business was defused and settled. Murray's visits home now became things everyone looked forward to, even though he was still up to his old tricks with stories and such.

Murray later wrote about the incident and incorporated what it had taught him into his training of other family therapists. He considered it the turning point in his career.

### Interpretation

Bowen's strategy in the experiment he conducted on his family was simple: he would make it impossible for any family member to make him take sides or hook him into any kind of alliance. He would also deliberately cause an emotional tempest to break up the stale family dynamic, particularly targeting June and his mother, that dynamic's centrifugal forces. He would make his family see things anew by getting them to talk about personal matters rather than avoiding them. He would work on himself to stay calm and rational, squelching any desire either to please or to run away from confrontation.

And in the midst of this experiment, Bowen experienced an unbelievable feeling of lightness—a near euphoria. For the first time in his life, he felt connected to the family without being submerged by its emotional pulls. He could engage, argue, and banter without either regressing into childish tantrums or striving to be falsely agreeable. The more he dealt with the family this way, the easier it became.

Bowen also noticed the effect his behavior had on others. First, they could not interact in their usual way: June could not avoid him, his weak sister could not internalize all the family's problems, the mother could not use him as a crutch. Next, they found themselves drawn to him. His refusal to take sides made it easier to open up to him. The stale family dynamic of gossip, secret communications, and irritating alliances was broken up in one visit. And, according to Bowen, it stayed that way for the rest of his life.

Bowen transferred his theory and practice beyond the family. He thought about his workplace, which had a family-like group ego and emotional system that infected him every time he was there: people would pull him into alliances, criticize absent colleagues, make it impossible for him to stay detached. Avoiding these conversations solved nothing; it meant he was still affected by the group dynamic, just unable to deal with it. Listening patiently to people's gossip while wishing they would stop was equally frustrating. Bowen had to take some kind of action to disrupt the dynamic—and he found he could apply the same tactics he had used on his family, to great success. He purposefully stirred things up while staying free of alliances. And, as with his family, he noticed the tremendous power his autonomy gave him in the group.

No one can get far in life without allies. The trick, however, is to recognize the difference between false allies and real ones. A false alliance is created out of an immediate emotional need. It requires that you give up something essential about yourself and makes it impossible for you to make your own decisions. A true alliance is formed out of mutual self-interest, each side supplying what the other cannot get alone. It does not require you to fuse your own identity with that of a group or pay attention to everyone else's emotional needs. It allows you autonomy.

Throughout your life you will find yourself in groups that demand fusion, forcing you into all kinds of false alliances that command your emotions. You must find a way to the position of strength and power: able to interact and engage with people while staying autonomous. You deftly avoid false alliances by taking provocative actions that make it impossible for people to entrap you. You shake up the dynamic as much as possible, targeting the troublemakers and controllers. Once you are in a position where you are able to stay rational within the group, you can seem to join an alliance without worrying about your emotions running away with you. And you will find that as the person who is simultaneously autonomous and part of the group, you will become a center of gravity and attention.

*Enter into action under the cover of helping another's interests, only to further your own in the end. . . . This is the perfect stratagem and disguise for realizing your ambitions, for the advantages you seem to offer only serve as lures to influence the other person's will. They think their interests are being advanced when in truth they are opening the way for yours.*
*—Baltasar Gracian (1601–1658)*

THE LION AND THE WILD ASS

*A lion and a wild ass entered into an agreement to hunt wild beasts together. The lion was to use his great strength, while the ass would make use of his greater speed. When they had taken a certain number of animals, the lion divided up the spoils into three portions. "I'll take the first share because I am the king," he said. "The second share will be mine because I have been your partner in the chase," he said. "As for the third share," he said to the wild ass, "this share will be a great source of harm to you, believe me, if you do not yield it up to me. And, by the way, get lost!"*
It is suitable always to calculate your own strength, and not to enter into an alliance with people stronger than yourself.

*FABLES*, AESOP, SIXTH CENTURY B.C.

## KEYS TO WARFARE

To survive and advance at all in life, we find ourselves constantly having to use other people for some purpose, some need—to obtain resources we cannot get on our own, to give us protection of some sort, to compensate for a skill or talent we do not possess. As a description of human relationships, however, the word "use" has ugly connotations, and in any case we always like to make our actions seem nobler than they are. We prefer to think of these interactions as relationships of assistance, partnering, friendship.

This is not a matter of mere semantics; it is the source of a dangerous confusion that will harm you in the end. When you look for an ally, you have a need, an interest you want met. This is a practical, strategic matter upon which your success depends. If you allow emotions and appearances to infect the kinds of alliances you form, you are in danger. The art of forming alliances depends on your ability to separate friendship from need.

*The state of Jin, located in modern Shaanhsi, grew steadily in strength by swallowing small neighbors. There were two small states, Hu and Yu, to its south. In the spring of the nineteenth year under King Hui of Zhou (658 B.C), Duke Xian of Jin sent for a trusted minister, Xun Xi, and declared his intention to attack Hu. "We have little chance to gain advantage," observed Xun Xi after a pause. "Hu and Yu have always been very close. When we attack one of them, the other will surely come to its rescue. Pitched one to one, neither of them is our match, but the result is far from certain if we fight both of them at the same time." "Surely you are not saying we have no way to cope with these two small states?" asked the duke. Xun Xi thought for a while before replying. . . . "I have thought up a plan by which we will be able to subdue both Hu and Yu. For the first step we should present the Duke of Yu with handsome gifts and ask him to lend us a path by which we can attack Hu." The duke asked, "But we have just offered gifts to Hu and signed a friendly agreement with it. We can hardly make Yu believe that we want to attack Hu instead of Yu itself." "That is not so difficult to work out," replied Xun Xi. "We may secretly order our men on the border to make*

The first step is to understand that all of us constantly use other people to help and advance ourselves. (Bowen went so far as to use his own family in an experiment to solve a professional dilemma.) There is no shame in this, no need to ever feel guilty. Nor should we take it personally when we realize that someone else is using us; using people is a human and social necessity. Next, with this understanding in mind, you must learn to make these necessary alliances strategic ones, aligning yourself with people who can give you something you cannot get on your own. This requires that you resist the temptation to let your decisions about alliances be governed by your emotions; your emotional needs are what your personal life is for, and you must leave them behind when you enter the arena of social battle. The alliances that will help you most are those involving mutual self-interest. Alliances infected with emotions, or with ties of loyalty and friendship, are nothing but trouble. Being strategic with your alliances will also keep you from the bad entanglements that are the undoing of so many.

Think of your alliances as stepping-stones toward a goal. Over the course of your life, you will be constantly jumping from one stone to the next to suit your needs. When this particular river is crossed, you will leave them behind you. We will call this constant shifting yet advancing use of allies the "Alliance Game."

Many key principles of the Alliance Game originated in ancient China, which was composed of numerous states in continual flux—now weak, now powerful, now weak again. War was a dangerous affair, for a state that invaded another would stir up a lot of mistrust among the others and would often find itself losing ground in the long run. Meanwhile, a state that remained too loyal to an ally might find itself pulled into a war from which it could not break free and would go down in the process. The formation of proper alliances was in some ways a more important art than that of warfare itself, and the statesmen adept at this art were more powerful than military leaders.

It was through the Alliance Game that the state of Chin was able to slowly expand during the dangerous Warring States period of 403–221 B.C. Chin would make alliances with distant states and attack nearby ones; the nearby state that Chin had invaded could not get help from its outlying neighbor because that neighbor was now allied to Chin. If Chin faced an enemy that had a key ally, it would work first to disrupt the alliance—sowing dissension, spreading rumors, courting one of the two sides with money—until the alliance fell apart. Then Chin would invade first one of the two states, then the other. Gradually, bit by bit, it gobbled up neighboring states until, in the late third century B.C., it was able to unify China—a remarkable feat.

To play the Alliance Game right, today as in ancient China, you must be realistic to the core, thinking far ahead and keeping the situation as fluid as possible. The ally of today may be the enemy of tomorrow. Sentiment has no place in the picture. If you are weak but clever, you can slowly leapfrog into a position of strength by bouncing from one alliance to another. The opposite approach is to make a key alliance and stick with it, valuing trust and an established relationship. This can work well in stable times, but in periods of flux, which are more common, it can prove to be your undoing: differences in interest will inevitably emerge, and at the same time it will become hard to disentangle yourself from a relationship in which so much emotion has been invested. It is safer to bank on change, to keep your options open and your alliances based on need, not loyalty or shared values.

In the golden age of Hollywood, actresses had almost the least amount of power of anyone. Careers were short; even a great star would be replaced in a few years by someone younger. An actress would stay loyal to her studio, then watch helplessly as the roles dried up. The actress who best bucked the trend was Joan Crawford, who played her own version of the Alliance Game. In 1933, for instance, she met the screenwriter Joseph Mankiewicz, then a timid young man just starting out on what would be an illustrious career. Crawford recognized his talent immediately and went out of her way to befriend him, much to his amazement. He went on to write nine screenplays for her, greatly lengthening her career.

Crawford would also court cameramen and photographers, who would then work overtime to light her well and make her look good. She might do the same with a producer who controlled a screenplay with a role in it she coveted. Crawford would often make alliances with up-and-coming young talent who valued a relationship with the star. Then she would gracefully break or forget the connection when it no longer served her needs. Nor would she stay loyal to the studio, or indeed to anyone—only to herself. Her unsentimental approach to her own shifting network of alliances allowed her to avoid the trap that most actresses found embedded in the system.

The key to playing the game is to recognize who can best advance your interests at that moment. This need not be the most obviously powerful person on the scene, the person who *seems* to be able to do most for you; alliances that meet specific needs or answer particular deficiencies are often more useful. (Grand alliances between two great powers are generally the least effective.) Because Louis XI had a weak army, the Swiss, though minor players on the European scene, were the allies he needed. Recognizing this years in advance, he cultivated an alliance that bewildered his enemies. As an ambitious young congressional assistant in Washington, Lyndon Johnson realized he lacked all kinds of powers and talents to get him to the top. He became a clever user of other people's talents. Realizing the importance of information in Congress, he made a point of befriending and allying himself with those at key positions—whether high or low—in the information chain. He was particularly good with older men who enjoyed the company of a lively young man and the role of the father figure giving advice. Slowly, from

*raids on Hu. When the men of Hu come to protest, we may use that as a pretext to attack them. In this way Yu will be convinced of our professed intention." The duke considered it a good plan. Before long, armed conflicts broke out along the Jin-Hu border to the south. Thereupon the duke asked, "Now we have good reason to convince Yu of our intention to attack Hu. But it will not lend the path to us unless it receives a good profit in return. So what shall we use to bribe the Duke of Yu?" Xun Xi replied, "Though the Duke of Yu is known to be very greedy, he will not be moved unless our gifts are extremely precious. So why not offer him fine horses from Qu and jade from Chuiji?" The duke looked reluctant. "But these are the best treasures I have. I can hardly bring myself to part with them." "I am not surprised by your doubts," said Xun Xi. "Nevertheless, we are bound to subdue Hu now that it has lost the shield of Yu. After Hu is conquered, Yu will not be able to survive on its own. Therefore, when you send these gifts to the Duke of Yu, you are simply consigning the jade to your external mansion and the horses to your external stable. . . ."*
*. . . When Xun Xi was ushered into the court of Yu and presented the gifts, the Duke of Yu's eyes bulged.*

*. . . "The men of Hu have repeatedly worked up disturbances along our border," [said Xun Xi]. "To protect our people from the calamity of war, we have exerted the highest restraint and concluded a peace treaty with Hu. Nevertheless, the impudent Hu takes our restraint for weakness and is now creating new troubles by making invidious charges against us. Therefore my lord was compelled to order a punitive expedition against Hu, and he dispatched me to ask your permission to let our troops pass through your land. This way, we can get around our border with Hu, where its defense is strong, and launch a surprise attack at its weak point. When we have defeated the men of Hu, we shall present you with splendid trophies to testify to our mutual alliance and friendship." . . . That summer the Jin troops attacked Hu by way of Yu. The Duke of Yu led a band of force in person to join in the expedition. They defeated the Hu army and captured Xiayang, one of Hu's two major cities. The Duke of Yu received his share of the booty and believed he had nothing to regret for. . . . . . . In autumn of the twenty-second year under King Hui of Zhou (655 B.C.), the Duke of Jin again sent an envoy to borrow a path from Yu [to Hu],*

being a poor kid from Texas with no connections, Johnson raised himself to the top, through his network of convenient alliances.

It is a common strategy in bicycle races not to go out in front but to stay right behind the leader, a position that cuts down wind resistance—the leader faces the wind for you and saves you energy. At the last minute, you sprint ahead. Letting other people cut resistance for you and waste their energy on your behalf is the height of economy and strategy.

One of the best stratagems in the Alliance Game is to begin by seeming to help another person in some cause or fight, only for the purpose of furthering your own interests in the end. It is easy to find such people: they have a glaring need, a temporary weakness that you can help them to overcome. Now you have put them under a subtle obligation to you, to use as you will—to dominate their affairs, to divert their energies in the direction you desire. The emotions you create with your offer of help will blind the other person to your ulterior purpose.

The artist Salvador Dalí was particularly adept at this version of the game: if someone needed to raise money, say, Dalí would come to the rescue, organizing a charity ball or other fund-raising event. The person in need could hardly resist: Dalí was friendly with royalty, Hollywood stars, socialites. Soon he would be ordering all kinds of elaborate props for the ball. For his infamous "Night in a Surrealist Forest" in Pebble Beach, California, in 1941, which was intended to benefit starving artists in war-torn Europe, Dalí requested a live giraffe, enough pine trees to create a fake forest, the largest bed in the world, a wrecked automobile, and thousands of pairs of shoes from which to serve the first course. In the end the party was a smash and got all kinds of publicity, but, as so often with Dalí, the bills far exceeded the receipts; no money was left over for the starving artists of Europe. And strangely enough, all of the publicity was focused on Dalí, increasing his fame and winning him more powerful allies.

A variation on the Alliance Game is to play the mediator, the center around which other powers pivot. While remaining covertly autonomous, you make those around you fight for your allegiance. This was essentially how Prince Klemens von Metternich, the Austrian foreign minister during the Napoleonic era and afterward, restored Austria as Europe's principal power. It helped that Austria is located in the center of Europe and so is strategically vital to the nations around it. Even during the reign of Napoleon, when Austria was at its weakest and Metternich had to cozy up to the French, he kept his country free of lasting entanglements. Without bonding Austria to France by any legal alliance, for example, he tied Napoleon to him emotionally by arranging for the emperor to marry into the Austrian royal family. Keeping all of the great powers—England, France, Russia—at arm's length, he made everything revolve around Austria, even though Austria itself was no longer a great military power.

The brilliance of this variation is that merely by assuming a central position, you can wield tremendous power. For instance, you place yourself at a critical point in the information chain, giving you access to and control over it. Or you produce something other people depend on, giving you incredible leverage. Or you play the mediator everyone needs to resolve a dispute. Whatever it is, you can maintain power in this central position only by keeping yourself unentangled and courted by all. The moment you enter into any kind of lasting alliance, your power is greatly reduced.

A key component of the Alliance Game is the ability to manipulate other people's alliances and even destroy them, sowing dissension among your opponents so that they fight among themselves. Breaking your enemy's alliances is as good as making alliances yourself. When Hernán Cortés landed in Mexico in 1519, he faced hundreds of thousands of Aztecs with 500 men. Knowing that many smaller Mexican tribes resented the powerful Aztec Empire, he slowly worked to peel them away from their alliances with the Aztecs. By filling a tribal leader's ears with horrible stories about the Aztec emperor's plans, for example, he might bait the man into arresting the Aztec envoys on their next visit. That of course would infuriate the emperor, and now the tribe would be isolated and in danger—and would appeal to Cortés for protection. On and on Cortés went with this negative version of the Alliance Game, until the Aztecs' allies had become his.

Your focus here is on stirring up mistrust. Make one partner suspicious of the other, spread rumors, cast doubts on people's motives, be friendly to one ally to make the other jealous. Divide and conquer. In this way you will create a tide of emotions, hitting first this side, then that, until the alliance totters. Now former members of the alliance will feel vulnerable. Through manipulation or outright invitation, make them turn to you for protection.

In facing an enemy that is composed of allies, no matter how large or formidable, do not be afraid. As Napoleon said, "Give me allies to fight." In war, allies generally have problems of command and control. The worst kind of leadership is divided leadership; compelled to debate and agree before they act, allied generals usually move like snails. When fighting large groups of allies, as he often did, Napoleon always attacked first the weak link, the junior partner. Collapse here could make the whole fabric of the alliance fall apart. He would also seek quick victory in battle, even a small one, for no force is more easily discouraged by a defeat than an allied one.

Finally, you will of course be attacked for playing the Alliance Game. People will accuse you of being feckless, amoral, treacherous. Remember: these charges are strategic themselves. They are part of a moral offensive (see chapter 25). To advance their own interests, your accusers are trying to make you feel guilty or look bad. Do not let them get to you. The only real danger is that your reputation will eventually keep people from making alliances with you—but self-interest rules the world. If you are seen to have benefited others in the past and as capable of doing the same

*and again the Duke of Yu gave his consent. . . . . . . In the eighth month, the Duke of Jin led six hundred war chariots and proceeded by way of Yu to attack Hu. They laid siege to Shangyang, the capital of Hu. . . . The city, after holding out for nearly four months, finally yielded. The Duke of Hu fled . . . and Hu as a feudal state was destroyed. On their way back, the Jin troops halted at Yu. The Duke of Yu came to welcome them, receiving the Duke of Jin into the capital. The Jin troops seized the chance to storm into the city. Taken totally off guard, the Yu army submitted with little resistance, and the Duke of Yu was taken prisoner. Duke Xian of Jin was extremely pleased when Xun Xi returned to present him with the horses and jade as well as the captured Duke of Yu.*

THE WILES OF WAR: 36 MILITARY STRATEGIES FROM ANCIENT CHINA, TRANSLATED BY SUN HAICHEN, 1991 in the present, you will have suitors and playing partners. Besides, you are loyal and generous, as long as there is mutual need. And when you show that you cannot be had by the false lure of permanent loyalty and friendship, you will actually find yourself treated with greater respect. Many will be drawn to your realistic and spirited way of playing the game.

**Image:** Stepping-stones. The stream runs fast and dangerous, but you must cross it at some point. There lie some stones in a haphazard line that can get you to the other side. If you linger too long on one stone, you will lose your balance. If you go too fast or skip one, you will slip. Instead you must jump lightly from one stone to the next and never look back.

**Authority:** Beware of sentimental alliances where the consciousness of good deeds is the only compensation for noble sacrifices. *—Otto von Bismarck (1815–1898)*

## REVERSAL

If you play the Alliance Game, so will those around you, and you cannot take their behavior personally—you must keep dealing with them. But there are some types with whom any kind of alliance will harm you. You can often recognize them by their overeagerness to pursue you: they will make the first move, trying to blind you with alluring offers and glittering promises. To protect yourself from being used in a negative way, always look at the tangible benefits you will gain from this alliance. If the benefits seem vague or hard to realize, think twice about joining forces. Look at your prospective allies' past for signs of greed or of using people without giving in return. Be wary of people who speak well, have apparently charming personalities, and talk about friendship, loyalty, and selflessness: they are most often con artists trying to prey on your emotions. Keep your eye on the interests involved on both sides, and never let yourself be distracted from them.

# 28

# GIVE YOUR RIVALS ENOUGH ROPE TO HANG THEMSELVES

## THE ONE-UPMANSHIP STRATEGY

*Life's greatest dangers often come not from external enemies but from our supposed colleagues and friends, who pretend to work for the common cause while scheming to sabotage us and steal our ideas for their gain. Although, in the court in which you serve, you must maintain the appearance of consideration and civility, you also must learn to defeat these people. Work to instill doubts and insecurities in such rivals, getting them to think too much and act defensively. Bait them with subtle challenges that get under their skin, triggering an overreaction, an embarrassing mistake. The victory you are after is to isolate them. Make them hang themselves through their own self-destructive tendencies, leaving you blameless and clean.*

*Life is war against the malice of men.*

BALTASAR GRACIAN, 1601–58

## THE ART OF ONE-UPMANSHIP

Throughout your life you will find yourself fighting on two fronts. First is the external front, your inevitable enemies—but second and less obvious is the internal front, your colleagues and fellow courtiers, many of whom will scheme against you, advancing their own agendas at your expense. The worst of it is that you will often have to fight on both fronts at once, facing your external enemies while also working to secure your internal position, an exhausting and debilitating struggle.

The solution is not to ignore the internal problem (you will have a short life if you do so) or to deal with it in a direct and conventional manner, by complaining, acting aggressively, or forming defensive alliances. Understand: internal warfare is by nature unconventional. Since people theoretically on the same side usually do their best to maintain the appearance of being team players working for the greater good, complaining about them or attacking them will only make you look bad and isolate you. Yet at the same time, you can expect these ambitious types to operate underhandedly and indirectly. Outwardly charming and cooperative, behind the scenes they are manipulative and slippery.

You need to adopt a form of warfare suited to these nebulous yet dangerous battles, which go on every day. And the unconventional strategy that works best in this arena is the art of one-upmanship. Developed by history's savviest courtiers, it is based on two simple premises: first, your rivals harbor the seeds of their own self-destruction, and second, a rival who is made to feel defensive and inferior, however subtly, will tend to act defensive and inferior, to his or her detriment.

People's personalities often form around weaknesses, character flaws, uncontrollable emotions. People who feel needy, or who have a superiority complex, or are afraid of chaos, or desperately want order, will develop a personality—a social mask—to cover up their flaws and make it possible for them to present a confident, pleasant, responsible exterior to the world. But the mask is like the scar tissue covering a wound: touch it the wrong way and it hurts. Your victims' responses start to go out of control: they complain, act defensive and paranoid, or show the arrogance they try so hard to conceal. For a moment the mask falls.

When you sense you have colleagues who may prove dangerous—or are actually already plotting something—you must try first to gather intelligence on them. Look at their everyday behavior, their past actions, their mistakes, for signs of their flaws. With this knowledge in hand, you are ready for the game of one-upmanship.

Begin by doing something to prick the underlying wound, creating doubt, insecurity, and anxiety. It might be an offhand comment or something that your victims sense as a challenge to their position within the court. Your goal is not to challenge them blatantly, though, but to get under their skin: they feel attacked but are not sure why or how. The result is a vague, troubling sensation. A feeling of inferiority creeps in.

You then follow up with secondary actions that feed their doubts.

Here it is often best to work covertly, getting other people, the media, or simple rumor to do the job for you. The endgame is deceptively simple: having piled up enough self-doubt to trigger a reaction, you stand back and let the target self-destruct. You must avoid the temptation to gloat or get in a last blow; at this point, in fact, it is best to act friendly, even offering dubious assistance and advice. Your targets' reaction will be an overreaction. Either they will lash out, make an embarrassing mistake, or reveal themselves too much, or they will get overly defensive and try too hard to please others, working all too obviously to secure their position and validate their self-esteem. Defensive people unconsciously push people away.

At this point your opening action, especially if it is only subtly aggressive, will be forgotten. What will stand out will be your rivals' overreaction and humiliation. Your hands are clean, your reputation unsullied. Their loss of position is your gain; you are one up and they are one down. If you had attacked them directly, your advantage would be temporary or nonexistent; in fact, your political position would be precarious: your pathetic, suffering rivals would win sympathy as your victims, and attention would focus on you as responsible for their undoing. Instead they must fall on their swords. You may have given them a little help, but to whatever extent possible in their own eyes, and certainly in everyone else's, they must have only themselves to blame. That will make their defeat doubly galling and doubly effective.

To win without your victim's knowing how it happened or just what you have done is the height of unconventional warfare. Master the art and not only will you find it easier to fight on two fronts at the same time, but your path to the highest ranks will be that much smoother.

*Never interfere with an enemy that is in the process of committing suicide.*
*—Napoleon Bonaparte (1769–1821)*

## HISTORICAL EXAMPLES

**1.** John A. McClernand (1812–1900) watched with envy as his friend and fellow lawyer Abraham Lincoln rose to the U.S. presidency. McClernand, a lawyer and congressman from Springfield, Illinois, had had this ambition himself. Shortly after the outbreak of the Civil War, in 1861, he resigned his congressional seat to accept a commission as a brigadier general in the Union army. He had no military experience, but the Union needed leadership of any kind it could get, and if he proved himself in battle, he could rise fast. He saw this army position as his path to the presidency.

McClernand's first post was at the head of a brigade in Missouri under the overall command of General Ulysses S. Grant. Within a year he was promoted to major general, still under Grant. But this was

*First of all, a complete definition of the technical term "one-upmanship" would fill, and in fact has filled, a rather large encyclopedia. It can be defined briefly here as the art of placing a person "one-down." The term "one-down" is technically defined as that psychological state which exists in an individual who is not "one-up" on another person. . . . To phrase these terms in popular language, at the risk of losing scientific rigor, it can be said that in any human relationship (and indeed among other mammals) one person is constantly maneuvering to imply that he is in a "superior position" to the other person in the relationship. This "superior position" does not necessarily mean superior in social status or economic position; many servants are masters at putting their employers one-down. Nor does it imply intellectual superiority as any intellectual knows who has been put "one-down" by a muscular garbage collector in a bout of Indian wrestling. "Superior position" is a relative term which is continually being defined and redefined by the ongoing relationship. Maneuvers to achieve superior position may be crude or they may be infinitely subtle. For example, one is not usually in a superior position if he must ask another person for something. Yet he can ask for it in such a way that he is implying, "This is, of course, what I deserve."*

THE STRATEGIES OF PSYCHOTHERAPY, JAY HALEY, 1963

not good enough for McClernand, who needed a stage for his talents, a campaign to run and get credit for. Grant had talked to him of his plans for capturing the Confederate fort at Vicksburg, on the Mississippi River. The fall of Vicksburg, according to Grant, could be the turning point in the war. McClernand decided to sell a march on Vicksburg as his own idea and use it as a springboard for his career.

In September 1862, on leave in Washington, D.C., McClernand paid a visit to President Lincoln. He was "tired of furnishing brains" for Grant's army, he said; he had proved himself on the battlefield and was a better strategist than Grant, who was a little too fond of his whiskey. McClernand proposed to go back to Illinois, where he was well known and could recruit a large army. Then he would follow the Mississippi River south to Vicksburg and capture the fort.

Vicksburg was technically in Grant's department, but Lincoln was not sure the general could lead the audacious attack necessary. He took McClernand to see Secretary of War Edwin Stanton, another former lawyer, who commiserated with his two visitors on the difficulties of dealing with military brass. Stanton listened to and liked McClernand's plan. That October the onetime congressman left Washington with confidential orders, giving him approval for his march on Vicksburg. The orders were a little vague, and Grant was not informed of them, but McClernand would make the best of them.

McClernand quickly recruited more soldiers than he had promised Lincoln he would. He sent his recruits to Memphis, Tennessee, where he would soon join them to march on Vicksburg. But when he arrived in Memphis, in late December 1862, the thousands of men he had recruited were not there. A telegram from Grant—dated ten days earlier and waiting for him in Memphis—informed him that the general was planning to attack Vicksburg. If McClernand arrived in time, he would lead the attack; if not, his men would be led by General William Tecumseh Sherman.

McClernand was livid. The situation had clearly been orchestrated to make it impossible for him to arrive in time to lead his own recruits; Grant must have figured out his plan. The general's polite telegram covering his bases made the whole affair doubly infuriating. Well, McClernand would show him: he would hurry downriver, catch up with Sherman, take over the campaign, and humiliate Grant by winning the credit and honor for capturing Vicksburg.

McClernand did catch up with Sherman, on January 2, 1863, and immediately assumed command of the army. He made an effort to be charming to Sherman, who, he learned, had been planning to raid Confederate outposts around Vicksburg to soften up the approach to the fort. The idea was heaven-sent for McClernand: he would take over these raids, win battles without Grant's name over his, earn himself some publicity, and make his command of the Vicksburg campaign a fait accompli. He followed Sherman's plan to the letter, and the campaign was a success.

At this triumphant point, out of the blue, McClernand received a telegram from Grant: he was to halt operations and wait for a meeting with the general. It was time for McClernand to play his trump card, the president; he wrote Lincoln requesting more explicit orders, and specifically an independent command, but he got no reply. And now vague doubts began to trouble McClernand's peace of mind. Sherman and other officers seemed cool; somehow he had rubbed them the wrong way. Perhaps they were conspiring with Grant to get rid of him. Grant soon appeared on the scene with detailed plans for a campaign against Vicksburg under his own direction. McClernand would lead a corps, which, however, was stationed at the faraway outpost of Helena, Arkansas. Grant made a point of treating him politely, but everything together added up to a humiliating setback.

How to be one up—*how to make the other man feel that something has gone wrong, however slightly. The Lifeman is never caddish himself, but how simply and certainly, often, he can make the other man feel a cad, and over prolonged periods.*

*The Complete Upmanship*, Stephen Potter, 1950

Now McClernand exploded, writing letter after letter to Lincoln and Stanton to remind them of their earlier rapport and of the support they had once given him, and complaining bitterly about Grant. After days of fuming and writing, McClernand finally received a response from Lincoln—and, to his shock and dismay, the president had somehow turned against him. There had been too many family quarrels among his generals, wrote Lincoln; for the sake of the Union cause, McClernand should subordinate himself to Grant.

McClernand was crushed. He could not figure out what he had done or how it had all gone wrong. Bitter and frustrated, he continued to serve under Grant but questioned his boss's abilities to anyone who would listen, including journalists. In June 1863, after enough negative articles had been printed, Grant finally fired him. McClernand's military career was over, and with it his dreams of personal glory.

### Interpretation

From the moment he met John McClernand, General Grant knew he had a troublemaker on his hands. McClernand was the type of man who thought only of his own career—who would steal other people's ideas and plot behind their backs for the sake of personal glory. But Grant would have to be careful: McClernand was popular with the public, a charmer. So when Grant figured out on his own that McClernand was trying to beat him to Vicksburg, he did not confront him or complain. Instead he took action.

Knowing that McClernand had an oversensitive ego, Grant recognized that it would be relatively easy to push the man's buttons. By taking over his subordinate's recruits (technically in his department anyway) while apparently covering his bases in the telegram, he forced McClernand into a rash response that seemed like insubordination to other military men and made it clear how far he was using the war for personal purposes. Once McClernand had rushed to take his troops back from Sherman, Grant stood aside. He knew that a man like this—vain and obnoxious—would irritate the hell out of his brother officers;

*There are other ways to fray nerves. During the Gulf War, President Bush kept pronouncing the name of the Iraqi leader as "SAD-am," which loosely means "shoeshine boy." On Capitol Hill, the ritual mispronunciation of a member's name is a time-tested way to rattle opponents or haze newcomers. Lyndon Johnson was a master of the practice. When Johnson was Senate majority leader, writes J. McIver Weatherford, he applied it with junior members who voted the wrong way: "While slapping the young chap on the back and telling him he understood, Johnson would break his name into shreds as a metaphorical statement of what would happen if the disloyalty persisted."*

THE ART OF POLITICAL WARFARE, JOHN PITNEY, JR., 2000

they would inevitably complain about him to Grant, who, as a responsible officer, would have to pass the complaints upward, apparently without personal feelings in play. Treating McClernand politely while indirectly checkmating him, Grant finally got him to overreact in the worst possible way, with his letters to Lincoln and Stanton. Grant knew that Lincoln was tired of squabbling within the Union high command. While Grant could be seen working quietly to perfect his plans for taking Vicksburg, McClernand was acting petty and throwing tantrums. The difference between the two men was all too clear. With this battle won, Grant repeated it, letting McClernand hang himself with his unwise complaints to the press.

You will often come across McClernands in your daily battles—people who are outwardly charming but treacherous behind the scenes. It does no good to confront them directly; they are proficient at the political game. But a subtle one-up campaign can work wonders.

Your goal is to get these rivals to put their ambition and selfishness on display. The way to do this is to pique their latent but powerful insecurities—make them worry that people do not like them, that their position is unstable, that their path to the top is not clear. Perhaps, like Grant, you can take action that thwarts their plans in some way while hiding your own beneath a veneer of politeness. You are making them feel defensive and disrespected. All the dark, ugly emotions they strive so hard to hide will boil up to the surface; they will tend to lash out, overplaying their hand. Work to make them grow emotional and lose their habitual cool. The more they reveal of themselves, the more they will alienate other people, and isolation will be their doom.

**2.** The Académie Française, founded by Cardinal Richelieu in 1635, is a highly select body of France's forty most learned scholars, whose task it is to oversee the purity of the French language. It was customary in the early years of the academy that when a seat became empty, potential members would petition to fill it, but on the occasion of a vacant seat in 1694, King Louis XIV decided to go against protocol and nominated the bishop of Noyon. Louis's nomination certainly made sense. The bishop was a learned man, well respected, an excellent orator, and a fine writer.

The bishop, however, had another quality as well: an incredible sense of self-importance. Louis was amused by this failing, but most in the court found it downright insufferable: the bishop had a way of making almost everyone feel inferior, in piety, erudition, family pedigree—whatever they had.

Because of his rank, for instance, the bishop was accorded the rare privilege of being able to have his coach drive up to the front door of the royal residence, while most others had to get out and walk from the entrance doors of the driveway. One time the archbishop of Paris was walking along the driveway when the bishop of Noyon passed. From his carriage the bishop waved and signaled for the archbishop to approach him. The archbishop expected him to alight and accompany him to the palace on foot. Instead Noyon had the carriage slow down and continued his drive to the front door, leading the archbishop through the window by the arm, as if he were a dog on a leash, meanwhile chatting away superciliously. Then, once the bishop did get out of the carriage and the two men started up the grand staircase, Noyon dropped the archbishop as if he were nobody. Almost everyone in the court had a story like this one to tell, and they all nursed secret grudges against the bishop.

With Louis's approval, however, it was impossible to not vote Noyon into the academy. The king further insisted that his courtiers attend the inauguration of the bishop, since this was his first nominee to the illustrious institution. At the inauguration, customarily, the nominee would deliver a speech, which would be answered by the academy's director—who at the time was a bold and witty man called the abbé de Caumartin. The abbé could not stand the bishop but particularly disliked his florid style of writing. De Caumartin conceived the idea of subtly mocking Noyon: he would compose his response in perfect imitation of the bishop, full of elaborate metaphors and gushing praise for the newest academician. To make sure he could not get into trouble for this, he would show his speech to the bishop beforehand. Noyon was delighted, read the text with great interest, and even went so far as to supplement it with more effusive words of praise and high-flying rhetoric.

On the day of the inauguration, the hall of the academy was packed with the most eminent members of French society. (None dared incur the king's displeasure by not attending.) The bishop appeared before them, monstrously pleased to command this prestigious audience. The speech he delivered had a flowery pomposity exceeding any he had given previously; it was tiresome in the extreme. Then came the abbé's response. It started slowly, and many listeners began to squirm. But then it gradually took off, as everyone realized that it was an elaborate yet subtle parody of the bishop's style. De Caumartin's bold satire captivated everyone, and when it was over, the audience applauded, loudly and gratefully. But the bishop—intoxicated by the event and the attention—thought that the applause was genuine and that in applauding the abbé's praise of him, the audience was really applauding him. He left with his vanity inflated beyond all proportion.

Soon Noyon was talking about the event to one and all, boring everyone to tears. Finally he had the misfortune to brag about it to the archbishop of Paris, who had never gotten over the carriage incident. The archbishop could not resist: he told Noyon that the abbé's speech was a joke on him and that everyone in the court was laughing at the bishop's expense. Noyon could not believe this, so he visited his friend and confessor Père La Chaise, who confirmed that it was true.

Now the bishop's former delight turned to the most bitter rage. He complained to the king and asked him to punish the abbé. The king

WHEN TO GIVE ADVICE

*In my own view (but compare Motherwell) there is only one correct time when the gamesman can give advice: and that is when the gamesman has achieved a* useful *though not necessarily a* winning *lead. Say three up and nine to play at golf, or, in billiards, sixty-five to his opponent's thirty. Most of the accepted methods are effective. E.g. in billiards, the old phrase serves. It runs like this:*

*Gamesman: Look . . . may I say something?*

*Layman: What?*

*Gamesman: Take it easy.*

*Layman: What do you mean?*

*Gamesman: I mean—you know how to make the strokes, but you're stretching yourself on the rack all the time. Look. Walk up to the ball. Look at the line. And make your stroke. Comfortable. Easy. It's as simple as that.*

*In other words, the advice* must be vague, *to make certain it is not helpful. But, in general, if properly managed, the mere giving of advice is sufficient to place the gamesman in a practically invincible position.*

*THE COMPLETE UPMANSHIP,* STEPHEN POTTER, 1950

tried to defuse the problem, but he valued peace and quiet, and Noyon's almost insane anger got on his nerves. Finally the bishop, wounded to the core, left the court and returned to his diocese, where he remained for a long time, humiliated and humbled.

THE LION, THE WOLF AND THE FOX

*A very old lion lay ill in his cave. All of the animals came to pay their respects to their king except for the fox. The wolf, sensing an opportunity, accused the fox in front of the lion: "The fox has no respect for you or your rule. That's why he hasn't even come to visit you." Just as the wolf was saying this, the fox arrived, and he overheard these words. Then the lion roared in rage at him, but the fox managed to say in his own defence: "And who, of all those who have gathered here, has rendered Your Majesty as much service as I have done? For I have traveled far and wide asking physicians for a remedy for your illness, and I have found one." The lion demanded to know at once what cure he had found, and the fox said: "It is necessary for you to flay a wolf alive, and then take his skin and wrap it around you while it is still warm." The wolf was ordered to be taken away immediately and flayed alive. As he was carried off, the fox turned to him with a smile and said: "You should have spoken well of me to His Majesty rather than ill."*

*FABLES*, AESOP, SIXTH CENTURY B.C.

Interpretation

The bishop of Noyon was not a harmless man. His conceit had made him think his power had no limits. He was grossly unaware of the offense he had given to so many people, but no one could confront him or bring his behavior to his attention. The abbé hit upon the only real way to bring such a man down. Had his parody been too obvious, it would not have been very entertaining, and the bishop, its poor victim, would have won sympathy. By making it devilishly subtle, and making the bishop complicit in it as well, de Caumartin both entertained the court (always important) and let Noyon dig his own grave with his reaction—from the heights of vanity to the depths of humiliation and rage. Suddenly aware of how people saw him, the bishop lost his balance, even alienating the king, who had once found his vanity amusing. Finally he had to absent himself from court, to many people's relief.

The worst colleagues and comrades are often the ones with inflated egos, who think everything they do is right and worthy of praise. Subtle mockery and disguised parody are brilliant ways of one-upping these types. You seem to be complimenting them, your style or ideas even imitating theirs, but the praise has a sting in its tail: Are you imitating them to poke fun at them? Does your praise hide criticism? These questions get under their skin, making them vaguely insecure about themselves. Maybe you think they have faults—and maybe that opinion is widely shared. You have disturbed their high sense of self, and they will tend to respond by overreacting and overplaying their hand. This strategy works particularly well on those who fancy themselves powerful intellectuals and who are impossible to best in any kind of argument. By quoting their words and ideas back at them in slightly grotesque form, you neutralize their verbal strengths and leave them self-doubting and insecure.

**3.** Toward the middle of the sixteenth century, a young samurai, whose name history has left behind, developed a novel way of fighting: he could wield two swords with equal dexterity in his right and left hands at the same time. This technique was formidable, and he was eager to use it to make a name for himself, so he decided to challenge the most famous swordsman of his time, Tsukahara Bokuden, to a duel. Bokuden was now middle-aged and in semiretirement. He answered the young man's challenge with a letter: a samurai who could use a sword in his left hand with the same effectiveness as his right had an unfair advantage. The young swordsman could not understand what he meant. "If you think my using a sword with my left is unfair," he wrote back, "renounce the match." Instead Bokuden sent off ten more letters, each repeating in slightly different words the charge about the left hand. Each letter only made the challenger more annoyed. Finally, however, Bokuden agreed to fight.

The young samurai was used to fighting on instinct and with great speed, but as the duel began, he could not stop thinking about his left hand and Bokuden's fear of it. With his left hand—he found himself calculating—he would stab here, slash there. His left hand could not fail; it seemed possessed of its own power. . . . Then, suddenly, out of nowhere, Bokuden's sword cut deeply across the challenger's *right* arm. The duel was over. The young samurai recovered physically, but his mind was forever unhinged: he could not fight by instinct anymore. He thought too much, and he soon gave up the sword.

In 1605, Genzaemon, head of the renowned Yoshioka family of Kyoto swordsmen, received the strangest challenge of his life. An unknown twenty-one-year-old samurai named Miyamoto Musashi, dressed like a beggar in dirty, ragged clothes, challenged him to a duel so haughtily that Musashi must have thought himself the more famous swordsman. Genzaemon did not feel he had to pay attention to this youth; a man as illustrious as he could not go through life accepting challenges from every bumpkin who crossed his path. Yet something about Musashi's arrogance got under his skin. Genzaemon would enjoy teaching this youth a lesson. The duel was set for five o'clock the following morning in a suburban field.

Genzaemon arrived at the appointed time, accompanied by his students. Musashi was not there. Minutes turned into an hour. The young man had probably gotten cold feet and skipped town. Genzaemon sent a student to look for the young samurai at the inn where he was staying. The student soon returned: Musashi, he reported, had been asleep when he arrived and, when awakened, had rather impertinently ordered him to send Genzaemon his regards and say he would be there shortly. Genzaemon was furious and began to pace the field. And Musashi still took his time. It was two more hours before he appeared in the distance, sauntering toward them across the field. He was wearing, too, a scarlet headband, not the traditional white headband that Genzaemon wore.

Genzaemon shouted angrily at Musashi and charged forward, impatient to have done with this irritating boor. But Musashi, looking almost bored, parried one blow after another. Each man was able to slash at the other's forehead, but where Genzaemon's white headband turned red with blood, Musashi's stayed the same color. Finally, frustrated and confused, Genzaemon charged forward yet again—right into Mushashi's sword, which struck his head and knocked him to the ground unconscious. Genzaemon would later recover, but he was so humiliated by his defeat that he left the world of swordsmanship and entered the priesthood, where he would spend his remaining years.

[Christy] *Mathewson in his later years recounted a knock-down incident from the first game of the 1911 World Series, which he won for the Giants, defeating the Philadelphia Athletics 2 to 1. Charles Albert "Chief" Bender started for the Athletics, and Bender was throwing harder that day than Mathewson had ever seen him throw. Twice Bender drilled Fred Snodgrass, the Giants' young center fielder. When Snodgrass came to bat for the third time—in a "pinch"—Bender smiled at him. "Look out, Fred-die," he said, "you don't get hit this time." Then he threw a fast ball at Snodgrass' head. Snodgrass ducked. Ball one. "If you can't throw better than that," Snodgrass shouted, "I won't need to get a hit." Bender continued to smile. ("He had perfect teeth," Mathewson remembered.) Then he threw a fast-ball strike that overpowered Snodgrass. "You missed that a mile," Bender said, grinning again. Snodgrass set his jaw in anger and began overswinging. "Grinning chronically," in Mathewson's phrase, Bender struck out Snodgrass with a curve that broke down into the dirt. Snodgrass was not afraid of Chief Bender's pitches. He was a solid hitter who finished with a lifetime average of .275. What happened, Mathewson said, was that a combination, the knockdown pitches, the sarcastic banter, the condescending grin, distracted Snodgrass. Then, having struck out his man, Bender pushed the needle deeper and twisted it. "You ain't a batter, Freddie. You're a backstop. You can never get anywhere without being hit!" Although beaten that day, Chief Bender won two other games. The Athletics won the World Series, 4 games to 2. Across six games, the rattled Fred Snodgrass, a .294 hitter all season, batted .105. But as Mathewson interpreted the episode, he was a victim of gamesmanship, obviously and distinctly quite different from being terrorized. "Chief took Fred's mind right out of the game," Mathewson said.*

THE HEAD GAME, ROGER KAHN, 2001

Interpretation

For a samurai, losing a duel could mean death or public humiliation. Swordsmen sought out any advantage—physical dexterity, a superior sword, the perfect technique—to avoid that fate. But the greatest samurais, the Bokudens and Musashis, sought their advantage in being able to subtly push the opponent off his game, messing with his mind. They might try to make him self-conscious, a little too aware of his technique and style—a deadly trap for anyone who must react in the moment. They might trick him into focusing on the wrong thing—the left hand, the scarlet headband. Particularly with conventional-minded opponents, they might show up late, sparking a frustration that would upset their timing and concentration. In all of these cases, a change in the enemy's focus or mood would lead to a mistake. To try to repair this mistake in the heat of the moment would lead to another, until the one-upped fighter might literally walk into the other man's sword.

Understand: what will yield the greatest effects in the game of one-upmanship is a subtle disturbance in your opponents' mood and mindset. Be too direct—make an insulting comment, an obvious threat—and you wake them to the danger you represent, stir their competitive juices, bring out the best in them. Instead you want to bring out the worst. A subtle comment that makes them self-conscious and gets under their skin will turn them inward, get them lost in the labyrinth of their own thoughts. A seemingly innocent action that stirs an emotion like frustration, anger, or impatience will equally cloud their vision. In both cases they will tend to misfire and start making mistakes.

This works particularly well against rivals who must perform in some way—deliver a speech, say, or present a project: the fixating thought or bad emotion you create in them makes them lose touch with the moment and messes up their timing. Do this right, too, and no one will be aware of your involvement in the bad performance, not even the rival you have one-upped.

**4.** In January 1988, Senator Robert Dole of Kansas could smell victory in his quest to become president of the United States. His main opponent for the Republican nomination was George H. W. Bush, the incumbent vice president in the administration of Ronald Reagan. In the Iowa caucuses, the first test in the primary season, Bush had been lackluster and had finished a distant third, behind Dole and televangelist Pat Robertson. Dole's aggressive campaigning had won him much attention—he had the momentum and was clearly the front-runner.

To Dole, however, there was one blemish to his great victory in Iowa. Lee Atwater, Bush's thirty-six-year-old campaign strategist, had spread to the media a story that questioned the integrity of the senator's wife, former secretary of transportation Elizabeth Dole. The senator was an elected politician of nearly three decades' standing and had developed the necessary thick skin, but attacks on his wife, he felt, were beyond the pale. He had a temper that his advisers worked hard to keep under wraps, and when the story broke, he lashed out at reporters—giving Atwater the opportunity to say, "He can dish it out, but if someone hits him back, he starts whining." Then Atwater sent Dole a ten-page letter enumerating the many times the Kansas senator had gone negative in the campaign, and this letter, too, made its way into the media. Dole was furious. Despite his victory in Iowa, he could not get over seeing his wife dragged into the dirt. He would get back at the Bush folk and Atwater.

Silence.—*The way of replying to a polemical attack the most unpleasant for both parties is to get annoyed and stay silent: for the attacker usually interprets the silence as a sign of contempt.*

FRIEDRICH NIETZSCHE, 1844–1900

Next up was the New Hampshire primary. Victory here would put Dole well on his way, and he was ahead in the polls, but this time Bush came out fighting and the race tightened up. The weekend before the vote, the Bush people ran an ad portraying Dole as a "straddler," a man with two faces whose senate votes depended on expediency, not sincere belief. Humorous, deceptive, bitingly negative, the ad had Atwater's fingerprints all over it. And the timing was perfect—too late for Dole to respond with an ad of his own. The ad helped propel Bush into the lead and, a few days later, to victory.

Shortly after the results of the New Hampshire primary were in, NBC newsman Tom Brokaw caught up with Bush and asked if he had any message for his rival. "Naw," he replied with a smile, "just wish him well." Then Brokaw found Dole and asked the same question. "Yeah," said Dole with a bitter scowl. "Stop lying about my record."

In the days to come, Dole's answer was rerun again and again on television and discussed in the papers. It made him look like a sore loser. The press began to pile on, and Dole was ungracious—he seemed whiny. A few weeks later, he went down to a crushing defeat in South Carolina and shortly thereafter an even worse string of losses in the Super Tuesday primaries throughout the South. Somewhere along the line, Dole's campaign had crashed and burned. Little did he suspect that it had all begun in Iowa.

### Interpretation

Lee Atwater believed that adults could be divided into two groups: the overly mature and the childlike. The overly mature are inflexible and overserious, making them highly vulnerable in politics, particularly in the age of television. Dole was clearly the mature type, Atwater the child.

It didn't take Atwater much research to see that Dole was hypersensitive about attacks on his wife. Replaying old charges against her in Iowa, Atwater was able to get under the senator's skin. He kept Dole's blood boiling with the letter that accused him of starting the dirty campaigning, and he upped the pressure with the perfectly timed ad that mocked Dole's record for New Hampshire voters. Although Atwater was the one pushing buttons, Dole's outburst to Brokaw focused all attention on him and his unsportsmanlike behavior. Atwater, a genius at

Glaciation . . . *is the name for the set of gambits which are designed to induce an awkward silence, or at any rate a disinclination to talk, on the part of possible opponents. The "freezing" effects of these gambits is sometimes of immense power: . . . If someone else tells a funny story, do not, whatever happens, tell your own funny story in reply, but listen intently and not only refrain from laughing or smiling, but make no response, change of expression or movement whatever. The teller of the funny story, whatever the nature of his joke, will then suddenly feel that what he has said is in bad taste. Press home your advantage. If he is a stranger, and has told a story about a man with one leg, it is no bad thing to pretend that one of your own legs is false, or at any rate that you have a severe limp. This will certainly silence Opponent for the rest of the evening. . . . . . . If, for instance, someone is being really funny or witty and there is a really pleasant atmosphere of hearty and explosive laughter, then (a) join in the laughter at first. Next (b) gradually become silent. Finally (c) at some pause in the conversation be overheard whispering, "Oh for some real talk."*

THE COMPLETE UPMANSHIP, STEPHEN POTTER 1950

one-upmanship, now stood back. Dole could only respond with more sourness, compounding the problem and leading to electoral suicide.

The easiest types to one-up are those who are rigid. Being rigid does not necessarily mean being humorless or charmless, but it does mean being intolerant of anything that breaks their code of acceptable behavior. Being the target of some anarchic or unconventional antic will trigger an overreaction that makes them look sour, vindictive, unleaderlike. The calm exterior of the mature adult is momentarily blown away, revealing something rather peevish and puerile.

Do not discourage such targets from getting personal: the more bitterly they protest and criticize you, the worse they look. They forget that the real issue is how they are perceived by the people around them or, in an electoral race, by the public. Inflexible to the core, they can be induced to make mistake after mistake with the slightest push.

**5.** In 1939, Joan Crawford (1904–77) talked her way into a relatively minor role in the film *The Women*: the lower-class perfume salesgirl who steals the husband of an elegant woman played by Norma Shearer. Crawford and Shearer were also bitter rivals in real life. Shearer was the wife of the movie producer Irving Thalberg, who always managed to get her the best parts. Crawford hated her for that, and for her haughty manner. Thalberg had died in 1936, but, to Crawford's disgust, the studio was still pampering Shearer. Everyone in Hollywood knew of their mutual dislike and was waiting for the showdown. But Crawford was the consummate professional on the set, and she kept matters civil.

The Crawford and Shearer characters in *The Women* share only one scene: the climax of the movie, when Shearer finally confronts Crawford about the affair with her husband. The rehearsal went well, as did the master shot showing the two actresses performing together. Then it came time for close-ups. Of course Norma Shearer went first. Crawford sat in a chair off camera, delivering her lines to Shearer. (Many actors would have an assistant or the director feed the lines while they retired to their dressing rooms, but Crawford always insisted on reading them herself.)

Crawford was knitting an afghan at the time, and as she said her lines, she knitted furiously, then stopped when it was time for Shearer to respond. She never looked Shearer in the eye. The needles made a loud clicking sound that began to drive Shearer crazy. Straining to stay polite, Shearer said, "Joan, darling, I find your knitting distracting." Pretending not to hear, Crawford kept knitting. Finally Shearer, a woman famous for her elegance, lost control: she *screamed* at Crawford, ordering her off the set and back to her dressing room. As Crawford walked away, still not looking at Shearer, the film's director, George Cukor, ran to her side, but Shearer commanded him to come back. Her voice had a bitter tone that no one there had heard before and few would forget—it was so unlike her. Or was it?

In 1962, Crawford and Bette Davis, longtime stars who had never appeared in the same movie, were finally to costar, in Robert Aldrich's film *What Ever Happened to Baby Jane?* Crawford and Davis had never been thought to like each other too much, but Crawford had encouraged the pairing—as good publicity, it would help to extend their careers. Once again their behavior was civil on set, but after the film came out, it was Davis, not Crawford, who was nominated for a Best Actress Oscar. Worse, she immediately started crowing about it, proudly announcing that she would be the first actress to win three Oscars. Crawford had only one.

Davis was the center of attention at the Oscars. Backstage before the event, she was unusually gracious to Crawford—after all, she could afford to be; this was her night. (Only three other actresses were nominated, and everyone expected Davis to get it.) Crawford was equally polite. During the ceremony, however, as Davis stood in the wings, waiting, she hoped, to accept the award, she got a shock: she lost. Anne Bancroft won for her role in *The Miracle Worker.* And there was more: as Davis stood taking it in, she felt a hand on her arm. "Excuse me," said Crawford, and she strode past the stunned Davis to accept the award on Bancroft's behalf. (The Oscar winner could not be there that night.) On what was supposed to be Davis's night of glory, Crawford had somehow stolen the limelight, an unbearable affront.

### Interpretation

A Hollywood actress has to be thick-skinned, and Joan Crawford was the quintessence of the Hollywood actress: she had a huge capacity to absorb and deal with insults and disrespect. Whenever she could, though, she plotted to get the last laugh on her various nemeses, leaving them humiliated. Crawford knew that people thought of her as somewhat of a bitch, a tough, even unpleasant woman. She felt this was unfair—she had been kind to many—but she could live with it. What annoyed her was how Shearer got away with playing the elegant lady when in fact, Crawford believed, she was a nasty specimen beneath her charming exterior. So Crawford maneuvered to get Shearer to expose a side of herself that few had seen. Just that glimmer was memorable to the Hollywood community and humiliating to Shearer.

With Davis it was all in the timing: Crawford ruined her night of glory (which she had been gloating about for months) without even saying a mean word. Crawford knew that Bancroft would be unable to attend and learned from inside information that she would win, so she happily volunteered to accept the prize on her behalf.

You will often find yourself nursing the desire to revenge yourself on those who have mistreated you. The temptation is to be direct, to say something honest and mean, to let people know how you feel—but words are ineffective here. A verbal spat lowers you to the other person's level and often leaves you with a bad feeling. The sweeter revenge is an action that gives you the last laugh, leaving your victims

*Inevitably a patient entering analysis begins to use ploys which have placed him one-up in previous relationships (this is called a "neurotic pattern"). The analyst learns to devastate these maneuvers of the patient. A simple way, for example, is to respond inappropriately to what the patient says. This places the patient in doubt about everything he has learned in relationships with other people. The patient may say, "Everyone should be truthful," hoping to get the analyst to agree with him and thereby follow his lead. He who follows another lead is one-down. The analyst may reply with silence, a rather weak ploy in this circumstance, or he may say, "Oh?" The "Oh?" is given just the proper inflection to imply, "How on earth could you have ever conceived such an idea?" This not only places the patient in doubt about his statement, but in doubt about what the analyst means by "Oh?" Doubt is, of course, the first step toward one-downness. When in doubt the patient tends to lean on the analyst to resolve the doubt, and we lean on those who are superior to us. Analytic maneuvers designed to arouse doubt in a patient are instituted early in analysis. For example, the analyst may say,*

*"I wonder if that's* really *what you're feeling." The use of "really" is standard in analytic practice. It implies the patient has motivations of which he is not aware. Anyone feels shaken, and therefore one-down, when this suspicion is placed in his mind.*

STRATEGIES OF PSYCHOTHERAPY, JAY HALEY, 1963

with a sense of vague but corrosive inferiority. Provoke them into exposing a hidden, unpleasant side to their character, steal their moment of glory—but make this the battle's last maneuver. That gives you the double delight of showing you are no one to mess with and inflicting a wound that sticks around. As they say, revenge is a dish best served cold.

**Image:**
The Mask. Every performer on the crowded stage is wearing a mask—a pleasant, appealing face to show the audience. Should an apparently innocent bump from a fellow performer make a mask fall, a far less pleasant look will be revealed, and one that few will forget even after the mask is restored.

**Authority:** We often give our rivals the means of our own destruction.
—*Aesop (sixth century* B.C.*)*

## REVERSAL

Sometimes outright war is best—when, for example, you can crush your enemies by encirclement. In the ongoing relationships of daily life, though, one-upmanship is usually the wiser strategy. It may sometimes seem therapeutic to outfight your rivals directly; it may sometimes be appealing to send an overtly intimidating message. But the momentary gains you may earn with a direct approach will be offset by the suspicions you arouse in your colleagues, who will worry that someday you will strong-arm them, too. In the long run, it is more important to secure good feelings and maintain appearances. Wise courtiers always seem to be paragons of civilized behavior, encasing their iron fist in a velvet glove.

# 29

# TAKE SMALL BITES

## THE FAIT ACCOMPLI STRATEGY

*If you seem too ambitious, you stir up resentment in other people; overt power grabs and sharp rises to the top are dangerous, creating envy, distrust, and suspicion. Often the best solution is to take small bites, swallow little territories, playing upon people's relatively short attention spans. Stay under the radar and they won't see your moves. And if they do, it may already be too late; the territory is yours, a fait accompli. You can always claim you acted out of self-defense. Before people realize it, you have accumulated an empire.*

## PIECEMEAL CONQUEST

On June 17, 1940, Winston Churchill, prime minister of England, received a surprise visit from the French general Charles de Gaulle. The Germans had begun their blitzkrieg invasion of the Low Countries and France a mere five weeks earlier, and they had advanced so far so fast that not only France's military but its government as well had already collapsed. The French authorities had fled, either to parts of France not yet occupied by the Germans or to French colonies in North Africa. None, however, had fled to England—but here was General de Gaulle, a solitary exile seeking refuge and offering his services to the Allied cause.

The two men had met before, when de Gaulle had briefly served as France's undersecretary of state for war during the weeks of the blitzkrieg. Churchill had admired his courage and resolution at that difficult moment, but de Gaulle was a strange fellow. At the age of fifty, he had a somewhat undistinguished military record and could hardly be considered an important political figure. But he always acted as if he were at the center of things. And here he was now, presenting himself as the man who could help rescue France, although many other Frenchmen could be considered more suitable for the role. Nevertheless, de Gaulle might be someone whom Churchill could mold and use for his purposes.

Within hours of de Gaulle's arrival in England, the French military sued for peace with the Germans. Under the agreement the two nations worked out, the unoccupied parts of France were to be ruled by a French government friendly to the invaders and based in Vichy. That same evening de Gaulle presented Churchill with a plan: Broadcasting on BBC Radio, he would address all Frenchmen still loyal to a free France and would urge them to not lose heart. He would also call on any who had managed to get to England to contact him. Churchill was reluctant: he did not want to offend the new French government, with which he might have to deal. But de Gaulle promised to say nothing that could be read as treachery to the Vichy government, and at the last minute he was given permission.

De Gaulle delivered the speech much as he had outlined it—except that he ended it with the promise he would be back on the air the next day. This was news to Churchill, yet once the promise had been made, it might look bad to keep de Gaulle off the air, and anything that would hearten the French during these dark days seemed worthwhile.

In the next broadcast, de Gaulle was decidedly bolder. "Any Frenchman who still has weapons," he announced, "has the absolute duty to continue the resistance." He even went so far as to instruct his fellow generals still in France to disobey the enemy. Those who rallied to him in England, he said, would form part of a nation without territory to be called Free France and of a new army to be called Fighting France, the spearhead of an eventual liberation of mainland France from the Germans.

Occupied with other matters and believing de Gaulle's audience to be small, Churchill overlooked the general's indiscretions and allowed him to continue his broadcasts—only to find that each new program made it harder to pull the plug. De Gaulle was transforming himself into a celebrity. The performance of the French military and government during the blitzkrieg had been widely seen as a disgrace, and in the aftermath no one had stepped forward to alter this perception of cowardice—except de Gaulle. His voice radiated confidence, and his face and tall figure stood out in photographs and newsreels. Most important, his appeals had effect: his Fighting France grew from a few hundred soldiers in July 1940 to several thousand a month later.

Soon de Gaulle was clamoring to lead his forces on a campaign to liberate French colonies in Central and Equatorial Africa from the Vichy government. The area was mostly desert and rain forest and was far from the more strategic regions of North Africa on the Mediterranean, but it contained some seaports that might be useful, and so Churchill gave de Gaulle his backing. The French forces were able to take Chad, Cameroon, the French Congo, and Gabon with relative ease.

When de Gaulle returned to England late in 1940, he now had thousands of square miles of territory under his control. His command meanwhile had swelled to close to 20,000 soldiers, and his bold venture had captured the imagination of the British public. No longer the low-order general who had sought refuge months before, he was now a military and political leader. And de Gaulle was equal to this change in status: he was now making demands of the English and acting in a rather aggressive manner. Churchill was beginning to regret giving him so much leeway.

The following year British intelligence discovered that de Gaulle had been making important contacts among the growing French Resistance movement. The Resistance, which was dominated by communists and socialists, had started off chaotic, lacking a coherent structure. De Gaulle had personally chosen an official in the prewar socialist government, Jean Moulin, who had come to England in October 1941, to help unify this underground force. Of all de Gaulle's maneuverings, this was the one that could benefit the Allies most directly; an efficient Resistance would be invaluable. So, with Churchill's blessing, Moulin was parachuted into southern France in early 1942.

By the end of that year, the increasingly imperious de Gaulle had so offended many within the Allied governments and armies—particularly U.S. President Franklin D. Roosevelt—that a plan was discussed to replace him with someone more malleable. The Americans believed they had found the perfect man for the job: General Henri Giraud, one of France's most respected military officials, a man with a record far more distinguished than de Gaulle's. Churchill approved, and Giraud was named commander in chief of French forces in North Africa. Sensing the allied plot, de Gaulle requested a personal meeting with Giraud to dis-

Chien/Development (Gradual Progress)

*This hexagram is made up of Sun (wood, penetration) above, i.e., without, and Ken (mountain, stillness) below, i.e., within. A tree on a mountain develops slowly according to the law of its being and consequently stands firmly rooted. This gives the idea of a development that proceeds gradually, step by step. The attributes of the trigrams also point to this: within is tranquility, which guards against precipitate actions, and without is penetration, which makes development and progress possible.*

*THE I CHING*, CHINA, CIRCA EIGHTH CENTURY B.C.

cuss the situation; after much bureaucratic wrangling, he was granted permission and arrived in Algiers in May 1943.

The two men were at each other's throats almost immediately, each making demands to which the other could never agree. Finally de Gaulle compromised: proposing a committee that would prepare to lead a postwar France, he drafted a document naming Giraud as commander in chief of the armed forces and copresident of France with de Gaulle. In return de Gaulle got the committee to be expanded in size and cleansed of officials with Vichy connections. Giraud was satisfied and signed on. Shortly thereafter, however, Giraud left Algiers for a visit to the United States, and de Gaulle, in his absence, filled the expanded committee with Gaullist sympathizers and Resistance members. Upon Giraud's return he discovered that he had been stripped of much of his political power. Isolated on a committee that he had helped to form, he had no way to defend himself, and in a matter of months de Gaulle was named sole president, then commander in chief. Giraud was quietly retired.

Roosevelt and Churchill watched these developments with increasing alarm. They tried to intervene, making various threats, but in the end they were powerless. Those BBC broadcasts that had started out so innocently were now listened to avidly by millions of Frenchmen. Through Moulin, de Gaulle had gained almost complete control of the French Resistance; a break with de Gaulle would put the Allies' relationship with the Resistance in jeopardy. And the committee that de Gaulle had helped form to govern postwar France was now recognized by governments around the world. To take on the general in any kind of political struggle would be a public-relations nightmare destructive to the war effort.

Somehow this once undistinguished general had forged a kind of empire under his control. And there was nothing anyone could do about it.

### Interpretation

When General Charles de Gaulle fled to England, he had one goal: to restore the honor of France. He intended to do this by leading a military and political organization that would work to liberate France. He wanted his country to be seen as an equal among the Allies, rather than as a vanquished nation dependent on others to regain its freedom.

Had de Gaulle announced his intentions, he would have been seen as a dangerous mix of delusion and ambition. And had he grabbed for power too quickly, he would have shown those intentions. Instead, supremely patient and with an eye on his goal, he took one small bite at a time. The first bite—always the most important—was to gain himself public exposure with first one BBC broadcast, then, through clever maneuvering, an ongoing series. Here, exploiting his keen dramatic instincts and hypnotic voice, he quickly established a larger-than-life presence. This allowed him to create and build up his military group Fighting France.

He took his next bite by bringing those African territories under the control of Fighting France. His control over a large geographical area, no matter how isolated, gave him unassailable political power. Then he insinuated himself into the Resistance, taking over a group that had been a communist bastion. Finally he created—and, bite by bite, gained complete control of—a committee to govern the free France of the future. Because he proceeded in such a piecemeal fashion, no one really noticed what he was up to. When Churchill and Roosevelt realized how far he had insinuated himself into the Resistance, and into the minds of the British and American publics as France's destined postwar leader, it was too late to stop him. His preeminence was a fait accompli.

It is not easy to make one's way in this world, to strive with energy to get what you want without incurring the envy or antipathy of others who may see you as aggressive and ambitious, someone to thwart. The answer is not to lower your ambitions but rather to disguise them. A piecemeal approach to conquest of anything is perfect for these political times, the ultimate mask of aggression. The key to making it work is to have a clear sense of your objective, the empire you want to forge, and then to identify the small, outlying areas of the empire that you will first gobble up. Each bite must have a logic in an overall strategy but must be small enough that no one senses your larger intentions. If your bites are too big, you will take on more than you are ready for and find yourself overwhelmed by problems; if you bite too fast, other people will see what you are up to. Let the passage of time masterfully disguise your intentions and give you the appearance of someone of modest ambition. By the time your rivals wake up to what you have consumed, they risk being consumed themselves if they stand in your way.

*Ambition can creep as well as soar.*
*—Edmund Burke (1729–1797)*

## KEYS TO WARFARE

At first glance we humans might seem hopelessly violent and aggressive. How else to account for history's endless series of wars, which continue into the present? But in fact this is somewhat of an illusion. Standing out dramatically from daily life, war and conflict compel disproportionate attention. The same can be said of those aggressive individuals in the public realm who are constantly grabbing for more.

The truth is that most people are conservative by nature. Desperate to keep what they have, they dread the unforeseen consequences and situations that conflict inevitably brings. They hate confrontation and try to avoid it. (That is why so many people resort to passive aggression to get what they want.) You must always remember this fact of human nature as you plot your way through life. It is also the foundation for any fait accompli strategy.

The strategy works as follows: Suppose there is something you want or need for your security and power. Take it without discussion or warning and you give your enemies a choice, either to fight or to accept the loss and leave you alone. Is whatever you have taken, and your unilateral action in taking it, worth the bother, cost, and danger of waging war? Which costs more, the war (which might easily escalate into something large) or the loss? Take something of real value and they will have to choose carefully; they have a big decision to make. Take something small and marginal, though, and it is almost impossible for your opponents to choose battle. There are likely to be many more reasons for leaving you alone than for fighting over something small. You have played to your enemy's conservative instincts, which are generally stronger than their acquisitive ones. And soon your ownership of this property becomes a fait accompli, part of the status quo, which is always best left alone.

Sooner or later, as part of this strategy, you will take another small bite. This time your rivals are warier; they are starting to see a pattern. But what you have taken is once again small, and once again they must ask themselves if fighting you is worth the headache. They didn't to do it before—why now? Execute a fait accompli strategy subtly and well, as de Gaulle did, and even though a time may come when your goal becomes clear, and when they regret their previous pacifism and consider war, by that time you will have altered the playing field: you are neither so small nor so easy to defeat. To take you on now entails a different kind of risk; there is a different, more powerful reason for avoiding conflict. Only nibble at what you want and you never spark enough anger, fear, or mistrust to make people overcome their natural reluctance to fight. Let enough time pass between bites and you will also play to the shortness of people's attention spans.

The key to the fait accompli strategy is to act fast and without discussion. If you reveal your intentions before taking action, you will open yourself to a slew of criticisms, analyses, and questions: "How dare you think of taking that bite! Be happy with what you have!" It is part of people's conservatism to prefer endless discussion to action. You must bypass this with a rapid seizure of your target. The discussion is foreclosed. No matter how small your bite, taking it also distinguishes you from the crowd and earns you respect and weight.

When Frederick the Great became king of Prussia in 1740, Prussia was a minor European power. Frederick's father had built up the Prussian army, at great expense, but had never really used it; the minute he put the army in play, he knew, the other European powers would have united against him, fearing any threat to the status quo. Frederick, though massively ambitious, understood what had kept his father in check.

The same year he took the throne, however, an opportunity presented itself. Prussia's great nemesis was Austria, where a new leader,

Maria Theresa, had recently become empress. There were many who questioned her legitimacy, though, and Frederick decided to exploit this political instability by moving his army into the small Austrian province of Silesia. Maria Theresa, wanting to prove her toughness, decided to fight to take it back. The war lasted several years—but Frederick had judged the moment well; he finally threatened to take more territory than Silesia alone, and in the end the empress sued for peace.

*All the conceptions born of impatience and aimed at obtaining speedy victory could only be gross errors. . . . It was necessary to accumulate thousands of small victories to turn them into a great success.*

GENERAL VO NGUYEN GIAP, 1911–

Frederick would repeat this strategy again and again, taking over small states here and there that weren't worth fighting for, at least not hard. In this way, almost before anyone noticed, he made Prussia a great power. Had he begun by invading some larger territory, he would have shown his ambitions too clearly and brought down upon himself an alliance of powers determined to maintain the status quo. The key to his piecemeal strategy was an opportunity that fell into his lap. Austria was at a weak moment; Silesia was small, yet by incorporating this neighboring state, Prussia enriched its resources and put itself in position for further growth. The two combined gave him momentum and allowed him space to slowly expand from small to large.

The problem that many of us face is that we have great dreams and ambitions. Caught up in the emotions of our dreams and the vastness of our desires, we find it very difficult to focus on the small, tedious steps usually necessary to attain them. We tend to think in terms of giant leaps toward our goals. But in the social world as in nature, anything of size and stability grows slowly. The piecemeal strategy is the perfect antidote to our natural impatience: it focuses us on something small and immediate, a first bite, then how and where a second bite can get us closer to our ultimate objective. It forces us to think in terms of a process, a sequence of connected steps and actions, no matter how small, which has immeasurable psychological benefits as well. Too often the magnitude of our desires overwhelms us; taking that small first step makes them seem realizable. There is nothing more therapeutic than action.

In plotting this strategy, be attentive to sudden opportunities and to your enemies' momentary crises and weaknesses. Do not be tempted, however, to try to take anything large; bite off more than you can chew and you will be consumed with problems and disproportionately discouraged if you fail to cope with them.

The fait accompli strategy is often the best way to take control of a project that would be ruined by divided leadership. In almost every film Alfred Hitchcock made, he had to go through the same wars, gradually wresting control of the film from the producer, the actors, and the rest of the team. His struggles with screenwriters were a microcosm of the larger war. Hitchcock always wanted his vision for a film to be exactly reflected in the script, but too firm a hand on his writer's neck would get him nothing except resentment and mediocre work. So instead he moved slowly, starting out by giving the writer room to work loosely off his notes, then asking for revisions that shaped the script his way. His control became obvious only gradually, and by that time the writer was emotionally tied to the project and, however frustrated, was working for his approval. A very patient man, Hitchcock let his power plays unfold over time, so that producer, writer, and stars understood the completeness of his domination only when the film was finished.

To gain control of any project, you must be willing to make time your ally. If you start out with complete control, you sap people's spirit and stir up envy and resentment. So begin by generating the illusion that you're all working together on a team effort; then slowly nibble away. If in the process you make people angry, do not worry. That's just a sign that their emotions are engaged, which means they can be manipulated.

Finally, the use of the piecemeal strategy to disguise your aggressive intentions is invaluable in these political times, but in masking your manipulations you can never go too far. So when you take a bite, even a small one, make a show of acting out of self-defense. It also helps to appear as the underdog. Give the impression your objectives are limited by taking a substantial pause between bites—exploiting people's short attention spans—while proclaiming to one and all that you are a person of peace. In fact, it would be the height of wisdom to make your bite a little larger upon occasion and then giving back some of what you have taken. People see only your generosity and your limited actions, not the steadily increasing empire you are amassing.

**Image:**
The Artichoke.
At first glance it
seems unappetizing,
even forbidding, with the
meager edible matter in its hard
exterior. The reward, however,
comes in taking it apart,
devouring it leaf by leaf. Its
leaves slowly become more
tender and tastier, until you
arrive at the succulent
heart.

**Authority:** To multiply small successes is precisely to build one treasure after another. In time one becomes rich without realizing how it has come about. *—Frederick the Great (1712–1786)*

## REVERSAL

Should you see or suspect that you yourself are being attacked bite by bite, your only counterstrategy is to prevent any further progress or faits accomplis. A quick and forceful response will usually be enough to discourage the nibblers, who often resort to this strategy out of weakness and cannot afford many battles. If they are tougher and more ambitious, like Frederick the Great, that forceful response becomes more crucial still. Letting them get away with their bites, however small, is too dangerous—nip them in the bud.

# 30

## PENETRATE THEIR MINDS

## COMMUNICATION STRATEGIES

*Communication is a kind of war, its field of battle the resistant and defensive minds of the people you want to influence. The goal is to advance, to penetrate their defenses and occupy their minds. Anything else is ineffective communication, self-indulgent talk. Learn to infiltrate your ideas behind enemy lines, sending messages through little details, luring people into coming to the conclusions you desire and into thinking they've gotten there by themselves. Some you can trick by cloaking your extraordinary ideas in ordinary forms; others, more resistant and dull, must be awoken with extreme language that bristles with newness. At all cost, avoid language that is static, preachy, and overly personal. Make your words a spark for action, not passive contemplation.*

*The most superficial way of trying to influence others is through talk that has nothing real behind it. The influence produced by such mere tongue wagging must necessarily remain insignificant.*

THE *I CHING*, CHINA, CIRCA EIGHTH CENTURY B.C.

## VISCERAL COMMUNICATION

To work with the film director Alfred Hitchcock for the first time was generally a disconcerting experience. He did not like to talk much on the sets of his movies—just the occasional sardonic and witty remark. Was he deliberately secretive? Or just quiet? And how could someone direct a film, which entails ordering so many people about, without talking a lot and giving explicit instructions?

This peculiarity of Hitchcock's was most troublesome for his actors. Many of them were used to film directors coddling them, discussing in detail the characters they were to play and how to get into the role. Hitchcock did none of this. In rehearsals he said very little; on the set, too, actors would glance over at him for his approval only to find him napping or looking bored. According to the actress Thelma Ritter, "If Hitchcock liked what you did, he said nothing. If he didn't, he looked like he was going to throw up." And yet somehow, in his own indirect way, he would get his actors to do precisely what he wanted.

On the first day of shooting for *The 39 Steps* in 1935, Hitchcock's two leads, Madeleine Carroll and Robert Donat, arrived on the set a little tense. That day they were to act in one of the movie's more complex scenes: playing relative strangers who, however, had gotten handcuffed together earlier in the plot and, still handcuffed, were forced to run through the Scottish countryside (actually a sound stage) to escape the film's villains. Hitchcock had given them no real sign of how he wanted them to act the scene. Carroll in particular was bothered by the director's behavior. This English actress, one of the most elegant film stars of the period, had spent much of her career in Hollywood, where directors had treated her like royalty; Hitchcock, on the other hand, was distant, hard to figure out. She had decided to play the scene with an air of dignity and reserve, the way she thought a lady would respond to the situation of being handcuffed to a strange man. To get over her nervousness, she chatted warmly with Donat, trying to put both him and herself in a collaborative mood.

When Hitchcock arrived on set, he explained the scene to the two actors, snapped a pair of handcuffs on them, and proceeded to lead them through the set, across a dummy bridge and among other props. Then, in the middle of this demonstration, he was suddenly called away to attend to a technical matter. He would return soon; they should take a break. He felt in his pockets for the key to the handcuffs—but no, he must have mislaid it, and off he hurried, ostensibly to find the key. Hours went by. Donat and Carroll became increasingly frustrated and embarrassed; suddenly they had no control, a most unusual feeling for two stars on set. While even the humblest crew members were free to go about their business, the two stars were shackled together. Their forced intimacy and discomfort made their earlier banter impossible. They could not even go to the bathroom. It was humiliating.

Hitchcock returned in the afternoon—he had found the key. Shooting began, but as the actors went to work, it was hard for them to get over the experience of that day; the movie stars' usual cool unflappability was gone. Carroll had forgotten all her ideas about how to play the scene. And yet, despite her and Donat's anger, the scene seemed to flow with unexpected naturalness. Now they knew what it was like to be tied together; they had *felt* the awkwardness, so there was no need to act it. It came from within.

Four years later Hitchcock made *Rebecca*, with Joan Fontaine and Laurence Olivier. Fontaine, at twenty-one, was taking her first leading role and was horribly nervous about playing opposite Olivier, who was widely recognized as an actor of genius. Another director might have eased her insecurities, but Hitchcock was seemingly doing the opposite. He chose to pass along gossip from the rest of the cast and crew: no one thought she was up to the job, he told her, and Olivier had really wanted his wife, Vivien Leigh, to get her part. Fontaine felt terrified, isolated, unsure—exactly the qualities of her character in the film. She hardly needed to act. And her memorable performance in *Rebecca* was the start of a glorious career.

When Hitchcock made *The Paradine Case*, in 1947, his leading lady, Ann Todd, was appearing in her first Hollywood movie and found it hard to relax. So in the silence on set before the director called, "Action!" Hitchcock would tell her a particularly salacious story that would make her laugh or gasp in shock. Before one scene in which she had to lie on a bed in an elegant nightgown, Hitchcock suddenly jumped on her, yelling, "Relax!" Antics like this made it easy for her to let go of her inhibitions and be more natural.

When cast and crew were tired on set, or when they'd gotten too casual and were chatting rather than concentrating on their work, Hitchcock would never yell or complain. Instead he might smash a lightbulb with his fist or throw his teacup against a wall; everyone would quickly sober up and recover his or her focus.

Clearly Hitchcock mistrusted language and explanation, preferring action to words as a way of communicating, and this preference extended to the form and content of his films. That gave his screenwriters a particularly hard time; after all, putting the film into words was their job. In story meetings Hitchcock would discuss the ideas he was interested in—themes like people's doubleness, their capacity for both good and evil, the fact that no one in this world is truly innocent. The writers would produce pages of dialog expressing these ideas elegantly and subtly, only to find them edited out in favor of actions and images. In *Vertigo* (1958) and *Psycho* (1960), for example, Hitchcock inserted mirrors in many scenes; in *Spellbound* (1945) it was shots of ski tracks and other kinds of parallel lines; the murder in *Strangers on a Train* (1951) was revealed through its reflection in a pair of glasses. For Hitchcock, evidently, images like these revealed his ideas of the doubleness in the human soul better than words did, but on paper this seemed somewhat contrived.

On set, the producers of Hitchcock's films often watched in bewilderment as the director moved the camera, not the actors, to stage his

*When you are trying to communicate and can't find the point in the experience of the other party at which he can receive and understand, then, you must create the experience for him.*
*I was trying to explain to two staff organizers in training how their problems in their community arose because they had gone outside the experience of their people: that when you go outside anyone's experience not only do you not communicate, you cause confusion. They had earnest, intelligent expressions on their faces and were verbally and visually agreeing and understanding, but I knew they really didn't understand and that I was not communicating. I had not got into their experience. So I had to give them an experience.*

*Rules for Radicals*, Saul D. Alinsky, 1971

*The letter set Cyrus thinking of the means by which he could most effectively persuade the Persians to revolt, and his deliberations led him to adopt the following plan, which he found best suited to his purpose. He wrote on a roll of parchment that Astyages had appointed him to command the Persian army; then he summoned an assembly of the Persians, opened the roll in their presence and read out what he had written. "And now," he added, "I have an order for you: every man is to appear on parade with a billhook." . . . The order was obeyed. All the men assembled with their billhooks, and Cyrus' next command was that before the day was out they should clear a certain piece of rough land full of thorn bushes, about eighteen or twenty furlongs square. This too was done, whereupon Cyrus issued the further order that they should present themselves again on the following day, after having taken a bath. Meanwhile Cyrus collected and slaughtered all his father's goats, sheep, and oxen in preparation for entertaining the whole Persian army at a banquet, together with the best wine and bread he could procure. The next day the guests assembled, and were told to sit down on the grass and enjoy themselves. After the meal*

scenes. It seemed to make no sense, as if he loved the technical side of filmmaking more than dialog and the human presence. Nor could editors fathom his obsession with sounds, colors, the size of the actors' heads within the frame, the speed with which people moved—he seemed to favor these endless visual details over the story itself.

And then the film would be a finished product, and suddenly everything that had seemed peculiar about his method made perfect sense. Audiences often responded to Hitchcock's films more deeply than they did to the work of any other director. The images, the pacing, the camera movements, swept them along and got under their skin. A Hitchcock film was not just seen, it was experienced, and it stayed in the mind long after the viewing.

### Interpretation

In interviews Hitchcock often told a story about his childhood: When he was around six, his father, upset at something he had done, sent him to the local police station with a note. The officer on duty read the note and locked little Alfred in a cell, telling him, "This is what we do to naughty boys." He was released after just a few minutes, but the experience marked him indelibly. Had his father yelled at him, as most boys' fathers did, he would have become defensive and rebellious. But leaving him alone, surrounded by frightening authority figures, in a dark cell, with its unfamiliar smells—that was a much more powerful way to communicate. As Hitchcock discovered, to teach people a lesson, to really alter their behavior, you must alter their experience, aim at their emotions, inject unforgettable images into their minds, shake them up. Unless you are supremely eloquent, it is hard to accomplish this through words and direct expression. There are simply too many people talking at us, trying to persuade us of this or that. Words become part of this noise, and we either tune them out or become even more resistant.

To communicate in a deep and real way, you must bring people back to their childhood, when they were less defensive and more impressed by sounds, images, actions, a world of preverbal communication. It requires speaking a kind of language composed of actions, all strategically designed to effect people's moods and emotions, what they can least control. That is precisely the language Hitchcock developed and perfected over the years. With actors he wanted to get the most natural performance out of them, in essence get them *not* to act. To tell them to relax or be natural would have been absurd; it would only have made them more awkward and defensive than they already were. Instead, just as his father had gotten him to feel terror in a London police station, he got them to *feel* the emotions of the movie: frustration, isolation, loss of inhibition. (Of course he hadn't mislaid the handcuffs' key somewhere on the set of *The 39 Steps*, as Donat later found out; the supposed loss was a strategy.) Instead of prodding actors with irritating words, which come from the outside and are pushed away, Hitchcock made these feelings part of their inner experience—and this communicated immediately on-screen. With audiences, too, Hitchcock never preached a message. Instead he used the visual power of film to return them to that childlike state when images and compelling symbols had such a visceral effect.

It is imperative in life's battles to be able to communicate your ideas to people, to be able to alter their behavior. Communication is a form of warfare. Your enemies here are defensive; they want to be left alone with their preexisting prejudices and beliefs. The more deeply you penetrate their defenses, the more you occupy their mental space, the more effectively you are communicating. In verbal terms, most people wage a kind of medieval warfare, using words, pleas, and calls for attention like battle-axes and clubs to hit people over the head. But in being so direct, they only make their targets more resistant. Instead you must learn to fight indirectly and unconventionally, tricking people into lowering their defenses—hitting their emotions, altering their experience, dazzling them with images, powerful symbols, and visceral sensory cues. Bringing them back to that childlike state when they were more vulnerable and fluid, the communicated idea penetrates deep behind their defenses. Because you are not fighting the usual way, you will have an unusual power.

*The priest Ryokan . . . asked Zen master Bukkan . . . for an explanation of the four Dharma-worlds. . . . [Bukkan] said: "To explain the four Dharma-worlds should not need a lot of chatter." He filled a white tea cup with tea, drank it up, and smashed the cup to pieces right in front of the priest, saying, "Have you got it?" The priest said: "Thanks to your here-and-now teaching, I have penetrated right into the realm of Principle and Event."*

*—Trevor Leggett,* Samurai Zen: The Warrior Koans *(1985)*

## THE MASTERMIND

In 1498 the twenty-nine-year-old Niccolò Machiavelli was appointed secretary of Florence's Second Chancery, which managed the city's foreign affairs. The choice was unusual: Machiavelli was of relatively low birth, had no experience in politics, and lacked a law degree or other professional qualification. He had a contact in the Florentine government, however, who knew him personally and saw great potential in him. And indeed, over the next few years, Machiavelli stood out from his colleagues in the Chancery for his tireless energy, his incisive reports on political matters, and his excellent advice to ambassadors and ministers. He won prestigious assignments, traveling around Europe on diplomatic missions—to various parts of northern Italy to meet with Cesare Borgia, to ferret out that ruthless statesman's intentions on Florence; to France to meet with King Louis XII; to Rome to confer with Pope Julius II. He seemed to be at the start of a brilliant career.

Not all was well, however, in Machiavelli's professional life. He complained to his friends about the Chancery's low pay; he also described doing all the hard work in various negotiations, only to see some

*Cyrus asked them which they preferred—yesterday's work or today's amusement; and they replied that it was indeed a far cry from the previous day's misery to their present pleasures. This was the answer which Cyrus wanted; he seized upon it at once and proceeded to lay bare what he had in mind. "Men of Persia," he said, "listen to me: obey my orders, and you will be able to enjoy a thousand pleasures as good as this without ever turning your hands to menial labour; but, if you disobey, yesterday's task will be the pattern of innumerable others you will be forced to perform. Take my advice and win your freedom. I am the man destined to undertake your liberation, and it is my belief that you are a match for the Medes in war as in everything else. It is the truth I tell you. Do not delay, but fling off the yoke of Astyages at once."*

*The Persians had long resented their subjection to the Medes. At last they had found a leader, and welcomed with enthusiasm the prospect of liberty.*

*The Histories,*
Herodotus,
484–432 B.C.

*Even more foolish is one who clings to words and phrases and thus tries to achieve understanding. It is like trying to strike the moon with a stick, or scratching a shoe because there is an itchy spot on the foot. It has nothing to do with the Truth.*

ZEN MASTER MUMON, 1183–1260

powerful senior minister brought on board at the last moment to finish the job and take the credit. Many above him, he said, were stupid and lazy, appointed to their positions by virtue of birth and connections. He was developing the art of dealing with these men, he told his friends, finding a way to use them instead of being used.

Before Machiavelli's arrival in the Chancery, Florence had been ruled by the Medici family, who, however, had been unseated in 1494, when the city became a republic. In 1512, Pope Julius II financed an army to take Florence by force, overthrow the republic, and restore the Medicis to power. The plan succeeded, and the Medicis took control, well in Julius's debt. A few weeks later, Machiavelli was sent to prison, vaguely implicated in a conspiracy against the Medicis. He was tortured but refused to talk, whether about his own involvement or that of others. Released from prison in March 1513, he retired in disgrace to a small farm owned by his family a few miles outside Florence.

Machiavelli had a close friend in a man called Francesco Vettori, who had managed to survive the change in government and to ingratiate himself with the Medicis. In the spring of 1513, Vettori began to receive letters in which Machiavelli described his new life. At night he would shut himself up in his study and converse in his mind with great figures in history, trying to uncover the secrets of their power. He wanted to distill the many things he himself had learned about politics and statecraft. And, he wrote to Vettori, he was writing a little pamphlet called *De principatibus*—later titled *The Prince*—"where I dive as deep as I can into ideas about this subject, discussing the nature of princely rule, what forms it takes, how these are acquired, how they are maintained, how they are lost." The knowledge and advice imparted in this pamphlet would be more valuable to a prince than the largest army—perhaps Vettori could show it to one of the Medicis, to whom Machiavelli would gladly dedicate the work? It could be of great use to this family of "new princes." It could also revive Machiavelli's career, for he was despondent at his isolation from politics.

Vettori passed the essay along to Lorenzo de' Medici, who accepted it with much less interest than he did two hunting dogs given to him at the same time. Actually, *The Prince* perplexed even Vettori: its advice was sometimes starkly violent and amoral, yet its language was quite dispassionate and matter-of-fact—a strange and uncommon mix. The author wrote the truth, but a little too boldly. Machiavelli also sent the manuscript to other friends, who were equally unsure what to make of it. Perhaps it was intended as satire? Machiavelli's disdain for aristocrats with power but no brains was well known to his circle.

Soon Machiavelli wrote another book, later known as *The Discourses*, a distillation of his talks with friends since his fall from grace. A series of meditations on politics, the book contained some of the same stark advice as the earlier work but was more geared toward the constitution of a republic than to the actions of a single prince.

Over the next few years, Machiavelli slowly returned to favor and was allowed to participate in Florentine affairs. He wrote a play, *Mandragola*, which, though scandalous, was admired by the pope and staged at the Vatican; he was also commissioned to write a history of Florence. *The Prince* and *The Discourses* remained unpublished, but they circulated in manuscript among the leaders and politicians of Italy. Their audience was small, and when Machiavelli died in 1527, the former secretary to the republic seemed destined to return to the obscurity from which he came.

After Machiavelli's death, however, those two unpublished works of his began to circulate outside Italy. In 1529, Thomas Cromwell, the crafty minister to Henry VIII of England, somehow got hold of a copy of *The Prince* and, unlike the flightier Lorenzo de' Medici, read it closely and carefully. To him the book's historical anecdotes made for a lively and entertaining read. The plain language was not bizarre but refreshing. Most important, the amoral advice was in fact indispensable: the writer explained not only what a leader had to do to hold on to power but how to present his actions to the public. Cromwell could not help but adapt Machiavelli's counsel in his advice to the king.

Published in several languages in the decades after Machiavelli's death, *The Prince* slowly spread far and wide. As the centuries passed, it took on a life of its own, in fact a double life: widely condemned as amoral, yet avidly read in private by great political figures down the ages. The French minister Cardinal Richelieu made it a kind of political bible. Napoleon consulted it often. The American president John Adams kept it by his bedside. With the help of Voltaire, the Prussian king Frederick the Great wrote a tract called *The Anti-Machiavel*, yet he shamelessly practiced many of Machiavelli's ideas to the letter.

As Machiavelli's books reached larger audiences, his influence extended beyond politics. Philosophers from Bacon to Hegel found in his writings confirmation for many of their own theories. Romantic poets such as Lord Byron admired the energy of his spirit. In Italy, Ireland, and Russia, young revolutionaries discovered in *The Discourses* an inspiring call to arms and a blueprint for a future society.

Over the centuries millions upon millions of readers have used Machiavelli's books for invaluable advice on power. But could it possibly be the opposite—that it is Machiavelli who has been using his readers? Scattered through his writings and through his letters to his friends, some of them uncovered centuries after his death, are signs that he pondered deeply the strategy of writing itself and the power he could wield after his death by infiltrating his ideas *indirectly* and *deeply* into his readers' minds, transforming them into unwitting disciples of his amoral philosophy.

Interpretation

Once retired to his farm, Machiavelli had the requisite time and distance to think deeply about those matters that concerned him most. First, he

*Yoriyasu was a swaggering and aggressive samurai. . . . In the spring of 1341 he was transferred from Kofu to Kamakura, where he visited Master Toden, the 45th teacher at Kenchoji, to ask about Zen.*
*The teacher said, "It is to manifest directly the Great Action in the hundred concerns of life. When it is loyalty as a samurai, it is the loyalty of Zen. 'Loyalty' is written with the Chinese character made up of 'centre' and 'heart,' so it means the lord in the centre of the man. There must be no wrong passions. But when this old priest looks at the samurai today, there are some whose heart centre leans towards name and money, and others where it is towards wine and lust, and with others it is inclined towards power and bravado. They are all on those slopes, and cannot have a centred heart; how could they have loyalty to the state? If you, Sir, wish to practise Zen, first of all practise loyalty and do not slip into wrong desires."*
*The warrior said, "Our loyalty is direct Great Action on the battlefield. What need have we for sermons from a priest?"*
*The teacher replied, "You, Sir, are a hero in strife, I am a gentleman of peace—we can have nothing to say to each other."*
*The warrior then drew his sword and said,*

*"Loyalty is in the hero's sword, and if you do not know this, you should not talk of loyalty." The teacher replied, "This old priest has the treasure sword of the Diamond King, and if you do not know it, you should not talk of the source of loyalty." The samurai said, "Loyalty of your Diamond Sword—what is the use of that sort of thing in actual fighting?" The teacher jumped forward and gave one Katzu! shout, giving the samurai such a shock that he lost consciousness. After some time the teacher shouted again and the samurai at once recovered. The teacher said, "The loyalty in the hero's sword, where is it? Speak!" The samurai was overawed; he apologized and took his departure.*

SAMURAI ZEN: *THE WARRIOR KOANS*, TREVOR LEGGETT, 1985

slowly formulated the political philosophy that had long been brewing in his mind. To Machiavelli the ultimate good was a world of dynamic change in which cities or republics were reordering and revitalizing themselves in perpetual motion. The greatest evil was stagnation and complacency. The agents of healthy change were what he called "new princes"—young, ambitious people, part lion, part fox, conscious or unconscious enemies of the established order. Second, Machiavelli analyzed the process by which new princes rose to the heights of power and, often, fell from it. Certain patterns were clear: the need to manage appearances, to play upon people's belief systems, and sometimes to take decidedly amoral action.

Machiavelli craved the power to spread his ideas and advice. Denied this power through politics, he set out to win it through books: he would convert readers to his cause, and *they* would spread his ideas, witting or unwitting carriers. Machiavelli knew that the powerful are often reluctant to take advice, particularly from someone apparently beneath them. He also knew that many of those not in power might be frightened by the dangerous aspects of his philosophy—that many readers would be attracted and repelled at the same time. (The powerless want power but are afraid of what they might have to do to get it.) To win over the resistant and ambivalent, Machiavelli's books would have to be strategic, indirect and crafty. So he devised unconventional rhetorical tactics to penetrate deep behind his readers' defenses.

First, he filled his books with indispensable advice—practical ideas on how to get power, stay in power, protect one's power. That draws in readers of all kinds, for all of us think first of our own self-interest. Also, no matter how much a reader resists, he or she realizes that ignoring this book and its ideas might be dangerous.

Next, Machiavelli stitched historical anecdotes throughout his writing to illustrate his ideas. People like to be shown ways to fancy themselves modern Caesars or Medicis, and they like to be entertained by a good story; and a mind captivated by a story is relatively undefended and open to suggestion. Readers barely notice that in reading these stories—or, rather, in reading Machiavelli's cleverly altered versions of them—they are absorbing ideas. Machiavelli also quoted classical writers, adjusting the quotations to suit his purposes. His dangerous counsels and ideas would be easier to accept if they seemed to be emerging from the mouth of a Livy or a Tacitus.

Finally, Machiavelli used stark, unadorned language to give his writing movement. Instead of finding their minds slowing and stopping, his readers are infected with the desire to go beyond thought and take action. His advice is often expressed in violent terms, but this works to rouse his readers from their stupor. It also appeals to the young, the most fertile ground from which new princes grow. He left his writing open-ended, never telling people exactly what to do. They must use their own ideas and experiences with power to fill in his writing, becoming complicit partners in the text. Through these various devices, Machiavelli gained power over his readers while disguising the nature of his manipulations. It is hard to resist what you cannot see.

Understand: you may have brilliant ideas, the kind that could revolutionize the world, but unless you can express them effectively, they will have no force, no power to enter people's minds in a deep and lasting way. You must focus not on yourself or on the need you feel to express what you have to say but on your audience—as intently as a general focuses on the enemy he is strategizing to defeat. When dealing with people who are bored and have short attention spans, you must entertain them, sneaking your ideas in through the back door. With leaders you must be careful and indirect, perhaps using third parties to disguise the source of the ideas you are trying to spread. With the young your expression must be more violent. In general, your words must have movement, sweeping readers along, never calling attention to their own cleverness. You are not after personal expression, but power and influence. The less people consciously focus on the communicative form you have chosen, the less they realize how far your dangerous ideas are burrowing into their minds.

> *For some time I have never said what I believed, and never believed what I said, and if I do sometimes happen to say what I think, I always hide it among so many lies that it is hard to recover it.*
> *—Niccolò Machiavelli, letter to Francesco Guicciardini (1521)*

*The Lydian King Croesus had had Miltiades much in this thoughts so when he learned of his capture, he sent a command to the people of Lampsacus to set him at liberty; if they refused, he was determined, he added, to "cut them down like a pine-tree." The people of the town were baffled by Croesus' threat, and at a loss to understand what being cut down like a pine-tree might mean, until at last the true significance of the phrase dawned upon a certain elderly man: the pine, he explained, was the only kind of tree which sent up no new shoots after being felled—cut down a pine and it will die off completely. The explanation made the Lampsacenes so frightened of Croesus that they let Miltiades go.*

THE HISTORIES, HERODOTUS, 484–432 B.C.

## KEYS TO WARFARE

For centuries people have searched for the magic formula that would give them the power to influence others through words. This search has been mostly elusive. Words have strange, paradoxical qualities: offer people advice, for instance, no matter how sound, and you imply that you know more than they do. To the extent that this strikes at their insecurities, your wise words may merely have the effect of entrenching them in the very habits you want to change. Once your language has gone out into the world, your audience will do what they want with it, interpreting it according to their own preconceptions. Often when people appear to listen, nod their heads, and seem persuaded, they are actually just trying to be agreeable—or even just to get rid of you. There are simply too many words inundating our lives for talk to have any real, long-lasting effect.

This does not mean that the search for power through language is futile, only that it must be much more strategic and based on knowledge of fundamental psychology. What really changes us and our behavior is not the actual words uttered by someone else but our own experience, something that comes not from without but from within. An event occurs that shakes us up emotionally, breaks up our usual patterns of look-

*That same day Jesus went out of the house and sat beside the sea. And great crowds gathered about him, so that he got into a boat and sat there; and the whole crowd stood on the beach. And he told them many things in parables, saying: "A sower went out to sow. And as he sowed, some seeds fell along the path, and the birds came and devoured them. Other seeds fell on rocky ground, where they had not much soil, and immediately they sprang up, since they had no depth of soil, but when the sun rose they were scorched; and since they had no root they withered away. Other seeds fell upon thorns, and the thorns grew up and choked them. Other seeds fell on good soil and brought forth grain, some a hundred-fold, some sixty, some thirty. He who has ears, let him hear." Then the disciples came and said to him, "Why do you speak to them in parables?" And he answered them, "To you it has been given to know the secrets of the kingdom of heaven, but to them it has not been given. For to him who has will more be given, and he will have abundance; but from him who has not, even what he has will be taken away. This is why I speak to them in parables, because seeing they do not see, and*

ing at the world, and has a lasting impact on us. Something we read or hear from a great teacher makes us question what we know, causes us to meditate on the issue at hand, and in the process changes how we think. The ideas are internalized and felt as personal experience. Images from a film penetrate our unconscious, communicating in a preverbal way, and become part of our dream life. Only what stirs deep within us, taking root in our minds as thought and experience, has the power to change what we do in any lasting way.

The historical figure who most deeply pondered the nature of communication was surely Socrates, the great philosopher of classical Athens. Socrates' goal was simple: he wanted to make people realize that their knowledge of the world was superficial, if not downright false. Had he tried to say this conventionally and directly, though, he would only have made his audience more resistant and would have strengthened their intellectual smugness. And so, pondering this phenomenon, and through much trial and error, Socrates came up with a method. First came the setup: he would make a show of his own ignorance, telling his audience of mostly young men that he himself knew little—that any wisdom he was reputed to have was just talk. Meanwhile he would compliment his listeners, feeding their vanity by praising their ideas in an offhand way. Then, in a series of questions constituting a dialog with a member of his audience, he would slowly tear apart the very ideas he had just praised. He would never directly say anything negative, but through his questions he would make the other person see the incompleteness or falsity of his ideas. This was confusing; he had just professed his own ignorance, and he had sincerely praised his interlocutors. Yet he had somehow raised a lot of doubts about what they had claimed to know.

The dialog would lie in the minds of Socrates' targets for several days, leading them to question their ideas about the world on their own. In this frame of mind, they would now be more open to real knowledge, to something new. Socrates broke down people's preconceptions about the world by adopting what he called a "midwife" role: he did not implant his ideas, he simply helped to deliver the doubts that are latent in everyone.

The success of the Socratic method was staggering: a whole generation of young Athenians fell under his spell and were permanently altered by his teachings. The most famous of these was Plato, who spread Socrates' ideas as if they were gospel. And Plato's influence over Western thought is perhaps greater than that of anyone else. Socrates' method was highly strategic. He began by tearing himself down and building others up, a way of defusing his listeners' natural defensiveness, imperceptibly lowering their walls. Then he would lure them into a labyrinth of discussion from which they could find no exit and in which everything they believed was questioned. According to Alcibiades, one of the young men whom Socrates had bewitched, you never knew what he really believed or what he really meant; everything he said was a rhetorical stance, was ironic. And since you were unsure what he was doing, what came to the surface in these conversations were your own confusion and doubt. He altered your experience of the world from within.

Think of this method as *communication-in-depth.* Normal discourse, and even fine writing and art, usually only hits people on the surface. Our attempts to communicate with them become absorbed in all of the noise that fills their ears in daily life. Even if something we say or do somehow touches an emotional chord and creates some kind of connection, it rarely stays in their minds long enough to alter how they think and act. A lot of the time, these surface communications are fine; we cannot go through life straining to reach everyone—that would be too exhausting. But the power to reach people more deeply, to alter their ideas and unpleasant behavior, is sometimes critical.

What you need to pay attention to is not simply the content of your communication but the form—the way you lead people to the conclusions you desire, rather than telling them the message in so many words. If you want people to change a bad habit, for example, much more effective than simply trying to persuade them to stop is to show them—perhaps by mirroring their bad behavior in some way—how annoying that habit *feels* to other people. If you want to make people with low self-esteem feel better about themselves, praise has a superficial effect; instead you must prod them into accomplishing something tangible, giving them a real experience. That will translate into a much deeper feeling of confidence. If you want to communicate an important idea, you must not preach; instead make your readers or listeners connect the dots and come to the conclusion on their own. Make them internalize the thought you are trying to communicate; make it seem to emerge from their own minds. Such indirect communication has the power to penetrate deep behind people's defenses.

In speaking this new language, learn to expand your vocabulary beyond explicit communication. Silence, for instance, can be used to great effect: by keeping quiet, not responding, you say a lot; by not mentioning something that people expect you to talk about, you call attention to this ellipsis, make it communicate. Similarly, the details—what Machiavelli calls *le cose piccole* (the little things)—in a text, speech, or work of art have great expressive power. When the famous Roman lawyer and orator Cicero wanted to defame the character of someone he was prosecuting, he would not accuse or rant; instead he would mention details from the life of the accused—the incredible luxury of his home (was it paid for out of illegal means?), the lavishness of his parties, the style of his dress, the little signs that he considered himself superior to the average Roman. Cicero would say these things in passing, but the subtext was clear. Without hitting listeners over the head, it directed them to a certain conclusion.

In any period it can be dangerous to express ideas that go against the grain of public opinion or offend notions of correctness. It is best to seem to conform to these norms, then, by parroting the accepted wis-

*hearing they do not hear, nor do they understand. With them indeed is fulfilled the prophecy of Isaiah which says: 'You shall indeed hear but never understand, and you shall indeed see but never perceive.'"*

MATTHEW 13:1–15

dom, including the proper moral ending. But you can use details here and there to say something else. If you are writing a novel, for instance, you might put your dangerous opinions in the mouth of the villain but express them with such energy and color that they become more interesting than the speeches of the hero. Not everyone will understand your innuendos and layers of meaning, but some certainly will, at least those with the proper discernment; and mixed messages will excite your audience: indirect forms of expression—silence, innuendo, loaded details, deliberate blunders—make people feel as if they were participating, uncovering the meaning on their own. The more that people participate in the communication process, the more deeply they internalize its ideas.

In putting this strategy into practice, avoid the common mistake of straining to get people's attention by using a form that is shocking or strange. The attention you get this way will be superficial and short-lived. By using a form that alienates a wide public, you narrow your audience; you will end up preaching to the converted. As the case of Machiavelli demonstrates, using a conventional form is more effective in the long run, because it attracts a larger audience. Once you have that audience, you can insinuate your real (and even shocking) content through details and subtext.

In war almost everything is judged by its result. If a general leads his army to defeat, his noble intentions do not matter; nor does the fact that unforeseen factors may have thrown him off course. He lost; no excuse will do. One of Machiavelli's most revolutionary ideas was to apply this standard to politics: what matters is not what people say or intend but the results of their actions, whether power is increased or decreased. This is what Machiavelli called the "effective truth"—the real truth, in other words, what happens in fact, not in words or theories. In examining the career of a pope, for instance, Machiavelli would look at the alliances he had built and the wealth and territory he had acquired, not at his character or religious proclamations. Deeds and results do not lie. You must learn to apply the same barometer to your attempts at communication, and to those of other people.

If a man says or writes something that he considers revolutionary and that he hopes will change the world and improve mankind, but in the end hardly anyone is affected in any real way, then it is not revolutionary or progressive at all. Communication that does not advance its cause or produce a desired result is just self-indulgent talk, reflecting no more than people's love of their own voice and of playing the role of the moral crusader. The effective truth of what they have written or said is that nothing has been changed. The ability to reach people and alter their opinions is a serious affair, as serious and strategic as war. You must be harsher on yourself and on others: failure to communicate is the fault not of the dull-witted audience but of the unstrategic communicator.

Irony.—*Irony is in place only as a pedagogic tool, employed by a teacher in dealing with any kind of pupil: its objective is humiliation, making ashamed, but of that salutary sort which awakens good resolutions and inspires respect and gratitude towards him who treats us thus of the kind we feel for a physician. The ironist poses as unknowing, and does so so well that the pupils in discussion with him are deceived, grow bold in their belief they know better and expose themselves in every way; they abandon circumspection and reveal themselves as they are—up to the moment when the lamp they have been holding up to the face of the teacher sends its beams very humiliatingly back on to themselves.—Where such a relationship as that between teacher and pupil does not obtain, irony is ill-breeding, a vulgar affectation. All ironical writers depend on the foolish species of men who together with the author would like to feel themselves superior to all others and who regard the author as the mouthpiece of their presumption.—Habituation to irony, moreover, like habituation to sarcasm, spoils the character, to which it gradually lends the quality of a malicious and jeering superiority: in the end one comes to resemble a snapping dog which has learned how to laugh but forgotten how to bite.*

*Human, All too Human*, Friedrich Nietzsche, 1878

**Image:** The Stiletto. It is long and tapered to a point. It requires no sharpening. In its form lies its perfection as an instrument to penetrate cleanly and deeply. Whether thrust into the flank, the back, or through the heart, it has a fatal effect.

**Authority:** I cannot give birth to wisdom myself and the accusation that many make against me, that while I question others, I myself bring nothing wise to light due to my lack of wisdom, is accurate. The reason for this is as follows: God forces me to serve as a midwife and prevents me from giving birth. —*Socrates (470–399 B.C.)*

## REVERSAL

Even as you plan your communications to make them more consciously strategic, you must develop the reverse ability to decode the subtexts, hidden messages, and unconscious signals in what other people say. When people speak in vague generalities, for example, and use a lot of abstract terms like "justice," "morality," "liberty," and so on, without really ever explaining the specifics of what they are talking about, they are almost always hiding something. This is often their own nasty but necessary actions, which they prefer to cover up under a screen of righteous verbiage. When you hear such talk, be suspicious.

Meanwhile people who use cutesy, colloquial language, brimming with clichés and slang, may be trying to distract you from the thinness of their ideas, trying to win you over not by the soundness of their arguments but by making you feel chummy and warm toward them. And people who use pretentious, flowery language, crammed with clever metaphors, are often more interested in the sound of their own voices than in reaching the audience with a genuine thought. In general, you must pay attention to the forms in which people express themselves; never take their content at face value.

# 31

## DESTROY FROM WITHIN

## THE INNER-FRONT STRATEGY

*A war can only really be fought against an enemy who shows himself. By infiltrating your opponents' ranks, working from within to bring them down, you give them nothing to see or react against—the ultimate advantage. From within, you also learn their weaknesses and open up possibilities of sowing internal dissension. So hide your hostile intentions. To take something you want, do not fight those who have it, but rather join them—then either slowly make it your own or wait for the moment to stage a coup d'état. No structure can stand for long when it rots from within.*

*Athene now inspired Prylis, son of Hermes, to suggest that entry should be gained into Troy by means of a wooden horse; and Epeius, son of Panopeus, a Phocian from Parnassus, volunteered to build one under Athene's supervision. Afterwards, of course, Odysseus claimed all the credit for this stratagem . . .*

*. . . [Epeius] built an enormous hollow horse of fir planks, with a trap-door fitted into one flank, and large letters cut on the other which consecrated it to Athene: "In thankful anticipation of a safe return to their homes, the Greeks dedicate this offering to the Goddess." Odysseus persuaded the bravest of the Greeks to climb fully armed up a rope-ladder and through the trap-door into the belly of the horse. . . . Among them were Menelaus, Odysseus, Diomedes, Sthenelus, Acamas, Thoas, and Neoptolemus. Coaxed, threatened, and bribed, Epeius himself joined the party. He climbed up last, drew the ladder in after him and, since he alone knew how to work the trap-door, took his seat beside the lock.*

*At nightfall, the remaining Greeks under Agamemnon followed Odysseus's instructions, which were to burn their camp, put out to sea and wait off Tenedos and the Calydnian*

## THE INVISIBLE ENEMY

Late in 1933, Adolf Hitler appointed the forty-six-year-old Rear Admiral Wilhelm Canaris chief of the Abwehr, the secret intelligence and counterespionage service of the German General Staff. Hitler had recently won dictatorial powers as the ruler of Germany, and, with an eye on future conquests in Europe, he wanted Canaris to make the Abwehr an agency as efficient as the British Secret Service. Canaris was a slightly odd choice for the position. He came from the aristocracy, was not a member of the Nazi Party, and had not had a particularly outstanding military career. But Hitler saw traits in Canaris that would make him a superior spymaster: cunning in the extreme, a man made for intrigue and deception, he knew how to get results. He would also owe his promotion exclusively to Hitler.

In the years to come, Hitler would have reason to feel proud of his choice. Canaris rigorously reorganized the Abwehr and extended its spy networks throughout Europe. Then, in May 1940, he provided exceptional intelligence for the blitzkrieg invasion of France and the Low Countries early in World War II. And so, in the summer of that same year, Hitler gave Canaris his most important task to date: providing intelligence for Operation Sealion, a plan to conquer England. After the blitzkrieg and the evacuation of the Allied army at Dunkirk, the British seemed deeply vulnerable, and knocking them out of the war at this point would ensure Hitler's conquest of Europe.

A few weeks into the job, however, Canaris reported that the Germans had underestimated the size of the English army and air force. Sealion would require resources much larger than the Führer had anticipated; unless Hitler was willing to commit many more troops, it could turn into a mess. This was highly disappointing news for Hitler, who had wanted to knock out England in one quick blow. With his eye on an imminent invasion of Russia, he was unwilling to commit large numbers to Sealion or to spend years subduing the British. Having come to trust Canaris, he abandoned the planned invasion.

That same summer General Alfred Jodl came up with a brilliant plan to damage England in another way: using Spain as a base of operations, he would invade the British-owned island of Gibraltar, cutting off England's sea routes through the Mediterranean and the Suez Canal to its empire in India and points east—a disastrous blow. But the Germans would have to act fast, before the English caught on to the threat. Excited by the prospect of ruining England in this indirect way, Hitler once again asked Canaris to assess the plan. The Abwehr chief went to Spain, studied the situation, and reported back. The moment a German army moved into Spain, he said, the English would see the plan, and Gibraltar had elaborate defenses. The Germans would also need the cooperation of Francisco Franco, dictator of Spain, who Canaris believed would not be sufficiently helpful. In short: Gibraltar was not worth the effort.

There were many around Hitler who believed that taking Gibraltar was eminently realizable and could mean overall victory in the war against Britain. Shocked at Canaris's report, they vocally expressed their doubts about the intelligence he had been providing all along. His enigmatic nature—he spoke little and was impossible to read—only fueled their suspicions that he was not to be trusted. Hitler heard his staff out, but a meeting with Generalissimo Franco to discuss the Gibraltar plan indirectly corroborated everything Canaris had said. Franco was difficult and made all kinds of silly demands; the Spanish would be impossible to deal with; the logistics were too complicated. Hitler quickly lost interest in Jodl's plan.

In the years that followed, German officials in increasing numbers would come to suspect Canaris of disloyalty to the Third Reich, but no one could pin anything concrete on him. And Hitler himself had great faith in the Abwehr chief and sent him on critical top-secret missions. One such assignment occurred in the summer of 1943, when Marshal Pietro Badoglio, the former chief of the Italian General Staff, arrested Benito Mussolini, dictator of Italy and Hitler's staunchest ally. The Germans feared that Badoglio might secretly open talks with General Dwight D. Eisenhower for Italy's surrender—a devastating blow to the Axis that Hitler could forestall, if necessary, by sending an army to Rome, arresting Badoglio, and occupying the capital. But was it necessary?

Hitler's armies were needed elsewhere, so Canaris was dispatched to assess the likelihood of Italy's surrender. He met with his counterpart in the Italian government, General Cesare Amé, then arranged for a meeting between high-ranking members of both countries' intelligence services. At the meeting, Amé emphatically denied that Badoglio had any intention of betraying Germany; in fact, the marshal was fiercely loyal to the cause. And Amé was very convincing. Hitler accordingly left Italy alone. A few weeks later, however, Badoglio indeed surrendered to Eisenhower, and the valuable Italian fleet moved into Allied hands. Canaris had been fooled—or was it Canaris who had done the fooling?

General Walter Schellenberg, chief of the foreign intelligence branch of the SS, began to investigate the Badoglio fiasco and found two men in Amé's service who had listened in on one of Canaris's talks with their boss. Canaris, they reported, had known of Badoglio's intentions to surrender all along and had collaborated with Amé to deceive Hitler. Surely this time the Abwehr chief had been caught in the act and would pay with his life. Schellenberg accumulated a thick dossier of other actions that cast more doubts on Canaris. He presented it to Heinrich Himmler, head of the SS, who, however, told his subordinate to keep quiet—he would present the dossier to Hitler when the time was right. Yet, to Schellenberg's dismay, months went by and Himmler did nothing, except eventually to retire Canaris with honors from the service.

Shortly after Canaris's retirement, his diaries fell into the hands of the SS. They revealed that he had conspired against Hitler from the

*Islands until the following evening. . . . . . . At the break of day, Trojan scouts reported that the camp lay in ashes and that the Greeks had departed, leaving a huge horse on the seashore. Priam and several of his sons went out to view it and, as they stood staring in wonder, Thymoetes was the first to break the silence. "Since this is a gift to Athene," he said, "I propose that we take it into Troy and haul it up to her citadel." "No, no!" cried Capys. "Athene favoured the Greeks too long; we must either burn it at once or break it open to see what the belly contains." But Priam declared: "Thymoetes is right. We will fetch it in on rollers. Let nobody desecrate Athene's property." The horse proved too broad to be squeezed through the gates. Even when the wall had been breached, it stuck four times. With enormous efforts the Trojans then hauled it up to the citadel; but at least took the precaution of repairing the breach behind them. . . . At midnight . . . Odysseus ordered Epeius to unlock the trapdoor. . . . Now the Greeks poured silently through the moonlit streets, broke into the unguarded houses, and cut the throats of the Trojans as they slept.*

*The Greek Myths*, vol. 2,
Robert Graves,
1955

beginning of his service as Abwehr chief, even plotting to assassinate the Führer in schemes that had only barely misfired. Canaris was sent to a concentration camp, where, in April 1945, he was tortured and killed.

### Interpretation

Wilhelm Canaris was a devoutly patriotic and conservative man. In the earliest days of the Nazi Party's rise to power, he had come to believe that Hitler would lead his beloved Germany to destruction. But what could he do? He was just one man, and to raise his voice against Hitler would get him no more than a little publicity and an early death. Canaris cared only about results. So he kept quiet, and, when offered the job of Abwehr chief, he seized his opportunity. At first he bided his time, gaining credibility by his work in the Abwehr and getting to understand the inner workings of the Nazi government. Meanwhile he secretly organized a group of like-minded conspirators, the Schwarze Kapelle (Black Orchestra), who would hatch several plots to kill Hitler. From his position in the Abwehr, Canaris was to some extent able to protect the Schwarze Kapelle from investigation. He also quietly gathered intelligence on the dirtiest secrets of high-ranking Nazis like Himmler and let them know that any move against him would result in revelations that would ruin them.

Assigned to prepare for Operation Sealion, Canaris doctored the intelligence to make England look much more formidable than it was. Assigned to investigate an invasion of Gibraltar, he secretly told the Spanish that to let Germany use their country would spell disaster: Germany would never leave. Hence Franco's alienating treatment of Hitler. In both of these cases, Canaris exploited Hitler's impatience for quick and easy victories to discourage him from ventures that could have easily and irrevocably turned the war in his favor. Finally, in the case of Badoglio, Canaris understood Hitler's weak spot—a paranoid concern with the loyalty of others—and coached Amé on how to appeal to this weakness and make a show of Italy's devotion to the Axis cause. The results of Canaris's work from the inside are astounding: one man played a major role in saving England, Spain, and Italy from disaster, arguably turning the tide of the war. The resources of the German war machine were essentially at his disposal, to disrupt and derail its efforts.

As the story of Canaris demonstrates, if there is something you want to fight or destroy, it is often best to repress your desire to act out your hostility, revealing your position and letting the other side know your intentions. What you gain in publicity, and perhaps in feeling good about expressing yourself openly, you lose in a curtailment of your power to cause real damage, particularly if the enemy is strong.

Instead the ultimate strategy is to seem to stay on the enemy side, burrowing deep into its heart. From there you can gather valuable information: weaknesses to attack, incriminating evidence to publicize. Here subtle maneuvers, like passing along false information or steering your opponent into a self-destructive policy, can have large effects—much larger than anything you could do from the outside. The enemy's powers become weapons you can use against it, a kind of turncoat armory at your disposal. It is hard for most people to imagine that someone who outwardly plays the part of a loyal supporter or friend can secretly be a foe. This makes your hostile intentions and maneuvers relatively easy to cloak. When you are invisible to the enemy, there is no limit to the destructive powers at your command.

*Speak deferentially, listen respectfully, follow his command, and accord with him in everything. He will never imagine you might be in conflict with him. Our treacherous measures will then be settled.*
—*Tai Kung,* Six Secret Teachings *(circa fourth century* B.C.*)*

## THE FRIENDLY TAKEOVER

In the summer of 1929, André Breton, the thirty-three-year-old leader of Paris's avant-garde surrealist movement, saw a private screening of a film called *Un Chien Andalou.* It was directed by a Spanish member of the group, Luis Buñuel, and its first image showed a man slicing open a woman's eye with a knife. This, Breton exclaimed, was the first surrealist film. *Un Chien Andalou* generated excitement in part because of the contribution to it of a new artist on the scene, Salvador Dalí, a friend and collaborator of Buñuel's. The director spoke highly to Breton of his fellow Spaniard, whose paintings, he said, could certainly be considered surrealist and whose personality was supremely peculiar. Soon others, too, were talking about Dalí, discussing what he called his "paranoid-critical" method of painting: he delved deep into his dreams and unconscious and interpreted the images he found there, no matter what their content, in delirious detail. Dalí still lived in Spain, but Breton was suddenly seeing his name everywhere he went. Then, in November 1929, the twenty-five-year-old Dalí had his first major show in a Paris gallery, and Breton was transfixed by the images. He wrote of the exhibition, "For the first time the windows of the mind had opened wide."

The late 1920s were a difficult period for Breton. The movement he had founded some five years earlier was stagnating, its members constantly bickering over ideological points that bored Breton to tears. In truth, surrealism was on the verge of becoming passé. Perhaps Dalí could offer the fresh blood it needed: his art, his ideas, and his provocative character might make surrealism something people talked about again. With all this in mind, Breton invited Dalí into the movement, and the Spaniard happily accepted. Dalí moved to Paris and established himself there.

For the next few years, Breton's strategy seemed to be working. Dalí's scandalous paintings were the talk of Paris. His exhibitions caused riots. Suddenly everyone was interested in surrealism again, even

*Throughout his revolutionary and missionary travels, Hasan [leader of the Nizari Ismailis] was searching for an impregnable fortress from which to conduct his resistance to the Seljuk empire. In about 1088, he finally chose the castle of Alamut, built on a narrow ridge on a high rock in the heart of the Elburz Mountains in a region known as the Rudbar. The castle dominated an enclosed cultivated valley thirty miles long and three miles across at its widest, approximately six thousand feet above sea level. Several villages dotted the valley, and their inhabitants were particularly receptive to the ascetic piety of Hasan. The castle was accessible only with the greatest difficulty through a narrow gorge of the Alamut River. . . . Hasan employed a careful strategy to take over the castle, which had been granted to its current Shiite owner, named Mahdi, by the Seljuk sultan Malikshah. First, Hasan sent his trusted* dai *Husayn Qai-ni and two others to win converts in the neighboring villages. Next, many of the residents and soldiers of Alamut were secretly converted to Ismailism. Finally, in September 1090, Hasan himself was secretly smuggled into*

younger artists. But by 1933, Breton was beginning to rue his inclusion of Dalí. He had begun to get letters from the Spaniard expressing great interest in Hitler as a source of paranoiac inspiration. Only the surrealists, Dalí felt, were capable of "saying pretty things on the subject" of Hitler; he even wrote of sexually charged dreams about Hitler. As news of Dalí's infatuation with the Führer spread within the movement, it provoked a great deal of argument. Many surrealists had communist sympathies and were disgusted by the Spanish artist's musings. To make matters worse, he included in one enormous painting an image of Lenin in a grotesque pose—exposing oversized buttocks (nine feet long), propped on a crutch. Many in the surrealist group admired Lenin; was Dalí being deliberately provocative? After Breton told Dalí he disliked this rendition of the human buttocks and anus, a delirious profusion of anus images suddenly began to populate the artist's paintings.

By early 1934, Breton could stand no more, and he issued a statement, cosigned by several members, proposing Dalí's expulsion from the surrealist group. The movement was split down the middle; Dalí had both supporters and enemies. Finally a meeting was called to debate the issue. Dalí had a fever and a sore throat; he came to the meeting wearing half a dozen layers of clothing and with a thermometer in his mouth. As Breton paced the room, listing the reasons for his banishment, Dalí began to take off and put on his overcoat, jacket, and sweaters, trying to regulate his temperature. It was hard for anyone to pay attention to Breton.

Finally Dalí was asked to respond. "I had painted both Lenin and Hitler on the basis of dreams," he said, the thermometer in his mouth making him spit many of his words. "Lenin's anamorphic buttock was not insulting, but the very proof of my fidelity to surrealism." He continued to put on and take off clothing. "All taboos are forbidden, or else a list has to be made of those to be observed, and let Breton formally state that the kingdom of surrealist poetry is nothing but a little domain used for the house arrest of those convicted felons placed under surveillance by the vice squad or the Communist Party."

The members of the circle were perplexed to say the least: Dalí had turned their meeting into a kind of surrealist performance, both making fun of the creative freedom they advocated and claiming it for himself. He had also made them laugh. A vote to exclude him would only confirm the accusations he had leveled at them. For the time being, they decided to leave him alone, but in the meeting's aftermath it was clear that the surrealist movement was now more divided than ever.

At the end of that year, Dalí disappeared to New York. Word came back to Paris that he had completely conquered the art world in America, making surrealism the hottest movement around. In the years to come, he would actually emigrate to the United States, and his face would grace the cover of *Time* magazine. From New York his fame spread far, wide, and around the world. Meanwhile the surrealists themselves faded quietly from public view, marginalized by other art movements. In 1939, Breton, disgusted by his lack of control over Dalí, finally expelled the Spanish artist from the group, but by then it hardly mattered: Dalí himself had become synonymous with surrealism, and it would stay that way long after the surrealist movement had died.

*the castle. When Mahdi realized that Hasan had in fact quietly taken over his fortress, he left peacefully. . . .*

*THE TEMPLARS AND THE ASSASSINS*, JAMES WASSERMAN, 2001

## Interpretation

Salvador Dalí was an extremely ambitious man. Although he appeared eccentric to say the least, his diaries show the extent to which he applied strategy to get what he wanted. Languishing in Spain early on in his career, he saw the importance of capturing the Paris art world, the center of the modern-art movement, if he were to rise to the heights of fame. And if he were to make it in Paris, his name would have to be attached to some kind of movement—that would demonstrate his avant-garde status and give him free publicity. Considering the nature of his work and paranoiac-critical method, surrealism was the only logical choice. Of course it helped that Dalí's good friend Buñuel was already a member of the group and that his lover, Gala, was also the wife of Paul Eluard, one of surrealism's principal authors and thinkers. Through Buñuel, Gala, and a few others (people Dalí called "messengers" and "porters"), he spread his name strategically throughout Paris and aimed himself directly at Breton. In truth, Dalí despised any kind of organized group and actively disliked Breton, but both could be useful to him. By insinuating his presence through others and suggesting that he was a surrealist *avant la lettre,* he cleverly managed to get Breton to invite him into the group.

Now, as a true surrealist, an official insider, Dalí could continue to wage his insidious war. At first he made a show of being a loyal member of the group, the platform from which he spent several years winning over Paris with his striking paintings. The surrealists were grateful for the new life he had given them, but in reality he was using their name and presence to propel his career. Then, once his fame was secure, he proceeded to dynamite the group from the inside. The weaker the surrealists were internally, the more he could dominate them publicly. Dalí deliberately chose Hitler and Lenin as images he knew would disgust many in the group. That would both bring out Breton's totalitarian side and cause a major split among the members. Dalí's "performance" at the meeting to expel him was a surrealist masterpiece in itself, and a strategic blow to any vestiges of group unity. Finally, when the movement was riven with division, he scampered off to New York to complete his campaign. Appropriating the seductive name of surrealism for himself, he would go down in history as its most famous member, far more famous than Breton.

It is hard to make your way in the world alone. Alliances can help, but if you are starting out, it is hard to get the right people interested in an alliance with you; there is nothing in it for them. The wisest strategy is often to join the group that can best serve your long-term interests, or the one with which you have the most affinity. Instead of trying to conquer this group from the outside, you burrow your way into it. As an insider you can gather valuable information about how it functions and particularly about its members' hypocrisies and weaknesses—knowledge you can use to wage insidious intraorganizational warfare. From the inside you can divide and conquer.

Remember: your advantage here is that, unlike the other members, you have no sentimental attachment to the group; your only allegiance is to yourself. That gives you the freedom you need to make the manipulative and destructive maneuvers that will propel you to the fore at the others' expense.

*If you decide to wage a war for the total triumph of your individuality, you must begin by inexorably destroying those who have the greatest affinity with you.*
*—Salvador Dalí (1904–1989)*

## KEYS TO WARFARE

The most common form of defense in old-fashioned warfare was the fortress or walled city, and military leaders strategized for centuries about how to take such structures. The fortress presented a simple problem: it was designed to be impenetrable, to require such an effort to take it that unless doing so was strategically essential, an army would tend to pass it by. The conventional strategy against the fortress was to scale or breach its walls, using siege engines and battering rams. Often that meant first besieging it, creating around it circles known as "lines of circumvallation and contravallation" that would prevent supplies and reinforcements from coming in and the defenders from leaving. The city's inhabitants would slowly starve and weaken, making it possible eventually to breach the walls and take the castle. These sieges tended to be quite long and bloody.

Over the centuries, however, certain enlightened strategists hit upon a different way to bring down the walls. Their strategy was based on a simple premise: the apparent strength of the fortress is an illusion, for behind its walls are people who are trapped, afraid, even desperate. The city's leaders have essentially run out of options; they can only put their faith in the fortress's architecture. To lay siege to these walls is to mistake the appearance of strength for reality. If in fact the walls are hiding great weakness within, then the proper strategy is to bypass them and aim for the interior. This can be done literally, by digging tunnels beneath the walls, undermining their strength—a conventional military strategy. A better, more devious route is to infiltrate people inside them or to work with the city's disaffected inhabitants. This is known as "opening an inner front"—finding a group on the inside who will work on your behalf to spread discontent and will eventually betray the fortress into your hands, sparing you a long siege.

In late January 1968, the North Vietnamese launched the famous Tet Offensive against the South Vietnamese and American armies. Among their targets was Hue, the ancient capital of Vietnam and a city of great religious significance for the Vietnamese people. In the center of Hue is a massive fort called the Citadel, and within the Citadel is the Imperial Palace compound, the heart and soul of Hue. The Citadel has incredibly thick and high walls and is surrounded on all sides by water. In 1968 it was guarded by American soldiers and their allies. Yet the North Vietnamese were somehow able to take the Citadel with remarkable ease. They held it for several weeks, then disappeared from Hue as if by magic after a massive U.S. counterattack. The Citadel was unimportant to them as a physical or strategic possession; what they were after was the symbolism of being able to take it, showing the world that American invincibility was a myth.

The capture of the Citadel was a remarkable feat, and this is how it was done. Months before Tet, the North Vietnamese began to infiltrate men into the city and to organize those of their sympathizers who already lived in Hue and worked inside the Citadel. They got hold of detailed plans of the fortress, which allowed them to dig elaborate tunnels under its walls. They were also able to leave stockpiles of weapons at key points. During the Tet holiday, they infiltrated even more of their men into the city, dressed as peasants. Confederates inside the fortress helped them to overrun some of the guard posts and open the gates. Melting into the local population, they made it impossible for the Citadel's defenders to distinguish friend from foe. Finally, having reconnoitered the location of the concentrated command structure inside the Citadel, the North Vietnamese were able to take it out right away, leaving the defenders unable to communicate with one another. This created mass confusion, and in the process the defense of the Citadel collapsed.

The North Vietnamese called this strategy the "blooming lotus." It has deep roots in Asian military thinking, and its applications go far beyond war. Instead of focusing on the enemy's formidable front, on capturing key points in the periphery of its defenses and finding a way through them (the traditional Western approach), the lotus strategy aims first and foremost at the center—the soft and vulnerable parts within. The goal is to funnel soldiers and confederates into this central area by whatever means possible and to attack it first in order to spread confusion. Rather than trying to penetrate defenses, it infiltrates them. This includes the minds of the enemy soldiers and officers—strategizing to get under their skin, to unbalance their reasoning powers, to soften them up from within. As with the lotus flower, everything unfolds from the center of the target.

The basic principle here is that it is easiest to topple a structure—a wall, a group, a defensive mind—from the inside out. When something

To attack or to intervene.—*We often make the mistake of actively opposing a tendency or party or age because we happen to have seen only its external side, its deliquescence or the "faults of its virtues" necessarily adhering to it—perhaps because we ourselves have participated in them to a marked degree. Then we turn our back on them and go off in an opposite direction; but it would be better if we sought out their good and strong side instead or evolved and developed it in ourself. It requires, to be sure, a more penetrating eye and a more favorable inclination to advance what is imperfect and evolving than to see through it in its imperfection and deny it.*

*Human, All Too Human*, Friedrich Nietzsche, 1878

*A prince need trouble little about conspiracies when the people are well disposed, but when they are hostile and hold him in hatred, then he must fear everything and everybody.*

NICCOLÒ MACHIAVELLI, 1469–1527

begins to rot or fall apart from within, it collapses of its own weight—a far better way to bring it down than ramming yourself against its walls. In attacking any group, the lotus strategist thinks first of opening an inner front. Confederates on the inside will provide valuable intelligence on the enemy's vulnerabilities. They will silently and subtly sabotage him. They will spread internal dissension and division. The strategy can weaken the enemy to the point where you can finish him off with a penetrating blow; it can also bring down the enemy in and of itself.

A variation on the lotus strategy is to befriend your enemies, worming your way into their hearts and minds. As your targets' friend, you will naturally learn their needs and insecurities, the soft interior they try so hard to hide. Their guard will come down with a friend. And even later on, when you play out your treacherous intentions, the lingering resonance of your friendship will still confuse them, letting you keep on manipulating them by toying with their emotions or pushing them into overreactions. For a more immediate effect, you can try a sudden act of kindness and generosity that gets people to lower their defenses—the Trojan Horse strategy. (For ten long years, the Greeks battered the walls of Troy to no effect; the simple gift of a wooden horse let them sneak a few men into Troy and open the gates from within.)

The lotus strategy is widely applicable. When confronted by something difficult or thorny, do not be distracted or discouraged by its formidable outer appearance; think your way into the soft core, the center from which the problem blossoms. Perhaps the source of your problem is a particular person; perhaps it is yourself and your own stale ideas; perhaps it is the dysfunctional organization of a group. Knowing the problem's core gives you great power to change it from the inside out. Your first thought must always be to infiltrate to the center—whether in thought or in action—never to whale away at the periphery or just pound at the walls.

If there is someone on the inside whom you need to get rid of or thwart, the natural tendency is to consider conspiring with others in your group who feel the same way. In most conspiracies the goal is some large-scale action to topple the leader and seize power. The stakes are high, which is why conspiracies are so often difficult and dangerous. The main weakness in any conspiracy is usually human nature: the higher the number of people who are in on the plot, the higher the odds that someone will reveal it, whether deliberately or accidentally. As Benjamin Franklin said, "Three may keep a secret if two of them are dead." No matter how confident you may be of your fellow conspirators, you cannot know for certain what is going on in their minds—the doubts they may be having, the people they may be talking to.

There are a few precautions you can take. Keep the number of conspirators as small as possible. Involve them in the details of the plot only as necessary; the less they know, the less they have to blab. Revealing the schedule of your plan as late as possible before you all act will give them no time to back out. Then, once the plan is described, stick to it. Nothing sows more doubts in conspirators' minds than last-minute changes. Even given all this insurance, keep in mind that most conspiracies fail and, in their failure, create all kinds of unintended consequences. Even the successful plot to assassinate Julius Caesar led not to the restoration of the Roman Republic, as the conspirators intended, but eventually to the undemocratic regime of the emperor Augustus. Too few conspirators and you lack the strength to control the consequences; too many and the conspiracy will be exposed before it bears fruit.

In destroying anything from within, you must be patient and resist the lure of large-scale, dramatic action. As Canaris showed, the placement of little wrenches in the machinery is just as destructive in the long run, and safer because harder to trace. Consider the ability to dissuade your opponents from acting aggressively or to make their plans misfire as a kind of battlefield victory, even if your triumph is surreptitious. A few such victories and your enemy will fall apart from within.

Finally, morale plays a crucial part in any war, and it is always wise to work to undermine the morale of the enemy troops. The Chinese call this "removing the firewood from under the cauldron." You can attempt this from the outside, through propaganda, but that often has the opposite effect, reinforcing the cohesion of soldiers and civilians in the face of an alien force trying to win them over. It is much more effective to find sympathizers within their ranks, who will spread discontent among them like a disease. When soldiers see those on their own side having doubts about the cause they are fighting for, they are generally demoralized and vulnerable to more disaffection. If their leaders overreact to this threat by punishing grumblers, they play into your hands, representing themselves as unjust and heavy-handed; if they leave the problem alone, it will only spread; and if they start to see enemies everywhere around them, their paranoia will cloud their strategic abilities. Using an inner front to spread dissension is often enough to give you the advantage you need to overwhelm the enemy.

**Image:** The Termite. From deep within the structure of the house, the termite silently eats away at the wood, its armies patiently boring through beams and supports. The work goes unnoticed, but not the result.

**Authority:** The worst [military policy is] to assault walled cities. . . . If your commander, unable to control his temper, sends your troops swarming at the walls, your casualties will be one in three and still you will not have taken the city. . . . Therefore the expert in using the military subdues the enemy's forces without going to battle, takes the enemy's walled cities without launching an attack. *—Sun-tzu (fourth century B.C.)*

## REVERSAL

There are always likely to be disgruntled people in your own group who will be liable to turning against you from the inside. The worst mistake is to be paranoid, suspecting one and all and trying to monitor their every move. Your only real safeguard against conspiracies and saboteurs is to keep your troops satisfied, engaged in their work, and united by their cause. They will tend to police themselves and turn in any grumblers who are trying to foment trouble from within. It is only in unhealthy and decaying bodies that cancerous cells can take root.

# 32

# DOMINATE WHILE SEEMING TO SUBMIT

## THE PASSIVE-AGGRESSION STRATEGY

*Any attempt to bend people to your will is a form of aggression. And in a world where political considerations are paramount, the most effective form of aggression is the best-hidden one: aggression behind a compliant, even loving exterior. To follow the passive-aggressive strategy, you must seem to go along with people, offering no resistance. But actually you dominate the situation. You are noncommittal, even a little helpless, but that only means that everything revolves around you. Some people may sense what you are up to and get angry. Don't worry—just make sure you have disguised your aggression enough that you can deny it exists. Do it right and they will feel guilty for accusing you. Passive aggression is a popular strategy; you must learn how to defend yourself against the vast legions of passive-aggressive warriors who will assail you in your daily life.*

*Gandhi and his associates repeatedly deplored the inability of their people to give organized, effective, violent resistance against injustice and tyranny. His own experience was corroborated by an unbroken series of reiterations from all the leaders of India—that India could not practice physical warfare against her enemies. Many reasons were given, including weakness, lack of arms, having been beaten into submission, and other arguments of a similar nature. . . . . . . Confronted with the issue of what means he could employ against the British, we come to the other criteria previously mentioned; that the kind of means selected and how they can be used is significantly dependent upon the face of the enemy, or the character of his opposition. Gandhi's opposition not only made the effective use of passive resistance possible but practically invited it. His enemy was a British administration characterized by an old, aristocratic, liberal tradition, one which granted a good deal of freedom to its colonials and which always had operated on a pattern of using, absorbing, seducing, or destroying, through flattery or corruption, the revolutionary leaders who arose from the colonial ranks.*

## THE GUILT WEAPON

In December 1929 the group of Englishmen who governed India were feeling a little nervous. The Indian National Congress—the country's main independence movement—had just broken off talks over the proposal that Britain would gradually return autonomous rule to the subcontinent. Instead the Congress was now calling for nothing less than immediate and total independence, and it had asked Mahatma Gandhi to lead a civil-disobedience campaign to initiate this struggle. Gandhi, who had studied law in London years before, had invented a form of passive-resistant protest in 1906, while working as a barrister in South Africa. In India in the early 1920s, he had led civil-disobedience campaigns against the British that had created quite a stir, had landed him in prison, and had made him the most revered man in the country. For the British, dealing with him was never easy; despite his frail appearance, he was uncompromising and relentless.

Although Gandhi believed in and practiced a rigorous form of nonviolence, the colonial officers of the British Raj were fearful: at a time when the English economy was weak, they imagined him organizing a boycott of British goods, not to mention mass demonstrations in the streets of India's cities, a police nightmare.

The man in charge of the Raj's strategy in combating the independence movement was the viceroy of India, Lord Edward Irwin. Although Irwin admired Gandhi personally, he had decided to respond to him rapidly and with force—he could not let the situation get out of hand. He waited anxiously to see what Gandhi would do. The weeks went by, and finally, on March 2, Irwin received a letter from Gandhi—rather touching in its honesty—that revealed the details of the civil-disobedience campaign he was about to launch. It was to be a protest against the salt tax. The British held a monopoly on India's production of salt, even though it could easily be gathered by anyone on the coast. They also levied a rather high tax on it. This was quite a burden for the poorest of the poor in India, for whom salt was their only condiment. Gandhi planned to lead a march of his followers from his ashram near Bombay (present-day Mumbai) to the coastal town of Dandi, where he would gather sea salt left on the shore by the waves and encourage Indians everywhere to do the same. All this could be prevented, he wrote to Irwin, if the viceroy would immediately repeal the salt tax.

Irwin read this letter with a sense of relief. He imagined the sixty-year-old Gandhi, rather fragile and leaning on a bamboo cane, leading his ragtag followers from his ashram—fewer than eighty people—on a two-hundred-mile march to the sea, where he would gather some salt from the sands. Compared to what Irwin and his staff had been expecting, the protest seemed almost ludicrously small in scale. What was Gandhi thinking? Had he lost touch with reality? Even some members of the Indian National Congress were deeply disappointed by his choice of protest. In any event, Irwin had to rethink his strategy. It simply would not do to harass or arrest this saintly old man and his followers (many of them women). That would look very bad. It would be better to leave him alone, avoiding the appearance of a heavy-handed response and letting the crisis play out and die down. In the end the ineffectiveness of this campaign would somewhat discredit Gandhi, breaking his spell over the Indian masses. The independence movement might fracture or at least lose some momentum, leaving England in a stronger position in the long run.

*This was the kind of opposition that would have tolerated and ultimately capitulated before the tactic of passive resistance.*

*RULES FOR RADICALS*, SAUL D. ALINSKY, 1971

As Irwin watched Gandhi's preparations for the march, he became still more convinced that he had chosen the right strategy. Gandhi was framing the event as almost religious in quality, like Lord Buddha's famous march to attain divine wisdom, or Lord Rama's retreat in the Ramayana. His language became increasingly apocalyptic: "We are entering upon a life-and-death struggle, a holy war." This seemed to resonate with the poor, who began to flock to Gandhi's ashram to hear him speak. He called in film crews from all over the world to record the march, as if it were a momentous historical event. Irwin himself was a religious man and saw himself as the representative of a God-fearing, civilized nation. It would redound to England's credit to be seen to leave this saintly man untouched on his procession to the sea.

*It is impossible to win a contest with a helpless opponent since if you win you have won nothing. Each blow you strike is unreturned so that all you can feel is guilt for having struck while at the same time experiencing the uneasy suspicion that the helplessness is calculated.*

*STRATEGIES OF PSYCHOTHERAPY*, JAY HALEY, 1963

Gandhi and his followers left their ashram on March 12, 1930. As the group passed from village to village, their ranks began to swell. With each passing day, Gandhi was bolder. He called on students throughout India to leave their studies and join him in the march. Thousands responded. Large crowds gathered along the way to see him pass; his speeches to them grew more and more inflammatory. He seemed to be trying to bait the English into arresting him. On April 6 he led his followers into the sea to purify themselves, then collected some salt from the shore. Word quickly spread throughout India that Gandhi had broken the salt law.

Irwin followed these events with increasing alarm. It dawned on him that Gandhi had tricked him: instead of responding quickly and decisively to this seemingly innocent march to the sea, the viceroy had left Gandhi alone, allowing the march to gain momentum. The religious symbolism that seemed so harmless had stirred the masses, and the salt issue had somehow become a lightning rod for disaffection with English policy. Gandhi had shrewdly chosen an issue that the English would not recognize as threatening but that would resonate with Indians. Had Irwin responded by arresting Gandhi immediately, the whole thing might have died down. Now it was too late; to arrest him at this point would only add fuel to the fire. Yet to leave him alone would show weakness and cede him the initiative. Meanwhile nonviolent demonstrations were breaking out in cities and villages all over India, and to respond to them with violence would only make the demonstrators more sympathetic to moderate Indians. Whatever Irwin did, it seemed, would make things worse. And so he fretted, held endless meetings, and did nothing.

In the days to come, the cause rippled outward. Thousands of

Indians traveled to India's coasts to collect salt as Gandhi had. Large cities saw mass demonstrations in which this illegal salt was given away or sold at a minimal price. One form of nonviolent protest cascaded into another—a Congress-led boycott of British goods, for one. Finally, on Irwin's orders, the British began to respond to the demonstrations with force. And on May 4 they arrested Gandhi and took him to prison, where he would stay for nine months without trial.

Gandhi's arrest sparked a conflagration of protest. On May 21 a group of 2,500 Indians marched peacefully on the government's Dharasana Salt Works, which was defended by armed Indian constables and British officers. When the marchers advanced on the factory, they were struck down with steel-plated clubs. Instructed in Gandhi's methods of nonviolence, the demonstrators made no attempt to defend themselves, simply submitting to the blows that rained down on them. Those who had not been hit continued to march until almost every last one had been clubbed. It was a nauseating scene that got a great deal of play in the press. Similar incidents all over India helped to destroy the last sentimental attachment any Indians still had toward England.

To end the spiraling unrest, Irwin was finally forced to negotiate with Gandhi, and, on several issues, to give ground—an unprecedented event for an English imperialist viceroy. Although the end of the Raj would take several years, the Salt March would prove to be the beginning of the end, and in 1947 the English finally left India without a fight.

### Interpretation

Gandhi was a deceptively clever strategist whose frail, even saintly appearance constantly misled his adversaries into underestimating him. The key to any successful strategy is to know both one's enemy and oneself, and Gandhi, educated in London, understood the English well. He judged them to be essentially liberal people who saw themselves as upholding traditions of political freedom and civilized behavior. This self-image—though riddled with contradictions, as indicated by their sometimes brutal behavior in their colonies—was deeply important to the English. The Indians, on the other hand, had been humiliated by many years of subservience to their English overlords. They were largely unarmed and in no position to engage in an insurrection or guerrilla war. If they rebelled violently, as other colonies had done, the English would crush them and claim to be acting out of self-defense; their civilized self-image would suffer no damage. The use of nonviolence, on the other hand—an ideal and philosophy that Gandhi deeply valued and one that had a rich tradition in India—would exploit to perfection the English reluctance to respond with force unless absolutely necessary. To attack people who were protesting peacefully would not jibe with the Englishman's sense of his own moral purity. Made to feel confused and guilty, the English would be paralyzed with ambivalence and would relinquish the strategic initiative.

The Salt March is perhaps the quintessential example of Gandhi's

*Huang Ti, the legendary Yellow Emperor and reputed ancestor of the Chou dynasty, the historical paradigm of concord and civilization, is said to have brought harmony from chaos, tamed the barbarians and wild beasts, cleared the forests and marshes, and invented the "five harmonious sounds," not through an act of epic bloodshed, but through his superior virtue, by adapting and yielding to "natural conditions" and to the Will of Heaven. Confucianism henceforth repudiates as unworkable the idea of military solutions to human problems. Huang Ti's most notable heir, we are told, was Ti Yao, a gentleman who "naturally and without effort," embraced reverence, courteousness, and intelligence. Nevertheless, during his reign, the Deluge, mythology's universal symbol of anomie, threatened to inundate the land. Thus it fell upon him to appoint a successor to preserve the order of his own son. Ti Yao chose the most qualified man for the job, the venerable Shun, who had in various tests already demonstrated a capacity to harmonize human affairs through righteousness. . . .*

*. . . Shun in turn selected Yu the Sage to engineer an end to the flood. Because Yu refused wine and*

strategic brilliance. First, he deliberately chose an issue that the British would consider harmless, even laughable. To respond with force to a march about salt would have given an Englishman trouble. Then, by identifying his apparently trivial issue in his letter to Irwin, Gandhi made space for himself in which to develop the march without fear of repression. He used that space to frame the march in an Indian context that would give it wide appeal. The religious symbolism he found for it had another function as well: it heightened the paralysis of the British, who were quite religious themselves in their own way and could not countenance repressing a spiritual event. Finally, like any good showman, Gandhi made the march dramatically visual and used the press to give it maximum exposure.

Once the march gained momentum, it was too late to stop it. Gandhi had sparked a fire, and the masses were now deeply engaged in the struggle. Whatever Irwin did at this point would make the situation worse. Not only did the Salt March become the model for future protests, but it was clearly the turning point in India's struggle for independence.

Many people today are as ambivalent as the English were about having power and authority. They need power to survive, yet at the same time they have an equally great need to believe in their own goodness. In this context to fight people with any kind of violence makes you look aggressive and ugly. And if they are stronger than you are, in effect you are playing into their hands, justifying a heavy-handed response from them. Instead it is the height of strategic wisdom to prey upon people's latent guilt and liberal ambivalence by making yourself look benign, gentle, even passive. That will disarm them and get past their defenses. If you take action to challenge and resist them, you must do it morally, righteously, peacefully. If they cannot help themselves and respond with force, they will look and feel bad; if they hesitate, you have the upper hand and an opening to determine the whole dynamic of the war. It is almost impossible to fight people who throw up their hands and do not resist in the usual aggressive way. It is completely confusing and disabling. Operating in this way, you inflict guilt as if it were a kind of weapon. In a political world, your passive, moralistic resistance will paralyze the enemy.

*always acted appropriately, moving with and not resisting nature, the Way of Heaven* (T'ien Tao) *was revealed to him. He subsequently harnessed the river waters not by fighting against them with a dam, but by yielding to them and clearing for them a wider channel within which to run. Were it not for Yu, so the story goes, who herein personified the wisdom of both Confucius and Lao-tzu, the Taoist prophet, we would all be fish.*

RELIGIOUS MYTHOLOGY AND THE ART OF WAR, JAMES A. AHO, 1981

*I was a believer in the politics of petitions, deputations and friendly negotiations. But all these have gone to dogs. I know that these are not the ways to bring this Government round. Sedition has become my religion. Ours is a nonviolent war.*
*—Mahatma Gandhi (1869–1947)*

## PASSIVE POWER

Early in 1820 a revolution broke out in Spain, followed a few months later by one in Naples, which at that time was a city-state incorporated

*The devotion of his soldiers to him, affirmed in many stories, must be a fact. [Julius Caesar] could not have done what he did without it. The speech in which it is always said he quelled a mutiny with a single word, calling his men not fellow-soldiers as was his custom, but citizens, civilians, shows a great deal more about his methods than the mere clever use of a term. It was a most critical moment for him. He was in Rome after Pompey's defeat, on the point of sailing for Africa, to put down the powerful senatorial army there. In the city he was surrounded by bitter enemies. His whole dependence was his army, and the best and most trusted legion in it mutinied. They nearly killed their officer; they marched to Rome and claimed their discharge; they would serve Caesar no longer. He sent for them, telling them to bring their swords with them, a direction perfectly characteristic of him. Everything told of him shows his unconcern about danger to himself. Face to face with them, he asked them to state their case and listened while they told him all they had done and suffered and been poorly rewarded for, and demanded to be discharged. His speech in answer was also characteristic, very gentle, very brief,*

within the Austrian Empire. Forced to accept liberal constitutions modeled on that of revolutionary France some thirty years earlier, the kings of both countries had reason to fear that they also faced the same fate as the French king of that period, Louis XVI, beheaded in 1793. Meanwhile the leaders of Europe's great powers—England, Austria, and Prussia—quaked at the thought of unrest and radicalism spreading across their borders, which had only recently been stabilized by the defeat of Napoleon. They all wanted to protect themselves and halt the tide of revolution.

In the midst of this general unease, Czar Alexander I of Russia (1777–1825) suddenly proposed a plan that to many seemed a cure more dangerous than the disease. The Russian army was the largest and most feared in Europe; Alexander wanted to send it to both Spain and Naples, crushing the two rebellions. In exchange he would insist that the kings of both realms enact liberal reforms that would grant their citizens greater freedoms, making them more content and diluting their desire for revolution.

Alexander saw his proposal as more than a practical program to safeguard Europe's monarchies; it was part of a great crusade, a dream he had nurtured since the earliest days of his reign. A deeply religious man who saw everything in terms of good and evil, he wanted the monarchies of Europe to reform themselves and create a kind of Christian brotherhood of wise, gentle rulers with himself, the czar, at their helm. Although the powerful considered Alexander a kind of Russian madman, many liberals and even revolutionaries throughout Europe saw him as their friend and protector, the rare leader sympathetic to their cause. It was even rumored that he had made contacts with various men of the left and had intrigued with them.

The czar went further with his idea: now he wanted a conference of the major powers to discuss the future of Spain, Naples, and Europe itself. The English foreign minister, Lord Castlereagh, wrote letter after letter trying to dissuade him of the need for the meeting. It was never wise to meddle in the affairs of other countries, Castlereagh said; Alexander should leave England to help stop the unrest in Spain, its close ally, while Austria did the same for Naples. Other ministers and rulers wrote to Alexander as well, using similar arguments. It was critical to show a united front against his plan. Yet one man—the Austrian foreign minister, Prince Klemens von Metternich—responded to the czar in a much different fashion, and it was shocking to say the least.

Metternich was the most powerful and respected minister in Europe. The quintessential realist, he was always slow to take bold action or to involve Austria in any kind of adventure; security and order were his primary concerns. He was a conservative, a man who believed in the virtues of the status quo. If change had to come, it should come slowly. But Metternich was also something of an enigma—an elegant courtier, he spoke little yet always seemed to get his way. Now not only was he supporting Alexander's call for a conference, but he also seemed open to the czar's other ideas. Perhaps he had undergone a change of heart and was moving to the left in his later years? In any event, he personally organized the conference for October of that year in the Austrian-held city of Troppau, in the modern-day Czech Republic.

Alexander was delighted: with Metternich on his side, he could realize his ambitions and then some. When he arrived in Troppau for the conference, however, the representatives of the other powers in attendance were less than friendly. The French and the Prussians were cool; Castlereagh had refused to come altogether. Feeling somewhat isolated, Alexander was delighted again when Metternich proposed they hold private meetings to discuss the czar's ideas. For several days, and for hours on end, they holed themselves up together in a room. The czar did most of the talking; Metternich listened with his usual attentive air, agreeing and nodding. The czar, whose thinking was somewhat vague, strained to explain his vision of Europe as best he could, and the need for the leaders at the conference to display their moral unity. He could not help but feel frustrated at his inability to make his ideas more specific.

Several days into these discussions, Metternich finally confessed to the czar that he, too, saw a moral danger brewing in Europe. Godless revolution was the scourge of the time; giving in to the radical spirit, showing any sign of compromise, would eventually lead to destruction at the hands of these satanic forces. During the Troppau conference, a mutiny had broken out in a regiment of Russian guards; Metternich warned Alexander that this was the first symptom of a revolutionary infection attacking Russia itself. Thank God the czar, a pillar of moral strength, would not give in. Alexander would have to serve as the leader of this counterrevolutionary crusade. This was why Metternich had become so excited by the czar's ideas about Naples and Spain and how he had interpreted them.

The czar was swept up in Metternich's enthusiasm: together they would stand firm against the radicals. Somehow, though, the result of their conversation was not a plan for Russia to invade Naples and Spain; indeed, Alexander speculated instead that it might not be the time to press the kings of those countries to reform their governments—that would just weaken both monarchs. For the time being, the leaders' energy should go into halting the revolutionary tide. In fact, the czar began to repent of some of his more liberal ideas, and he confessed as much to Metternich. The conference ended with a statement of grand common purpose among the powers—much of its language the czar's—and an agreement that Austrian troops, not Russian ones, would return the king of Naples to full power, then leave him to pursue the policies of his choice.

After Alexander returned to Russia, Metternich wrote to praise him for leading the way. The czar wrote back in fervor: "We are engaged in a combat with the realm of Satan. Ambassadors do not suffice for this

*exactly to the point: "You say well, citizens. You have worked hard—you have suffered much. You desire your discharge. You have it. I discharge you all. You shall have your recompense. It shall never be said of me that I made use of you when I was in danger, and was ungrateful to you when danger was past." That was all, yet the legionaries listening were completely broken to his will. They cried out that they would never leave him; they implored him to forgive them, to receive them again as his soldiers. Back of the words was his personality, and although that can never be recaptured, something of it yet comes through the brief, bald sentences: the strength that faced tranquilly desertion at a moment of great need; the pride that would not utter a word of appeal or reproach; the mild tolerance of one who knew men and counted upon nothing from them.*

THE ROMAN WAY,
EDITH HAMILTON,
1932

task. Only those whom the Lord has placed at the head of their peoples may, if He gives His blessings, survive the contest . . . with this diabolic force." In fact, the czar wanted to go further; he had returned to the idea of marching his army into Spain to put down the revolution there. Metternich responded that that would not be necessary—the British were handling the situation—but a conference next year could readdress the issue.

In early 1821 another revolution broke out, this time in Piedmont, the one Italian state outside Austrian control. The king was forced to abdicate. In this instance Metternich welcomed Russian intervention, and 90,000 Russian troops became reserves in an Austrian army heading for Piedmont. A Russian military presence so close to their borders greatly dampened the spirits of the rebels and of their sympathizers throughout Italy—all those leftists who had seen the czar as their friend and protector. They thought that no more.

The Austrian army crushed the revolution within a few weeks. At Metternich's request, the Russians politely withdrew their forces. The czar was proud of his growing influence in Europe, but somehow he had embarked on the very opposite of his original plans for a crusade: instead of being in the forefront of the fight for progress and reform, he had become a guardian of the status quo, a conservative in the mold of Metternich himself. Those around him could not understand how this had happened.

*At times one has to deal with hidden enemies, intangible influences that slink into dark corners and from this hiding affect people by suggestion. In instances like this, it is necessary to trace these things back to the most secret recesses, in order to determine the nature of the influences to be dealt with. . . . The very anonymity of such plotting requires an especially vigorous and indefatigable effort, but this is well worth while. For when such elusive influences are brought into the light and branded, they lose their power over people.*

*THE I CHING,* CHINA, CIRCA EIGHTH CENTURY B.C.

### Interpretation

Prince Metternich may have been history's most effective public practitioner of passive aggression. Other diplomats sometimes thought him cautious, even weak, but in the end, as if by magic, he always got what he wanted. The key to his success was his ability to hide his aggression to the point where it was invisible.

Metternich was always careful to take the measure of his opponent. In the case of Czar Alexander, he was dealing with a man governed by emotion and subject to wild mood swings. Yet the czar, behind his moralistic Christian façade, was also aggressive in his own way, and ambitious; he itched to lead a crusade. In Metternich's eyes he was as dangerous as Napoleon had been: in the name of doing good for Europe, such a man might march his troops from one end of the continent to the other, creating untold chaos.

To stand in the way of Alexander's powerful army would be destructive in itself. But the canny Metternich knew that to try to persuade the czar that he was wrong would have the unintended effect of feeding his insecurities and pushing him to the left, making him more prone to take dangerous action on his own. Instead the prince would have to handle him like a child, diverting his energies to the right through a passive-aggressive campaign.

The passive part was simple: Metternich presented himself as compliant, going along with ideas that he actually disagreed with to the extreme. He accepted Alexander's request for a congress, for example, although he personally opposed it. Then, in his private discussions with the czar at Troppau, he at first just listened, then enthusiastically agreed. The czar believed in demonstrating moral unity? Then so did Metternich—although his own policies had always been more practical than moral; he was the master of realpolitik. He flattered personal qualities in the czar—moral fervor, for example—that he actually thought dangerous. He also encouraged the czar to go further with his ideas.

*In those days force and arms did prevail; but now the wit of the fox is everywhere on foot, so hardly a faithful or virtuous man may be found.*

QUEEN ELIZABETH I, 1533–1603

Having disarmed Alexander's suspicions and resistance this way, Metternich at the same time operated aggressively. At Troppau he worked behind the scenes to isolate the czar from the other powers, so that the Russian leader became dependent on him. Next he cleverly arranged those long hours of private meetings, in which he subtly infected the czar with the idea that revolution was far more dangerous than the status quo and diverted the Russian's radical Christian crusade into an attack on liberalism itself. Finally, having mirrored Alexander's energy, his moods, his fervor, and his language, Metternich managed to lure him into sending troops against the rebellion in Piedmont. That action both committed Alexander in deed to the conservative cause and alienated him from the liberals of Europe. No longer could he spout vague, ambiguous pronouncements on the left; he had finally taken action, and it was in the opposite direction. Metternich's triumph was complete.

Although the phrase "passive aggression" has negative connotations for most of us, as conscious strategy passive-aggressive behavior offers an insidiously powerful way of manipulating people and waging personal war. Like Metternich, you must operate on two fronts. You are outwardly agreeable, apparently bending to people's ideas, energy, and will, changing shape like Proteus himself. Remember: people are willful and perverse. Opposing them directly or trying to change their ideas will often have the contrary effect. A passive, compliant front, on the other hand, gives them nothing to fight against or resist. Going along with their energy gives you the power to divert it in the direction you want, as if you were channeling a river rather than trying to dam it. Meanwhile the aggressive part of your strategy takes the form of infecting people with subtle changes in their ideas and with an energy that will make them act on your behalf. Their inability to get what you are doing in focus gives you room to work behind the scenes, checking their progress, isolating them from other people, luring them into dangerous moves that make them dependent on your support. They think you are their ally. Behind a pleasant, compliant, even weak front, you are pulling the strings.

*This was the real achievement of Metternich's policy, that it had killed Russian liberalism and achieved a measure of domination over Austria's most dangerous rival in the guise of submitting to him.*

—*Henry Kissinger,* A World Restored *(1957)*

*In this postscript on the solution of Caesar's problem, it is not our intention to trace Octavian's rise to power from the time he arrived in Rome to claim his inheritance until, in 31 B.C., with the aid of Vipsanius Agrippa, he defeated Antony and Cleopatra at Actium and became master of the Roman world. Instead, it is to describe in brief how as such he solved Caesar's problem and established a peace which was to last for over 200 years. When he contemplated the empire he had won and its heterogeneous local governments and peoples, he realized that it was far too large and complex to be ruled by the council of a city state; that instead it demanded some form of one-man rule, and that his problem was how to disguise it. From the outset he decided not to tamper with the constitution of the Republic, or contemplate monarchy. . . . . . . Firstly, in 28 B.C. he declined all honours calculated to remind the Romans of the kingly power; adopted the title of princeps ("first citizen"), and called his system the Principate. Secondly, he accepted all the old conventions—consuls, tribunes, magistrates, elections, etc. Thirdly, instead of ignoring the Senate and insulting its members as Caesar had done, he went out of his way to consult it and placate them. Lastly,*

## KEYS TO WARFARE

We humans have a particular limitation to our reasoning powers that causes us endless problems: when we are thinking about someone or about something that has happened to us, we generally opt for the simplest, most easily digestible interpretation. An acquaintance is good or bad, nice or mean, his or her intentions noble or nefarious; an event is positive or negative, beneficial or harmful; we are happy or sad. The truth is that nothing in life is ever so simple. People are invariably a mix of good and bad qualities, strengths and weaknesses. Their intentions in doing something can be helpful and harmful to us at the same time, a result of their ambivalent feelings toward us. Even the most positive event has a downside. And we often feel happy and sad at the same time. Reducing things to simpler terms makes them easier for us to handle, but because it is not related to reality, it also means we are constantly misunderstanding and misreading. It would be of infinite benefit for us to allow more nuances and ambiguity into our judgments of people and events.

This tendency of ours to judge things in simple terms explains why passive aggression is so devilishly effective as a strategy and why so many people use it—consciously and unconsciously. By definition, people who are acting passive-aggressively are being passive and aggressive simultaneously. They are outwardly compliant, friendly, obedient, even loving. At the same time, they inwardly plot and take hostile action. Their aggression is often quite subtle—little acts of sabotage, remarks designed to get under your skin. It can also be blatantly harmful.

When we are the victims of this behavior, we find it hard to imagine that both things are happening at the same time. We can manage the idea that someone can be nice one day and nasty the next; that is just called being moody. But to be nasty and nice simultaneously—that confuses us. We tend to take these people's passive exterior for reality, becoming emotionally engaged with their pleasant, nonthreatening appearance. If we notice that something is not quite right, that while seeming friendly they might be doing something hostile, we are genuinely bewildered. Our confusion gives the passive-aggressive warrior great manipulative power over us.

There are two kinds of passive aggression. The first is conscious strategy as practiced by Metternich. The second is a semiconscious or even unconscious behavior that people use all the time in the petty and not-so-petty matters of daily life. You may be tempted to forgive this second passive-aggressive type, who seems unaware of the effects of his or her actions or helpless to stop, but people often understand what they are doing far better than you imagine, and you are more than likely being taken in by their friendly and helpless exterior. We are generally too lenient with this second variety.

The key to using passive aggression as a conscious, positive strategy is the front you present to your enemies. They must never be able to detect the sullen, defiant thoughts that are going on inside of you.

In 1802 what today is Haiti was a French possession riven by a revolt of the country's black slaves under the leadership of Toussaint-L'ouverture. That year an army sent by Napoleon to crush the rebellion managed to seize Toussaint through treachery and ship him off to France, where he would eventually die in prison. Among Toussaint's most-decorated generals was a man named Jean-Jacques Dessalines, who now surrendered to the French and even served in their army, helping them to put down isolated pockets of revolt and winning from them much appreciation. But it was all a ploy: as Dessalines squashed these remnants of the rebellion, he would hand over the weapons he captured to the French, but secretly he always kept some of them back, stashing them away until he had quite a large armory. Meanwhile he built up and trained a new rebel army in the remote areas where his assignment led him. Then, choosing a moment when an outbreak of yellow fever had decimated the French army, he resumed hostilities. Within a few years, he had defeated the French and liberated Haiti for good from colonial control.

Dessalines's use of passive aggression has deep roots in military strategy, in what can be called the "false surrender." In war your enemies can never read your thoughts. They must make your appearance their guide, reading the signs you give off to decipher what you are thinking and planning. Meanwhile the surrender of an army tends to be followed by a great flood of emotion and a lowering of everyone's guard. The victor will keep an eye on the beaten troops but, exhausted by the effort it took to win, will be hugely tempted to be less wary than before. A clever strategist, then, may falsely surrender—announce that he is defeated in body and spirit. Seeing no indication otherwise, and unable to read his mind, the enemy is likely to take his submission at face value. Now the false surrenderer has time and space to plot new hostilities.

In war as in life, the false surrender depends on the seamless appearance of submission. Dessalines did not just give in, he actively served his former enemies. To make this work, you must do likewise: play up your weakness, your crushed spirit, your desire to be friends—an emotional ploy with great power to distract. You must also be something of an actor. Any sign of ambivalence will ruin the effect.

In 1940, President Franklin D. Roosevelt faced a dilemma. He was nearing the end of his second term in office, and it was an unwritten tradition in American politics that no president would run for a third term. But Roosevelt had much unfinished business. Abroad, Europe was deep in a war that would almost certainly end up involving the United States; at home, the country had been going through difficult times, and Roosevelt wanted to bring his programs to remedy them to completion. If he revealed his desire for a third term, though, he would stir up opposition even within his own party. Many had already accused him of dictatorial tendencies. So Roosevelt decided to get what he wanted through a form of passive aggression.

*on January 13, 27 B.C., at a session of the Senate, he renounced all his extraordinary powers and placed them at the disposal of the Senate and the people. And when the senators begged him to resume them and not to abandon the Commonwealth he had saved, he yielded to their request and consented to assume proconsular authority over an enlarged province, which included Spain, Gaul, Syria, Cilicia, and Cyprus, while the Senate was left with the remaining provinces. Thus in semblance the sovereignty of the Senate and the people was restored; but in fact, because his enlarged province comprised the majority of the legions, and Egypt, over which he ruled as king . . . the basis of political power passed into his hands. Three days later the Senate decreed that the title "Augustus" (the Revered) should be conferred upon him.*

*Julius Caesar*, J.F.C. Fuller, 1965

*It is not an enemy who taunts me—*
*then I could bear it;*
*it is not an adversary who deals insolently with me—*
*then I could hide from him.*
*But it is you, my equal, my companion, my familiar friend. . . .*
*. . . My companion stretched out his hand against his friends,*
*he violated his covenant.*
*His speech was smoother than butter,*
*yet war was in his heart;*
*his words were softer than oil,*
*yet they were drawn swords.*

PSALMS, 55:12–15, 20–21

In the months leading up to the Democratic Convention, which was to choose which candidate the party would run in the race, Roosevelt repeatedly stated his lack of interest in a third term. He also actively encouraged others in the party to seek the nomination to replace him. At the same time, he carefully crafted his language so that he never completely closed the door on running himself, and he pushed enough candidates into the nomination race that no single one of them could come to the convention as the favorite. Then, as the convention opened, Roosevelt withdrew from the scene, making his large presence known by his absence: without him the proceedings were incomparably dull. Reports came back to him that people on the floor were beginning to clamor for him to appear. Letting that desire reach its peak, the president then had his friend Senator Alben Barkley insert into his own convention speech a message from Roosevelt: "The president has never had, and has not today, any desire or purpose to continue in the office of president, to be a candidate for that office, or to be nominated by the convention for that office." After a moment of silence, the convention floor began to ring with the delegates' cry: "WE WANT ROOSEVELT!" The appeal went on for an hour. The next day the delegates were to vote, and chants of "ROOSEVELT!" again filled the hall. The president's name was entered for the nomination, and he won by a landslide on the first ballot.

Remember: it is never wise to seem too eager for power, wealth, or fame. Your ambition may carry you to the top, but you will not be liked and will find your unpopularity a problem. Better to disguise your maneuvers for power: you do not want it but have found it forced upon you. Being passive and making others come to you is a brilliant form of aggression.

Subtle acts of sabotage can work wonders in the passive-aggressive strategy because you can camouflage them under your friendly, compliant front. That was how the film director Alfred Hitchcock would outmaneuver the meddlesome producer David O. Selznick, who used to alter the script to his liking, then show up on set to make sure it was shot the way he wanted it. On these occasions Hitchcock might arrange for the camera to malfunction or let it run without any film in it—by the time Selznick saw the edit, reshooting would be expensive and impossible. Meanwhile the director would make a show of being happy to see Selznick on set and bewildered if the camera didn't roll or rolled but recorded no film.

Passive aggression is so common in daily life that you have to know how to play defense as well as offense. By all means use the strategy yourself; it is too effective to drop from your armory. But you must also know how to deal with those semiconscious passive-aggressive types so prevalent in the modern world, recognizing what they are up to before they get under your skin and being able to defend yourself against this strange form of attack.

First, you must understand why passive aggression has become so omnipresent. In the world today, the expression of overt criticism or negative feelings toward others has become increasingly discouraged. People tend to take criticism far too personally. Furthermore, conflict is something to be avoided at all costs. There is great societal pressure to please and be liked by as many people as possible. Yet it is human nature to have aggressive impulses, negative feelings, and critical thoughts about people. Unable to express these feelings openly, without fear of being disliked, more and more people resort to a kind of constant, just-below-the-surface passive aggression.

*The idiom represents an archetype in world literature: a person with a smiling face and a cruel heart, dubbed a "smiling tiger" in Chinese folklore.*

*The Wiles of War*, translated by Sun Haichen, 1991

Most often their behavior is relatively harmless: perhaps they are chronically late, or make flattering comments that hide a sarcastic sting, or offer help but never follow through. These common tactics are best ignored; just let them wash over you as part of the current of modern life, and never take them personally. You have more important battles to fight.

There are, however, stronger, more harmful versions of passive aggression, acts of sabotage that do real damage. A colleague is warm to your face but says things behind your back that cause you problems. You let someone into your life who proceeds to steal something valuable of yours. An employee takes on an important job for you but does it slowly and badly. These types do harm but are excellent at avoiding any kind of blame. Their modus operandi is to create enough doubt that they were the ones who did the aggressive act; it is never their fault. Somehow they are innocent bystanders, helpless, the real victims in the whole dynamic. Their denials of responsibility are confusing: you suspect they have done something, but you cannot prove it, or, worse, if they are *really* skillful, you feel guilty for even thinking them at fault. And if in your frustration you lash out at them, you pay a high price: they will focus attention on your angry, aggressive response, your overreaction, distracting your thoughts from the passive-aggressive maneuvers that got you so irritated in the first place. The guilt you feel is a sign of the power they have over you. Indeed, you can virtually recognize the harmful variety of passive aggression by the strength of emotions it churns up in you: not superficial annoyance but confusion, paranoia, insecurity, and anger.

To defeat the passive-aggressive warrior, you must first work on yourself. This means being acutely aware of the blame-shifting tactic as it happens. Squash any feelings of guilt it might begin to make you feel. These types can be very ingratiating, using flattery to draw you into their web, preying on your insecurities. It is often your own weakness that sucks you into the passive-aggressive dynamic. Be alert to this.

Second, once you realize you are dealing with the dangerous variety, the smartest move is to disengage, at best to get the person out of your life, or at the least to not flare up and cause a scene, all of which plays into his hands. You need to stay calm. If it happens to be a partner in a relationship in which you cannot disengage, the only solution is to find a way to make the person feel comfortable in expressing any

*It is not pathological to attempt to gain control of a relationship, we all do this, but when one attempts to gain that control while denying it, then such a person is exhibiting symptomatic behavior. In any relationship that stabilizes, such as that between a husband and wife, the two people work out agreements about who is to control what area of the relationship. . . . A relationship becomes psychopathological when one of the two people will maneuver to circumscribe the other's behavior while indicating he is not. The wife in such a relationship will force her husband to take care of the house in such a way that she denies she is doing so. She may, for example, have obscure dizzy spells, an allergy to soap, or various types of attacks which require her to lie down regularly. Such a wife is circumscribing her husband's behavior while denying that* she *is doing this; after all, she cannot help her dizzy spells. When one person circumscribes the behavior of another while denying that he is doing so, the relationship begins to be rather peculiar. For example, when a wife requires her husband to be home every night because she has anxiety attacks when she is left alone, he cannot acknowledge that she is controlling his behavior because* she *is not*

negative feelings toward you and encouraging it. This may be hard to take initially, but it may defuse his or her need to be underhanded; and open criticisms are easier to deal with than covert sabotage.

The Spaniard Hernán Cortés had many passive-aggressive soldiers in the army with which he conquered Mexico, men who outwardly accepted his leadership but were inwardly treacherous. Cortés never confronted or accused these people, never lashed out at them at all; instead he quietly figured out who they were and what they were up to, then fought fire with fire, maintaining a friendly front but working behind the scenes to isolate them and bait them into attacks in which they revealed themselves. The most effective counterstrategy with the passive-aggressive is often to be subtle and underhanded right back at them, neutralizing their powers. You can also try this with the less harmful types—the ones who are chronically late, for instance: giving them a taste of their own medicine may open their eyes to the irritating effects of their behavior.

In any event, you must never leave the passive-aggressive time and space in which to operate. Let them take root and they will find all kinds of sly ways to pull you here and there. Your best defense is to be sensitive to any passive-aggressive manifestations in those around you and to keep your mind as free as possible from their insidious influence.

**Image:** The River. It flows with great force, sometimes flooding its banks and creating untold damage. Try to dam it and you only add to its pent-up energy and increase your risk. Instead divert its course, channel it, make its power serve your purposes.

**Authority:** As dripping water wears through rock, so the weak and yielding can subdue the firm and strong. —*Sun Haichen,* Wiles of War *(1991)*

## REVERSAL

The reversal of passive aggression is aggressive passivity, presenting an apparently hostile face while inwardly staying calm and taking no unfriendly action. The purpose here is intimidation: perhaps you know you are the weaker of the two sides and hope to discourage your enemies from attacking you by presenting a blustery front. Taken in by your appearance, they will find it hard to believe that you do not intend to do anything. In general, presenting yourself as the opposite of what you really are and intend can be a useful way of disguising your strategies.

*requiring him to be home—the anxiety is and her behavior is involuntary. Neither can he refuse to let her control his behavior for the same reason.*

STRATEGIES OF PSYCHOTHERAPY, JAY HALEY, 1963

# 33

# SOW UNCERTAINTY AND PANIC THROUGH ACTS OF TERROR

## THE CHAIN-REACTION STRATEGY

*Terror is the ultimate way to paralyze a people's will to resist and destroy their ability to plan a strategic response. Such power is gained through sporadic acts of violence that create a constant feeling of threat, incubating a fear that spreads throughout the public sphere. The goal in a terror campaign is not battlefield victory but causing maximum chaos and provoking the other side into desperate overreaction. Melting invisibly into the population, tailoring their actions for the mass media, the strategists of terror create the illusion that they are everywhere and therefore that they are far more powerful than they really are. It is a war of nerves. The victims of terror must not succumb to fear or even anger; to plot the most effective counterstrategy, they must stay balanced. In the face of a terror campaign, one's rationality is the last line of defense.*

*"Brothers," says an Ismaili poet, "when the time of triumph comes, with good fortune from both worlds as our companion, then by one single warrior on foot a king may be stricken with terror, though he own more than a hundred thousand horsemen."*

QUOTED IN *THE ASSASSINS*, BERNARD LEWIS, 1967

## THE ANATOMY OF PANIC

In Isfahan (in present-day Iran) toward the end of the eleventh century, Nizam al-Mulk, the powerful vizier to Sultan Malik Shah, ruler of the great Islamic empire of the period, became aware of a small yet irritating threat. In northern Persia lived a sect called the Nizari Ismailis, followers of a religion combining mysticism with the Koran. Their leader, the charismatic Hasan-i-Sabah, had recruited thousands of converts alienated by the tight control the empire exercised over religious and political practices. The influence of the Ismailis was growing, and what was most disturbing to Nizam al-Mulk was the utter secrecy in which they operated: it was impossible to know who had converted to the sect, for its members did so in private and kept their allegiance hidden.

The vizier monitored their activities as best he could, until finally he heard some news that roused him to action. Over the years, it seemed, thousands of these secret Ismaili converts had managed to infiltrate key castles, and now they had taken them over in the name of Hasan-i-Sabah. This gave them control over part of northern Persia, a kind of independent state within the empire. Nizam al-Mulk was a benevolent administrator, but he knew the danger of allowing sects like the Ismailis to flourish. Better to snuff them out early on than face revolution. So, in 1092, the vizier convinced the sultan to send two armies to bring down the castles and destroy the Nizari Ismailis.

The castles were strongly defended, and the countryside around them teemed with sympathizers. The war turned into a stalemate, and eventually the sultan's armies were forced to come home. Nizam al-Mulk would have to find some other solution, perhaps an occupying force for the region—but a few months later, as he was traveling from Isfahan to Baghdad, a Sufi monk approached the litter on which he was carried, pulled out a dagger from under his clothes, and stabbed the vizier to death. The killer was revealed to be an Ismaili dressed as a peaceful Sufi, and he confessed to his captors that Hasan himself had assigned him to do the job.

The death of Nizam al-Mulk was followed within weeks by the death, from natural causes, of Malik Shah. His loss would have been a blow at any time, but without his crafty vizier to oversee the succession, the empire fell into a period of chaos that lasted several years. By 1105, however, a degree of stability had been reestablished and attention again focused on the Ismailis. With one murder they had managed to make the entire empire tremble. They had to be destroyed. A new and vigorous campaign was launched against the sect. And soon it was revealed that the assassination of Nizam al-Mulk was not a single act of revenge, as it had seemed at the time, but an Ismaili policy, a strange and frightening new way of waging war. Over the next few years, key members of the administration of the new sultan, Muhammad Tapar, were assassinated in the same ritualistic fashion: a killer would emerge from a crowd to deliver a deadly blow with a dagger. The deed was most often done in public and in broad daylight; sometimes, though, it took place while the victim was in bed, a secret Ismaili having infiltrated his household staff.

*Losses to which we are accustomed affect us less deeply.*

JUVENAL, FIRST TO SECOND CENTURY A.D.

A wave of fear fanned out among the empire's hierarchy. It was impossible to tell who was an Ismaili: the sect's adherents were patient, disciplined, and had mastered the art of keeping their beliefs to themselves and fitting in anywhere. It did not help that when the assassins were captured and tortured, they would accuse various people within the sultan's inner circle of being either paid spies for the Ismailis or secret converts. No one could know for certain if they were telling the truth, but suspicion was cast on everyone.

Now viziers, judges, and local officials had to travel surrounded by bodyguards. Many of them began to wear thick, uncomfortable shirts of mail. In certain cities no one could move from house to house without a permit, which spread disaffection among the citizenry and made it easier for the Ismailis to recruit converts. Many found it hard to sleep at night or to trust their closest friends. All kinds of wild rumors were spread by those who had grown delirious with paranoia. Bitter divisions sprang up within the hierarchy, as some argued for a hard-line approach to Hasan, while others preached accommodation as the only answer.

Meanwhile, as the empire struggled to somehow repress the Ismailis, the killings went on—but they were highly sporadic. Months would pass without one, and then suddenly there would be two within a week. There was no real rhyme or reason to when it happened or which high administrator was singled out. Officials would talk endlessly about a pattern, analyzing every Ismaili move. Without their realizing it, this little sect had come to dominate their thoughts.

In 1120, Sanjar, the new sultan, decided to take action, planning a military campaign to capture the Ismaili castles with overwhelming force and turn the region around them into an armed camp. He took extra precautions to prevent any attempt on his life, changing his sleeping arrangements and allowing only those he knew well to approach him. By making himself personally secure, he believed he could stay free of the panic around him.

As preparations for the war got under way, Hasan-i-Sabah sent ambassador after ambassador to Sanjar offering to negotiate an end to the killings. They were all turned away. The tables seemed to have turned: now it was the Ismailis who were frightened.

Shortly before the campaign was to be launched, the sultan awoke one morning to find a dagger thrust neatly in the ground a few feet away from the position where his breast lay on the bed. How did it get there? What did it mean? The more he thought about it, the more he began to literally tremble with fear—it was clearly a message. He told no one about this, for whom could he trust? Even his wives were suspect. By the end of the day, he was an emotional wreck. That evening he received a message from Hasan himself: "Did I not wish the sultan well, that dagger which was struck into the hard ground would have been planted in his soft breast."

Sanjar had had enough. He could not spend another day like this. He was not willing to live in constant fear, his mind deranged by uncertainty and suspicion. It was better, he thought, to negotiate with this demon. He called off his campaign and made peace with Hasan.

Over the years, as the Ismailis' political power grew and the sect expanded into Syria, its killers became almost mythic. The assassins had never tried to escape; their killing done, they were caught, tortured, and executed to a man, but new ones kept on coming, and nothing seemed to deter them from completing their task. They seemed possessed, utterly devoted to their cause. Some called them *hashshashin,* from the Arabic word *hashish,* because they acted as if they were drugged. European crusaders to the Holy Land heard stories about these devilish *hashshashin* and passed them on, the word slowly transforming into "assassins," passing forever into the language.

### Interpretation

Hasan-i-Sabah had one goal: to carve out a state for his sect in northern Persia, allowing it to survive and thrive within the Islamic empire. Given his relatively small numbers and the powers arrayed against him, he could not hope for more, so he devised a strategy that was surely history's first organized terrorist campaign for political power. Hasan's plan was deceptively simple. In the Islamic world, a leader who had won respect was invested with considerable authority, and to the extent that he had authority, his death could sow chaos. Accordingly Hasan chose to strike these leaders, but in a somewhat random way: it was impossible to see any pattern in his choices, and the possibility of being the next victim was more disturbing than many could bear. In truth, except for the castles they held, the Ismailis were quite weak and vulnerable, but by patiently infiltrating his men deep into the heart of the sultan's administration, Hasan was able to create the illusion they were everywhere. Only fifty or so assassinations are recorded in his entire lifetime, and yet he won as much political power through them as if he had an enormous army.

This power could not come by merely making individuals feel afraid. It depended on the effect the killings would have on the entire social group. The weakest officials in the hierarchy were the ones who would succumb to paranoia and voice doubts and rumors that would spread and infect those who were less weak. The result was a ripple effect—wild swings of emotion, from anger to surrender, up and down the line. A group caught by this kind of panic cannot find its balance and can fall with the slightest push. Even the strongest and most determined will be infected in the end, as Sultan Sanjar was: his attempts at security, and the harsh life to which he subjected himself as protection, revealed that he was under the influence of this panic. One simple dagger in the ground was enough to push him over the edge.

*On their voyage Pisander and the others abolished the democracies in the [Greek] cities, as had been decided. From some places they also took hoplites to add to their forces, and so came to Athens. Here they found that most of the work had already been done by members of their [antidemocratic] party. Some of the younger men had formed a group among themselves and had murdered without being detected a certain Androcles, who was one of the chief leaders of the [democratic] party. . . . There were also some other people whom they regarded as undesirable and did away with secretly. . . . . . . [Athenians] were afraid when they saw their numbers, and no one now dared to speak in opposition to them. If anyone did venture to do so, some appropriate method was soon found for having him killed, and no one tried to investigate such crimes or take action against those suspected of them. Instead the people kept quiet, and were in such a state of terror that they thought themselves lucky enough to be left unmolested even if they had said nothing at all. They imagined that the revolutionary party was much bigger than it really was, and they lost all confidence in themselves, being unable to find out the*

Understand: we are all are extremely susceptible to the emotions of those around us. It is often hard for us to perceive how deeply we are affected by the moods that can pass through a group. This is what makes the use of terror so effective and so dangerous: with a few well-timed acts of violence, a handful of assassins can spark all kinds of corrosive thoughts and uncertainties. The weakest members of the target group will succumb to the greatest fear, spreading rumors and anxieties that slowly overcome the rest. The strong may respond angrily and violently to the terror campaign, but that only shows how influenced they are by the panic; they are reacting rather than strategizing—a sign of weakness, not strength. In normal circumstances individuals who become frightened in some way can often regain their mental balance over time, especially when they are around others who are calm. But this is almost impossible within a panicked group.

*facts because of the size of the city and because they had insufficient knowledge of each other. . . . Throughout the democratic party people approached each other suspiciously, everyone thinking that the next man had something to do with what was going on.*

*History of the Peloponnesian War*, Thucydides, circa 460–circa 399 B.C.

As the public's imagination runs wild, the assassins become something much larger, seeming omnipotent and omnipresent. As Hasan proved, a handful of terrorists can hold an entire empire hostage with a few well-calibrated blows against the group psyche. And once the group's leaders succumb to the emotional pull—whether by surrendering or launching an unstrategic counterattack—the success of the terror campaign is complete.

*Victory is gained not by the number killed but by the number frightened.*

*—Arab proverb*

## KEYS TO WARFARE

In the course of our daily lives, we are subject to fears of many kinds. These fears are generally related to something specific: someone might harm us, a particular problem is brewing, we are threatened by disease or even death itself. In the throes of any deep fear, our willpower is momentarily paralyzed as we contemplate the bad that could happen to us. If this condition lasted too long or were too intense, it would make life unbearable, so we find ways to avoid these thoughts and ease our fears. Maybe we turn to the distractions of daily life: work, social routines, activities with friends. Religion or some other belief system, such as faith in technology or science, might also offer hope. These distractions and beliefs become the ground beneath us, keeping us upright and able to walk on without the paralysis that fear can bring.

Under certain circumstances, however, this ground can fall away from under us, and then there is nothing we can do to steady ourselves. In the course of history, we can track a kind of madness that overcomes humans during certain disasters—a great earthquake, a ferocious plague, a violent civil war. What troubles us most in these situations is not any specific dreadful event that happened in the recent past; we have a tremendous capacity to overcome and adapt to anything horrible. It is the uncertain future, the fear that more terrible things are coming and that we might soon suffer some unpredictable tragedy—that is what unnerves us. We cannot crowd out these thoughts with routines or religion. Fear becomes chronic and intense, our minds besieged by all kinds of irrational thoughts. The specific fears become more general. Among a group, panic will set in.

In essence, this is terror: an intense, overpowering fear that we cannot manage or get rid of in the normal way. There is too much uncertainty, too many bad things that can happen to us.

During World War II, when the Germans bombed London, psychologists noted that when the bombing was frequent and somewhat regular, the people of the city became numb to it; they grew accustomed to the noise, discomfort, and carnage. But when the bombing was irregular and sporadic, fear became terror. It was much harder to deal with the uncertainty of when the next one would land.

It is a law of war and strategy that in the search for an advantage, anything will be tried and tested. And so it is that groups and individuals, seeing the immense power that terror can have over humans, have found a way to use terror as a strategy. People are crafty, resourceful, and adaptable creatures. The way to paralyze their will and destroy their capacity to think straight is to consciously create uncertainty, confusion, and an unmanageable fear.

Such strategic terror can take the form of exemplary acts of destruction. The masters of this art were the Mongols. They would level a few cities here and there, in as horrible a manner as possible. The terrifying legend of the Mongol Horde spread quickly. At its very approach to a city, panic would ensue as the inhabitants could only imagine the worst. More often than not, the city would surrender without a fight—the Mongols' goal all along. A relatively small army far from home, they could not afford long sieges or protracted wars.

This strategic terror can also be used for political purposes, to hold a group or nation together. In 1792 the French Revolution was spinning out of control. Foreign armies were on the verge of invading France; the country was hopelessly factionalized. The radicals, led by Robespierre, confronted this threat by initiating a war against the moderates, the Reign of Terror. Accused of counterrevolution, thousands were sent to the guillotine. No one knew who would be next. Although the radicals were relatively small in number, by creating such uncertainty and fear they were able to paralyze their opponents' will. Paradoxically, the Reign of Terror—which gives us the first recorded instance of the use of the words "terrorism" and "terrorist"—produced a degree of stability.

Although terror as a strategy can be employed by large armies and indeed whole states, it is most effectively practiced by those small in number. The reason is simple: the use of terror usually requires a willingness to kill innocent civilians in the name of a greater good and for a strategic purpose. For centuries, with a few notable exceptions such as

*Six at the top means: Shock brings ruin and terrified gazing around. Going ahead brings misfortune. If it has not yet touched one's own body But has reached one's neighbor first, There is no blame. One's comrades have something to talk about. When inner shock is at its height, it robs a man of reflection and clarity of vision. In such a state of shock it is of course impossible to act with presence of mind. Then the right thing is to keep still until composure and clarity are restored. But this a man can do only when he himself is not yet infected by the agitation, although its disastrous effects are already visible in those around him. If he withdraws from the affair in time, he remains free of mistakes and injury. But his comrades, who no longer heed any warning, will in their excitement certainly be displeased with him. However, he must not take this into account.*

THE *I CHING*, CHINA, CIRCA EIGHTH CENTURY B.C.

the Mongols, military leaders were unwilling to go so far. Meanwhile a state that inflicted mass terror on its own populace would unleash demons and create a chaos it might find hard to control. But small groups have no such problems. Being so few in number, they cannot hope to wage a conventional war or even a guerrilla campaign. Terror is their strategy of last resort. Taking on a much larger enemy, they are often desperate, and they have a cause to which they are utterly committed. Ethical considerations pale in comparison. And creating chaos is part of their strategy.

Terrorism was limited for many centuries by its tools: the sword, the knife, the gun, all agents of individual killing. Then, in the nineteenth century, a single campaign produced a radical innovation, giving birth to terrorism as we know it today.

In the late 1870s, a group of Russian radicals, mostly from the intelligentsia, had been agitating for a peasant-led revolution. Eventually they realized that their cause was hopeless: the peasants were unprepared to take this kind of action, and, more important, the czarist regime and its repressive forces were far too powerful. Czar Alexander II had recently initiated what became known as the White Terror, a brutal crackdown on any form of dissidence. It was almost impossible for the radicals to operate in the open, let alone spread their influence. Yet if they did nothing, the czar's strength would only grow.

And so from among these radicals, a group emerged bent on waging a terrorist war. They called themselves Narodnaya Volia, or "People's Will." To keep their organization clandestine, they kept it small. They dressed inconspicuously, melting into the crowd. And they began to make bombs. Once they had assassinated a number of government ministers, the czar became a virtual prisoner in his palace. Deranged with the desire to hunt the terrorists down, he directed all of his energies toward this goal, with the result that much of his administration became dysfunctional.

In 1880 the radicals were able to explode a bomb in the Winter Palace, the czar's residence in St. Petersburg. Then, finally, the following year another bomb killed Alexander himself. The government naturally responded with repression still harsher than the policy already in place, erecting a virtual police state. In spite of this, in 1888, Alexander Ulianov—the brother of Vladimir Lenin and a member of Narodnaya Volia—nearly succeeded in killing Alexander's successor, Czar Alexander III.

The capture and execution of Ulianov brought the activities of Narodnaya Volia to a close, but the group had already begun to inspire a wave of terrorist strikes internationally, including the anarchist assassinations of the American presidents James A. Garfield in 1881 and William McKinley in 1901. And with Narodnaya all the elements of modern terrorism are in place. The group thought bombs better than guns, being more dramatic and more frightening. They believed that if they killed

*The basis of Mongol warfare was unadulterated terror. Massacre, rapine and torture were the price of defeat, whether enforced or negotiated. . . . The whole apparatus of terror was remorselessly applied to sap the victim's will to resist, and in practical terms this policy of "frightfulness" certainly paid short-term dividends. Whole armies were known to dissolve into fear-ridden fragments at the news of the approach of the* toumans. . . . *Many enemies were paralysed . . . before a [Mongol] army crossed their frontiers.*

THE ART OF WARFARE ON LAND,
DAVID CHANDLER,
1974

*"This is what you should try for. An attempt upon a crowned head or on a president is sensational enough in a way, but not so much as it used to be. It has entered into a general conception of the existence of all chiefs of state. It's almost conventional—especially since so many presidents have been assassinated. Now let us take an outrage upon—say a church. Horrible enough at first sight, no doubt, and yet not so effective as a person of an ordinary mind might think. No matter how revolutionary and anarchist in inception, there would be fools enough to give such an outrage the character of a religious manifestation. And that would detract from the especial alarming significance we wish to give to the act. A murderous attempt on a restaurant or a theatre would suffer in the same way from the suggestion of non-political passion; the exasperation of a hungry man, an act of social revenge. All this is used up; it is no longer instructive as an object lesson in revolutionary anarchism. Every newspaper has ready-made phrases to explain such manifestations away. I am about to give you the philosophy of bomb throwing from my point of view; from the point of view you pretend to have been serving for the last eleven years. I will try not to talk above your head. The sensibilities of the class you*

enough ministers of the government, extending upward to the czar himself, the regime would either collapse or go to extremes to try to defend itself. That repressive reaction, though, would in the long run play into the radicals' hands, fomenting a discontent that would eventually spark a revolution. Meanwhile the bombing campaign would win the group coverage in the press, indirectly publicizing their cause to sympathizers around the world. They called this "the propaganda of the deed."

Narodnaya Volia aimed principally at the government but was willing to kill civilians in the process. The fall of the czarist government was worth a few lives lost, and in the end the bombs were less deadly than their alternative, which was civil war. At the very least, Narodnaya Volia would show the Russian people that the government was not the untouchable monolithic power that it made itself seem; it was vulnerable. The group's members understood that the regime was quite likely to be able to liquidate them in time, but they were willing to die for their cause.

Narodnaya Volia saw that it could use one relatively small event—a bomb blast—to set off a chain reaction: fear in the administration would produce harsh repression, which would win the group publicity and sympathy and heighten the government's unpopularity, which would lead to more radicalism, which would lead to more repression, and so on until the whole cycle collapsed in turmoil. Narodnaya Volia was weak and small, yet simple but dramatic acts of violence could give it a disproportionate power to sow chaos and uncertainty, creating the appearance of strength among police and public. In fact, its smallness and inconspicuousness gave it a tremendous edge: at enormous expense, a cumbersome force of thousands of police would have to search out a tiny, clandestine band that had the advantages of mobility, surprise, and relative invisibility. Besides giving the terrorists the chance to present themselves as heroic underdogs, the asymmetry of forces made them almost impossible to fight.

This asymmetry brings war to its ultimate extreme: the smallest number of people waging war against an enormous power, leveraging their smallness and desperation into a potent weapon. The dilemma that all terrorism presents, and the reason it attracts so many and is so potent, is that terrorists have a great deal less to lose than the armies arrayed against them, and a great deal to gain through terror.

It is often argued that terrorist groups like Narodnaya Volia are doomed to failure: inviting severe repression, they play into the hands of the authorities, who can effectively claim carte blanche to fight this threat—and in the end they bring about no real change. But this argument misses the point and misreads terrorism. Narodnaya Volia awakened millions of Russians to its cause, and its techniques were copied around the world. It also profoundly unbalanced the czarist regime, which responded irrationally and heavy-handedly, devoting resources to repression that could have been better applied to reforms that might have prolonged its stay in power. The repression also incubated a much more potent revolutionary group, the burgeoning communist movement.

In essence, terrorists kick a rock in order to start an avalanche. If no landslide follows, little is lost, except perhaps their own lives, which they are willing to sacrifice in their devotion to their cause. If mayhem and chaos ensue, though, they have great power to influence events. Terrorists are often reacting against an extremely static situation in which change by any route is blocked. In their desperation they can often break up the status quo.

It is a mistake to judge war by the rubric of victory or defeat: both states have shades and gradations. Few victories in history are total or bring about lasting peace; few defeats lead to permanent destruction. The ability to effect some kind of change, to attain a limited goal, is what makes terrorism so alluring, particularly to those who are otherwise powerless.

For instance, terrorism can be used quite effectively for the limited goal of gaining publicity for a cause. Once this is achieved, a public presence is established that can be translated into political power. When Palestinian terrorists hijacked an El Al plane in 1968, they captured the attention of the mass media all over the world. In the years to come, they would stage-manage other terrorist acts that played well on television, including the infamous attack on the 1972 Munich Olympics. Although such acts made them hated by most in non-Arab countries, they were willing to live with that—the publicity for their cause, and the power that came from it, was all they were after. As the writer Brian Jenkins notes, "Insurgents fought in Angola, Mozambique, and Portuguese Guinea for fourteen years using the standard tactics of rural guerrilla warfare. The world hardly noticed their struggle, while an approximately equal number of Palestinian commandos employing terrorist tactics have in a few years become a primary concern to the world."

In a world dominated by appearances, in which value is determined by public presence, terrorism can offer a spectacular shortcut to publicity—and terrorists accordingly tailor their violence to the media, particularly television. They make it too gruesome, too compelling, to ignore. Reporters and pundits can profess to be shocked and disgusted, but they are helpless: it is their job to spread the news, yet in essence they are spreading the virus that can only aid the terrorists by giving them such presence. The effect does not go unnoticed among the small and powerless, making the use of terrorism perversely appealing to a new generation.

Yet for all its strengths, terrorism also has limitations that have proved the death of many a violent campaign, and those opposing it must know and exploit this. The strategy's main weakness is the terrorists' lack of ties to the public or to a real political base. Often isolated, living in hiding, they are prone to lose contact with reality, overestimating their own

*are attacking are soon blunted. Property seems to them an indestructible thing. You can't count upon their emotions either of pity or fear for very long. A bomb outrage to have any influence on public opinion now must go beyond the intention of vengeance or terrorism. It must be purely destructive. It must be that, and only that, beyond the faintest suspicion of any other object. You anarchists should make it clear that you are perfectly determined to make a clean sweep of the whole social creation . . .*

*. . . What is one to say to an act of destructive ferocity so absurd as to be incomprehensible, inexplicable, almost unthinkable, in fact, mad? Madness alone is truly terrifying, inasmuch as you cannot placate it either by threats, persuasion, or bribes. Moreover, I am a civilized man. I would never dream of directing you to organize a mere butchery even if I expected the best results from it. But I wouldn't expect from a butchery the results I want. Murder is always with us. It is almost an institution. The demonstration must be against learning—science. But not every science will do. The attack must have all the shocking senselessness of gratuitous blasphemy. . . ."*

*The Secret Agent*, Joseph Conrad, 1857–1924

power and overplaying their hand. Although their use of violence must be strategic to succeed, their alienation from the public makes it hard for them to maintain a sense of balance. The members of Narodnaya Volia had a somewhat developed understanding of the Russian serfs, but more recent terrorist groups, such as the Weathermen in the United States and the Red Brigades in Italy, have been so divorced from the public as to verge on the delusional. Accentuating the terrorists' isolation and denying them a political base should be part of any effective counterstrategy against them.

Terrorism is usually born out of feelings of weakness and despair, combined with a conviction that the cause one stands for, whether public or personal, is worth both the inflicting and the suffering of any kind of damage. A world in which the faces of power are often large and apparently invulnerable only makes the strategy more appealing. In this sense terrorism can become a kind of style, a mode of behavior that filters down into society itself.

In the 1920s and '30s, the French psychoanalyst Jacques Lacan butted heads with the extremely conservative medical societies that dominated almost all aspects of psychoanalytic practice. Realizing the futility of taking on these authorities in a conventional way, Lacan developed a style that can fairly be described as terroristic. His sessions with his patients, for example, were often cut short before the usual fifty minutes were up; they could last any period of time that he saw fit and were sometimes as brief as ten minutes. This deliberate provocation to the medical establishment caused a great deal of scandal, setting off a chain reaction that shook the psychoanalytic community for years. (These sessions were also quite terrorizing for the patients, who could never be sure when Lacan would end them and so were forced to concentrate and make every moment count—all of which had great therapeutic value, according to Lacan.) Having gained much publicity this way, Lacan kept stirring the pot with new provocative acts, culminating in the creation of his own rival school and professional society. His books are written in a style to match this strategy: violent and arcane. It was as if he occasionally liked to throw little bombs into the world, thriving on the terror and attention they got him.

People who feel weak and powerless are often tempted into outbursts of anger or irrational behavior, which keeps those around them in suspense as to when the next attack will come. These fits of temper, like other, more serious kinds of terror, can have a chilling effect on their targets, sapping the will to resist; when the simplest dealings with these people are potentially so unpleasant, why fight? Why not just give in? A violent temper or outlandish act, volcanic and startling, can also create the illusion of power, disguising actual weaknesses and insecurities. And an emotional or out-of-control response to it just plays into the other person's hands, creating the kind of chaos and attention he or she thrives on. If you have to deal with a terroristic spouse or boss, it is best to fight back

*When Odawara Castle fell to the attackers in the Meiō period (the end of the fifteenth century), Akiko, who had been a maid in the service of Mori Fujiyori, the lord of the castle, escaped with a cat which had been her pet for years, and then the cat became a wild supernatural monster which terrorized the people, finally even preying on infants in the village. The local officials joined with the people in attempts to catch it, but with its strange powers of appearing and disappearing, the swordsmen and archers could find nothing to attack, and men and women went in dread day and night. Then in December of the second year of Eishō (1505), priest Yakkoku went up on to the dais at Hokokuji and drew the picture of a cat, which he displayed to the congregation with the words: "As I have drawn it, so I kill it with Katzu!, that the fears may be removed from the hearts of the people." He gave the shout, and tore to pieces the picture of the cat. On that day a woodcutter in the valley near the Takuma villa heard a terrible screech; he guided a company of archers to the upper part of the valley, where they found the body of the cat-monster, as big as a bear-cub, dead on a rock. The people* in a determined but dispassionate manner—the response such types least expect.

Although organized terrorism has evolved and technology has increased its capacity for violence, its essential makeup does not seem to have changed—the elements developed by Narodnaya Volia are still in effect. Yet the question many ask today is whether a new, more virulent kind of terrorism may be developing, one far surpassing the classical version. If terrorists could get hold of more potent armaments, for example—nuclear or biological weapons, say—and had the stomach to use them, their kind of war and the power it may bring them would make a qualitative leap into a new, apocalyptic form. But perhaps a new form of terrorism has already emerged that does not need the threat of dirty weapons to create a more devastating result.

On September 11, 2001, a handful of terrorists linked to the Islamic Al Qaeda movement produced the single deadliest terrorist action to date, in their attacks on the World Trade Center in New York City and the Pentagon outside Washington, D.C. The attack had many of the earmarks of classical terrorism: a small group, with extremely limited means, using the technology of the United States at their disposal, was able to strike with maximum effect. Here was the familiar asymmetry of forces in which smallness becomes an asset, being inconspicuous within the larger population and accordingly quite difficult to detect. The terror of the event itself set in motion a paniclike reaction from which the United States has still not fully recovered. The drama and symbolism of the Twin Towers themselves, not to mention the Pentagon, created a grotesquely compelling spectacle that gave the terrorists maximum exposure while incisively demonstrating the vulnerability of the United States, often described in recent years as the world's only remaining superpower. There were those around the world who had never imagined that America could be so quickly and seriously harmed but were delighted to find they were wrong.

Many deny that 9/11 was a new form of terrorism. It simply distinguished itself, they say, by the number of its victims; the change was quantitative, not qualitative. And, as in classical terrorism, these analysts continue, Al Qaeda is ultimately doomed to failure: the U.S. counterattack on Afghanistan destroyed their operational base, and they are now the targets of the unbending will of the American government, whose invasion of Iraq was a stage in a grand strategy to rid the region of terrorism in general. But there is another way to look at the attack, keeping in mind the chain reaction that is always the terrorist's goal.

The full economic impact of 9/11 is hard to measure, but the ripple effect of the attack is by any standard immense and undeniable: substantial increases in security costs, including the funding of new government programs for that purpose; enormous military expenditures on the invasions of two separate nations; a depressive effect on the stock market (always particularly susceptible to the psychology of panic) and a

*agreed that this had been the result of the master's Katzu!*

*Tests*
*(1) How can tearing up a picture with a Katzu! destroy a living monster?*
*(2) That devil-cat is right now rampaging among the people, bewitching and killing them. Kill it quickly with a Katzu! Show the proof!*

*Samurai Zen: The Warrior Koans*, Trevor Leggett, 1985

*When a man has learned within his heart what fear and trembling mean, he is safeguarded against any terror produced by outside influences. Let the thunder roll and spread terror a hundred miles around: he remains so composed and reverent in spirit that the sacrificial rite is not interrupted. This is the spirit that must animate leaders and rulers of men—a profound inner seriousness from which all outer terrors glance off harmlessly.*

*The I Ching*, China, circa eighth century B.C.

consequent injury to consumer confidence; hits on specific industries, such as travel and tourism; and the reverberating effect of all these on the global economy. The attack also had tremendous political effects—in fact, the American elections of 2002 and 2004 were arguably determined by it. And as the chain reaction has continued to play out, a growing rift has emerged between the United States and its European allies. (Terrorism often implicitly aims to create such splits in alliances and in public opinion as well, where hawks and doves line up.) September 11 has also had a definite and obvious impact on the American way of life, leading directly to a curtailment of the civil liberties that are the distinguishing mark of our country. Finally—though this is impossible to measure—it has had a depressive and chilling effect on the culture at large.

*"It appears to me that this mystery is considered insoluble, for the very reason which should cause it to be regarded as easy of solution—I mean for the* outré *character of its features. The police are confounded by the seeming absence of motive—not for the murder itself—but for the atrocity of the murder. . . . They have fallen into the gross but common error of confounding the unusual with the abstruse. But it is by these deviations from the plane of the ordinary, that reason feels its way, if at all, in its search for the true. In investigations such as we are now pursuing, it should not be so much asked 'what has occurred,' as 'what has occurred that has never occurred before.' In fact, the facility with which I shall arrive, or have arrived, at the solution of this mystery, is in the direct ratio of its apparent insolubility in the eyes of the police."*

AUGUSTE DUPIN IN "THE MURDERS IN THE RUE MORGUE," EDGAR ALLAN POE, 1809–1849

Perhaps the strategists of Al Qaeda neither intended all this nor even imagined it; we will never know. But terrorism is by its nature a throw of the dice, and the terrorist always hopes for the maximum effect. Creating as much chaos, uncertainty, and panic as possible is the whole idea. In this sense the 9/11 attack must be considered a success to such an extent that it does indeed represent a qualitative leap in terrorism's virulence. It may not have been as physically destructive as the explosion of a nuclear or biological weapon could be, but over time its reverberating power has far surpassed that of any terror attack before it. And this power comes from the altered nature of the world. Given the deep interconnections of the new global scene, whether commercial, political, or cultural, a powerful attack at a single point can have a chain-reactive effect that terrorists of earlier years could never have imagined. A system of interconnected markets that thrives on open borders and networks is intensely vulnerable to this intense ripple effect. The kind of panic that once might stir in a crowd or through a city can now spread over the world, fed spectacularly by the media.

To consider the 9/11 attack a failure because it did not achieve Al Qaeda's ultimate goal of pushing the United States out of the Middle East or spurring a pan-Islamic revolution is to misread their strategy and to judge them by the standards of conventional warfare. Terrorists quite often have a large goal, but they know that the chances of reaching it in one blow are fairly negligible. They just do what they can to start off their chain reaction. Their enemy is the status quo, and their success can be measured by the impact of their actions as it plays out over the years.

To combat terrorism—classical or the new version on the horizon—it is always tempting to resort to a military solution, fighting violence with violence, showing the enemy that your will is not broken and that any future attacks on their part will come with a heavy price. The problem here is that terrorists by nature have much less to lose than you do. A counterstrike may hurt them but will not deter them; in fact, it may even embolden them and help them gain recruits. Terrorists are often willing to spend years bringing you down. To hit them with a dramatic counterstrike is only to show your impatience, your need for immediate results, your vulnerability to emotional responses—all signs not of strength but of weakness.

Because of the extreme asymmetry of forces at play in the terrorist strategy, the military solution is often the least effective. Terrorists are vaporous, spread out, linked not physically but by some radical and fanatic idea. As a frustrated Napoleon Bonaparte said when he was struggling to deal with German nationalist groups resorting to acts of terror against the French, "A sect cannot be destroyed by cannonballs."

The French writer Raymond Aron defines terrorism as an act of violence whose psychological impact far exceeds its physical one. This psychological impact, however, then translates into something physical—panic, chaos, political division—all of which makes the terrorists seem more powerful than they are in reality. Any effective counterstrategy must take this into consideration. In the aftermath of a terrorist blow, what is most essential is stopping the psychological ripple effect. And the effort here must begin with the leaders of the country or group under attack.

In 1944, near the end of World War II, the city of London was subjected to a fierce campaign of terror from Germany's V-1 and V-2 rockets, an act of desperation that Hitler hoped would spread internal division and paralyze the will of the British public to continue the war. Over six thousand people were killed, many more were injured, and millions of homes were damaged or destroyed. But instead of allowing despondency and worry to set in, Prime Minister Winston Churchill turned the bombing campaign to his advantage as an opportunity to rally and unify the British people. He designed his speeches and policies to calm panic and allay anxiety. Instead of drawing attention to the V-1 attacks, or to the more dreaded V-2s, he emphasized the need to stay resolved. The English would not give Germany the satisfaction of seeing them bow to such terror.

In 1961, when President Charles de Gaulle of France faced a vicious right-wing terror campaign by French forces in Algeria opposed to his plan to grant the colony its independence, he used a similar strategy: he appeared on television to say that the French could not surrender to this campaign, that the costs in lives were relatively small compared to what they had recently suffered in World War II, that the terrorists were few in number, and that to defeat them the French must not succumb to panic but must simply unite. In both these cases, a leader was able to provide a steadying influence, a ballast against the latent hysteria felt by the threatened citizenry and stoked by the media. The threat was real, Churchill and de Gaulle acknowledged; security measures were being taken; but the important thing was to channel public emotions away from fear and into something positive. The leaders turned the attacks into rallying points, using them to unite a fractured public—a crucial issue, for polarization is always a goal of terrorism. Instead of trying to mount a dramatic counterstrike, Churchill and de Gaulle included the public in their

*We can no longer conceive of the idea of a symbolic calculation, as in poker or the potlatch: minimum stake, maximum result. This is exactly what the terrorists have accomplished with their attack on Manhattan, which illustrates rather well the theory of chaos: an initial shock, provoking incalculable consequences.*

*The Spirit of Terrorism*, Jean Baudrillard, 2002

*In general, the most effective response to unconventional provocation is the least response: do as little as possible and that cunningly adjusted to the arena. Do no harm. Deny one's self, do less rather than more. These are uncongenial to Americans who instead desire to deploy great force, quickly, to achieve a swift and final result. What is needed is a shift in the perception of those responsible in Washington: less can be more, others are not like us, and a neat and tidy world is not worth the cost.*

*Dragonwars*, J. Bowyer Bell, 1999

*And it is this uncontrollable chain reaction effect of reversals that is the true power of terrorism. This power is visible in the obvious and less obvious aftereffects of the event—not only in the economic and political recession throughout the system, and the psychological recession that comes out of that, but also in the recession in the value system, in the ideology of liberty, the freedom of movement, etc., which was the pride of the Western world and the source of its power over the rest of the world.*

*It has reached a point where the idea of liberty, one that is relatively recent, is in the process of disappearing from our customs and consciences, and the globalization of liberal values is about to realize itself in its exact opposite form: a globalization of police forces, of total control, of a terror of security measures. This reversal moves towards a maximum of restrictions, resembling those of a fundamentalist society.*

THE SPIRIT OF TERRORISM, JEAN BAUDRILLARD, 2002

strategic thinking and made the citizenry active participants in the battle against these destructive forces.

While working to halt the psychological damage from an attack, the leader must do everything possible to thwart a further strike. Terrorists often work sporadically and with no pattern, partly because unpredictability is frightening, partly because they are often in fact too weak to mount a sustained effort. Time must be taken to patiently uproot the terrorist threat. More valuable than military force here is solid intelligence, infiltration of the enemy ranks (working to find dissidents from within), and slowly and steadily drying up the money and resources on which the terrorist depends.

At the same time, it is important to occupy the moral high ground. As the victim of the attack, you have the advantage here, but you may lose it if you counterattack aggressively. The high ground is not a minor luxury but a critical strategic ploy: world opinion and alliances with other nations will prove crucial in isolating the terrorists and preventing them from sowing division. All this requires the willingness to wage the war over the course of many years, and mostly behind the scenes. Patient resolve and the refusal to overreact will serve as their own deterrents. Show you mean business and make your enemies feel it, not through the blustery front used for political purposes—this is not a sign of strength—but through the cool and calculating strategies you employ to corner them.

In the end, in a world that is intimately interlinked and dependent on open borders, there will never be perfect security. The question is, how much threat are we willing to live with? Those who are strong can deal with a certain acceptable level of insecurity. Feelings of panic and hysteria reveal the degree to which the enemy has triumphed, as does an overly rigid attempt at defense, in which a society and culture at large are made hostage to a handful of men.

**Image:** The Tidal Wave. Something disturbs the water far out at sea—a tremor, a volcano, a landslide. A wave a few inches high begins to ripple, cresting into a larger wave and then a larger one still, the depth of the water giving it momentum, until it breaks on shore with an unimaginable destructive force.

**Authority:** There is no fate worse than being continuously under guard, for it means you are always afraid.
*—Julius Caesar (100–44 B.C.)*

## REVERSAL

The reverse of terrorism would be direct and symmetrical war, a return to the very origins of warfare, to fighting that is up-front and honest, a simple test of strength against strength—essentially an archaic and useless strategy for modern times.

# SELECTED BIBLIOGRAPHY


Alinsky, Saul D. *Rules for Radicals.* New York: Vintage Books, 1972.

Beer, Sir Gavin de. *Hannibal.* New York: Viking, 1969.

Brown, Anthony Cave. *Bodyguard of Lies.* New York: Bantam Books, 1976.

Chambers, James. *The Devil's Horsemen: The Mongol Invasion of Europe.* New York: Atheneum, 1979.

Chandler, David G. *The Art of Warfare on Land.* London: Penguin Books, 1974.

———. *The Campaigns of Napoleon.* New York: Macmillan, 1966.

Clausewitz, Carl von. *On War.* Michael Howard and Peter Paret, eds. and trs. New York: Everyman's Library, 1993.

Cohen, Eliot A. and John Gooch. *Military Misfortunes: The Anatomy of Failure in War.* New York: Vintage Books, 1991.

Creveld, Martin van. *Command in War.* Cambridge, MA: Harvard University Press, 1985.

Douglass, Frederick. *My Bondage and My Freedom.* New York: Penguin Books, 2003.

Dupuy, Colonel T. N. *A Genius for War: The German Army and General Staff, 1807–1945.* Englewood Cliffs, NJ: Prentice-Hall Inc., 1977.

Foote, Shelby. *The Civil War: A Narrative* (3 volumes). New York: Vintage Books, 1986.

Green, Peter. *The Greco-Persian Wars.* Berkeley: University of California Press, 1998.

Haley, Jay. *Strategies of Psychotherapy.* New York: Grune and Stratton, 1963.

Hammond, Grant T. *The Mind of War: John Boyd and American Security.* Washington, D.C.: Smithsonian Institution Press, 2001.

Hart, B. H. Liddell. *Strategy.* New York: A Meridian Book, 1991.

Kissinger, Henry. *A World Restored.* Boston: Houghton Mifflin Co., 1957.

Kjetsaa, Geir. *Fyodor Dostoyevsky: A Writer's Life.* Siri Hustvedt and David McDuff, trs. New York: Viking, 1987.

Lawrence, T. E. *Seven Pillars of Wisdom.* New York: Anchor Books, 1991.

Leonard, Maurice. *Mae West: Empress of Sex.* New York: A Birch Lane Press Book, 1992.

Lewis, Bernard. *The Assassins: A Radical Sect in Islam.* New York: Oxford University Press, 1987.

Madariaga, Salvador de. *Hernán Cortés: Conqueror of Mexico.* Garden City, NY: Anchor Books, 1969.

Mansfield, Harvey C. *Machiavelli's Virtue.* Chicago: University of Chicago Press, 1998.

Morris, Donald R. *The Washing of the Spears: The Rise and Fall of the Zulu Nation.* New York: Da Capo Press, 1998.

Musashi, Miyamoto. *The Way to Victory: The Annotated Book of Five Rings.* Translated and commentary by Hidy Ochiai. Woodstock, NY: Overlook Press, 2001.

Nietzsche, Friedrich. *Ecce Homo.* R. J. Hollingdale, tr. London: Penguin Books, 1992.

Picq, Colonel Ardant du. *Battle Studies: Ancient and Modern Battle.* Colonel John N. Greely and Major Robert C. Cotton, trs. New York: Macmillan, 1921.

Poole, H. John. *Phantom Soldier: The Enemy's Answer to U.S. Firepower.* Emerald Isle, NC: Posterity Press, 2001.

Potter, Stephen. *The Complete Upmanship.* New York: Holt, Rinehart and Winston, 1971.

Schmitt, Carl. *The Concept of the Political.* Chicago: University of Chicago Press, 1996.

Spoto, Donald. *The Dark Side of Genius: The Life of Alfred Hitchcock.* New York: Da Capo Press, 1999.

Sugawara, Makoto. *The Lives of Master Swordsmen.* Tokyo: The East Publications, 1985.

Sun-tzu. *The Art of Warfare.* Translated and with commentary by Roger T. Ames. New York: Ballantine Books, 1993.

*Sword and the Mind, The.* Translated and with introduction by Hiroaki Sato. Woodstock, NY: Overlook Press, 1986.

Tomkins, Calvin. *Duchamp: A Biography.* New York: Henry Holt and Co., 1996.

Tsunetomo, Yamamato. *Hagakure: The Book of the Samurai.* William Scott Wilson, tr. Tokyo: Kodansha International, 1983.

Wilden, Anthony. *Man and Woman, War and Peace: The Strategist's Companion.* London: Routledge & Kegan Paul, 1987.

Wilhelm, Richard. *The I Ching* (or *Book of Changes*). Princeton, NJ: Princeton University Press, 1977.

*Wiles of War: 36 Military Strategies from Ancient China, The.* Compiled and translated by Sun Haichen. Beijing: Foreign Languages Press, 1991.

*Xenophon's Anabasis: The March Up Country.* W.H.D. Rouse, tr. New York: A Mentor Classic, 1959.

Young, Desmond. *Rommel.* London: Collins, 1950.

# INDEX

Grateful acknowledgment is made for permission to reprint excerpts from the following copyrighted works:

*Religious Mythology and the Art of War: Comparative Religious Symbolisms of Military Violence* by James A. Aho. Copyright © 1981 by James A. Aho. Reprinted by permission of Greenwood Publishing Group, Inc., Westport, Connecticut.

*Dragonwars: Armed Struggle and the Conventions of Modern War* by J. Bowyer Bell. Copyright © 1999 by Transaction Publishers. Reprinted by permission of Transaction Publishers.

*Roosevelt: The Lion and the Fox* by James MacGregor Burns. Copyright © 1956 by James MacGregor Burns. Copyright renewed 1984 by James MacGregor Burns. Reprinted by permission of Harcourt, Inc.

*The Years of Lyndon Johnson: The Path to Power* by Robert A. Caro. Copyright © 1982 by Robert A. Caro. Used by permission of Alfred A. Knopf, a division of Random House, Inc.

*Journey to Ixtlan: The Lessons of Don Juan* by Carlos Castaneda. Copyright © 1972 by Carlos Castaneda. Abridged and used by permission of Simon & Schuster Adult Publishing Group.

*The Art of War in World History: From Antiquity to the Nuclear Age,* edited by Gerard Chaliand. Copyright © 1994 by The Regents of the University of California. By permission of the University of California Press.

*Titan: The Life of John D. Rockefeller, Sr.* by Ron Chernow. Copyright © 1998 by Ron Chernow. Reprinted by permission of Random House, Inc.

*Clausewitz on Strategy: Inspiration and Insight from a Master Strategist,* edited by Tiha von Ghyczy, Bolko von Oetinger, and Christopher Bassford (John Wiley & Sons). Copyright © 2001 by The Boston Consulting Group, Inc. Reprinted by permission of The Strategy Institute.

*On War* by Carl von Clausewitz, edited and translated by Michael Howard and Peter Paret. Copyright © 1976 by Princeton University Press, renewed 2004 by Princeton University Press. Reprinted by permission of Princeton University Press.

*Command in War* by Martin van Creveld. Copyright © 1985 by the President and Fellows of Harvard College. Reprinted by permission of Harvard University Press, Cambridge, Mass.

*The Generalship of Alexander the Great* by J.F.C. Fuller. Copyright © 1960 by J.F.C. Fuller. Reprinted by permission of Rutgers University Press.

*Grant and Lee: A Study in Personality and Generalship* by J.F.C. Fuller (Indiana University Press). Copyright © 1957 by J.F.C. Fuller. By permission of David Higham Associates.

*Julius Caesar: Man, Soldier and Tyrant* by J.F.C. Fuller. Copyright © 1965 by J.F.C. Fuller. Reprinted by permission of Rutgers University Press.

*The Greco-Persian Wars* by Peter Green. Copyright © 1996 by Peter Green. Reprinted by permission of University of California Press.

*Strategies of Psychotherapy* by Jay Haley (Triangle Press). Copyright © 1967 by Jay Haley. By permission of the author.

*Masters of War: Classic Strategic Thought* by Michael I. Handel (Frank Cass Publishers). Copyright © 1992 by Michael I. Handel. Reprinted by permission of Taylor & Francis Books.

*Iliad* by Homer, translated by Stanley Lombardo. Copyright © 1997 by Hackett Publishing Company, Inc. Reprinted by permission of Hackett Publishing Company, Inc. All rights reserved.

*The Head Game: Baseball Seen from the Pitcher's Mound* by Roger Kahn. Copyright © 2000 Hook Slide, Inc. Reprinted by permission of Harcourt, Inc.

*A World Restored: Metternich, Castlereagh and the Problems of Peace 1812–1822* by Henry Kissinger (Boston: Houghton Mifflin, 1957). Reprinted by permission of the publisher.

*The Anatomy of the Zulu Army: From Shaka to Cetshwayo 1818–1879* by Ian Knight. Copyright © 1995 by Ian Knight. Reprinted by permission of Greenhill Books, London.

*Samurai Zen: The Warrior Koans* by Trevor Leggett (Routledge). Copyright © 2002 by The Trevor Leggett Adhyatma Yoga Trust. Reprinted by permission of Taylor & Francis Books.

*The Art of Maneuver: Maneuver-Warfare Theory and Airland Battle* by Robert R. Leonhard. Copyright © 1991 by Robert R. Leonhard. Used by permission of Presidio Press, an imprint of The Ballantine Publishing Group, a division of Random House, Inc.

*Hitter: The Life and Turmoils of Ted Williams* by Ed Linn. Copyright © 1993 by Edward A. Linn. Reprinted by permission of Harcourt, Inc.

*The Ramayana of R. K. Narayan* by R. K. Narayan. Copyright © R. K. Narayan, 1972. Used by permission of Viking Penguin, a member of Penguin Group (USA) Inc.

*The Gay Science* by Friedrich Nietzsche, edited by Bernard Williams, translated by Josefine Nauckhoff. Copyright © 2001 by Cambridge University Press. Reprinted by permission of Cambridge University Press.

*Human, All Too Human: A Book of Free Spirits* by Friedrich Nietzsche, translated by R. J. Hollingdale. Copyright © 1986, 1996 by Cambridge University Press. Reprinted by permission of Cambridge University Press.

*The Art of Political Warfare* by John J. Pitney, Jr. Copyright © 2000 by University of Oklahoma Press. Reprinted by permission of University of Oklahoma Press.

*The Tao of Spycraft: Intelligence Theory and Practice in Traditional China* by Ralph D. Sawyer. Copyright © 1998 by Ralph D. Sawyer. Reprinted by permission of Westview Press, a member of Perseus Books, LLC.

*The Art of War* by Sun-tzu, translated by Ralph D. Sawyer. Copyright © 1994 by Ralph D. Sawyer. Reprinted by permission of Westview Press, a member of Perseus Books, LLC.

*Sun-tzu: The Art of Warfare* by Sun-tzu, translated by Roger T. Ames. Copyright © 1993 by Roger T. Ames. Used by permission of Ballantine Books, a division of Random House, Inc.

*Mao: A Biography* by Ross Terrill. Copyright © 1999 by Ross Terrill. All rights reserved. Used by permission of Stanford University Press.

*The Templars and the Assassins: The Militia of Heaven* by James Wasserman. Copyright © 2001 by James Wasserman. By permission of Destiny Books.

*The I Ching or Book of Changes* (third edition), translated by Richard Wilhelm. Copyright © 1950 by Bollingen Foundation, Inc. New material copyright © 1967 by Bollingen Foundation. Copyright renewed 1977 by Princeton University Press. Reprinted by permission of Princeton University Press.